AF855083

Electronic Tagging and Tracking in Marine Fisheries

Reviews: Methods and Technologies in Fish Biology and Fisheries

VOLUME 1

Series editor:

Jennifer L. Nielsen

U.S. Geological Survey,
Biological Resources Division,
Alaska Biological Science Center,
Anchorage, Alaska

Electronic Tagging and Tracking in Marine Fisheries

Proceedings of the Symposium on Tagging and Tracking Marine Fish with Electronic Devices, February 7-11, 2000, East-West Center, University of Hawaii

Edited by

John R. Sibert
University of Hawaii, Honolulu

and

Jennifer L. Nielsen
*U.S. Geological Survey,
Biological Resources Division,
Alaska Biological Science Center,
Anchorage, Alaska*

KLUWER ACADEMIC PUBLISHERS
DORDRECHT / BOSTON / LONDON

A C.I.P. Catalogue record for this book is available from the Library of Congress.

ISBN 1-4020-0125-8

Published by Kluwer Academic Publishers,
P.O. Box 17, 3300 AA Dordrecht, The Netherlands.

Sold and distributed in North, Central and South America
by Kluwer Academic Publishers,
101 Philip Drive, Norwell, MA 02061, U.S.A.

In all other countries, sold and distributed
by Kluwer Academic Publishers,
P.O. Box 322, 3300 AH Dordrecht, The Netherlands.

Printed on acid-free paper

All Rights Reserved
© 2001 Kluwer Academic Publishers
No part of the material protected by this copyright notice may be reproduced or utilized in any form or by any means, electronic or mechanical, including photocopying, recording or by any information storage and retrieval system, without written permission from the copyright owner.

Printed in the Netherlands.

General Series Preface

Reviews: Methods and Technology in Fish Biology and Fisheries published by Kluwer Academic Publishers is a book series dedicated to the publication of information on advanced, forward-looking methodologies, technologies, or perspectives in fish and fisheries. This series is especially dedicated to relevant topics addressing global, international concern in fish and fisheries. Humans continue to challenge our environments with new technologies and technological applications. The dynamic creativity of our own species often tends to place the greatest burden on our supporting ecosystems. This is especially true for aquatic networks of creeks, lakes, rivers and ocean environments. We also frequently use our conceptual powers to balance conflicting requirements and demands on nature and continue to develop new approaches and tools to provide sustainable resources as well as conserve what we hold most dear on local and global scales. This book series will provide a window into the developing dynamic among humans, aquatic ecosystems (both freshwater and marine), and the organisms that inhabit aquatic environments.

There are many reasons to doubt the increasing social and economic value technology has gained over the last two centuries. Science and technology represent stages in human development. I agree with Ernst Mayer when he said in *Toward a New Philosophy of Biology* (1988) that "endeavors to solve all scientific problems by pure logic and refined measurements are unproductive, if not totally irrelevant." Living aquatic systems are extremely complex environments that appear alien to most human beings. We only "go there" in a limited capacity and for relatively short periods of time. To reduce these biological systems to simple physio-chemical processes that can be controlled by technology, and made subject to arbitrary management decisions or broad arm-waving policies is to deny their unique nature. That is certainly not what is meant by "*Methods and Technology*" in this series. Rather, I want to provide a forum for discussions on the living systems themselves and the organisms inhabiting them through new applications of science and technology with special emphasis on aspects of fish and fisheries under-represented in the current literature.

Our understanding of aquatic biology in freshwater and marine environments demands a careful and protracted approach ranging broadly from studies of genes, regulatory processes, isolating mechanisms, individual behavior, population structure, biodiversity, to interactive ecosystems. The tools and technologies that allow these investigations change rapidly, always removing old uncertainties and creating new ones. The intent of this series is to monitor that change and document perspectives and developments that mark a fundamental re-evaluation of nature in aquatic habitats and its role in relationship to human society and resource management.

Dr. Jennifer L. Nielsen, Series Editor
Reviews: Methods and Technology in Fish Biology and Fisheries
US Geological Survey, Biological Resource Division
Alaska Biological Science Center
Anchorage, Alaska

CONTENTS

Series Editor's Preface

The developing technology of microcomputers and electronic tagging of organisms in marine environments has opened up a new field of fisheries science. The natural history of pelagic and near-shore organisms inhabiting ocean habitats has been difficult to ascertain until this technology became available to marine scientists. Inference gained from archival and pop-up tags in ocean environments relies on important technical and statistical developments that are new to the field of fisheries. This volume marks the first compilation of the history of this technology, past and present technological and developmental problems, recent case histories and applications in marine environments, and general assessments of its impacts on fisheries management. The various forms of electronic tags currently available and the diversity of their applications documented by this volume demonstrates the unique capacity this tool provides for improving our dialogue with ocean dwelling organisms. This volume also demonstrates the ability of marine scientists to quickly grasp a new technology and their unique creativity in its application in a wide array of questions, many of which were unanswerable prior to the new technology. With this tool scientists let the fish speak for themselves by documenting preferred activity, seasonal behavior and habitat selection independent of human activity and affiliated fisheries. In the future, applications of this technology will continue to grow and the science supporting interpretation of these types of data should become an integral part of fisheries management and marine science. I think most readers will enjoy speculating on potential future applications this technology may provide for fish and fisheries throughout the next 50 years.

Dr. Jennifer L. Nielsen, Series Editor
Reviews: Methods and Technology in Fish Biology and Fisheries
US Geological Survey, Biological Resource Division
Alaska Biological Science Center
Anchorage, Alaska

Preface

John Sibert
Jennifer Nielsen

The Symposium on Tagging and Tracking Marine Fish with Electronic Devices was held February 7-11, 2000, at the East-West Center in Honolulu and hosted by the University of Hawaii Pelagic Fisheries Research Program. Over 100 scientists and electronics manufacturers from a dozen countries assembled to present the current state of the art in this rapidly developing field. Results from all of the oceans of the planet were presented. The species discussed included tropical cephalopods (squids, octopii, and cuttlefish), salmon, tunas, marlins, cod, reef fishes, sea basses and sharks from habitats ranging from shallow inshore waters to the deep ocean.

This volume presents a subset of the symposium presentations supplemented by some specifically commissioned papers and documents the state of the art of electronic tagging as it was in early 2000. The majority of the papers selected for the volume are "case studies" rather than primary reports of current research results. Case studies synthesize results of research projects strongly motivated by practical fishery management concerns and also display the wide range of problems that can be addressed with electronic tagging. The technology is changing rapidly and is far from mature. Consequently, there are a number of unresolved and controversial problems such as geolocation by daylight and how to apply tracking data to whole populations. Approaches to these problems are also included in the volume.

We thank the symposium organizing committee, Geoff Arnold, Rich Brill, Laurent Dagorn, John Gunn, Kim Holland, and Molly Lutcavage, for their ideas and support. Diane Henderson and May Izumi from Publication Services of School of the Ocean and Earth Science and Technology, University of Hawaii, deserve special thanks for their meticulous professionalism in preparing this book for publication. Dodie Lau of the Pelagic Fisheries Research Program provided invaluable logistical and world wide web support through all phases of this project from planning and hosting the symposium though coordination of the publication. Preparation of this volume was supported by the University of Hawaii Pelagic Fisheries Research Program under Cooperative Agreement number NA67R-J0154 from the National Oceanic and Atmospheric Administration.

Electronic Tagging and Tracking in Marine Fisheries

Introduction to the Proceedings

John Sibert
Pelagic Fisheries Research Program, Joint Institute for Marine and Atmospheric Research, University of Hawai'i at Mānoa

1. INTRODUCTION

Sea water is nearly opaque to electromagnetic radiation. It is, therefore, difficult to track marine animals visually over significant distances, and radio tracking is rendered impractical by antennae and power requirements. Marine biologists have been laboring for over 40 years to develop alternative electronic systems to understand the behavior and physiology of marine animals in their natural environment. The pace of development accelerated in the 1990s motivated by new fishery management imperatives and propelled by rapid advances in electronics. By the late 1990s, a wide variety of electronic tagging and tracking devices were being applied in every ocean, in all marine habitats, and to an astonishing array of animals. Behaviorists, physiologists, engineers, modelers, fishery managers, and equipment manufacturers converged in Honolulu in February 2000 to participate in the *Symposium on Tagging and Tracking Marine Fish with Electronic Devices*. This volume reviews the history and presents the state-of-the-art of electronic tags used in the marine environment, as described in presentations at the symposium.

2. ELECTRONIC TAGS

Electronic systems used to track marine animals fall roughly into two categories: acoustic telemetry and "archival" or data storage devices. Acous-

J.R. Sibert and J.L. Nielsen (eds.), Electronic Tagging and Tracking in Marine Fisheries, 1–6.
© 2001 *Kluwer Academic Publishers. Printed in the Netherlands.*

tic tags can be relatively simple devices attached to an animal that emit a high-frequency, ultrasonic "ping" that is detected by a hydrophone. Hydrophones may be mounted on vessels ranging in sophistication from a simple kayak to modern, fully instrumented research ships. Hydrophones can also be deployed on moored automated recorders deployed singly or in arrays. Acoustic devices are capable of yielding extremely detailed, minute-by-minute descriptions of movement over time scales dependent on the endurance of the tracking vessel (hours to weeks). Longer term tracking and monitoring have become practical with the development of automated and remotely monitored recorders. Additional sensors can be incorporated in tags and their output encoded in the telemetry signal.

Archival tags are miniature computers containing a clock integrated with a variety of sensors. The sensors are sampled at pre-set intervals and the data are stored in the memory of the computer for periods up to several years. Archival tags typically record ambient light intensity (for estimating geographic position), water pressure (for estimating depth), water temperature, and internal body temperature. The first generation of archival tags had to be recovered to obtain the data stored in the tag. High tag costs and low tag return rates restricted the application of archival tags to heavily exploited, high-value fisheries. Newer devices help to overcome the recovery problem. "Pop-up" tags detach themselves from the fish after a pre-set date, rise to the sea-surface, and transmit stored data to the researcher via satellite. In a new and interesting convergence of technologies, acoustic devices have been married to archival tags. "Chat" tags are acoustically interrogated when they come within the range of a moored "listening" station and transmit stored data to the station.

Acoustic telemetry systems were the first electronic fish tagging systems to be widely applied in the ocean and have been under continuous development since the 1950s. In contrast, the development and use of archival tags is relatively recent. Arnold and Dewar review the development of electronic tags over the last 30 years and assess the impact their use has had on marine fisheries research.

3. FISHERY MANAGEMENT

The most promising current application of electronic tags to fishery management is elucidation of habitat use. Preservation of habitat is an essential component of plans to conserve and rebuild fishery resources, and identification of "essential fish habitat" has become a standard requirement of fishery management policy in some countries. This volume contains many "case studies" in which tags of various kinds, both electronic and conventional, have been applied to single species over extended periods and where electronic tags have been used to monitor multiple species in a single habitat. The habitats considered range from near-shore reef and estuarine to open ocean pelagic, and the fish include small reef fishes, sharks, tunas and even some invertebrates.

Different tagging techniques applied to a single species over time (Block et al.; Eklund et al.; Holland et al.) demonstrate modes of habitat usage, define home range, and document changes in habitat use at different life history stages. Applications of a single technique to several species in a single location (O'Dor et al.; Meyer and Holland) reveal how different features of the habitat mediate the structure of fish communities. In the future, such studies will constitute the core of ecosystem based management policies.

The relationship between the dynamics of individuals on small time and space scales to the dynamics of whole populations on large time and space scale is a major preoccupation of general ecology (Levin, 1992). Electronic tags report on the behavior of individuals, whereas fisheries management concerns populations. Applications of tracking data to population modeling require explicit approaches to the general ecological problem of changing scale. Dagorn et al. and Kirby explore the interplay between field work and theory in attempting to move across scales. Newlands and Lutcavage, and Sibert and Fournier present potential models for integrating tracks into larger scale models of population movement.

4. LIMITATIONS TO TECHNOLOGY

Electronic devices deployed in the marine environment can fail owing to loss of battery power, flooding, and other causes. Deployment on free-ranging animals exacerbates the causes of these failures. Some of the larger tropical predators, such as tunas and marlins, make frequent excursions from warm, 25°C, surface water to depths exceeding 500 m where the water temperature may be less than 5°C and where the pressure exceeds 50 atmospheres. The thermal expansion and contraction combined with the pressure changes increase the likelihood of flooding. If the device is attached externally, it may be bitten by a predator or simply fall off. If the device is implanted internally, it may be expelled by the animal. Animals tagged with acoustic transmitters may swim beyond the detection range of the monitors. These problems complicate the use of electronic tags in applications where estimation of mortality is a research issue. Tag manufactures are well aware of the problems and are rapidly implementing improvements. Newer pop-off devices incorporate more robust pressure resistant flotation materials and are equipped with pressure sensitive "fail-safe" releases. "Mortality" sensors that sense orientation (ventral side up) or physiological activity could be incorporated into tags, but the logical dilemma of discriminating between electronic malfunction and mortality remains.

The ability to determine position from data collected by archival tags depends on the deceptively simple concept of geolocation by light intensity. If the light intensity is known at every instant, the times of sunrise and sunset can be determined. Latitude is related to day length, and longitude is related to the time of local noon. Small errors in estimation of times of sunrise and sunset translate to large errors in position estimates and define the fundamental "limits to the accuracy" of geolocation by light (Metcalfe). Under ideal conditions, the best that can be expected is an uncertainty in geographic position of slightly less than 1 degree (Hill and Braun). Under field conditions, biofouling of light sensors, crepuscular diving behavior, cloud cover, and other factors combine to increase geolocation uncertainty. As with any new and rapidly developing field, the debate surrounding the subject of geolocation by light intensity has been lively, and there is no consensus on computation algorithms. Papers presented in this volume explore geolocation errors in detail (Musyl et al.), offer improved geolocation algorithms (Welch and Eveson; Hill and Braun), and propose statistical methods for estimating the most probable track from a time series of position estimates (Sibert and Fournier).

5. FUTURE DEVELOPMENTS

It is feasible to incorporate a variety of sensors into electronic tags. The possibility of improving geolocation estimates by including clues from the Earth's geomagnetic field had been discussed in the literature. Internally deployed archival tags are capable of recording internal body temperature and the information has been used in feeding (Gunn et al.) and thermoregulation studies. Tags that measure tail beat (Webber et al.) and heart rate have been developed and applied. Other physiological sensors, perhaps adopted from medical telemetry, can be envisaged and could be incorporated into either acoustic or archival tags. While equipment manufacturers appear to be willing to collaborate in *ad hoc* projects, they do not anticipate that tags with exotic sensors will become "off-the-shelf" products. Researchers, on the other hand, are demanding higher levels of accuracy and precision in geolocation estimates from archival tags than can be obtained from geolocation by daylight alone.

Methods of obtaining information from electronic tags will continue to improve. The RAPT system discussed by O'Dor et al. may become the model for future monitoring methods. It is possible to envisage arrays of "listening stations" that receive information from electronic tags, both archival and otherwise, and that store and relay the information to a central station. Large-scale arrays can be envisaged, and preliminary planning has begun for an array of listening stations on the Pacific continental shelf of North America that will extend from Mexico to Alaska (David Welch, personal communication).

The data produced by electronic tags can be voluminous. An archival tag, for instance, is capable of storing measurements of swimming depth, body temperature, and ambient water temperature every fifteen minutes for two years. Acoustic telemetry arrays are capable of generating comparable volumes of data on the minute by minute positions of individual animals. Simply organizing such data for easy storage and access is a formidable task. Researchers who choose to employ electronic tagging systems will need to allocate resources for data management and analysis. Large data sets such as these often contain information relevant to scientific questions not envisaged by the researchers who originally collected the data. The prospects for collaborative analysis are improving. Some researchers have already begun to place data in the public domain. The Lowestoft Laboratory has published data from archival tags deployed on North Sea plaice in a series of technical reports (Buckley 1994, 1995, 1999). Musyl et al. have made raw data from their archival tag mooring experiment available on the World Wide Web. The CSIRO Division of Marine Research will make software for manipulating data from archival tags available for research use by arrangement. Some consideration could be given to creation of a repository for electronic tagging data. Such a repository would maintain the data in electronic format, be accessible from the World Wide Web, and contain ancillary information documenting the source of the data, and how they were collected.

The current state of the art in electronic fish tagging is the product of a 30-year collaboration between researchers, engineers, and electronic manufacturers. It has been the beneficiary of the amazing development of electronic technology. Trends towards decreasing size and increasing functionality of electronic devices are likely to continue. The imagination and creativity of marine biologists will undoubtedly stimulate the development of new ways to apply electronic tagging technology to research problems. We can expect more information from these devices delivered to the researchers more rapidly. The results will be more refined descriptions of habitat use by fisheries resources, new ideas on which develop ecosystem-based fishery management policy, more intimate understanding of the behavior and physiology of marine animals in their natural environment, and quantitative links between individuals and populations.

REFERENCES

Buckley, A. A. 1994. Open sea fish tracks: plaice fitted with compass tags. Data Report, MAFF Directorate of Fisheries Research, Lowestoft (35), 60 pp.

Buckley, A. A. 1995. Open sea fish tracks: cod fitted with transponding acoustic tags. Data Report, MAFF Directorate of Fisheries Research, Lowestoft (37), 107 pages.

Buckley, A. A. 1999. Open sea fish tracks: plaice fitted with transponding acoustic tags. Science Series Data Report, CEFAS, Lowestoft, (39), 81 pp.

Levin, S. A. 1992. The problem of pattern and scale in ecology. *Ecology* **73**, 1843-1967.

Electronic Tags in Marine Fisheries Research:
A 30-Year Perspective

Geoff Arnold[1] and Heidi Dewar[2]
[1]*The Centre for Environment, Fisheries and Aquaculture Science, Lowestoft Laboratory, Pakefield Road, Lowestoft, Suffolk, NR33 0HT, UK*
[2] *Pfleger Institute of Environmental Research, 1400 N Pacific Street, Oceanside, CA 92054, USA.*

Key words: acoustic telemetry, archival tags, satellite telemetry, fisheries

Abstract: Remote sensing technologies have been used to study the behaviour and movements of fish in the sea for about 30 years. The first relatively simple devices have given rise to a suite of sophisticated tools that have spawned a whole new field of fisheries science. The oldest method, acoustic telemetry, has now advanced to the level at which we can study detailed behaviour in the field in relation to the immediate environment, and monitor physiological variables, as well as document larger-scale movement patterns. Archival tags, available for the last 7-8 years, provide a powerful tool to obtain detailed information on a range of variables for large number of individual fish over long periods of time. Their primary limitation is that they must be recovered to obtain the data. The most recent development, the pop-up satellite tag, requires no recovery and can thus be used on any fish of sufficient size. Being fully independent of the fisheries, this type of tag is ideal for studying large, highly migratory fish and dramatically expands the scope of questions that can be addressed. The pace of technical development will undoubtedly continue to accelerate over the next few decades, producing new tools that can be expected to solve many currently unanswered questions in fisheries biology. Solutions to these questions should help to achieve improved management of the seas and their living resources.

1. INTRODUCTION

Efforts to study the movements of individual fish have been in progress for over a century since Atlantic salmon tagged in the Penobscot River,

J.R. Sibert and J.L. Nielsen (eds.), Electronic Tagging and Tracking in Marine Fisheries, 7–64.
© 2001 *British Crown copyright. Printed in the Netherlands.*

Maine, were first recovered in fair numbers in 1873 (Everhart et al., 1975). For nearly 75 years after this early success, simple mark-recapture experiments provided the only direct way to document fish movements. Then in the mid-1950s the first electronic tags marked the beginning of the technological age. Acoustic tags provided marine fisheries biologists with a tool that allowed them to follow individual fish over several days, describe their behaviour in great detail and simultaneously collect data about the immediate physical environment. The pace of technical development began to accelerate in the 1980s, in parallel with developments in microchip technology. And, in less than a decade in the 1990s, we have seen the introduction of archival tags, followed rapidly by pop-up tags that detach themselves from the fish at a predetermined time, float to the surface and transmit data to satellite. These latest electronic tags, which use low power memory chips and micro-controllers, can record a range of environmental, behavioural and physiological data over long periods from large numbers of individual fish. Most significantly, the data recorded by these 'computer' tags are increasingly independent of the fisheries.

This paper sets out to review the development of electronic tags over the last 30 years, to assess the impact their use has had on marine fisheries research and consider briefly possibilities for the future. Researchers can now tackle far more complicated questions covering a broader range of species than ever before possible, and the availability of new tools, which allow us to address the critical issues now facing global fisheries, is likely to prove timely.

Our review is not intended to be exhaustive. Further information can be found in Stasko and Pincock (1977), Priede (1992), Moore et al. (2000) and Voegeli et al. (2001), as well as the reports of a recent Concerted Action 'Improvements of Tagging Methods for Stock Assessment and Research in Fisheries' (Anon, 2000, 2001) and an earlier workshop 'Electronic Tags in Fisheries Research and Management' (Arnold and Lundgren, 2001), both funded by the European Commission. Further information about Concerted Action (CATAG) can be found at the web site http://www.hafro.is/catag. Our goals here are to survey the types of electronic tags that are now available and to review the biological knowledge that has been acquired recently by the application of new technology. Because the symposium in Honolulu was concerned with marine fish, our review concentrates on acoustic tags, archival tags, and tags that can transmit data by radio to a variety of listening stations—shore-stations, research vessels and satellites—relevant to the marine environment. We have not attempted to deal with the freshwater environment, where radio tags are widely used because they are little affected by aeration and can be readily detected through the air-water interface. We do not discuss methods of tag attachment, fish welfare, or possible effects on fish behaviour, which are discussed in Anon (2000, 2001) or Arnold and Lundgren (2001).

2. ACOUSTIC TAGS

Acoustic telemetry, which dates back to the 1950s, is one of the oldest tools used to study the behaviour of individual fish in their natural environment. The acoustic tag, which is usually attached to the fish externally, but may also be placed internally, emits a high-frequency sound that is inaudible to the fish but allows the tag to be located by sonar, or a hydrophone on a mobile vessel, or static listening station.

Most acoustic tags emit ultrasonic frequencies (30 to 300 kHz), which are produced by driving an annular ceramic transducer at its resonant frequency. Sounds within this frequency range propagate well through seawater, whereas the electromagnetic waves used in radio telemetry are rapidly absorbed. Tag size is determined largely by the size of the transducer, whose diameter is inversely proportional to frequency (Priede, 1986). Range also varies with frequency. A large diameter 32 kHz tag may have a range of as much as 2.5 km (Klimley et al., 1998); a small 300 kHz tag usually has a range of less than 400 m. Frequencies of 30-50 kHz are commonly used to track large pelagic fish in the open ocean; frequencies of 60-80 kHz are most common in coastal and estuarine waters (Priede, 1992). Higher frequencies (150-300 kHz) are used where a small tag is required (Priede, 1992) or where specialised high-frequency imaging sonar is available (Arnold et al., 1990). The physical constraints on acoustic signal propagation and detection, including refraction and reflection, spreading and absorption losses, and ambient noise, are discussed by Priede (1986, 1992) and Voegeli and Pincock (1996), among others.

Acoustic tags are available as simple *pingers*, which transmit regularly and continuously, and *transponders*, which only transmit on receipt of an external signal from a sonar. Tags can be individually coded by modulating either frequency, or pulse repetition rate; this permits groups of fish to be followed at the same time. Pulse rate and frequency modulation are also used to transmit data recorded from sensors on the tag. Coding with pseudo random (PN) numbers, a recent development, allows large numbers of fish to be tracked simultaneously (Cote et al., 1998; Voegeli et al., 1998, 2001).

3. TRACKING SYSTEMS

Acoustic telemetry encompasses a wide range of tracking systems with varying levels of complexity. The simplest system, which provides mobility equal to that of the fish, consists of a directional hydrophone, a receiver and headphones. It allows the operator to follow the tagged fish and obtain an approximate estimate of its position relative to the tracking boat (e.g., Holland et al., 1985). . Weather permitting, a large research vessel can track an individual fish until the tag batteries are exhausted, usually after 10-15 days at most. With a small boat, track duration is primarily limited by human endurance (Nelson, 1978), although autonomous vehicles may perhaps overcome this limitation in the future (Goudey et al., 1998). A sonar operating in passive mode may also be used (Døving et al., 1985). Accurate location, however, requires a hydrophone array, or active sonar. Fixed hydrophone arrays (e.g., Urquhart and Smith, 1992; Smith et al., 1998) have traditionally been used in sheltered waters, where it is possible to run cables ashore. Modern systems, which log data or transmit them by radio, can be used in the open sea (e.g., Engås et al., 1998), as well as estuaries (e.g., Moore, 1996).

3.1 Fixed Hydrophone Arrays and Sonar Buoys

Fixed hydrophone arrays and sonar buoys are used to study movements and behaviour on small spatial scales. High resolution tracking systems with a fixed hydrophone array connected by cables to an onshore listening station have been successful in confined coastal waters (e.g., Glass et al., 1992). The concept has been extended to the open sea with sonar buoys, which can convert acoustic signals to radio signals and transmit them to a ship, or shore-based listening station (Konagaya, 1982). To obtain two-dimensional paths at least three buoys are required, moored 400-600 m apart in a triangular configuration; four buoys can provide three-dimensional information on the position of the fish (e.g., Voegeli and McKinnon, 1996). Sonar buoys have been used to study activity and feeding behaviour in cod, whiting, ling (Engås et al., 1996; Løkkeborg, 1998; Løkkeborg and Fernö, 1999; Løkkeborg et al., 2000) and edible crabs (Skajaa et al., 1998). They are also being used to study predator-prey interactions of white sharks and elephant seals (Klimley et al., 1998, 2001) (Fig. 1). The method has been applied to describe the natural behaviour of cod and whiting and the reactions of these fish to an approaching trawler (Engås et al., 1996, 1998). Commercial sonar buoy systems are manufactured by Vemco Ltd. (Shad Bay, Nova Scotia, Canada) (e.g., O'Dor et al., 1998) and Lotek Wireless Inc. (Newmarket, Ontario, Canada) (e.g., Cote et al., 1998).

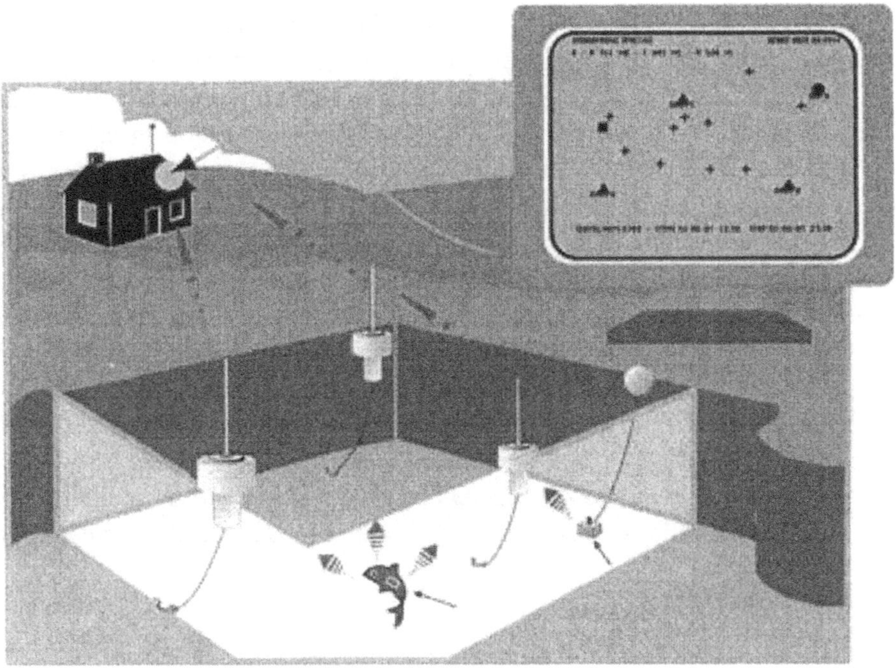

Figure 1. Fixed hydrophone array with radio transmission to shore station. This system has been used to monitor the predator-prey interactions between elephant seals and white sharks (Klimley et al., 1998)

Chains of moored sonar buoys have been used to monitor the movements of individual fish over larger spatial scales and this system has proved to be very effective for examining the movements of diadromous fish through estuaries in the United Kingdom (Solomon and Potter, 1988). Tracks are reconstructed simply from the time at which the fish passes each buoy. Radio transmissions from several buoys are received by an automatic listening station (ALS), which records the date and time at which each signal is received, the identity of the sonar buoy and an audio recording of the pulse rate of the tag. The system has been used to track the movements of adult Atlantic salmon (*Salmo salar*) and sea trout (*S. trutta*) returning to spawn in freshwater (e.g., Potter, 1988; Potter et al., 1992; Mee et al., 1996) using combined acoustic and radio (CART) tags inserted in the stomach (Solomon and Storeton-West, 1983). CART tags switch from acoustic to radio transmission after a pre-set time interval (Solomon and Potter, 1988), or in direct response to a change in salinity (Deary et al., 1998; Niezgoda et al., 1998). Wild smolts emigrating to sea have been tracked in a number of estuaries in England and Wales using small 300 kHz acoustic tags surgically implanted in the peritoneum (Moore et al., 1995, 1998a, b). These tags, which weigh only 0.35 g in water, have been shown to affect neither survival nor swimming performance (Moore et al., 1990). Klimley and Holloway (1999) used a simple system of non-overlapping receivers to record the arrival of tagged yellowfin tuna at fish aggregating devices (FADs) off the coast of Hawaii. They were able to document some level of school cohesion in tagged tuna that arrived simultaneously at the FADs on numerous occasions. Ohta et al. (this volume) describe a similar study.

A similar system (Fig. 2) has been used to investigate the seaward movements of wild Atlantic salmon smolts in coastal waters in eastern Canada (Lacroix and McCurdy, 1996; Voegeli et al., 1998; Lacroix and Voegeli, 2000). Smolts marked with small (single-chip) acoustic pingers and released in freshwater, are detected by chains of self-contained automated underwater receivers moored in tidal passages between Passamaquoddy Bay and the Bay of Fundy, and recently across the whole width of the bay itself. In its latest version the system uses single channel receivers, designed for deployment in various configurations (lines or grids) and miniature coded transmitters.

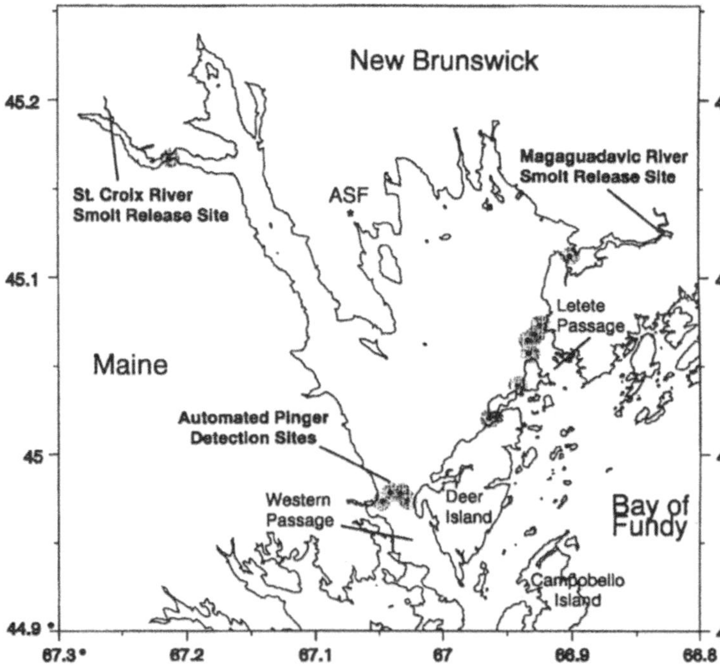

Figure 2. Map of Passamaquoddy Bay showing locations of underwater automated acoustic receivers, smolt (*Salmo salar*) release sites and Atlantic Salmon Federation (ASF) hatchery. (Voegeli et al., 1998).

The transmitters, which contain a micro-controller, transmit a complex six-pulse code capable of identifying 4096 different fish on the same frequency, although in practice only 30 are currently used (Voegeli et al., 1998; Lacroix and Voegeli, 2000). The system, which has proved to be very successful in protected coastal waters, has the potential to be used in the open sea, but has yet to be fully tested in this environment.

3.2 Mobile Hydrophone Systems

Mobile hydrophone systems are ideal for obtaining continuous tracks of individual fish, especially those that move long distances. A number of different tags and receiving systems have been developed over the years. Most marine fish tracking has been carried out with small boats using a directional hydrophone (e.g., Lawson and Carey, 1972) attached to a long retractable pole, which allows the hydrophone to be deployed at a depth of 1.5 to 3 m. The hydrophone is usually contained in a streamlined housing, which permits acoustic tags to be detected at speeds up to 7 knots (e.g., Stasko and Polar, 1973; Holland et al., 1985). The retractable pole can be fitted to a block clamped to the gunwale (Nelson, 1978), or mounted in a special bracket attached to the side of the boat (Holland et al., 1985, 1992). Another option is to use a fared shaft on the bow (Stasko and Polar, 1973; Jolley and Irby, 1979; Carey and Robison, 1981). Bow mounted hydrophones are, however, more susceptible to movement than those mounted on the side of the boat and may come out of the water as the boat pitches, leading to loss of the tag signal (Jolley and Irby, 1979). The hydrophone can be fixed facing forwards parallel to the keel, in which case the tag is located by turning the boat (Jolley and Irby, 1979; Holland et al., 1985). Alternatively, the hydrophone may be rotated to estimate the direction of maximum signal strength (Lawson and Carey, 1971; Carey and Lawson, 1973). Some systems have incorporated a remote, recording direction indicator (e.g., Young et al., 1976). Rotating hydrophones have also been used on larger research vessels (~25-50 m), attached to the side of the vessel (Ogura and Ishida, 1992) or mounted under the hull (Yuen, 1970; Tesch, 1972, 1974; Block et al., 1992a,b). Ogura and Ishida's (1992) system, which was used on a 48-m research vessel, employed one receiver and three directional hydrophones submerged 3 m below the surface. The hydrophones, which were mounted on a single pole, pointed down at angles of 10, 45 and 80°, respectively, and the receiver was switched between them as the fish changed depth.

More recently Vemco has simplified tracking by developing a multi-hydrophone system, which eliminates the need to rotate either vessel or hydrophone. The VR28 Tracking System (Vemco Ltd., Nova Scotia, Canada), has four line arrays spaced at 90° intervals, ahead, astern, port and starboard. A micro-controller receiver and four independent acoustic receivers are required to track the fish. Data from the receiver are interfaced with PC-based tracking software, which displays the range and bearing of the fish and provides a display of signal strength and background noise. The system, which can operate at speeds up to 7 knots, can be attached directly to the tracking vessel, or towed off the fore-deck using a V-fin, or other streamlined body, and a fared conducting cable. It can be used with CHAT tags (see Section 6.1), as well as pingers and transponders. Block et al. (1997), who used the system to track yellowfin tuna (*Thunnus albacares*) in the eastern Pacific, reported detection ranges of 0.8 to 1 nautical mile with 34 and 50 kHz transmitters.

After a fish is released the tracking boat remains within audible range of the acoustic tag and follows it, closing range every 10-20 min to take a position fix. The boat may be as close as 15 m when the fix is taken (e.g., LaBar et al., 1978) or as much as 200 m distant (Holland et al., 1985). Although errors in track reconstruction are introduced, it is assumed that the fish is at the same position as the boat. Prior to the use of GPS a number of techniques were used to determine the ship's location. Close to shore, the position of the boat could be fixed by triangulation using visual reference points and a compass (e.g., Groot et al., 1975), or sextant (Madison et al., 1972; LaBar et al., 1978). Flashing beacons (Young et al., 1976) or radar could be used at night and radar was also useful when visibility was poor by day (Madison et al., 1972; LaBar et al., 1978). Offshore, radio-navigation systems, such as Decca (e.g., Greer Walker et al., 1971, 1978; Tesch, 1972; Westerberg, 1979) or Loran C (e.g., Lawson and Carey, 1971; Laurs et al., 1977; Holland et al., 1985; Carey and Scharold, 1990), which generally gave less accurate positions than GPS, were used. Arnold et al. (1994) and McCleave and Arnold (1999) discuss the errors associated with the use of the Decca Navigator system for tracking fish in the North Sea.

Mobile hydrophone systems have been used extensively over the last 30 years to study the vertical and small-scale horizontal movements of a wide range of oceanic teleosts, including salmon (e.g. Soeda et al., 1987; Ogura and Ishida, 1992, 1995; Ogura, 1997), anguillid eels (e.g. Tesch, 1978, 1989) and various large pelagic species, including billfish (e.g., Carey and Robison, 1981; Block et al., 1992a,b; Brill et al., 1993) and tuna (e.g., Yuen, 1970; Holland et al., 1990a; Block et al., 1997; Brill et al., 1999; Josse et al., 1998; Dagorn et al., 2000; Lutcavage et al., 2000). The tuna studies have included several investigations of behaviour associated with fish aggregating devices (FADs) (Holland et al., 1990b; Cayré, 1991; Marsac and Cayré, 1998; Dagorn et al., 2001). Ultrasonic telemetry has also been used in attempts to pinpoint the open ocean spawning grounds of both European (*Anguilla anguilla*) and Japanese (*A. japonica*) eels. Artificially matured silver eels were released in probable spawning locations in the Sargasso Sea (Fricke and Kaese, 1995) and in the Pacific, west of the Mariana Islands (Aoyama et al., 1999), respectively. Applications of telemetry in the open ocean have also included studies of the behaviour of several cephalopods, including *Nautilus belauensis* (Carlson et al., 1984; Ward et al., 1984), the flying squid *Ommastrephes bartrami* (Nakamura, 1991, 1993), and the squid *Loligo forbesi* (O'Dor et al., 1994; Sauer et al., 1997).

There have been as many, if not more, ultrasonic tracking studies in shelf and coastal waters. As in the ocean —a lot of work has been done with salmon (e.g. Madison et al., 1972; Stasko et al., 1973, 1976; Døving et al., 1975; Groot et al., 1975; Stasko, 1975; Smith et al., 1981; Westerberg, 1982a,b; Potter, 1985; Shimamura et al., 1987; Soeda et al., 1987; Quinn, 1988, Quinn et al., 1989; Ruggerone et al., 1990; Olson and Quinn, 1992; Candy and Quinn, 1999) and anguillid eels (e.g. Tesch, 1972, 1974, 1979; Stasko and Rommel, 1974; Tesch et al., 1991; Westerberg, 1979). There have also been several studies of cod (*Gadus morhua*) (e.g. Hawkins et al., 1974; Clarke and Green, 1990; Wroblewski et al., 1994, 1995, 2000; Green and Wroblewski, 2000) and a number of other teleosts, including shad (*Alosa sapidissima*) (Dodson and Leggett, 1973, 1974), lingcod (*Ophiodon elongatus*) (Yamanka and Richards, 1993), yellowtail rockfish (*Sebastes flavidus*) (Pearcy, 1992), deepwater rockfishes (*Sebastes chlorostictus* and *S. paucispinis*) (Starr et al., 2000) and white goatfish (*Mulloides flavolineatus*) (Holland et al., 1993a). The coral trout (*Plectropomus leopardus)* has also been studied extensively, using ultrasonic telemetry to describe patterns of movement, space utilisation and the dynamics of spawning aggregations (Zeller, 1997, 1998a,b). A substantial amount of work has also been done with commercially exploited crustacea, using both acoustic tags (e.g. Monan and Thorne, 1973; Chapman and Rice, 1975; Hernnkind, 1980; Jensen et al., 1994; Hines et al., 1995; van der Meeren, 1997; Freire and González-Gurriarán, 1998; Tremblay et al., 1999; Watson et al., 1999) and electromagnetic tracking systems (e.g. Jernakoff, 1987; Jernakoff and Phillips, 1988; Smith et al., 1998, 1999; Collins et al., 2000).

Mobile tracking systems have been used to study the behaviour of several species of sharks in lagoons, and on reefs, banks and seamounts in both temperate and tropical waters. Studies have encompassed space utilisation and diel patterns of activity (Standora and Nelson, 1977; Sciarrotta et al., 1977; Tricas et al., 1981; Gruber et al., 1988; Klimley et al., 1988; Carey and Clark, 1995; Nelson et al., 1997; Goldman and Anderson, 1999; Gunn et al., 1999; Ackerman et al., 2000; Wetherbee et al., this volume), social groupings (McKibben and Nelson, 1986) and bioenergetics (Nelson, 1990; Matern et al., 2000; Lowe and Goldman, 2001), as well as homing, orientation and navigation (Carey and Scharold, 1990; Klimley, 1993; Holland et al., 1999).

Observations have been made with both juveniles (Medved and Marshall, 1983; Holland et al., 1992, 1993b; Holts and Bedford, 1993) and adults. Ultrasonic telemetry studies of sharks and rays have recently been reviewed by Sundström et al. (2001) and Voegeli et al. (2001).

3.3 Sector Scanning Sonar

High-frequency, high-resolution sector scanning sonar (Voglis and Cook, 1966), used in combination with transponding acoustic tags (Mitson and Storeton West, 1971; Greer Walker et al., 1971, 1978), provides another method for examining the movements of fish in the open sea. With this technique the position of the fish can be determined precisely and its track reconstructed more accurately than with a hydrophone system. The Lowestoft Laboratory first used this method to track fish in the open sea in 1970 and has used the technology almost continuously ever since.

The original version of the sonar, which was developed for mine hunting by the Admiralty Research Laboratory, was installed in RV 'Clione' (42.5 m; Mitson and Cook, 1971) and transferred to RV 'Corystes' (53 m) in 1988. The current system uses digital signal processing. The sector scanning sonar produces acoustic pictures of the seabed and other underwater targets, such as trawl gear (Cushing and Harden Jones, 1966). It insonifies a 30° x 10° beam (Fig. 3a) with a 2 ms pulse of sound at rates of 2 or 4 pulses s^{-1}, which correspond to operating ranges of 200 and 400 m. The sonar beam is scanned electronically at 10 000 Hz. Range and bearing resolution are 20 cm and 0.33°, respectively, although during tracking the position of the fish is estimated to the nearest 5 m and 1°. The acoustic picture is usually presented on a rectangular (B-scan) display with range (200 or 400 m maximum) up the long axis and bearing (±15° of the centre line) across the short axis (Greer Walker et al., 1978). The transducer is hydraulically stabilised against roll, pitch and yaw and can be steered in azimuth (0 to 220° in either direction from dead ahead) to observe all round the ship. Tilt angle can be varied continuously from vertical to horizontal. Azimuth control is referenced to the navigational gyrocompass; a second gyrocompass provides a vertical reference. The sonar can be operated in vertical mode (Fig. 3b) with the same control over azimuth and tilt angles; switching between horizontal and vertical modes takes only a few seconds. The depth of targets can be measured in vertical mode by increasing the tilt of the transducer until first the target and then the sea bed echo are brought onto the centre line of the display (Greer Walker et al., 1978). Depth estimates are quite good (± 0.5 m) at close range but increase significantly at greater distances (Arnold and Greer Walker, 1992).

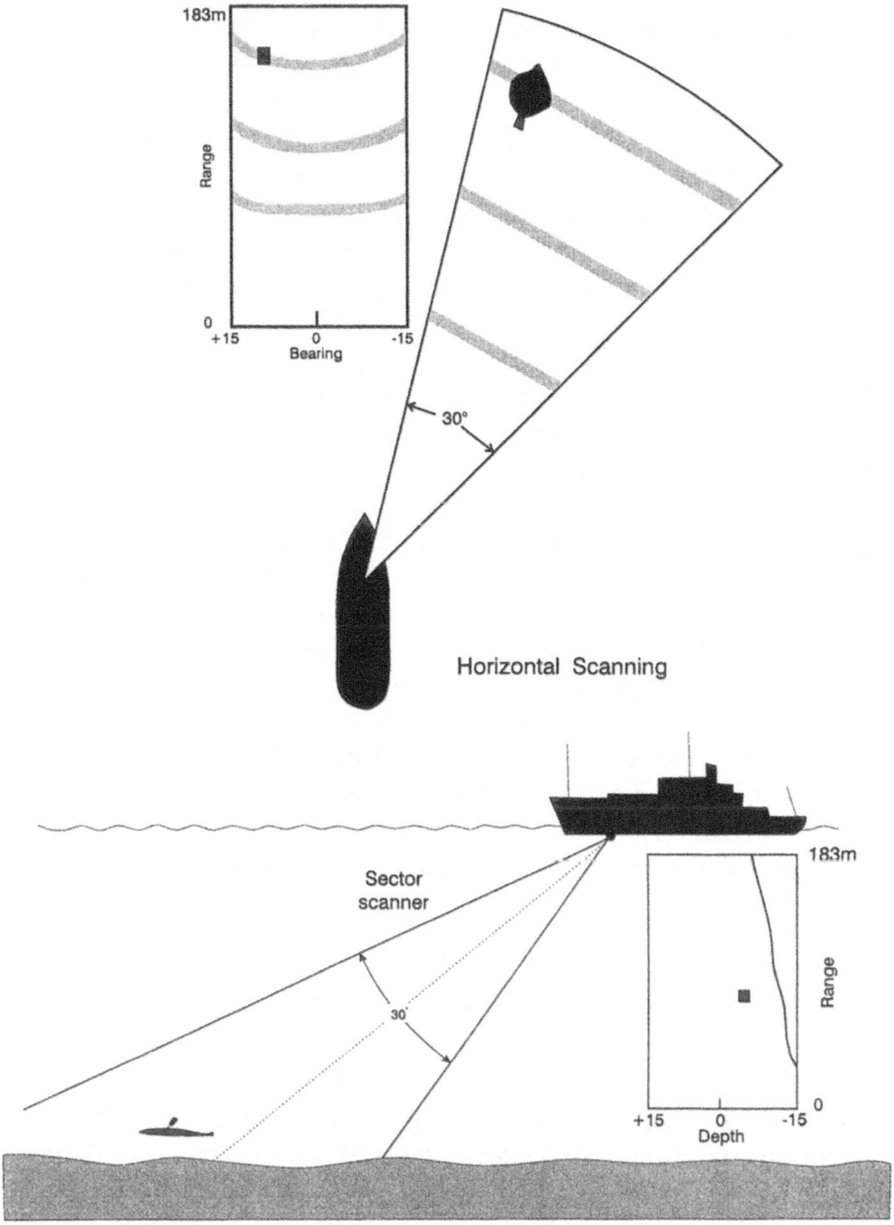

Figure 3. Tracking fish in the open sea using transponding acoustic tags and 300 kHz sector scanning sonar in horizontal (top) and vertical (bottom) mode. The range and bearing of the fish are measured in horizontal mode. Depths are estimated in vertical mode from slant range and the tilt angle of the transducer, when first the tag signal and then the bottom signal are brought on to the centre line of the B-scan display (see lower insert). For further details see Greer Walker et al. (1978), Arnold and Greer Walker (1992) and McCleave and Arnold (1999).

Transponding acoustic tags, which transmit a 3 ms pulse of sound when insonified by the sonar (Mitson and Storeton West, 1971), can be detected out to ranges of about 600 m under ideal conditions. However, fish are normally tracked at ranges between 120 and 250 m (mean 175 ± S.D. 30 m) (Arnold et al., 1994). The fish track is plotted electronically, using the range and bearing of the fish from the ship estimated from the sonar display and the position of the ship determined by GPS (Buckley and Arnold, this volume). The original 300 kHz acoustic tag, whose dimensions were 50 x 10 mm (Mitson and Storeton West, 1971), had negligible drag (Arnold and Holford, 1978) but an operating life of only about 2-3 days. The tag currently in use is larger (60 x 15 mm), has bigger batteries and a life of about 10 days, which can be increased by transmitting alternate pulses of sound with the sector scanning sonar. Tracks of up to 12.5 days have been achieved in this way (Buckley and Arnold, this volume).

The Lowestoft Laboratory used sector scanning sonar and transponding acoustic tags to measure the efficiency of the Granton trawl (Harden Jones et al., 1977; Harden Jones and Arnold, 1982), at the time one of the most important fishing gears in the North Sea. The same technique was used to study the migratory behaviour of plaice (*Pleuronectes platessa*), sole (*Solea solea*), cod (*G. morhua*), dogfish (*Scyliorhinus canicula*), eels (*A. anguilla*) (Greer Walker et al., 1978; Greer Walker et al., 1980; Arnold and Greer Walker, 1992; Arnold et al., 1994; McCleave and Arnold, 1999; Buckley and Arnold, this volume) and salmon (*S. salar*). Other projects have involved telemetering physical (depth and compass heading) and physiological data (heart rate) from free-swimming fish, including plaice, cod and salmon (Storeton West et al., 1978; Harden Jones and Arnold, 1982; Potter, 1985; Arnold et al., 1990; Metcalfe et al., 1993).

3.4 Deep-Sea Tracking Systems

Scientists at Aberdeen University in Scotland have developed an autonomous free-fall acoustic tracking vehicle capable of following the movements of abyssal scavengers, such as grenadiers (*Coryphaenoides* spp.), in the deep ocean (e.g., Priede et al., 1990; Armstrong et al., 1992). In its latest version, the AUDOS (Aberdeen University Deep Ocean Submersible) vehicle, which is capable of operating at depths down to 6000 m, carries a scanning directional sonar, as well as a camera, flash unit, electromagnetic current meter, compass, micro-controller and batteries (Bagley and Priede, 1996). The scanning sonar, which has a 75 mm diameter ceramic disc transducer resonant at 77 kHz, is rotated mechanically through 360° in alternate directions on successive scans. The electric motor that rotates the transducer is programmed to stop 52 times during each full rotation to allow tag interrogation.

Fish are tagged with ingestible code-activated transponding (CAT) tags, also operating at 77 kHz (Bagley, 1992; Bagley et al., 2000); up to three tags have been deployed at any one time. The tags (65 x 12.5 mm), which transmit a single return pulse on receipt of an individually recognised pulse code, can be detected out to a range of 500 m with a precision of 0.5 m. The activation code consist of two 10 ms acoustic pulses separated by a programmable interval of 100, 200 or 300 ms.

4. TELEMETRY OF PHYSICAL AND PHYSIOLOGICAL DATA

Early transmitters (e.g., Trefethen, 1956; Johnson, 1960; Bass and Raskovich, 1965) functioned solely as acoustic beacons, but within a few years sensors were added to acoustic tags to telemeter physical or physiological data. With smaller fish it is still generally necessary to use single channel devices. Multiplex systems have, however, been used with sharks (e.g., Ferrel et al., 1974; Carey et al., 1982; Nelson, 1990) and large pelagic teleosts (Lawson and Carey, 1971; Block et al., 1992a) for 30 years to transmit data sequentially from multiple sensors. Variables are identified by a series of synchronisation pulses and values by the subsequent pulse repetition rate.

4.1 Physical Data

No picture of marine fish behaviour would be complete without detailed information on vertical movement, which constitutes a key component in the life history of most species, regardless of whether they are nominally described as demersal or pelagic. Vertical movements are to a large extent dictated by the physical environment, either directly through a reaction of the fish to light intensity or temperature, or indirectly through the responses of prey organisms to the same environmental factors (e.g., Carey and Robinson, 1981; Block et al., 1992a; Josse et al., 1998; Marsac and Cayré, 1998). Much effort has therefore been invested in producing tags that transmit physical data, such as pressure and temperature. Pressure sensors, which measure depth, have been in widespread use since the first devices were built in the early 1970s (Lawson and Carey, 1972; Luke et al., 1973). They are now an essential component of all tracking studies, except those with small fish (e.g., salmon smolts, Moore et al., 1990) unable to carry a telemetry tag, or those in which the depth of the fish can be measured directly by sonar (e.g., Arnold and Greer Walker, 1992). Temperature sensitive tags were developed simultaneously and both sensors are now usually incorporated into the same device (Lawson and Carey, 1972; Carey and Lawson, 1973).

Tags capable of measuring other physical factors, such as salinity (Priede, 1982), dissolved oxygen concentration (Priede et al., 1988) and light intensity (Standora and Nelson, 1977; Nelson, 1978), have generally been custom built for specific applications, as have devices for measuring swimming speed and direction and locomotor activity. Direct measurements of swimming speed have usually been made with mechanical devices (wands, drogues or propellers), whose size has limited application to sharks (Nelson, 1976, 1978; Standora and Nelson, 1977; Sundström and Gruber, 1998), and large pelagic teleosts, such as tuna and marlin (Block et al., 1992a). The biomorph sensor used with brown trout (Young et al., 1972) and the rolling ball device developed by Holand (1974, 1987) were exceptions. Several non-mechanical tail beat frequency sensors have, however, been developed for smaller fish (Stasko and Horrall, 1976; Ross et al, 1981; Voegeli and Pincock, 1981) and juvenile sharks (Lowe et al., 1998). Recently, differential pressure measurements have been used to telemeter tail beat frequencies from European sea bass (*Dicentrarchus labrax*) (Webber et al., this volume). Mechanical compasses have been developed for use with both large sharks (Nelson 1976, 1978) and small teleosts (Mitson et al., 1982; Pearson and Storeton-West, 1987).

4.2 Physiological Data

Many investigators have developed specialised sensors to study the physiological ecology of free-swimming fish. Many studies have involved acoustic telemetry of one or more correlates of metabolism (Lucas and Johnstone, 1993), such as heart or respiration rate (Oswald, 1978). Heart rate has been recorded in a variety of freshwater species, which have included salmon (Kanwisher et al., 1974), trout (Priede and Tytler, 1977) and pike (Armstrong et al., 1989). In the sea measurements have been made with plaice (Kanwisher et al., 1974; Storeton-West et al., 1978), sole and bass (Sureau and Lagardère, 1991), cod (Kanwisher et al., 1974; Wardle and Kanwisher, 1974), mackerel (Kanwisher et al., 1974), and lemon sharks (Scharold and Gruber, 1991). Tail beat frequency and measures of swimming speed (see previous section) have also been used to estimate energetic costs, and similar studies have been conducted with cephalopods, using acoustic tags to telemeter measurements of swimming jet pressure (e.g., Webber and O'Dor, 1986; O'Dor et al., 1994; Aitken et al., 2000). Temperature has been measured in tunas to assess their capacity for thermoregulation and document the extent to which body temperatures are elevated above ambient (Carey and Lawson, 1973; Marcenik et al., 2001). Similar studies have been carried out with other endotherms, such as swordfish (Carey and Robison, 1981), blue marlin (Block et al., 1992a) and sharks (Carey et al., 1982; Carey and Gibson, 1987; Lowe and Goldman, 2001). Using a combination of depth and muscle temperature transmitters, Holland and colleagues were able to demonstrate that elevated body temperatures in bigeye tuna resulted from physiologically controlled short-term changes in whole-body thermal conductivity (Holland et al., 1992; Holland and Sibert, 1994). Visceral temperatures have been investigated in several species of sharks in relation to feeding physiology (Carey et al., 1981; McCosker, 1987). As with other telemetry studies, these physiological measurements have been made almost entirely with tags developed for specific projects. Exceptionally, field studies, which began with custom-built acoustic tags (Rogers et al., 1984; Wolcott and Hines, 1989) have led to commercial production of EMG telemetry tags, albeit so far only with radio transmitters for use in freshwater (Demers et al., 1996; Kaseloo et al., 1996). Dewar et al. (1999) have, however, recently described tank tests of an acoustic telemetry tag designed to measure EMGs in free-swimming marine fish.

In addition to direct measurement of physiological variables, there have been at least two attempts to investigate sensory physiology by altering the local magnetic field around a free-swimming fish. In each case a magnetic field generator has been combined with an electronic tag such that it could be switched on and off remotely under controlled conditions (Westerberg, 1982; Yano et al., 1996, 1997).

5. ARCHIVAL (DATA STORAGE) TAGS

Archival, or data storage, tags record data that are subsequently downloaded when the fish is recaptured and the tag recovered. These tags were first deployed in the early 1990s. CSIRO (Gunn et al., 1994; Anon, 1999) deployed 28 tags on southern bluefin tuna (*Thunnus maccoyii*) in Australia in 1993 and in the same year CEFAS (Metcalfe et al., 1994; Metcalfe and Arnold, 1997) released 50 tagged plaice (*P. platessa*) in the North Sea. In the following year, the Institute of Freshwater Fisheries, Reykjavik, released 60 tagged Atlantic salmon (*S. salar*) in the sea off the coast of Iceland. We estimate that, since then, several thousand archival tags have been deployed on a wide range of fish. Rates of return have approached 50% in some over-exploited species of finfish, such as cod, and have been even higher (~70%) for some decapod crustacea (Freire et al., 1999). As with conventional tagging, the method still depends on the return of tags by fishermen. The ermen. The data (Fig. 4) are, however, independent of the fisheries and have the potential to solve some major ecological questions that are unanswerable with data from either commercial fisheries, or research vessel surveys. Although there are already substantial databases of archival tag data, few definitive reports have, as yet, appeared. We have therefore drawn on reports of ICES Annual Science Conferences, and similar events, to produce an overall summary. The picture is an exciting one, as can be seen from the following vignettes, which show that important insights can be gained from a relatively small number of tag returns.

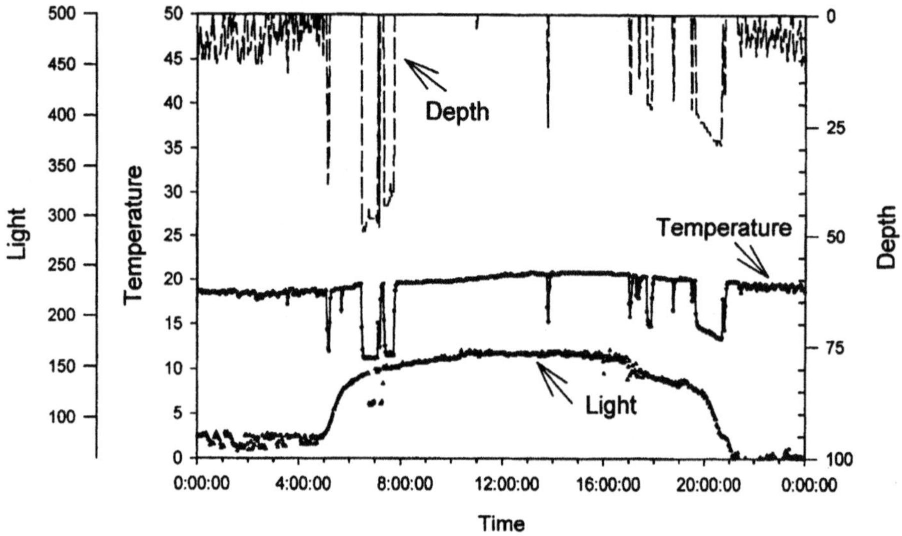

Figure 4. Depth, ambient temperature and light intensity for a 24-h period from an archival tag recovered from a small white shark (*Charcharodon carcharias*). The data were sampled every 2 minutes and are typical of records obtained from large pelagic species (H. Dewar, unpublished data).

5.1 Types of Tags and Sensors

Designed to address specific questions, the archival tags developed for the CSIRO and CEFAS research programmes were sophisticated multichannel devices that could record several physical factors (e.g., pressure, temperature, light intensity) simultaneously at frequent intervals (Gunn et al., 1994; Metcalfe and Arnold, 1997). Smaller, simpler devices have only subsequently become available (Walker et al, 2000). Several manufacturers now produce a variety of tags of different sizes and capabilities. Some are only suitable for use with large, highly migratory pelagic species, such as sharks, tuna and billfish, others are simple temperature-depth recorders (TDRs) that can be used with smaller species. Costs range from a few hundred dollars for a small TDR to $1300, or more, for a fully programmable multi-channel device with a recording life of several years (Table 1). Despite their high price, archival tags offer a very cost-effective option when compared to research vessels, which have limited range and endurance, are vulnerable to bad weather and costly to operate. Archival tags allow the detailed behaviour of individual fish to be observed over long periods, regardless of where they go. Observations can also be replicated.

5.2 Early Technical Problems

Although archival tags have already provided a large amount of high-quality data, and some tags have performed flawlessly over long periods, all manufacturers have experienced technical difficulties and tag failures. Some problems (e.g., battery passivation and pressure sensor failure) have been purely technical; others have resulted from unexpected behaviour on the part of the fish, which has subjected the tags to physical conditions outside their design specifications. Bluefin tuna, for example, have been found to dive rapidly to depths >1000 m (Block et al., this volume), well in excess of the pressure rating of most first-generation archival tags. Several types of tag have suffered from intrusion of water, which has damaged the electronics. Solar heating has been a problem with Star-Oddi and Lotek tags, both of which have translucent cases (Westerberg et al., 1999b). The circuit boards and electronic components absorb visible light and heat up, giving an erroneous measurement of ambient water temperature. This may be critical when surface temperatures are used to help estimate the position of the fish, although the problem can apparently be overcome by providing an efficient heat bridge to connect the thermistor with the water.

Rigid light stalks have proved to be a weak point in the design of some surgically implanted tags, often failing at the point of entry into the tag. Block et al. (this volume), for example, report that all first generation Mk7F tags (Wildlife Computers, Redmond Washington, USA) used on Atlantic bluefin tuna in 1997 failed during the first year of the mission, or shortly into the second year. Despite strengthening, stalk failures were still a problem with second generation tags, possibly because of increased stiffness, and giant bluefin tuna still appear to be able to break the stalk of third generation Mk7 tags during both feeding and spawning. The thin flexible light stalk of the NMT tag has proved to be much more resilient, although the slender, cylindrical, stainless steel case has sometimes migrated into the body cavity, rendering the light stalk inoperable (see Section 5.4.3).

Tag recovery has proved to be a problem for several archival tagging programmes and continuous publicity is needed to ensure that tags are removed from the fish and returned to the relevant laboratory. Despite explicit markers on the outside of the fish, surgically implanted tags have frequently been overlooked when tuna have been gutted at sea, for example. In one case the light stalk was cut off and returned with the conventional tag, leaving the archival tag inside the fish (Block et al., this volume). Atlantic tuna tagging programmes pay a high reward ($1000 US) for the return of each archival tag, but even this is not sufficient to ensure the return of all recaptured tags. Externally placed tags, which are more obvious, have generally presented less of a problem.

5.3 Light-Based Geolocation

For several centuries before the advent of GPS, human navigators used a succession of instruments, from astrolabe to sextant, to measure the altitude of the sun and determine latitude. With the advent of accurate marine chronometers in the late eighteenth century, it became possible to measure the time of local noon relative to Greenwich, and thus calculate longitude. Although far less precise, the zenith angle of the sun can also be used to determine geographical position in a much simpler way. This entails measuring surface light intensity with time using a recording device that incorporates an accurate real-time clock. If the animal dives, the measurements must be corrected for the attenuation of light with depth. Critical periods occur during twilight, when the zenith angle of the sun is between 90.83° (sunset and sunrise) and 108° (astronomical twilight). To estimate geographical position the observed light intensity must be related to a specific zenith angle, although not necessarily to a specific event, such as sunset, or civil (96°) or nautical (102°) twilight. Longitude is calculated from the time of noon, which occurs midway between the times at which light intensity reaches the level corresponding to the chosen zenith angle. Latitude is calculated from the length of the intervening day. The choice of angle can make a significant difference to the expected variability in latitude, especially around the equinoxes (Hill and Braun, this volume). There is an optimum zenith angle, which will minimise errors in estimating latitude, but this varies with both season and latitude. The optimum angle can be selected if the latitude of the tag is known, otherwise it is necessary to associate a series of points on the light intensity curve with a corresponding series of zenith angles, as is done in Wildlife Computers' tags. These points can then be used to produce a series of day lengths, one for each zenith angle, which can be compared with predictions to identify the latitude that gives the best fit to the data. Fuller treatments of the subject are given by De Long et al. (1992), Wilson et al. (1992), Hill (1994) and Welch and Eveson (1999) and by Metcalfe, Hill and Braun, Musyl et al. (this volume).

While the theory behind light-based geolocation is relatively straightforward, the actual estimation of the sun's zenith angle is complicated by a number of factors. Because of changes in light refraction in the atmosphere, the greatest accuracy with which latitude and longitude can be calculated is 0.7° and 0.32°, respectively (Hill and Braun, this volume). In practice, accuracy of geolocation is also affected by physical factors, such as weather, biofouling and turbidity, the behaviour of the fish (primarily diving and large daily movements) and the characteristics of the light sensor. For estimates of longitude it is critical that the real-time clock in the tag is accurate and can be corrected for drift at the end of the deployment

There have been several attempts to determine the accuracy of geolocation by archival tags, the most recent by Musyl et al. (this volume), who assessed the performance of both NMT (now marketed by Lotek Wireless) and Wildlife Computers (WC) tags (Table 1). The tags were suspended from a stationary mooring over a range of depths for almost a year. A number of methods for calculating latitude and longitude were compared for the WC tags, which, unlike NMT tags, record the original light intensity measurements. The results showed that longitude can be determined with accuracy greater than 0.5°. The mean absolute longitude errors were less than 0.35 and 0.25° throughout the year for NMT and WC tags, respectively. As expected from theory, estimates of latitude were an order of magnitude less accurate and mean absolute errors for the NMT tags ranged from 1.5 to 5.5°. The equivalent errors for the WC tags ranged from 0.78 to 3.5° and were influenced by the method chosen to define 'dawn' and 'dusk'. For comparison, Welch and Eveson (1999) reported errors of 0.9° for longitude and 1.2° for latitude for archival tags secured both to moorings and a moving vessel. Gunn et al. (1994), similarly, reported errors of 0.54° for longitude and 1.5° for latitude for tuna towed around the southeastern tip of Australia in a pen.

5.4 Applications with Pelagic and Semi-Pelagic Species

5.4.1 Southern bluefin tuna

CSIRO scientists in Hobart have obtained valuable new information about the movement, behaviour and physiology of juvenile southern bluefin tuna (*T. maccoyii*), using archival tags made by Zelcon Technic Pty (Gunn et al., 1994) and Wildlife Computers. Only 2 (7%) of the first batch of 28 tags were recovered, but the recapture rate increased to 10% for releases in 1994 (148 tags) and 31% for releases in 1995 (137 fish). The overall recovery rate to September 2000 was 85 tags (16.5%) from 514 releases (Anon, 1999). The tags, which were implanted surgically beneath the caecum and stomach, incorporated pressure and internal temperature sensors. An external stalk with light and temperature sensors penetrated the body wall and trailed beneath the fish. The CSIRO programme has shown that 3- to 6-year-old

Table 1. Capabilities of archival tags produced by various manufacturers

Manufacturer	Corporation Devices	Alec Electronics	Lotek Wireless			Star-Oddi		Vemco	Wildlife Computers
Model	RL-05T	MDS	LTD-100	LTD-1000	NMT-750	DST 200	DST 300	Minilog	Mk 7 TDR
Dimensions (mm)	40 x 23 x 8	18 x 109[1]	18 x 57	23 x 64	16 x 100	18 x 48	13 x 46	16 x 71	20 x 10 x 87.5
"	-	18 x 109[1]	-	-	-	-	-	22 x 95[10]	20 x 10 x 100
"	-	18 x 127[2]	-	-	-	-	-	-	20 x 20 x 70
Sensor stalk (mm)	-	-	-	-	>30 x 2.2	-	-	-	≤300 x 3.5
Weight (g) in air (water)	9.5 (3.3)	28.4 (3.5)[1] 33.6 (5.0)[2]	16 (1)	30 (6.1)	52 (-)	12 (1)	8 (1.5)	-	32 (15)
Memory capacity	-	32 kb	1.5 Mb	1.5 Mb	256 kb	-	-	8 kb [11]	2 Mb
No. measurements	8192	32 000	750 000	750 000	-	32 676	32 676	8128[12]	2 000 000
Depth: (m)									
max.	>1000	500	173	1250	-	600	900	150% fsd	120% fsd
full scale (fsd)	-	200, 500	100	1000	510[5], 700[6]	500	800	17, 34, 68, 136, 204, 340, 680	500, 1000 or 2000
accuracy	-	±1, ±2	±1	±10	-	-	-	±2% fsd	±1% reading
resolution	-	-	0.05	0.5	1 (3) [7]	2	3.2	1% fsd	0.25, 0.5, or 1[13]
Temperature: (°C)									
range	-5 to 30	0 to 30	2 to 35[3] -20 to 35[4]	2 to 35[3] -20 to 35[4]	-40 to 85 -5 to 35[8]	-25 to 50	-25 to 50	-4 to 20, -5 to 35 -30 to 40	-40 to 60
accuracy	0.2	0.15	0.06	0.06	-	-	-	0.2, 0.3, 0.5	0.1
resolution	-	-	0.03	0.03	0.2	0.1	0.1	0.1, 0.2, 0.3	0.05[14]
Salinity: (psu)									
range	-	-	-	-	-	5 to 37	-	-	-
resolution	-	-	-	-	-	0.2 to 2[9]	-	-	-

(table continued next page)

(Table 1 continued)

Manufacturer / Model	Corporation Devices RL-05T	Alec Electronics MDS	Lotek Wireless LTD-100	Lotek Wireless LTD-1000	Lotek Wireless NMT-750	Star-Oddi DST 200	Star-Oddi DST 300	Vemco Minilog	Wildlife Computers Mk 7 TDR
Tilt angle: (°)									
range	-	-	-	-	-	-	-45 to 45	-	-
resolution	-	-	-	-	-	-	<5	-	-
Light sensor:									
location	-	Internal	internal	internal	external	-	-	-	int/external
wavelength (nm)	-		-	-	450 ± 50	-	-	-	500 ± 60
Sampling interval	14 s and multiples	1 s plus 1, 10, 20 min	1 s to 99 days	1 s to 99 days	128 s			1 s to 6 h	1 s to 55 min
Data channels	1	1	8	8	4	2 or 3	2 or 3	1 or 2	4
Housing	clear ure-thane	Polyester resin	poly-carbonate	poly-carbonate	stainless steel	poly-carbonate	poly-carbonate	epoxy	epoxy

Notes: (1) MDS-T (temperature) and MDS-L (light) recorders; (2) MDS-D (depth) recorder; (3) silver oxide battery; (4) lithium battery; (5) time series; (6) histogram; (7) 1 m from 0-126 m, then 3 m; (8) range for full accuracy; (9) 0.2 psu from 5 to 16 psu; (10) with replaceable battery; (11) standard memory 8 k bytes per sensor - 16 and 64 k bytes also available; (12) per sensor with standard memory; (13) only 252 values are reported - the actual resolution varies with depth, depending on the reporting range selected; (14) resolution varies with temperature, depending on the reporting range selected.

bluefin, which spend the summer in the warm (17-21°C) waters of the Great Australian Bight, make previously unknown feeding migrations into the southern Indian Ocean and Tasman Seas in winter at latitudes between 30 and 40°S. On their feeding grounds, the larger juveniles appear to move regularly between daytime feeding areas in cold water and a night-time thermal refuge (Gunn and Young, 2000). Off eastern Tasmania, tuna move offshore at night to the oligotrophic East Australia Current, which originates in the tropics, and can be 8°C warmer than the productive waters on the continental shelf. In the Indian Ocean, juvenile SBT dive into deep (>300 m), cold (7-8°C) water to feed on squid by day and return to the surface at night, the reverse of the pattern younger fish exhibit in the Great Australian Bight (Gunn and Young, 2000).

By using internally implanted tags, CSIRO scientists have also been able to quantify the cyclical changes of internal temperature that accompany feeding and that were originally described in large Atlantic bluefin tuna (*T. thynnus*) by Carey et al. (1984), using ultrasonic tags inserted in the stomach. Juvenile SBT have been shown to maintain their basal visceral temperature at 2-3°C above that of the surrounding water, regardless of seasonal changes of 6-7°C in ambient temperature. There are strong relationships between food intake and the extent of visceral warming and with clupeoid prey the time taken for the viscera to reach maximum heat provides an easy, robust measure of the quantity of food consumed (Gunn et al., this volume).

5.4.2 Pacific bluefin tuna

During the winters of 1995 and 1996, Japanese scientists tagged 105 young Pacific bluefin tuna (*T. orientalis*) in the Tsushima Strait between the Japan Sea and the East China Sea (Kittigawa et al., 2000; Inagake, et al., 2001). Fifteen of the 24 recaptured fish stayed in the East China Sea until the following summer. Although making frequent feeding dives to depths of 120 m during the day, they remained predominantly in the surface mixed layer. Average swimming depths were much shallower during summer (0.8–6.3 m by day, 2.4-5.2 m at night) than in winter (27-70 m by day, 23-32 m at night), when there was no thermal stratification (Kittigawa et al., 2000). Five fish moved into the North Pacific, either south of Kyushu or through the Japan Sea, and were recaptured after periods of 611 to 1635 days at liberty. Two fish showed trans-Pacific migrations. One made a round trip to the North American coast and subsequently returned to Japan; the other was caught close to the US coast at 31° 48'N, 117° 18'W. Two of the three remaining tuna, and possibly the third, showed clockwise annual migration circuits closely related to oceanographic conditions, perhaps making selective use of ocean currents. These fish moved west in spring around the Kuroshio Extension and north in summer in the warm water spreading from the crest of the Kuroshio Extension. In autumn, they moved east along the south of the Oyashio Front and the Sub-arctic Boundary, and in early winter south to the Kuroshio Extension (Inagake, et al., 2001).

5.4.3 Atlantic bluefin tuna

US scientists began tagging Atlantic bluefin tuna (*T. thynnus*) in 1996 to investigate migration and spawning site fidelity and test the current ICCAT management hypothesis that there are discrete western and eastern stocks in the North Atlantic (Block et al., 1998a). Since 1996 the "Tag-A-Giant" (TAG) programme, a collaborative effort between the Monterey Bay Aquarium, Stanford University, and the National Marine Fisheries Service (NMFS), has deployed 279 archival tags, of which 39 (14%) had been reported recaptured up to October 2000 (Block et al., this volume). All of the tags have been deployed in the recreational tag-and-release fishery off Cape Hatteras and Cape Lookout, North Carolina, where large winter aggregations of bluefin have occurred from 1994 onwards, together with large concentrations of menhaden and other baitfish. Medium (61-141 kg) and giant (>141 kg) tuna caught on rod-and-line with heavy tackle are landed through a door in the stern of a small (<14 m) sports fishing vessel and tagged internally, using archival tags made by NMT and Wildlife Computers. Because cylindrical tags can migrate into the body cavity (see Section 5.2), recent tags have been sutured in place and restrained by a monofilament line anchored in the ventral muscle (Block et al., this volume).

The 39 recaptures from this programme have been reported from the western, central and eastern North Atlantic, the Gulf of Mexico and the Mediterranean Sea after periods of 160 to 1310 days at liberty (Block et al., this volume), the majority from the western Atlantic. Thirty tags have been returned and sufficient data obtained to describe a number of stereotyped patterns of behaviour. The latitude of the fish is only known at the release and recapture positions, but estimates of longitude from daily measurements of times of sunset and sunrise—or regular deep dives associated with these events—are sufficient to identify general patterns of movement.

Atlantic bluefin prefer the top 200 m of the water column and spend more than half their time in the upper 40 m (Block et al., this volume), although at times they can dive rapidly to depths of more than 1000 m. In spring, when the fish have migrated off the continental shelf, deep diving is probably associated with feeding. In the Gulf of Mexico, where deep dives are quite common during the spawning season, the function may be thermoregulatory. Most of the bluefin tagged off North Carolina between 1997 and 1999 have remained in the western North Atlantic. They have moved offshore in spring and travelled northeast with the Gulf Stream before returning to the continental shelf in the summer. Feeding occurs off New England in summer and early fall and continues into winter, by which time the fish have returned to North Carolina. A number of fish have shown this pattern for several years, before moving much greater distances towards the two known spawning areas in the Gulf of Mexico and western Mediterranean. Three large fish, for example, which were sufficiently large to be Mediterranean spawners, were caught in summer months near the Straits of Sicily after crossing the Atlantic from west to east. There are already sufficient data to show that Atlantic bluefin tuna tagged off the east coast of the USA are vulnerable to fishing on both sides of the North Atlantic, with obvious implications for management. With additional tag returns it should be possible to estimate the extent to which fish that spawn in the Gulf of Mexico mix with fish that breed in the Mediterranean. Archival tag data will probably also be able to resolve whether bluefin spawn annually and if they demonstrate spawning site fidelity.

Archival tags have also been used in the Mediterranean. Japanese scientists released 60 juvenile bluefin (70-90 cm fork length) in the Adriatic Sea near Zada, Croatia, in November 1999. The fish were caught by purse seine and kept in cages for about 4 months before release. Seven tags (12%) had been recovered by September 2000, all near the release area. The fish dived extensively, initially in the upper 60-70 m of the water column, but subsequently to depths of 150 m several times a day during the winter. In April and May two fish occasionally dived to at least 765 m, the limit of the pressure sensor (Yamashita and Miyabe, 2000).

5.4.4 Juvenile yellowtails

Small numbers of single-channel archival tags have been used to investigate distribution and movements of pre-recruit yellowtails (*Seriola quinqueradiata*) in Sagami Bay, Japan (Kasai et al., 2000). Temperature or depth data were recorded from four 0-group and two 1-group fish for periods of 5-69 days. The small temperature variations ($<0.5°C\ d^{-1}$) around ambient sea surface temperature (SST) and the depths (~10 m) recorded by the tags on the 0-group fish showed that this year-class remained within the immediate coastal area. In contrast, the tags on the 1-group fish recorded systematic diel changes in temperature ($3°C\ d^{-1}$) and depth (10-30 m), which, together with frequent daytime dives to 40 m, indicated that the older fish were moving offshore to feed during the day.

5.5 Applications with Salmonids

Archival tags have been used quite widely on salmonids in Iceland, the Baltic and Japan. In Iceland the main aim has been to study the marine migrations of Atlantic salmon (*Salmo salar*) (Sturlaugsson, 1995; Sturlaugsson and Thorisson, 1997), sea trout (*S. trutta*) (Sturlaugsson and Jóhansson, 1996) and arctic char (*Salvelinus alpinus*) (Sturlaugsson et al., 1998). Salmon were captured in the estuaries of their home rivers, tagged and released at several sites up to 200 km away; recapture rates were 40-65%. The Baltic studies (Karlsson *et al.*, 1996, 1999) also focused on homing migration in salmon and incorporated an appraisal of tag performance (Westerberg et al., 1999b). Fish were, however, tagged much farther away from their spawning rivers, at an earlier stage of migration. From 1995 to 1997, 211 salmon were tagged and 75 (36%) have been recovered, most in the year of tagging. The tags were positively buoyant, a point that all archival tag users and manufacturers could usefully keep in mind. About 21% of the recoveries were found either on shore or floating in the water, after becoming detached from the fish. Recoveries of archival tags were also nearly twice as high as those of the conventional Carlin tags used as controls, probably because of the higher reward and greater conspicuousness (Karlsson et al., 1999). Walker et al. (2000) observed similar effects.

The Icelandic and Baltic studies confirm that migrating Atlantic salmon (and sea trout) spend much of their time within a few metres of the surface, allowing satellite measurements of SST to be used to deduce location. In both locations salmon migrated along the coast rather than further offshore. In 1995 and 1996 Baltic salmon migrated up the eastern side of the Gulf of Bothnia, remaining within the narrow band of warmer water inshore of the strong temperature front on the Finnish side of the Gulf. These fish swam at a mean speed of approximately 30 km d^{-1} and at a median depth of 2 m, with no indication of any periodic change in depth (Westerberg et al., 1999a). Although the Icelandic salmon swam at similar depths ($\leq$ 2 m for 90% of the time), there was evidence of a diurnal rhythm of vertical movement, with fish swimming deeper at night. The salmon were closest to the surface at noon and the majority of deepest dives were related to sunrise and sunset. Most dives were rapid and shallow, but some penetrated the thermocline at 20-40 m depth and one fish dived to a maximum depth of 110 m (Sturlaugsson and Thorisson, 1997).

Japanese studies (Ogura, 1997; Ishida et al., 1998; Tanaka et al., 1998), in which chum salmon (*Oncorhynchus keta*) were tagged with TDRs, have shown the opposite pattern of vertical movement to the Icelandic study, with Pacific salmon deeper by day. The Japanese data also revealed seasonal differences in swimming depth with fish remaining in cool, deep (100-200 m) water during the autumn, but swimming at shallower depths in winter after the surface waters had cooled down (Ishida et al., 1998). Recapture rates were 44-70% from releases of 10-27 tagged fish (Ogura, 1997). Repeated diving in the top 60 m during the day was also a pronounced feature of the behaviour of 3 chum salmon migrating from the Bering Sea (where 25 fish were tagged) to Hokkaido in the summer of 1998 (Wada and Ueno, 1999). The fish, which migrated distances of 2500 to 3000 km over periods of 47 to 77 days, remained near the surface at night, where they were probably feeding. Toward the end of the migration, each fish dived to depths of 150 to 300 m for a single day as it reached, and crossed, the edge of the continental shelf. Subsequently all three fish returned to the surface, where they spent the last few days and nights, as they migrated through the shallow waters around Hokkaido. Wada and Ueno (1999) suggest that the pronounced vertical movements during migration in the open ocean may be associated with location and orientation, possibly involving detection of solar elevation. The deep, daylong dives at the shelf-edge may similarly be associated with transition to the olfactory homing system thought to be used by salmonids in coastal waters.

Similar patterns of vertical movement have been deduced from long-term records obtained from 8 Pacific salmonids tagged with small temperature loggers (Walker et al., 2000). Five chum salmon (22%) were recovered in Japan 62-117 days after being tagged in the Bering Sea in July 1998 with another 18 fish of the same species. Three single fish (a steelhead trout, a pink and a coho salmon) were recaptured in Alaska 21-52 days after being tagged as part of a batch of 12 mixed salmonids in the Gulf of Alaska in the same month. After an apparent refractory period of variable length, the records all showed large regular fluctuations in temperature consistent with the fish moving towards the surface at night, possible to feed, and descending periodically to deeper, cooler water during the day.

5.6 Applications with Demersal Species

Archival tags have been used extensively with demersal species in the North Atlantic and adjacent seas, to study both vertical and horizontal movement of individual fish. Rates of tag recovery have generally been high (17% to >40%) and the data have provided important new insights about seasonal distribution, as well as behaviour during feeding, migration, and spawning. Most programmes have concentrated on plaice (*P. platessa*) and cod (*G. morhua*), but there has also been some work with small elasmobranchs and invertebrates.

5.6.1 Plaice

The Lowestoft Laboratory has been using archival tags on demersal fish since 1993 (Metcalfe et al., 1994; see above). Since then 754 tagged plaice have been released in the North Sea. Return rates from 302 fish tagged with hemispherical Mk1 tags and 452 tagged with the smaller, cylindrical Mk3 (Lotek LTD-100) tags have been 17% (50 tags) and 29% (131 tags), respectively. The 181 tags returned by 30 November 2000 have yielded 22,513 days of data.

Using a variety of methods it has been possible to define movements between tag and recapture on a scale that would not be possible with light-based geolocation, even if the water were sufficiently clear to make the necessary measurements of light intensity at sunrise and sunset. In areas of fast, bi-directional tidal streams in the southern North Sea and English Channel, plaice exhibit a tidal pattern of vertical migration known as selective tidal stream transport (Greer Walker et al., 1978). The vertical movements associated with this pattern of behaviour can be used to reconstruct the geographical track of the fish (Arnold and Holford, 1995), which can often be independently validated by the temperature measurements recorded by the tags (Metcalfe et al., 1994). The sinusoidal pressure cycle, which is recorded by the tag whenever the fish remains on the seabed for one (13 h), or more, tidal cycles (Fig. 5) can also be used in the same way (Metcalfe and Arnold, 1997). Plaice usually remain on the bottom during the day when spawning and feeding. They behave in the same way during migration in areas, such as the German Bight, where tidal streams are slow and there appears to be no energetic advantage in selective tidal stream transport. In these instances it is possible to interpolate the geographical location of the fish using the tidal range and the time of local high tide recorded by the tag and the intersecting grid of co-tidal and co-range lines on the appropriate navigation chart. The large body of data is shedding new light on the spatial dynamics of plaice in the North Sea (Metcalfe et al., 1999).

5.6.2 Cod

Archival tags have been deployed on cod at Iceland (Thorsteinsson, 1995; Thorsteinsson and Marteinsdottir, 1998), the Faeroe Islands (Steingrund, 1999) and northern Norway (Godø and Michalsen, 2000), as well as in the North Sea and Irish Sea (Righton et al., 2000, 2001). Almost 800 tags have been used to date, yielding some very interesting results and demonstrating the great potential of the technique with a heavily exploited species.

Godø and Michalsen (2000) tagged north-east Arctic cod (56-87 cm) at Lofoten (mature fish) and Finmark (immature fish) using 200 Star Oddi

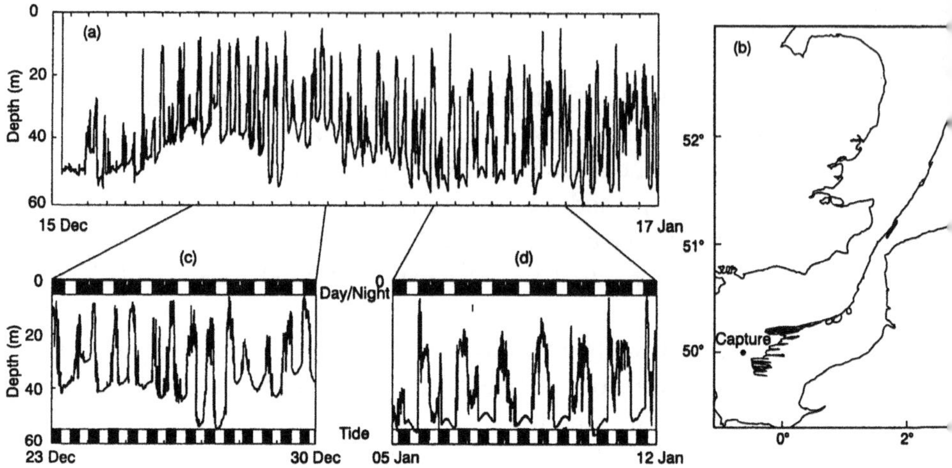

Figure 5. The reconstructed track of a 48 cm maturing female plaice fitted with a data storage tag, which recorded depth (pressure) at 10-min and water temperature at 24-h intervals. The track (b) was reconstructed from the vertical movements of the fish (a) and a computer simulation model of the tidal streams (Arnold and Holford, 1995). During the first 7 days the fish showed selective tidal stream transport (c) and moved rapidly from the Southern Bight into the English Channel. During the remainder of the track the fish showed a diel pattern of vertical migration (d) and swam in midwater at night. It remained within the plaice spawning area in the eastern English Channel, where it was caught by a French trawler (Arnold, 2000).

(DST 100) tags attached externally in front of the dorsal fin. The overall recovery rate was 16% and fish were recaptured in the Barents Sea after periods of up to 383 days. Data records showed seasonal changes in mean depth and temperature, but limited (<10 m) short-term vertical movements. Rapid movements of up to 250 m within 2 h were recorded sporadically and some fish showed distinct diel rhythms of vertical migration for restricted periods. Rates of vertical movement were such that the cod had to be negatively buoyant for most of the time (Arnold and Greer Walker, 1992).

Steingrund (1999) tagged somewhat smaller cod (55-70 cm) on Faeroe Plateau in autumn, at or very close to a spawning area. Eighty-eight fish were tagged internally with Star Oddi DST 300 tags sutured to the body wall; the rate of return was 47% over 22 months. The cod remained in warm (>5°C) water at nearly constant depth (90-130 m) throughout the winter. Most appear to have remained on the bottom on the plateau, although clear diel patterns of vertical migration, possibly associated with feeding, were seen in two fish for short periods. A few of the 41 recaptures showed a progressive change in depth from late February onwards, possibly indicating the onset of spawning.

Between 1995 and 1998 the Institute of Marine Research in Reykjavik tagged 398 sexually mature cod on spawning grounds in relatively shallow water (40-70 and 80-110 m) off southwest Iceland (Thorsteinsson and Mar-

Marteinsdottir, 1998). The tags (Star Oddi, DST 200 and 300), which were surgically implanted in the peritoneum, were programmed to record at 4 or 6 h intervals for periods of 6 days, followed by 40 or 60 min intervals for the ensuing 24 hours. The tags could record maximum depths of either 420 or 700 m and the maximum recording period was 390 days. By June 1998, 125 tags (31%) had been recovered. Most fish (predominantly males) remained in shallow water until May, after which they progressively moved off the continental shelf and descended to depths of 200-300 m and more. The records of 11 tags, which spanned two successive spawning seasons, showed that the cod returned to the continental shelf between February and April the following year, remaining in shallow water for periods of 14 to 33 days. Most fish showed extensive vertical movements in deep water, but much reduced activity on the continental shelf during the spawning season.

The Lowestoft Laboratory has recovered 31 tags (46%) from 68 cod tagged with Lotek LTD-100 tags in the southern North Sea in spring and summer of 1999. The fish, which were at liberty for periods of up to 278 days, switched between several distinct patterns of vertical movement. These patterns appeared to be associated with the post-spawning migration in April and May, when the fish remained in midwater for most of the time, and a low-activity recovery period in June and July, when the fish mostly remained on the sea bed. This was followed by a 'feeding' period in August and September, when the fish moved into midwater at night, prior to the autumnal pre-spawning migration (Righton et al., 2000). Comparable data from the Irish Sea, where one tagged fish was at liberty for 474 days, suggest that these patterns of vertical movement vary systematically between areas and may reflect regional differences in marine productivity (Righton et al., 2001).

5.6.3 Elasmobranchs

The Lowestoft Laboratory has also used archival tags to study the movements of thornback rays (*Raja clavata*) on and off the coast in both the Irish Sea and Thames Estuary. In the Irish Sea 18% of the 96 tags deployed were recovered (Metcalfe, pers.com.). Much higher rates of return (42%) have been recorded in the Thames Estuary, for which a fishery closure has been proposed to conserve a species that is now regarded as endangered. Ninety-seven rays were tagged in the Thames Estuary between October and November 1999 and 41 tags had been returned by 31 October 2000, producing nearly 5,000 days of data (Buckley pers. com.). The longest record to date is 345 days. As with conventional tagging experiments, on which the proposed conservation measure was based, 81% of the fish were recaptured within the Estuary. Depth records, however, show that 82% of the recaptured rays left the Estuary during the winter and some had moved into localised areas of deep water offshore.

Depth data have similarly allowed CSIRO scientists to deduce the position of school sharks (*Galeorhinus galeus*) on the continental shelf off South Australia. West and Stevens (2001) recovered 9 archival tags from 30 sharks tagged in late 1997. The fish, which were at liberty for periods from 8 days to 18 months, spent about 80% of their time on the continental shelf. They often moved into midwater for several hours each night but appeared to swim close to the bottom during the day. In deep water the sharks typically descended to depths of up to 600 m at dawn and returned to shallow water at dusk. The regularity of these movements suggested that they could be used to determine longitude when the water was too deep for the tags to record light intensity. On the continental shelf longitude could be estimated from the light intensity data and latitude from the maximum daily depth assuming the fish was then on the bottom.

5.6.4 Crustacea

The spider crab (*Maja squinado*), which attains sexual maturity after a terminal moult, is a good candidate for archival tag studies, particularly where there is high fishing mortality (Freire and González-Gurriarán, 1998). The crabs migrate offshore in autumn to mate and spawn, before returning to shallow, warmer water to incubate their eggs (González-Gurriarán and Freire, 1994). During recent investigations to assess the utility of electronic tags in the Galician fishery, Freire et al. (1999) double-tagged 22 females with Star-Oddi DST-100 data storage tags and acoustic tags, and released them in shallow, inshore water. Fifteen individuals (68%) were recaptured after periods of up to 150 days at liberty. Freire at al. (1999) were able to characterise three patterns of crab behaviour based on rate of change of depth. A 'stationary' phase in October was characterised by a low rate of change of depth (1 m d^{-1}) and bottom depths of less than 10 m. Starting in late October to early November, the females entered a 'migratory' phase, which lasted on average about 6 days. During this phase they migrated rapidly offshore to an average depth of 54 m (range 20-94 m), increasing their depth at a mean rate of approximately 9 m d^{-1}. Three crabs returned to warmer, shallow water soon after mating, decreasing their depth at an average rate of 12 m d^{-1}. Further large-scale studies are underway to extend these observations, investigate differences in behaviour between the sexes and explore the effect of temperature on behaviour and distribution.

6. REMOTE DATA RETRIEVAL

Tracking fish at sea is an expensive, time-consuming operation and replication of observations is difficult, sometimes impossible. Archival tags offer a more cost-effective method when recapture rates in commercial, or recreational fisheries are sufficiently high (e.g., Metcalfe and Arnold, 1997). The tags are expensive, however, and for many species recapture rates are too low to justify their use. In these circumstances, or where fishermen are reluctant to return recaptured tags, fishery-independent methods of data retrieval are needed. Automated acoustic listening stations deployed at places where fish regularly aggregate (e.g., Klimley et al., 1998, 2001) may provide a better alternative in some circumstances. 'Pop-up' tags that detach themselves from the fish at a predetermined time are probably more appropriate for a wider range of species, especially large pelagic ones.

6.1 Automated Acoustic Data Retrieval Stations

Fully automated acoustic systems capable of routinely downloading data and reprogramming the tag have so far had limited application. One system currently in use in a small number of studies is the Communicating History Acoustic Transponder or CHAT tag (Vemco Ltd., Nova Scotia). CHAT tags (Voegeli et al., 2001) can be tracked with moored listening stations or a mobile hydrophone system. When interrogated by the receiver at 27 kHz the tag transmits either raw or summarised data at 32.8 kHz. Calculation of maximum, minimum and average values, at intervals specified by the user, reduces the amount of data to be downloaded. In addition to data, one of 4096 unique ID codes is also transmitted. The time between the query and response can be used to calculate the distance from the receiver. The tags are relatively large (32 mm) in diameter and only suitable for use with large animals at present. CHAT tags are currently being used to examine the movements of tiger sharks (*Galeocerdo cuvier*) around the Hawaiian Islands (K. Holland, pers. com.).

CHAT tags will be most applicable for use on species that return predictably to a given location. Tuna, for example, make repeated visits to FADs (Klimley and Holloway, 1999). Additionally, seasonal aggregations are not uncommon in marine fish, occurring, for example, in coho salmon (*O. nerka*), yellowfin tuna (*T. albacares*) and several species of shark (Klimley et al., 1998). Regular returns to precise locations probably also occur in a number of other migratory species; the bass *(Dicentrarchus labrax)* (Pawson et al., 1987) is a good example.

6.2 Satellite Data Retrieval

Polar orbiting satellites offer a more versatile way of obtaining information from telemetry tags than stationary listening stations. The Argos satellite system, which has been used to monitor the movements of marine animals for almost two decades (Priede 1984; Taillade, 1992; Merrick 1994), estimates a location with relatively high accuracy and allows for limited data acquisition (32 bytes per transmission). The high frequency radio signal (401.65 MHz) is rapidly attenuated by seawater and the tag antenna must therefore be clear of the water for successful transmission. Most early work was conducted with marine mammals and turtles (e.g., Fancy et al., 1988; Stewart et al., 1989), which proved to be ideal platforms because of their need to surface regularly to breathe, and their size, which enabled them to carry a large transmitter package.

6.2.1 Towed tags

Prior to the advent of the pop-up satellite tag in the late 1990s, the only successful use of satellite telemetry with fish involved large sharks that swim at the surface. Tags were either towed behind the shark on a long tether, or secured to the first dorsal fin. Although bulky, the tags provided real-time data with good estimates of geolocation. In the early 1980s Priede (1984) used satellite telemetry to track a basking shark (*Cetirhinus maximus*) off the west coast of Scotland. The tag, which was 60 cm long, positively buoyant and self-righting, was towed behind the shark on a 10-m tether. Priede tracked a shark for 17 days and obtained 15 location fixes from Argos before the tether parted. Seven fixes were obtained on one cloudless day, when the animal appeared to spend most of its time at the surface. In 1994, Eckert and Stewart (2001) initiated similar studies on whale sharks (*Rhincodon typus*)—also large filter feeders—in the Sea of Cortez, Mexico. The transmitter, which was also towed on a long tether, had a maximum dimension of 42 cm. Eckert and Stewart tracked 15 animals between 1994 and 1998, using tags that provided depth and temperature histograms in addition to location. One shark retained its tag for 37 months, during which it moved nearly 13,000 km out into the western North Pacific. Satellite telemetry has also been applied to smaller sharks using a dorsal fin saddle, rather than a towed tag. Carey developed the first of these tags in 1993 to monitor the movements of blue sharks (*Prionace glauca*) in the Atlantic, and deployed three tags in 1994. The longest track, which lasted 36 days, involved a 2000 km migration across the Atlantic, and swimming depths of up to 500 m (Kingman, 1996). In 1999, a similar device was used to follow a small white shark around Australia (Bruce and Stevens pers. com.).

Fuelled by the success of satellite tags on sharks, Dewar and Block (unpublished data.) tried to develop a similar towed tag for blue marlin. While marlin are large predatory teleosts, acoustic tracks have shown that they spend as much as 50% of their time at the surface and swim relatively slowly (Block et al, 1992b). Coupled with anecdotal reports of basking, this information suggested that a towed tag might work for marlin, as it had for large sharks. A 0.5-W transmitter, developed by Microwave Telemetry (Columbia, MD, USA) for birds, was potted in a 29 cm long, positively buoyant, self-righting, fusiform body. Five of these tags were deployed on blue marlin in the summer of 1995 off the Kona coast of Hawaii (Dewar and Block, unpublished data). The tags were secured to a 1.5-m long tether and anchored in the dorsal muscles of the fish with a titanium dart. The first tag was heard from only once, and location could not be determined. The second tag indicated that the marlin had moved 26 km southwest in 7 hours after release but was not heard from again. A third tag was recovered from Kauai with the antennae severely bent, possible from impacting the fish. No information was received from any of the other tags. Clearly, a large towed tag was not suitable for active pelagic fish, for which a different tool would be required.

6.2.2 Single-point pop-up tags

Following Dewar and Block's failures with large towed tags, Paul Howey (Microwave Telemetry) suggested using a single-point, pop-up satellite tag that could be secured to the fish and released at a predetermined time. The tag would then float to the surface and communicate with the Argos satellite system. The fish would not need to spend any time at the surface and its location could be determined without the need for recapture.

The pop-up satellite tag concept appears to have arisen independently a number of times. In the early 1970s it was suggested that pop-up satellite tags should be used to track migrating gray whales. Because these whales spend only a short time at the surface, Goodman et al. (1973) recommended using a series of expendable satellite transmitters to be released at predetermined intervals. Pop-up locations could then be used to reconstruct the migratory path of the whale. Nelson (1978) proposed a similar system for sharks to overcome the limitations of acoustic telemetry. He suggested using multiple single-point radio tags that could be located with a radio direction finder in an aircraft and described the "ultimate" tracking system as one that replaced aircraft with satellites. A decade later, Hunter et al. (1986) discussed a larger number of methods for obtaining critical information on tuna movements and concluded that pop-up tags would provide an ideal solution. These tags would have higher recovery rates, be fisheries independent, and provide more detailed information on movement patterns than any other system. The idea was discussed again in 1994 at a meeting to review the development of archival tags, which was organised by the Inter American Tropical Tuna Commission, and held in conjunction with the 45th Annual Tuna Conference at Lake Arrowhead, California (Klimley et al., 1994). A year later, when the TAB (Tagging Atlantic Bluefin) workshop was held in Miami, Florida (Restrepo, 1996), a pop-up tag for large pelagic fish had become a realistic possibility. And, in fact, around this time, the NMFS in Honolulu was soliciting manufacturers to produce a pop-up archival tag that would release itself on a pre-specified date, float to the surface and transmit stored data via an Argos satellite link. An electrolytic breakaway link had existed for a long time (Nelson, 1976; Nelson and McKibben 1981) and miniature data storage capabilities had recently become available (Gunn et al. 1994; Metcalfe et al., 1994). All that remained, therefore, to set the stage for the construction of a practical pop-up tag, which would not need the large battery pack required for real-time uplinks, was a miniature satellite transmitter. Once this had been developed, Priede's 60-cm tag could be replaced by one only about 15 cm long.

By 1996 there were two parallel efforts to build a pop-up tag. Baba and Ukai (1996) described an intelligent tag and recovery system, which would record temperature and depth in 2 Mb of flash memory. They proposed using a Toyocom T-2050, 0.125 W transmitter (1.8 x 2.5 x 3.5 cm, including batteries) primarily to facilitate tag recovery. The description was published whilst the tag was still in development and we know of no reports that describe the actual use of this device. In the same year, Telemetry 2000 (the development arm of Microwave Telemetry) produced the first pop-up satellite tag to be proven at sea. The PTT-100 tag, which is still in use, consists of a small carbon fibre tube (~20 mm dia.) with a nose cone on the leading end and a float (~40 mm dia.) and antenna at the trailing end. The total length with a quarter-wavelength antenna (15.875 cm) is approximately 34 cm. The carbon fibre tubing houses a microprocessor, thermistor, batteries and a satellite transmitter. The nose cone contains the release mechanism, which consists of a small plastic eye loop secured to the tip of the nose cone with fine gauge stainless-steel wire. The tag is attached to the fish with a monofilament leader, which passes through the eye loop. At a pre-programmed time, a current is passed between the stainless steel wire and a "ground" electrode, corroding the stainless steel wire and releasing the eye loop and leader from the tag. Once released, the float, which is constructed from syntactic foam, provides the buoyancy necessary to bring the tag to the surface and allow it to float vertically. During the initial development phase, the stability of the tag was improved by modifying the shape of the float and maximising the distance between the centres of buoyancy and mass. Tests of float design were conducted at the Tuna Research and Conservation Center in Monterey, California. Once at the surface the tag transmits a radio signal every 60 s. Successive signals allow the Argos system to calculate the latitude and longitude of the tag by measuring the frequency change (Doppler shift) resulting from the rapid transit of the satellite. The first generation PTT-100 tag transmitted for approximately 30 days, which was more than sufficient for the satellite to receive all the recorded data several times to allow checks for accuracy.

The first pop-up satellite tags were deployed in 1997, in conjunction with a large-scale effort to study the long-term movements of Atlantic bluefin tuna (Block et al., 1998b). These early tags recorded 61 hourly, or daily temperature measurements. Initial deployments were in two phases. During phase one, whose goal was to test the concept and examine satellite transmission quality, nine tags were deployed for 3-14 days. All 9 tags reported, although two had relatively low transmission quality, and not all data were obtained. In phase two, a further 28 tags were deployed for periods of 60 to 90 days. The release dates were established to test long-term tag function and attachment and to determine where the bluefin were located in spring and summer during the expected spawning period. Success was phenomenal: signals from 26 (93%) of the tags were received by satellite and all 61 temperature measurements were obtained from 24 tags. Most fish were located over the continental shelf north of the release point, or offshore in association with the edge of the Gulf Stream or its rings. No tags surfaced on the known spawning grounds in the Mediterranean or Gulf of Mexico. Two fish had crossed to the east of the stock boundary established by ICCAT at 45° W.

Lutcavage et al. (1999) initiated a parallel programme with similar goals in the same year, although the bluefin were larger than those tagged by Block et al. (1998b) off North Carolina. Twenty PTT-100 pop-up tags were deployed off New England and set to detach from the fish after 3-9 months. Seventeen tags (85%) popped up successfully and transmitted data to satellite. Twelve tags detached between May and July, the known spawning period, but none surfaced in the Mediterranean, or Gulf of Mexico. Five fish had crossed to the east of 45° W, suggesting, with the results from the parallel investigations by Block et al. (1998b), that there might be a case for ICCAT to re-examine the stock boundary.

PTT-100 tags have been used in a third study of Atlantic bluefin, which commenced in the Mediterranean in 1998 (De Metrio et al., 1999, 2000). Pop-up rates have been much lower than those in the North American studies. Reasons probably include the difficulty of tagging large tuna underwater in traps, high rates of exploitation leading to non-reported recapture of fish prior to the specified pop-up date, and problems of detecting tags against background noise in European waters. Despite these difficulties, several tags have surfaced on schedule in the Mediterranean or eastern North Atlantic. No tuna have crossed the North Atlantic, but one tag, which was attached to a giant bluefin in southern Spain in July 1998, was located in the Greenland Sea (75.1°N 01.1°E) on the scheduled day in March 1999. Another was located close to the presumed southern limit of bluefin in the eastern North Atlantic, 180 days after release.

Since 1997, PTT-100 tags have been successfully deployed on several other large pelagic species, including yellowfin tuna, albacore, blue and striped marlin and white, thresher and salmon sharks (Graves et al., in press; Block et al., unpublished data; D. Holts, pers. com.; Gunn and Block, 2001).

6.2.3 Pop-up archival tags

The PTT-100 tag represented a critical step in proving the pop-up concept (Block et al., 1998b; Lutcavage et al., 1999; Graves et al., 2001). The aim of the next phase of development has been to increase the amount of information that can be obtained by including data on location, temperature and depth preference throughout the deployment. We are rapidly approaching this goal and commercially built pop-up archival tags are now available. There are currently two manufacturers, Microwave Telemetry and Wildlife Computers whose tags have different designs, although similar external shape and configuration. In addition, both have been used successfully on a number of species of sharks, tuna and billfish and have released successfully after being attached to the fish for up to a year.

Microwave Telemetry's archival pop-up tag is an improved version of the single-point PTT-100 described above. The tag now records temperature at hourly intervals throughout the deployment and also measures depth at the same frequency. Times of sunrise and sunset for geolocation are estimated from measurements of light intensity, using a fuzzy logic data processor. Rather than compressing information into histograms, raw data are transmitted through the Argos system. Geolocation errors are reported to be approximately $\pm$ 1.0° for longitude and $\pm$ 2.0° for latitude (Lutcavage et al., 2000). The tag incorporates software that limits transmission to periods when a satellite is likely to be in receiving range and with four Argos satellites now operational, the tag can transmit a year's worth of data in about 10 days at mid-latitudes. The data stream incorporates additional values, which include a calibration of the real time clock necessary for geolocation, as well as an indicator of data quality. Both the single-point and archival versions of the PTT-100 tag have an integrating inclinometer, which indicates whether the tag remained horizontal on the fish for the duration of the deployment, or detached prematurely and floated vertically. Tag cases are pressure rated to withstand depths down to 2000 m, and an interface is under development, which will allow the user to set the release date.

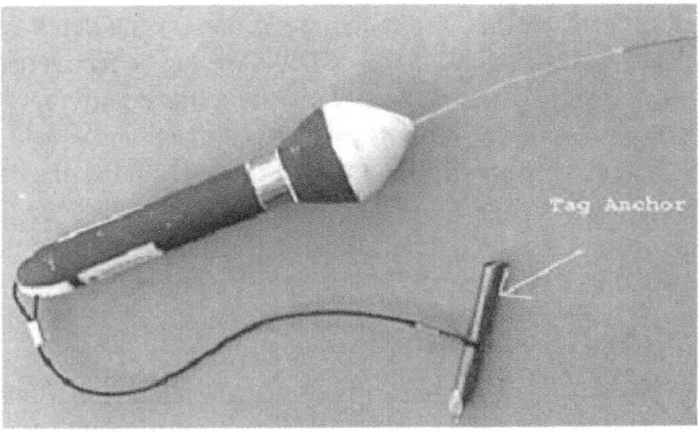

Figure 6. Pop-up archival transmitting (PAT) tag (Wildlife Computers, Redmond, Washington, USA.

The PAT tag constructed by Wildlife Computers (Fig. 6) collects temperature, depth and light intensity. Sunrise and sunset can be detected at depths down to 300 m, if the water is sufficiently clear. Data are stored in 1 to 2 Mb of non-volatile (flash) memory. The collected data are processed onboard the PAT, so that only data critical to understanding behaviour are transmitted. The tag is fully programmable and a software interface is provided for setting the release date, structure of histograms, intervals at which data will be saved into the archive, and transmission characteristics. The tag, which has a power output of 0.5 W, is capable of 10,000 transmissions (32-byte). Data for transmission are compressed into four different message types: depth and temperature histograms, which summarise the time the animal spent within specified depth and temperature ranges; depth-temperature profiles, which document the thermal character of the water column; status messages, which contain data to correct for clock drift and other diagnostic information; and location messages, which contain the calculated longitude for each day, together with light levels recorded around sunrise and sunset. Longitude is calculated in the tag from light and depth measurements made at 2-min intervals. Latitude is estimated from processed data transmitted with the location message because Argos is unable to transmit all the recorded light and depth data at present. Longitude is determined with an accuracy of ± 0.5 latitude with an accuracy of ±1° to ±10°, depending upon position and time of year.

Each location message also includes an indicator of the quality of the estimated longitude, the turbidity of the water, and an assessment of the quality of the light level data for determining latitude. The case can withstand pressures down to depths of 2000 m, although at present the depth sensor only records to 1000 m (resolution 0.5 m). Should the tag be recovered before or after pop-off, the entire archival record can be retrieved, giving data identical to that recorded with simple archival tags (see Section 5).

One challenge with pop-up tags has been determining the reasons for non-reports by tags through the Argos system. Non-reports are generally a result of tag failure, fish or fishing mortality, and other unpredictable problems, such as human interference. Microwave Telemetry and Wildlife Computers have both recently announced new features, which should obviate some of these problems. Both manufacturers include software, which will initiate transmission if the tag becomes detached prematurely, thus reducing the risk of losing tags floating on the surface prior to their programmed pop-off date. Both manufacturers have also developed release mechanisms that will trigger if the tag goes below a predetermined depth. The Wildlife Computers tag can be programmed to pop off early should its depth sensor indicate little or no depth change over an extended period of time, as would happen if a fish died and sank to the sea floor. The Microwave Telemetry tag will similarly start transmitting, if it remains at the surface, or at a constant depth, for longer than a predetermined time.

Archival pop-up satellite tags from both manufacturers have been used successfully on a range of tuna and elasmobranchs. Given their recent development, few studies have been published. One report by Lutcavage et al. (2000) provides some information on the use of these second-generation tags on bluefin tuna in the Atlantic. In 1999 these investigators deployed 21 PTT-100 archival pop-up tags (minus depth sensors) manufactured by Microwave Telemetry. By September 2000, 15 out of 19 tags (79%) had surfaced, as programmed, and transmitted successfully by satellite. Bluefin left the New England feeding grounds within 10-30 days and headed south past North Carolina, after which they followed the subtropical convergence east into the mid-Atlantic. No tracks entered the Gulf of Mexico or the Mediterranean during the expected spawning period. In a second study two Wildlife Computers PAT tags (recording only temperature and depth) were deployed on bluefin tuna in the Pacific for 24 and 52 days (Marcenik et al., 2001). Both tags surfaced as expected. The data confirmed patterns obtained from acoustic tracks over the same time period and showed that bluefin spent more than 70% of the time above 20 m in waters warmer than 17°C. The

PAT tags did, however, document deeper dives (>350 m) than those seen during acoustic tracking when the fish remained above 100 m.

7. DISCUSSION

After 30 years of development electronic tags have now become an indispensable tool for fisheries biologists and the limitations of conventional tags have generally been overcome (Gunn, 2000). Most importantly, perhaps, information on the position of the fish is obtained independently of any fishery and data can be obtained on the movements of the fish throughout its track. As a result the complexity of question that can be addressed with tagging experiments has dramatically increased. The various types of tag are complementary in nature (e.g., Holland et al., this volume) and the choice of technique depends on the question to be addressed, as well as the species and size of fish. Acoustic tracking can provide frequent position fixes on time scales of hours or minutes and detailed descriptions of both vertical and horizontal movement. At the same time it is possible to record a range of physical and physiological variables, relate the behaviour of the fish to its surrounding environment and develop hypotheses about behavioural and other ecological mechanisms. The technique can also be used to assess survival after capture and release (e.g., Candy et al., 1996; Candy and Quinn, 1999) and sometimes the effects of the tracking process itself on fish behaviour (e.g., Dagorn et al., 2001). Mobile tracking systems suffer from several limitations, however. These include the high costs of research vessels, the limited number of fish that can be followed at any time, the short duration of tracks, and uncertainty about the length of time for which the fish may exhibit abnormal behaviour after tagging. Archival tags, which offer many of the benefits of acoustic tags, allow observations to be made over much longer periods and also replicated. The recorded data are fishery independent, although the tags must be recovered and therefore suffer from some of the disadvantages of conventional tags, such as low recapture rates and biased geographical sampling. The data may also allow behaviour associated with post-tagging trauma to be identified and the length of the recovery period estimated (e.g., Wada and Ueno, 1999; Walker et al., 2000). Estimates of geographical position, although useful on a regional scale, are, however, subject to large errors and cannot approach the high precision—of the order of metres—that can be achieved with GPS on a tracking vessel. Pop-up tags are truly fishery independent and promise to be very useful with species, such as billfish (NMFS, 1997), where tag recapture rates are only a few percent. These tags are, however, at present only suitable for use with large fish. They also record more data than can currently be transmitted via the

Argos satellite system and significant data compression is consequently needed. Full recovery of data is, however, possible if fish are recaptured, or stranded tags are recovered from the beach (P. Howey, R. Hill, B. Block, all pers. com.).

As technology continues to develop, the applications for the different tags will undoubtedly expand, placing fewer questions beyond our reach. As identified by the CATAG working group (Anon, 2000, 2001), the principal technical needs are smaller tags with longer life, more memory, improved sensors, and increased operating range. Further improvements are needed to acoustic systems for tracking or monitoring large numbers of individual fish and these should include improved coding systems, increased detection range, higher data transmission rates, and better software for processing and interrogating data. Sensors need to be smaller to enable archival tags to record a wider range of physical and physiological variables. Important physical variables include body attitude (tilt angle), compass heading, magnetic field strength, magnetic dip, tail-beat frequency and swimming speed or acceleration. Important physiological variables include heart rate and cardiac output, feeding rate, growth rate, gonad development and related levels of enzymes or hormones in the blood. Better methods of geolocation are needed for archival and pop-up tags, and direct (e.g., geomagnetic sensors) and indirect (e.g., sequentially released tags) methods are both possible. More reliable techniques are required for fishery-independent data retrieval and these might include using other satellites, or cellular telephone systems, as an alternative to Argos. Improved software is needed to interface data from tags with that obtained from other sources, such as remote sensing. Desirable technical improvements are not limited to engineering, however. Biologists need to carry out systematic studies to improve methods of attachment and minimise the effects of tags on behaviour and swimming performance. Underwater tagging is an attractive possibility, particularly for deep-water species and species with closed swimbladders, and one system has recently been developed by Star-Oddi. There is also a pressing need to develop models that can predict the behaviour of populations from the behaviour observed in individual fish (e.g., Metcalfe et al., 1999; Dagorn et al., Newlands and Lutcavage, Sibert and Fournier, all this volume).

Electronic tags have a major role to play in fisheries science in the 21st century and are likely to provide solutions to many currently intractable problems. Defining movements over both large and small scales is critical at a number of different levels. Marine protected areas are increasingly popular as a conservation measure. Knowledge of movements over smaller spatial scales will help determine the effectiveness of given boundaries for protecting the species of interest (e.g., Holland, et al., 1996; Walters and Maguire, 1996; Metcalfe et al., 1998; Zeller, 1999; Bolden, 2000; Polovina et al.,

2000; O'Dor et al., this volume). Similar principles apply to artificial reefs (e.g., Collins et al., 1997), where it is hoped to aggregate or enhance local fish populations. The high frequency, high precision observations afforded by acoustic tracking are appropriate at these scales and also enable feeding and spawning behaviour to be described. Knowledge of movements over medium to large scales is important when dealing with basic biological questions, such as stock structure, spawning site fidelity and migration routes, all of which are critical to effective fisheries management. Pop-up tags and archival tags are ideal for addressing questions at these larger scales. Information on small and large-scale movements, coupled with satellite imagery and physical data obtained from the tag itself (e.g., Inagake et al., 2001) will provide insight into habitat preferences and the influence of environmental parameters on movements and behaviour. These data should improve our ability to differentiate natural trends from those associated with the impacts of human activities.

Fisheries biologists generally have only just begun to realise the full potential of electronic tags in studying both fisheries ecology and applied fisheries questions. Among the latter, there are many questions that are tractable with the application of modern technology (Anon, 2001). Although electronic tags have already been applied with considerable success to the study of movement and migration, large-scale systematic studies are still required to describe and understand the migrations, distribution and stock identity of many of the most important commercially exploited species, in particular highly migratory ones. These studies need to be extended to juveniles, as well as adults, and particular attention needs to be given to the interaction between behaviour and the physical environment, with special emphasis on the role of currents as transporting, or possibly guiding, agents. Reactions of fish to research vessels and survey trawls can seriously bias fishery-independent estimates of abundance and systematic investigations of these effects deserve to be encouraged. Similar studies are needed of the 'natural behaviour' of fish beyond the influence of the vessel and should include systematic investigation of vertical migration and distribution in relation to swimbladder function, depth of neutral buoyancy, body attitude and acoustic target strength. Studies with species caught incidentally are also important for understanding the broader ecological implications of different gear types. Predator-prey interactions are likely to prove a fruitful area of investigation, and direct measurements of feeding rates of free-ranging fish should become possible, once the appropriate sensors have been developed. Studies of the physiology of free-ranging fish are required to understand how behaviour changes in response to metabolic condition and reproductive development. Direct measurements of growth in relation to temperature and food intake would also be valuable. A number of investigations have demonstrated the benefits of using electronic tracking systems to evaluate the environmental effects of man-made structures such as barrages (e.g., Mee et al., 1996), dams and oil rigs and further application of these methods should be encouraged.

In conclusion, this is a very exciting time for researchers studying the biology of marine fish and invertebrates. New technologies and an accelerating pace of development are rapidly reducing the number of unanswerable questions. Suites of variables, which could previously only be measured with large, expensive instruments, can now be recorded simultaneously in a package small enough to be carried by a medium size fish. We can remotely monitor key physiological variables and examine fish movements at levels ranging from the subtle behaviour of mating squid to the basin scale migrations of giant bluefin tuna. At the same time we can record key features of the physical environment at the precise location of the fish. The arrival of the new technologies is timely, given the large numbers of populations and habitats now threatened with over exploitation, or even destruction. Used intelligently, the new tools should help to minimise human impact on marine ecosystems and begin to repair the damage that has already been done.

ACKNOWLEDGEMENTS

This work was part-funded by the UK's Ministry of Agriculture, Fisheries and Food under its Marine Fisheries R&D Commission (Contracts MF0145 and MF0149) and the Pfleger Institute of Environmental Research.

REFERENCES

Ackerman, J.T., Kondratieff, M.C., Matern, S.A., and Cech, J.J. (2000). Tidal influence on spatial dynamics of leopard sharks, *Triakis semifasciata*, in Tomales Bay, California. *Envir. Biol. Fishes* **58**, 33-43.

Aitken, J., Webber, D., O'Dor, R. and Voegeli, F. (2000). Acoustic telemetry of integrated differential pressures: calibration and applications. *In* Moore, A., and Russell, I., *Advances in Fish Telemetry*. Proceedings of the Third Conference on Fish Telemetry in Europe, held in Norwich, England, 20-25 June 1999. CEFAS Lowestoft, pp. 21-27.

Anon (1999). Final Report. In Davis, T.L.O., (ed.), *Southern Bluefin Tuna Recruitment Monitoring Program (1993-1998)*. pp. 7-27, CSIRO Marine Research, Hobart, Australia.

Anon (2000). Improvements of tagging methods for stock assessment and research in fisheries. Scientific Report of EU Concerted Action, FAIR CT.96.1394, 220 pp.

Anon (2001). Tagging Methods for Stock Assessment and Research in Fisheries. *Tech. Rep. (79), Mar. Res. Inst.*, Reykjavik, Iceland (in press).

Aoyama, J., Hissmann, K., Yoshinaga, T., Sasai, S., Uto, T., and Ueda, H. (1999). Swimming depth of migrating silver eels *Anguilla japonica* released at seamounts of the West Mariana Ridge, their estimated spawning sites. *Mar. Ecol. Prog. Ser.* **186**, 265-269.

Armstrong, J.D., Lucas, M.C., Priede, I.G., and de Vera, L. (1989). An acoustic telemetry system for monitoring the heart rate of pike, *Esox lucius* L., and other fish in their natural environment. *J. Exp. Biol.* **143**, 549-552.

Armstrong, J.D., Bagley, P.M., and Priede, I.G. (1992). Tracking deep-sea fish using ingestible transmitters and an autonomous sea-floor instrument package. *In* Priede, I.G., and Swift, S.M. (eds.), *Wildlife Telemetry*, Ellis Horwood, New York, pp. 376-386.

Arnold, G.P. (2000). Acoustic tracking. *In* Hancock, D.A., Smith, D.C., and Koehn, J.D. (eds.), *Fish Movement and Migration*, Australian Society for Fish Biology Workshop Proceedings, Bendigo, Victoria, September 1999, Australian Society for Fish Biology, Sydney, pp. 44-54.

Arnold, G.P., and Lundgren, B. (eds.) (2001). Electronic Tags in Fisheries Research and Management. Sci. Ser., Tech. Rep., CEFAS, Lowestoft (in press).

Arnold, G.P., and Holford, B.H. (1978). The physical effects of an acoustic tag on the swimming performance of plaice and cod. *J. Cons. Int. Explor. Mer* **38**, 189-200.

Arnold, G.P., and Holford, B.H., (1995). A computer simulation model for predicting rates and scales of movement of demersal fish on the European continental shelf. *ICES J. Mar. Sci.*, **52**, 981-990.

Arnold, G.P., and Greer Walker, M. (1992). Vertical movements of cod (*Gadus morhua* L.) in the open sea and the hydrostatic function of the swimbladder. *ICES J. Mar. Sci.*, **49**, 357-72.

Arnold, G.P., Greer Walker, M., and Holford, B.H. (1990). Fish behaviour: achievements and potential of high-resolution sector scanning sonar. *Rapp. PV Réun. Cons. Int. Explor. Mer* **189**, 112-122.

Arnold, G.P., Greer Walker, M, Emerson, L.S., and Holford, B.H. (1994). Movements of cod (*Gadus morhua* L.) in relation to the tidal streams in the southern North Sea. *ICES J. Mar. Sci.* **51**, 207-232.

Baba, N., and Ukai, T. (1996). Intelligent tag and its recovery system for studying the behavior of free-ranging salmon in the ocean. *Bull. Natl. Res. Inst. Aquacult.*, Supplement (2), 29-32.

Bagley, P.M. (1992). A code-activated transponder for the individual identification and tracking of deep-sea fish. *In* Priede, I.G., and Swift, S.M. (eds.), *Wildlife Telemetry*, Ellis Horwood, New York, pp.111-119.

Bagley, P.M., and Priede, I.G. (1996). Investigations of the behaviour of deep-sea fish using an autonomous free-fall acoustic tracking vehicle operating at abyssal depths. *In* Baras, E., and Philippart, J.C. (eds.), *Underwater Biotelemetry*, Proceedings of the First Conference and Workshop on Fish Telemetry in Europe. University of Liège, Belgium, pp. 215-224.

Bagley, P.M., Bradley, S., Collins, M.A., Priede, I.G., and Gray, P. (2000). Miniature acoustic code activated transponder for tracking fish at abyssal depths using delayed activation to overcome reverberation. *In* Moore, A., and Russell, I., *Advances in Fish Telemetry.* Proceedings of the Third Conference on Fish Telemetry in Europe, held in Norwich, England, 20-25 June 1999. CEFAS Lowestoft, pp. 13-19.

Bass, G.A., and Rascovich, M. (1965). A device for sonic tracking of large fishes. *Zoologica* **50**, 75-82.

Block, B.A., Booth, D.T., and Carey, F.G. (1992a). Depth and temperature of the blue marlin, *Makaira nigricans*, observed by acoustic telemetry. *Mar. Biol.* **114**, 175-183.

Block, B.A., Booth, D, and Carey, F.G. (1992b). Direct measurement of swimming speeds and depth of blue marlin. *J. Exp. Biol.* **166**, 267-284.

Block, B.A., Keen, J.E., Castillo, B., Dewar, H., Freund, E.V., Marcinek, D.J., Brill, R.W., and Farwell, C. (1997). Environmental preferences of yellowfin tuna (*Thunnus albacares*) at the northern extent of its range. *Mar. Biol.* **130**, 119-132.

Block, B.A., Dewar, H., Williams, T., Prince, E.D. Farwell, C., and Fudge, D. (1998a). Archival tagging of Atlantic bluefin tuna (*Thunnus thynnus thynnus*). *Mar. Tech. Soc. J.* **32**, 37-46.

Block, B.A., Dewar, H., Farwell, C., and Prince, E.D. (1998b). A new satellite technology for tracking the movements of Atlantic bluefin tuna. *Proc. Natl. Acad. Sci. USA* **95**, 9384-9389.

Bolden, S.K. (2000). Long-distance movement of a Nassau grouper (*Epinephelus striatus*) to a spawning aggregation in the central Bahamas. *Fish. Bull. US* **98**, 642-645.

Brill, R.W., Holts, D.B., Chang, R.K.C., Sullivan, S., Dewar, H., and Carey, F.G. (1993). Vertical and horizontal movements of striped marlin (*Tetrapturus audax*) near the Hawaiian Islands, determined by ultrasonic telemetry, with simultaneous measurements of oceanic currents. *Mar. Biol.* **117**, 567-574.

Brill, R.W., Block, B.A., Boggs, C.H., Bigelow, K.A., Freund, E.V., and Marcinek, D.J. (1999). Horizontal movements and depth distribution of large adult yellowfin tuna (*Thunnus albacares*) near the Hawaiian Islands, recorded using ultrasonic telemetry: implications for the physiological ecology of pelagic fishes. *Mar. Biol.* **133**, 395-408.

Candy, J.R., and Quinn, T.P (1999). Behavior of adult chinook salmon (*Oncorhynchus tshawytscha* in British Columbia coastal waters determined from ultrasonic telemetry. *Can. J. Zool.* **77**, 1161-1169.

Candy, J.R., Carter, E.W., Quinn, T.P., and Riddell, B.E. (1996). Adult chinook salmon behavior and survival after catch and release from purse-seine vessels in Johnstone Strait, British Columbia. *N. Am. J. Fish. Mgmt.* **16**, 521-529.

Carey, F.G., and Lawson, K.D. (1973). Temperature regulation in free-swimming bluefin tuna. *Comp. Biochem. Physiol.* **44A**, 375-392.

Carey, F.G., and Robison, B.H. (1981). Daily patterns in the activity of swordfish, *Xiphias gladius*, observed by acoustic telemetry. *Fish. Bull. US* **79**, 277-292.

Carey, F.G., and Gibson, Q.H. (1987). Blood flow in the muscle of free-swimming fish. *Physiol. Zool.* **60,** 138-148.

Carey, F.G., and Scharold, J.V. (1990). Movements of blue sharks (*Prionace glauca*) in depth and course. *Mar. Biol.* **106**, 329-342.

Carey, F.G., and Clark, E. (1995). Depth telemetry from the sixgill shark, *Hexanchus griseus*, at Bermuda. *Envir. Biol. Fishes* **42**, 7-14.

Carey, F.G., Teal, J.M., and Kanwisher, J.W. (1981). The visceral temperatures of mackerel sharks (Lamnidae). *Physiol. Zool.* **54**, 334-344.

Carey, F.G., Kanwisher, J.W., Brazier, O., Gabrielson, G., Casey, J.G., and Pratt, H.L. (1982). Temperature and activities of a white shark, *Carcharodon carcharias*. *Copeia* **1982** (2), 254-260.

Carey, F.G., Kanwisher, J.W., and Stevens, E.D. (1984). Bluefin tuna warm their viscera during digestion. *J. Exp. Biol.* **109**, 1-20.

Carlson, B. A., McKibben, J.N., and deGruy, M.V. (1984). Telemetric investigation of vertical migration of *Nautilus belauensis* in Palau. *Pac. Sci.* **38**, 183-188.

Cayré, P. 1991. Behaviour of yellowfin tuna (*Thunnus albacares*) and skipjack tuna (*Katsuwonus pelamis*) around fish aggregating devices (FADs) in the Comoros Islands as determined by ultrasonic tagging. *Aquat. Living Resour.* **4**, 1-12.

Chapman, C.J., Johnstone, A.D.F., and Rice, A.L. (1975). The behaviour and ecology of the Norway lobster, *Nephrops norvegicus* (L.). *In* Barnes, H. (ed.), Proc. 9th Europ. Mar. Biol. Symp. (9), pp. 59-74.

Clark, D. S and Green, J. M. (1990). Activity and movement patterns of juvenile cod, *Gadus morhua*, in Conception Bay, Newfoundland, as determined by sonic telemetry. *Can. J. Zool.* **68**, 1434-1442.

Cote, D., Scruton, D.A., Niezgoda, G.H., McKinley, R.S., Rowsell, D.F., Lindstrom, R.T., Ollerhead, L.M.N., and Whitt, C.J. (1998). A coded acoustic telemetry system for high precision monitoring of fish location and movement: application to the study of nearshore nursery habitat of juvenile Atlantic cod (*Gadus morhua*). *Mar. Tech. Soc. J.* **32**, 54-62.

Collins, K., Jensen, A., and Smith, P. (1997). Tagging, tracking and telemetry in artificial reef research. *In* Jensen, A.C. (ed.), *European Artificial Reef Research*, Proceedings of the 1st EARRN conference, Ancona, Italy, March 1996. Southampton Oceanography Centre, United Kingdom.

Collins, K.J., Smith, I.P., Jensen, A.C., Addison, J.T., and Bannister, R.C.A. (2000). Measuring lobster (*Homarus gammarus* (L.)) catchability using electromagnetic telemetry. *In* Moore, A., and Russell, I., *Advances in Fish Telemetry.* Proceedings of the Third Conference on Fish Telemetry in Europe, held in Norwich, England, 20-25 June 1999. CEFAS Lowestoft, pp. 185-192.

Cushing, D.H., and Harden Jones, F.R. (1966). Sea trials with modulation sector scanning sonar. *J. Cons. Int. Explor. Mer* **30**, 324-345.

Dagorn, L., Bach, P., and Josse, E. (2000). Movement patterns of large bigeye tuna (*Thunnus obesus*) in the open ocean, determined using ultrasonic telemetry. *Mar. Biol.* **136**, 361-371.

Dagorn, L., Josse, E., and Bach, P. (2001). Association of yellowfin tuna (*Thunnus albacares*) with tracking vessels during ultrasonic telemetry experiments. *Fish. Bull. US* **99**, 40-48.

Deary, C., Scruton, D.A., Niezgoda, G.H., McKinley, S., Cote, D., Clarke, K.D., Perry, D., Lindstrom, T., and White, D. (1998). A dynamically switched combined acoustic and radio transmitting (CART) tag: an improved tool for the study of diadromous fishes. *Mar. Tech. Soc. J.* **32**, 63-69.

De Long, R.L., Stewart, B.S., and Hill, R.D. (1992). Documenting migrations of northern elephant seals using day length. *Mar. Mamm. Sci.* **8**, 155-159.

Demers, E., McKinley, R.S., Weatherley, A.H., and McQueen, D.J. (1996). Activity patterns of largemouth and smallmouth bass determined with electromyogram biotelemetry. *Trans. Am. Fish. Soc.* **125**, 434-439.

Dewar, H., Deffenbaugh, M., Thurmond, G., Lashkari, K., and Block, B.A. (1999). Development of an acoustic telemetry tag for monitoring electromyograms in free-swimming fish. *J. Exp. Biol.* **202**, 2693-2699.

De Metrio, G., Arnold, G., Cort, J.L., de la Serna, J.M., Yannopoulos, C., Megalofonou, P., and Labini, G.S. (1999). Bluefin tuna tagging using "pop-up tags": first experiments in the Mediterranean and eastern Atlantic. ICCAT, Coll. Vol. Sci. Pap., XLIX(1), 113-119.

De Metrio, G., Arnold, G.P., de la Serna, J.M., Yannopoulos, C., Megalofonou, P., Buckley, A.A., and Pappalepore, M. (2000). Further results of tagging Mediterranean bluefin tuna with pop-up satellite-detected tags. ICCAT, SCRS/00/109, 7 pp. (mimeo.).

Dodson, J.J., and Leggett, W.C. (1973). Behavior of adult American shad (*Alosa sapidissima*) homing to the Connecticut River from Long Island Sound. *J. Fish. Res. Bd Can.* **30**, 1847-1860.

Dodson, J.J., and Leggett, W.C. (1974). Role of olfaction and vision in the behavior of American shad (*Alosa sapidissima*) homing to the Connecticut River from Long Island Sound. *J Fish. Res. Bd Can.* **31**, 1607-1619.

Døving, K.B., Westerberg, H., and Johnsen, P.B. (1985). Role of olfaction in the behavioural and neuronal responses of Atlantic salmon, *Salmo salar*, to hydrographic stratification. *Can. J. Fish. Aquat Sci.* **42**, 1658-1667.

Eckert, S.A., and Stewart, B.S. (2001). Telemetry and satellite tracking of whale sharks, *Rhincodon typus,* in the Sea of Cortez, Mexico, and the North Pacific Ocean. *Envir. Biol. Fish* **60**, 299-308.

Engås, A., Løkkeborg, S., and Øvredal, J.T (1996). Natural and fishing-gear-induced behaviour of tagged fish, studied by means of a stationary positioning system. *In* Baras, E., and Philippart, J.C. (eds.), *Underwater Biotelemetry*, Proceedings of the First Conference and Workshop on Fish Telemetry in Europe. University of Liège, Belgium, pp.203-211.

Engås, A., Haugland, E.K., and Øvredal, J.T. (1998). Reactions of cod (*Gadus morhua* L.) in the pre-vessel zone to an approaching trawler under different light conditions. *Hydrobiologia*, 371/372. *In* Lagardère, J.-P., Bégout Anras, M.-L., and Claireaux, G. (eds.), *Advances in Invertebrates and Fish Telemetry*, Kluwer Academic Publishers, Dordrecht, The Netherlands, pp. 199-206.

Everhart, W.H., Eipper, A.W., and Youngs, W.D. (1975). *Principles of Fisheries Science.* Cornell University Press, London, 288 pp.

Fancy, S.G., Pank, L.F., Douglas, D.C., Curby, C.H., Gardner, G.W., Amstrup, S.C., and Regelin, W.L. (1988). Satellite telemetry: a new tool for wildlife research and management. U.S. Dep. Inter., Fish Wildl. Serv.Resour. Publ. 172, 54 pp.

Ferrel, D.W., Nelson, D.R., Sciarrotta, T.C., Standora, E.A., and Carter, H.C. (1974). A multichannel ultrasonic biotelemetry system for monitoring marine animal behavior at sea. *Trans. Instr. Soc. Am.* **13**, 120-131.

Freire, J., and González-Gurriarán, E. (1998). New approaches to the behavioural ecology of decapod crustaceans using telemetry and electronic tags. *Hydrobiologia*, 371/372. *In* Lagardère, J.-P., Bégout Anras, M.-L., and Claireaux, G. (eds.), *Advances in Invertebrates and Fish Telemetry*, Kluwer Academic Publishers, Dordrecht, The Netherlands, pp. 123-132.

Freire, J., González-Gurriarán, E., and Bernárdez, C. (1999). Migratory patterns in the spider crab *Maja squinado* using telemetry and electronic tags. *ICES Annual Science Conference*, CM 1999/AA:14, 11 pages (mimeo).

Fricke, H., and Kaese, R. (1995). Tracking of artificially matured eels (*Anguilla anguilla*) in the Sargasso Sea and the problem of the eel's spawning site. *Naturwissenschaften* **82**, 32-36.

Glass, C.W., Johnstone, A.D.F., Smith, G.W., and Mojsiewicz, W.R. (1992). The movements of saithe (*Pollachius virens* L.) in the vicinity of an underwater reef. *In* Priede, I.G., and Swift, S.M. (eds.), *Wildlife Telemetry*, Ellis Horwood, New York, pp. 328-341.

Godø, O.R., and Michalsen, K. (2000). Migratory behaviour of north-east Arctic cod studied by the use of data storage tags. *Fish. Res.* **48**, 127-140.

Goldman, K.J., and Anderson, S.D. (1999). Space utilization and swimming depth of white sharks, *Carcharadon carcharias*, at the South Farallon Islands, central California. *Envir. Biol. Fishes* **56**, 351-364.

Goodman, R.M., Margolin, J., and Kratzer, D. (1973). Animal tracking satellite systems study. Final Report F-C3482, NASA contract NASW-2407. The Franklin Institute Research Laboratories, Philadelphia, Pa., USA.

González-Gurriarán, E., and Freire, J. (1994). Movement patterns and habitat utilization in the spider crab *Maja squinado* (Herbst) (Decapoda, Majidae) measured by ultrasonic telemetry. *J. Exp. Mar. Biol. Ecol.* **184**, 269-291.

Goudey, C.A., Consi, T., Manley, J., Graham, M., Donovan, B., and Kiley, L. (1998). A robotic boat for autonomous fish tracking. *Mar. Tech. Soc. J.* **32**, 47-53.

Graves, J.E., Luckhurst, B.E., and Prince, E.D. (2001). An evaluation of pop-up satellite tag technology to estimate post-release survival of blue marlin (*Makaira nigricans*). *Fish. Bull. US* (in press).

Green, J.M., and Wroblewski, J.S. (2000). Movement patterns of Atlantic cod in Gilbert Bay, Labrador: evidence for bay residency and spawning site fidelity. *J. Mar. Biol. Assoc. UK* **80**, 1077-1085.

Greer Walker, M., Mitson, R.B., and Storeton-West, T. (1971). Trials with a transponding acoustic fish tag tracked with an electronic sector scanning sonar. *Nature* **229**, 196-198.

Greer Walker, M., Harden Jones, F.R., and Arnold, G.P. (1978). The movements of plaice (*Pleuronectes platessa* L.) tracked in the open sea. *J. Cons. Int. Explor. Mer* **38**, 58-86.

Greer Walker, M., Riley, J.D., and Emerson, L. (1980). On the movements of sole (*Solea solea*) and dogfish (*Scyliorhinus canicula*) tracked off the East Anglian coast. *Neth. J. Sea Res.* **14,** 66-77.

Groot, C., Simpson, K., Todd, I., Murray, P.D., and Buxton, G.A. (1975). Movements of sockeye salmon (*Oncorhynchus nerka*) in the Skeena River estuary as revealed by ultrasonic tracking. *J. Fish. Res. Bd Can.* **32**, 233-242.

Gruber, S.H., Nelson, D.R. and Morrissey, J. F. (1988). Patterns of activity and space utilization of lemon sharks, *Negaprion brevirostris*, in a shallow Bahamian lagoon. *Bull. Mar. Sci.* **43**, 61-76.

Gunn, J. (2000). From plastic darts to pop-up satellite tags. *In* Hancock, D.A., Smith, D.C., and Kochn, J.D (cds.), *Fish Movement and Migration*, Australian Society for Fish Biology Workshop Proceedings, Bendigo, Victoria, September 1999. Australian Society for Fish Biology, Sydney, pp. 55-60.

Gunn, J., and Young, J. (2000). Environmental determinants of the movement and migration of juvenile southern bluefin tuna. *In* Hancock, D.A., Smith, D.C., and Koehn, J.D (eds.), *Fish Movement and Migration*, Australian Society for Fish Biology Workshop Proceedings, Bendigo, Victoria, September 1999. Australian Society for Fish Biology, Sydney, pp. 123-128.

Gunn, J.S., and Block, B.A. (2001). Advances in acoustic, archival and satellite tagging of tunas. In Block, B.A., and Stevens, E.D., *Tuna: Physiological Ecology and Evolution*. Academic Press (in press).

Gunn, J.S., Stevens, J.D., Davis, T.L.O., and Norman, B.M. (1999). Observations on the short-term movements and behaviour of whale sharks (*Rhincodon typus*) at Ningaloo Reef, Western Australia. *Mar. Biol.* **135**, 553-559.

Gunn, J., Polacheck, T., Davis, T., Sherlock, M., and Betlehem, A. (1994). The development and use of archival tags for studying the migration, behaviour and physiology of southern bluefin tuna, with an assessment of the potential for transfer of the technology to groundfish research. *ICES Annual Science Conference*, CM 1994 Mini: 2.1, 23 pages (mimeo).

Harden Jones, F. R., and Arnold, G.P. (1982). Acoustic telemetry and the marine fisheries. *In* Cheeseman, C.L., and Mitson, R.B. (eds.), *Telemetric Studies of Vertebrates*. Symp. Zool. Soc. Lond. (49), Academic Press, London, pp.75-93.

Harden Jones, F. R., Margetts, A.R., Greer Walker, M., and Arnold, G.P. (1977). The efficiency of the Granton otter trawl determined by sector scanning sonar and acoustic transponding tags. *Rapp. PV Réun. Cons. Int. Explor. Mer* **170**, 45-51.

Hawkins, A. D., MacLennan, D.N., Urquhart, G.G., and Robb, C. (1974). Tracking cod *Gadus morhua* L. in a Scottish sea loch. *J. Fish Biol.* **6**, 225-236.

Hernnkind, W. F. (1980). Spiny lobsters: patterns of movement. *In* Cobb, J. S., and Phillips, B.F. (eds.), *The Biology and Management of Lobsters*. Vol. 1. Academic Press, New York, pp. 349-407.

Hill, R.D. (1994) Theory of geolocation by light levels. *In* Le Bouef, B.J., and Laws. R.M. (eds.) *Elephant Seals: Population Ecology, Behavior, and Physiology*. University of California Press, Berkley, CA, pp. 227-236.

Hines, A.H., Wolcott, T.G., González-Gurriarán, E., González-Escalente, J.L., and Freire, J. (1995). Movement patterns and migrations in crabs: telemetry of juvenile and adult behaviour in *Callinectes sapidus* and *Maja squinado*. *J. Mar. Biol. Assoc. UK* **75**, 27-42.

Holand, B. (1974). Devices and Results 1974. *Fish Telemetry Report* (5), STF48 A4049, SINTEF, N-7034, Trondheim, Norway, 78 pp.

Holand, B.A. (1987). Underwater telemetry as a tool in aquaculture research and development. *Modeling, Identification and Control* **8**, 11-18.

Holland, K.N. and Sibert, J.R. (1994). Physiological thermoregulation in bigeye tuna, *Thunnus obesus*. *Envir. Biol. Fishes* **40**, 319-327.

Holland, K., Brill, R., and Chang, R. K. C. (1990). Horizontal and vertical movements of Pacific blue marlin captured and released using sportfishing gear. *Fish. Bull. US* **88**, 397-402.

Holland, K.N., Brill, R.W., and Chang, R. K. C. (1990). Horizontal and vertical movements of yellowfin and bigeye tuna associated with fish aggregating devices. *Fish. Bull. US* **88**, 493-507.

Holland, K.N., Lowe, C.G., and Wetherbee, B.M. (1996). Movements and dispersal patterns of blue trevally (*Caranx melampygus*) in a fisheries conservation zone. *Fish. Res.* 25, 279-292.

Holland, K.N., Lowe, C.G., Peterson, J.D., and Gill, A. (1992). Tracking coastal sharks with small boats: hammerhead shark pups as a case study. *Aust. J. Mar. Freshw. Res.* **43**, 61-66.

Holland, K.N., Peterson, J.D., Lowe, C.G., and Wetherbee, B.M. (1993a). Movements, distribution and growth rates of the white goatfish *Mulloides flavolineatus* in a fisheries conservation zone. *Bull. Mar. Sci.* **52**, 982-992.

Holland, K.N., Wetherbee, B.M., Peterson, J.D., and Lowe, C.G. (1993b). Movements and distribution of hammerhead shark pups on their natal grounds. *Copeia* **1993**(2), 495-502.

Holland, K.N., Wetherbee, B.M., Lowe, C.G., and Meyer, C.G. (1999). Movements of tiger sharks (*Galeocerdo cuvier*) in coastal Hawaiian waters. *Mar. Biol.* **134**, 665-673.

Holland, K., Brill, R., Ferguson, S., Chang, R., and Yost, R. (1985). A small vessel technique for tracking pelagic fish. *Mar. Fish. Rev.* **47**, 26-32.

Holland, K.N., Brill, R.W., Chang, R.K.C., Sibert, J.R. and Fournier, D.A. (1992). Physiological and behavioural thermoregulation in bigeye tuna *Thunnus obesus*. *Nature* **358**, 410-412.

Holts, D.B., and Bedford, D.W. (1993). Horizontal and vertical movements of the shortfin Mako shark, *Isurus oxyrinchus*, in the Southern California Bight. *Aust. J. Mar. Freshw. Res.* **44**, 901-909.

Hunter, J.R., Argue, A.W., Bayliff, W.H., Dizon, A.E., Fonteneau, A., Goodman, D., and Seckel, G.R. (1986). The dynamics of tuna movements: an evaluation of past and future research. *FAO Fish. Tech. Pap.* (277), 78 pp.

Inagake, D., Yamada, H., Segawa, K., Okazaki, M., Nitta, A., and Itoh, T. (2001). Migration of young bluefin tuna, *Thunnus orientalis* Temminck et Schlegel, through archival tagging experiments and its relation with oceanographic conditions in the western North Pacific. *Bull. Nat. Res. Inst. Far Seas Fish.*, **38**, 53-81.

Ishida, Y., Yano, A., Ban, M., and Ogura, M. (1998). Vertical movement of chum salmon, *Oncorhynchus keta*, in the western North Pacific Ocean as determined by a depth-recording archival tag. *Nat. Res. Inst. Far Seas Fish., Salmon Report Series* (45), 199-211.

Jensen, A.C., Collins, K.J., Free, E.K., and Bannister, R.C.A. (1994). Lobster (*Homarus gammarus*) movement on an artificial reef: the potential use of artificial reefs for stock enhancement. *Crustaceana* **67**, 198-211.

Jernakoff, P. (1987). Foraging patterns of juvenile western rock lobsters *Panulirus cygnus* George. *J. Exp. Mar. Biol. Ecol.* **113**, 125-144.

Jernakoff, P., and Phillips, B.F. (1988). Effect of a baited trap on the foraging movements of juvenile western rock lobsters *Panulirus cygnus* George. *Aust. J. Mar. Freshw. Res.* **39**, 185-192.

Johnson, J.H. (1960). Sonic tracking of adult salmon at Bonneville Dam, 1957. *Fish. Bull. US* **60**, 471-485.

Jolley, J.W., and Irby, E.W. (1979). Survival of tagged and released Atlantic sailfish (*Istiophorus platypterus*: Istiophoridae) determined with acoustical telemetry. *Bull. Mar. Sci.* **29**, 155-169.

Josse, E., Bach, P., and Dagorn, L. (1998). Simultaneous observations of tuna movements and their prey by sonic tracking and acoustic surveys. *Hydrobiologia*, 371/372. *In* Lagardère, J.-P., Bégout Anras, M.-L., and Claireaux, G. (eds.), *Advances in Invertebrates and Fish Telemetry*, Kluwer Academic Publishers, Dordrecht, The Netherlands, pp. 61-69.

Kanwisher, J., Lawson, K., and Sundnes, G. (1974). Acoustic telemetry from fish. *Fish. Bull. US* **72**, 251-255.

Karlsson, L., Ikonen, E., Westerberg, H., and Sturlaugsson, J. (1996). Use of data storage tags to study the spawning migration of Baltic salmon (*Salmo salar* L.) in the Gulf of Bothnia. *ICES Annual Science Conference*, CM 1996/M:9, 15 pages (mimeo).

Karlsson, L., Ikonen, E., Westerberg, H., and Sturlaugsson, J. (1999). Data storage tag study of salmon (*Salmo salar*) migration in the Baltic: the spawning migration of wild and hatchery-reared fish and a comparison of tagging methods. *ICES Annual Science Conference*, CM 1999/AA:05, 17 pages (mimeo).

Kasai, A., Sakamoto, W. Mitsunaga, Y., and Yamamoto, S. (2000). Behaviour of immature yellowtails (*Seriola quinqueradiata*) observed by electronic data-recording tags. *Fish. Oceanogr.* **9**, 259-270.

Kaseloo, P.A., Weatherley, A.H., Ihssen, P.E., Anstey, D.A., and Gare, M.D. (1996). Electromyograms from radiotelemetry as indicators of reproductive activity in lake trout. *J. Fish Biol.* **48**, 664-674.

Kingman, A. (1996). Satellite tracking blue sharks. *Shark News*, (7), 6.

Kittigawa, T., Nakata, H., Kimura, S., Itoh, T., Tsuji, S., and Nitta, A. (2000). Effect of ambient temperature on the vertical distribution and movement of Pacific bluefin tuna *Thunnus thynnus orientalis*. *Mar. Ecol. Prog. Ser.* **206**, 251-260.

Klimley, A.P. (1993). Highly directional swimming by scalloped hammerhead sharks, *Sphyrna lewini*, and subsurface irradiance, temperature, bathymetry, and geomagnetic field. *Mar. Biol.* **117**, 1-22.

Klimley, A.P., and Holloway, C.F. (1999). School fidelity and homing synchronicity of yellowfin tuna, *Thunnus albacares*. *Mar. Biol.* **133**, 307-317.

Klimley, A.P, Butler, S.B., Nelson, D.R., and Stull, A.T. (1988). Diel movements of scalloped hammerhead sharks, *Sphyrna lewini* Griffith and Smith, to and from a seamount in the Gulf of California. *J. Fish Biol.* **33**, 751-761.

Klimley, A.P., Prince, E.D., Brill, R.W., and Holland, K. (1994). Archival Tags 1994: Present and Future. US Department of Commerce, *NOAA Technical Memorandum NMFS-SEFSC 357.*, 42 pp.

Klimley, A.P., Voegeli, F., Beavers, S.C., and Le Boeuf, B.J. (1998). Automated listening stations for tagged marine fishes. *Mar. Tech. Soc. J.* **32**, 94-101.

Klimley, A.P., Le Boeuf, B.J., Cantara, K.M., Richert, J.E., Davis, S.F., and Van Sommeran, S. (2001). Radio-acoustic positioning as a tool for studying site-specific behavior of the white shark and other large marine species. *Mar. Biol.* **138**, 429-446.

Konagaya, T. (1982). A new telemetric method of determining the positions of swimming fish. *Bull. Jpn Soc. Sci. Fish.* **48**, 1545-1550.

LaBar, G.W., McCleave, J.D., and Fried, S.M. (1978). Seaward migration of hatchery-reared Atlantic salmon (*Salmo salar*) smolts in the Penobscot River estuary, Maine: open-water movements. *J. Cons. Int. Explor. Mer* **38**, 257-269.

Lacroix, G.L., and McCurdy, P. (1996). Migratory behaviour of post-smolt Atlantic salmon during initial stages of seaward migration. *J. Fish Biol.* **49**, 1086-1101.

Lacroix, G.L., and Voegeli, F.A. (2000). Development of automated monitoring systems for ultrasonic transmitters. *In* Moore, A., and Russell, I., *Advances in Fish Telemetry.* Proceedings of the Third Conference on Fish Telemetry in Europe, held in Norwich, England, 20-25 June 1999. CEFAS Lowestoft, pp. 37-50.

Lawson, K.D., and Carey, F.G. (1972). An acoustic telemetry system for transmitting body and water temperature from free swimming fish. *Woods Hole Oceanographic Institution Technical Report*, WHOI-71-67, 25 pp.

Laurs, R.M., Yuen, H.S.H., and Johnson, J.H. (1977). Small-scale movements of albacore, *Thunnus alalunga*, in relation to ocean features as indicated by ultrasonic tracking and oceanographic sampling. *Fish. Bull. US* **75**, 347-355.

Løkkeborg, S. (1998). Feeding behaviour of cod, *Gadus morhua*: activity rhythm and chemically mediated food search. *Anim. Behav.* **56**, 371-378.

Løkkeborg, S., and Fernö, A. (1999). Diel activity pattern and food search behaviour in cod, *Gadus morhua. Envir. Biol. Fishes* **54**, 345-353.

Løkkeborg, L., Skajaa, K., and Fernö, A. (2000). A food search strategy in ling (*Molva molva* L.) *J. Exp. Mar. Biol. Ecol.* **247**, 195-208.

Lowe, C.G., Holland, K.N., and Wolcott, T.G. (1998). A new acoustic tailbeat transmitter for fishes. *Fish. Res.* **36**, 275-283.

Lowe, C.G., and Goldman, K.J. (2001). Thermal and bioenergetics of elasmobranchs: bridging the gap. *Envir. Biol. Fishes* **60**, 251-266.

Lucas, M.C., Johnstone, A.D.F. and Priede, I.G. (1993). Use of physiological telemetry as a method of estimating metabolism of fish in the natural environment. *Trans. Am. Fish. Soc.* **122**, 822-833.

Luke, D.McG., Pincock, D.G., and Stasko, A.B. (1973). Pressure-sensing ultrasonic transmitter for tracking aquatic animals. *J. Fish. Res. Bd Can.* **30**, 1402-1404.

Lutcavage, M.E., Brill, R.W., Skomal, G.B., Chase, B.C., and Howey, P.W. (1999). Results of pop-up satellite tagging of spawning size class fish in the Gulf of Maine: do North Atlantic bluefin tuna spawn in the mid-Atlantic? *Can. J. Fish. Aquat. Sci.* **56**, 173-177.

Lutcavage, M.E., Brill, R.W., Skomal, G.B., Chase, B.C, Goldstein, J.L., and Tutein, J. (2000). Tracking adult North Atlantic bluefin tuna (*Thunnus thynnus*) in the northwestern Atlantic using ultrasonic telemetry. *Mar. Biol.* **137**, 347-358.

Lutcavage, M., Brill, R., Porter, J., Howey, P., Murray, E., Mendillo, A., Chaprales, W., Genovese, M., and Rollins, T. (2000). Summary of pop-up satellite tagging of giant bluefin tuna in the joint US-Canadian program, Gulf of Maine and Canadian Atlantic. ICCAT, SCRS/00/95, 9 pp. (mimeo.).

Madison, D.M., Horrall, R.M., Stasko, A.B., and Hasler, A.D. (1972). Migratory movements of adult sockeye salmon (*Oncorhynchus nerka*) in coastal British Columbia as revealed by ultrasonic tracking. *J. Fish. Res. Bd Can.* **29**, 1025-1033.

Marcenik, D.J., ,Blackwell, S.B., Dewar, H., Freund, E.V., Farwell, C.F., Dau, D.J., Seitz, A., and Block, B.A. (2001). Depth and muscle temperatures of Pacific bluefin tuna examined with acoustic telemetry and pop-up satellite tags. *Mar. Biol.* (in press).

Marsac, F., and Cayré, P. (1998). Telemetry applied to behaviour analysis of yellowfin tuna (*Thunnus albacares*, Bonnaterre, 1788) movements in a network of fish aggregating devices. *Hydrobiologia*, 371/372. *In* Lagardère, J.-P., Bégout Anras, M.-L., and Claireaux, G. (eds.), *Advances in Invertebrates and Fish Telemetry*, Kluwer Academic Publishers, Dordrecht, The Netherlands, pp. 155-171.

Matern, S.A., Cech, J.J., and Hopkins, T.E. (2000). Diel movements of bat rays, *Myliobatis californica*, in Tomales Bay, California: evidence for behavioral thermoregulation? *Envir. Biol. Fishes* **58**, 173-182.

McCleave, J.D., and Arnold, G.P. (1999). Movements of yellow- and silver-phase European eels (*Anguilla anguilla* L.) tracked in the western North Sea. *ICES J. Mar. Sci.* **56**, 510-536.

McCosker, J.E. (1987). The white shark, *Carcharodon carcharias*, has a warm stomach. *Copeia* **1987**(1), 195-197.

McKibben, J.N., and Nelson, D.R. (1986). Patterns of movement and grouping of gray reef sharks, *Carcharhinus amblyrhynchos*, at Enewetak, Marshall Islands. *Bull. Mar. Sci.* **38**, 89-110.

Medved, R.J., and Marshall, J.A. (1983). Short-term movements of young sandbar sharks, *Carcharhinus plumbeus* (Pisces, Carcharhinidae). *Bull. Mar. Sci.* **33**, 87-93.

Mee, D.M., Kirkpatrick, A.J., and Stonehewer, R.O. (1996). Post impoundment fishery investigations on the Tawe Barrage, South Wales. *In* Burt, N., and Watts, J. (eds.), *Barrages engineering design and environmental impacts*, John Wiley & Sons, Chichester, pp. 395-408.

Meeren, G.I. van der (1997). Preliminary acoustic tracking of native and transplanted European lobsters (*Homarus gammarus*) in an open sea lagoon. *Mar. Freshw. Res.* **48**, 915-921.

Merrick, R.L., Loughlin, T.R., Antonelis, G.A., and Hill, R. (1994). Use of satellite-linked time telemetry to study Steller sea lion and northern fur seal foraging. *Polar Res.* **13**, 105-114.

Metcalfe, J.D., and Arnold, G.P. (1997). Tracking fish with electronic tags. *Nature* **387**, 665-666.

Metcalfe, J.D., Holford, B.H., and Arnold, G.P. (1993). Orientation of plaice (*Pleuronectes platessa*) in the open sea: evidence for the use of external directional clues. *Mar. Biol.* **117**, 559-566.

Metcalfe, J.D., Arnold, G.P., and Holford, B.H. (1994). The migratory behaviour of plaice in the North Sea as revealed by data storage tags. *ICES Annual Science Conference,* CM 1994/Mini:11, 13 pp. (mimeo).

Metcalfe, J.D., Arnold, G.P., and Hunter, E. (1998). The use of electronic data storage tags in modelling the effectiveness of marine protected areas. *ICES Annual Science Conference*, CM 1998/U:4, 8 pp. (mimeo).

Metcalfe, J.D., Hunter, E., Holford, B.H., and Arnold, G.P. (1999). The application of electronic data storage tags to spatial dynamics of fish populations. *ICES Annual Science Conference*, CM 1999/AA:01, 7 pp. (mimeo).

Mitson, R.B., and J.C. Cook (1971). Shipboard installation and trials of an electronic sector-scanning sonar. *Radio Electron. Eng.* **41**, 339-350.

Mitson, R.B., and Storeton West, T.J. (1971). A transponding acoustic fish tag. *Radio Electron. Eng.* **41**, 483-489.

Mitson, R.B., Storeton-West, T.J., and Pearson, N.D. (1982). Trials of an acoustic transponding fish tag compass. *Biotelem. Pat. Mon.* **9**, 69-79.

Monan, G. E., and Thorne, D.L. (1973). Sonic tags attached to Alaska king crab. *Mar. Fish. Rev.* **35**, 18-21.

Moore, A. (1996). Tracking Atlantic salmon (*Salmo salar* L.) in estuaries and coastal waters. *In* Baras, E., and Philippart, J.C. (eds.), *Underwater Biotelemetry.* Proceedings of the First

Conference and Workshop on Fish Telemetry in Europe. University of Liège, Belgium, pp. 93-100.

Moore, A., Russell, I.C., and Potter, E.C.E. (1990). The effects of intraperitoneally implanted dummy acoustic transmitters on the behaviour and physiology of juvenile Atlantic salmon, *Salmo salar* L. *J. Fish Biol.* **37**, 713-721.

Moore, A., Lacroix, G.L., and Sturlaugsson, J. (2000). Tracking Atlantic salmon post-smolts in the sea. *In* Mills, D. (ed.) *The Ocean Life of Atlantic Salmon.* Fishing News Books, Blackwell Science, Oxford, pp. 49-64.

Moore, A., Potter, E.C.E., Milner, N.J., and Bamber, S. (1995). The migratory behaviour of wild Atlantic salmon (*Salmo salar*) smolts in the estuary of the River Conwy, North Wales. *Can. J. Fish. Aquat. Sci.* **52**, 1923-1935.

Moore, A., Ives, M., Scott, M., and Bamber, S. (1998a). The migratory behaviour of wild sea trout (*Salmo trutta* L.) smolts in the estuary of the River Conwy, North Wales. *Aquaculture* **168**, 57-68.

Moore, A., Ives, S., Mead, T.A., and Talks, L. (1998b). The migratory behaviour of wild Atlantic salmon (*Salmo salar* L.) smolts in the River Test and Southampton Water, southern England. *Hydrobiologia*, 371/372. *In* Lagardère, J.-P., Bégout Anras, M.-L., and Claireaux, G. (eds.), *Advances in Invertebrates and Fish Telemetry*, Kluwer Academic Publishers, Dordrecht, The Netherlands, pp. 295-304.

Nakamura, Y. (1991). Tracking of the mature female of flying squid, *Ommastrephes bartrami*, by an ultrasonic transmitter. *Bull. Hokkaido Natl Fish. Res. Inst.* **55**, 205-207.

Nakamura, Y. (1993). Vertical and horizontal movements of mature females of *Ommastrephes bartramii* observed by ultrasonic telemetry. *In* Okutani, T., O'Dor, R.K., and Kubodera, T. (eds.), *Recent Advances in Fisheries Biology*, Tokai University Press, Tokyo, pp. 331-336.

Nelson, D.R. (1976). Ultrasonic telemetry of shark behavior. *J. Acoust. Soc. Am.* **59**, 1004-1007.

Nelson, D.R. (1978). Telemetering techniques for the study of free-ranging sharks. *In* Hodgson, E.S., and Mathewson, R.F. (eds.), *Sensory Biology of Sharks, Skates, and Rays.* Office of Naval Research, Department of the Navy, Arlington, Va., USA, pp. 419-482.

Nelson, D.R. (1990). Telemetry studies of sharks: a review, with applications in resource management. *In* Pratt, H.L., Gruber, S.H., and Taniuchi, T. (eds.), *Elasmobranchs as living resources: advances in the biology, ecology, systematics, and the status of the fisheries.* US Department of Commerce, *NOAA Technical Report NOAA-NMFS-SWFSC-90*, , pp. 239-256.

Nelson, D.R., and McKibben, J.N. (1981). Timed-release, recoverable, ultrasonic/radio transmitters for tracking pelagic sharks. *In* Long, F.M. (ed.) Proc. Third Int. Conf. Wildl. Biotelem. University of Wyoming, Laramie, pp. 90-104.

Nelson, D.R., McKibben, J.N., Strong, W.R., Lowe, C.G., Sisneros, J.A., Schroeder, D.M., and Lavenberg, R.J. (1997). An acoustic tracking of a megamouth shark, *Megachasma pelagios*: a crepuscular vertical migrator. *Envir. Biol. Fishes* **49**, 389-399.

Niezgoda, G.H., McKinley, R.S., White, D., Anderson, G., and Cote, D. (1998). A dynamic combined acoustic and radio transmitting tag for diadromous fish. *Hydrobiologia*, 371/372. *In* Lagardère, J.-P., Bégout Anras, M.-L., and Claireaux, G. (eds.), *Advances in Invertebrates and Fish Telemetry*, Kluwer Academic Publishers, Dordrecht, The Netherlands, pp. 47-52.

NMFS (1997). Billfish News Letter. U.S. National Marine Fisheries Service Southwest Fisheries Center, La Jolla CA.

O'Dor, R.K., Hoar, J.A., Webber, D.M., Carey, F.G., Tanaka, S., Martins, H.R., and Porteiro, F.M. (1994). Squid (*Loligo forbesi*) performance and metabolic rates in nature. *Mar. Freshw. Behav. Physiol.* **25**, 163-177.

O'Dor, R.K., Andrade, Y., Webber, D.M., Sauer, W.H.H., Roberts, M.J., Smale, M.J., and Voegeli, F.M. (1998). Applications and performance of Radio-Acoustic Positioning and Telemetry (RAPT) systems. *Hydrobiologia*, 371/372. *In* Lagardère, J.-P., Bégout Anras, M.-L., and Claireaux, G. (eds.), *Advances in Invertebrates and Fish Telemetry*, Kluwer Academic Publishers, Dordrecht, The Netherlands, pp. 1-8.

Ogura, M. (1997). Acoustic and archival tagging work on salmonids in Japan. In Boehlert, G.W., Application of acoustic and archival tags to assess estuarine, nearshore, and off-shore habitat utilization and movement by salmonids. *NOAA Technical Memorandum NMFS*, NOAA-TM-NMFS-SWFSC-236, 62 pp.

Ogura, M., and Ishida, Y. (1992). Swimming behavior of coho salmon, *Oncorhynchus kisutch*, in the open sea as determined by ultrasonic telemetry. *Can. J. Fish. Aquat. Sci.* **49**, 453-457.

Ogura, M., and Ishida, Y. (1995). Homing behavior and vertical movements of four species of Pacific salmon (*Oncorhynchus* spp.) in the central Bering Sea. *Can. J. Fish. Aquat. Sci.* **52**, 532-540.

Olson, A.F., and Quinn, T.P. (1993). Vertical and horizontal movements of adult chinook salmon *Oncorhynchus tshawytscha* in the Columbia River estuary. *Fish. Bull. US* **91**, 171-178.

Oswald, R.L. (1978). The use of telemetry to study light synchronization with feeding and gill ventilation rates in *Salmo trutta*. *J. Fish Biol.* **13**, 729-739.

Pawson, M.G., Kelley, D.F., and Pickett, G.D. (1987). The distribution and migrations of bass, *Dicentrarchus labrax* L., in waters around England and Wales as shown by tagging. *J. Mar. Biol. Assoc. UK* **67**, 183-217.

Pearcy, W.G. (1992). Movements of acoustically-tagged yellowtail rockfish *Sebastes flavidus* on Heceta Bank, Oregon. *Fish. Bull. US* **90**, 726-735.

Pearson, N.D., and Storeton West, T.J. (1987). The design of an acoustic transponding compass tag for free-swimming fish. *In Electronics for Ocean Technology*, IERE Conf. Proc. (Lond.) (72), pp. 83-92.

Polovina, J.J., Kobayashi, D.R., Parker, D.M., Seki, M.P., and Balazs, G.H. (2000). Turtles on the edge: movement of loggerhead turtles (*Caretta caretta*) along oceanic fronts, spanning longline fishing grounds in the central North Pacific, 1997-1998. *Fish. Oceanogr.* **9**, 71-82.

Potter, E.C.E. (1988). Movements of Atlantic salmon, *Salmo salar* L., in an estuary in south-west England. *J. Fish Biol.* **33**, Supplement A, 153-159.

Potter, E.C.E, Solomon, D.J., and Buckley, A.A. (1992). Estuarine movements of adult Atlantic salmon (*Salmo salar* L.) in Christchurch Harbour, southern England. *In* Priede, I.G., and Swift, S.M. (eds.), *Wildlife Telemetry*, Ellis Horwood, New York, pp. 400-409.

Priede, I.G. (1982). An ultrasonic salinity telemetry sensor for use on fish in estuaries. *Biotelem. Pat. Mon.* **9**, 1-9.

Priede, I.G. (1984). A basking shark (*Cetorhinus maximus*) tracked by satellite together with simultaneous remote sensing. *Fish. Res.* **2**, 201-216.

Priede, I.G. (1986). The limitations of fish tracking systems: acoustic and satellite techniques. US Department of Commerce, *NOAA Technical Memorandum NMFS*, NOAA-TM-NMFS-SWFC-61, 17 pages.

Priede, I.G. (1992). Wildlife telemetry: an introduction. *In* Priede, I.G., and Swift, S.M. (eds.), *Wildlife Telemetry*, Ellis Horwood, New York, pp. 3-25.

Priede, I.G., and Tytler, P. (1977). Heart rate as a measure of metabolic rate in teleost fishes; *Salmo gairdneri, Salmo trutta* and *Gadus morhua. J. Fish Biol.* **10**, 231--242.

Priede, I.G., Solbé, J.F. de L.G., and Nott, J.E. (1988). An acoustic oxygen telemetry transmitter for the study of exposure of fish to variations in environmental dissolved oxygen. *J. Exp. Biol.* **140**, 563-567.

Priede, I.G., Smith, K.L., and Armstrong, J.D. (1990). Foraging behavior of abyssal grenadier fish: inferences from acoustic tagging and tracking in the North Pacific Ocean. *Deep-Sea Res.* **37**, 81-101.

Quinn, T.P., terHart, B.A., and Groot, C. (1989). Migratory orientation and vertical movements of homing adult sockeye salmon, *Oncorhynchus nerka,* in coastal waters. *Anim. Behav.* **37**, 587-599.

Restrepo, V.R. (1996). Report of the TAB workshop: a research and planning workshop for Atlantic bluefin tuna tagging studies. ICCAT Coll. Vol. Sci. Pap., XLV(2), 228-244.

Righton, D., Turner, K., and Metcalfe, J.D. (2000). Behavioural switching in North Sea cod: implications for foraging strategy? *ICES Annual Science Conference*, CM 2000/ Q:09, 14 pp. (mimeo).

Righton, D., Metcalfe, J., and Connolly, P. (2001). Different behaviour of North and Irish Sea cod. *Nature*, **411**, 156.

Rogers, S.C., Church, D.W., Weatherley, A.H., and Pincock, D.G. (1984). An automated ultrasonic telemetry system for the assessment of locomotor activity in free-ranging rainbow trout, *Salmo gairdneri*, Richardson. *J. Fish Biol.* **25**, 697-710.

Ross, L.G., Watts, W., and Young, A.H. (1981). An ultrasonic biotelemetry system for the continuous monitoring of tail-beat rate from free-swimming fish. *J. Fish Biol.* **18**, 479-490.

Ruggerone, G.T., Quinn, T.P., McGregor, I.A., and Wilkinson, T.D. (1990). Horizontal and vertical movements of adult steelhead trout, *Oncorhynchus mykiss*, in the Dean and Fisher Channels, British Columbia. *Can. J. Fish. Aquat. Sci.* **47**, 1963-1969.

Sauer, W.H.H., Roberts, M.J., Lipinski, M.R., Smale, M.J., Hanlon, R.T., Webber, D.M., and O'Dor, R.K. (1997). Choreography of the squid's "nuptial dance". *Biol. Bull.* **192**, 203-207.

Scharold, J., and Gruber, S.H. (1991). Telemetered heart rate as a measure of metabolic rate in the lemon shark, *Negaprion brevirostris. Copeia* **1991**(4), 942-953.

Sciarrotta, T.C., and Nelson, D.R. (1977). Diel behavior of the blue shark, *Prionace glauca*, near Santa Catalina Island, California. *Fish. Bull. US* **75**, 519-28.

Shimamura, T., Yoza, K. Hasewaga, E., and Furuta, M. (1987). On the swimming behaviour of chum salmon in early migratory season off the coast of Abashiri, Hokkaido. *Nippon Suisan Gakkaishi* **53**, 411-415.

Skajaa, K., Fernö, A., Løkkeborg, S., and Haugland, E.K. (1998). Basic movement pattern and chemo-orientated search towards baited pots in edible crab (*Cancer pagurus* L.). *Hydrobiologia*, 371/372. *In* Lagardère, J.-P., Bégout Anras, M.-L., and Claireaux, G. (eds.), *Advances in Invertebrates and Fish Telemetry*, Kluwer Academic Publishers, Dordrecht, The Netherlands, pp. 143-153.

Smith, G.W., Hawkins, A.D., Urquhart, G.G., and Shearer, W.M. (1981). Orientation and energetic efficiency in the offshore movements of returning Atlantic salmon (*Salmo salar* L.). *Scottish Fisheries Research Report*, (21), 22 pp.

Smith, G.W., Urquhart, G.G., MacLennan, D.N., and Sarno, B. (1998). A comparison of theoretical estimates of the errors associated with ultrasonic tracking using a fixed hydrophone array and field measurements. *Hydrobiologia*, 371/372. *In* Lagardère, J.-P., Bégout Anras, M.-L., and Claireaux, G. (eds.), *Advances in Invertebrates and Fish Telemetry*, Kluwer Academic Publishers, Dordrecht, The Netherlands, pp. 9-17.

Smith, I.P., Collins, K.J., and Jensen, A.C. (1998). Movement and activity patterns of European lobster (*Homarus gammarus*) revealed by electromagnetic telemetry. *Mar. Biol.* **132**, 611-623.

Smith, I.P., Collins, K.J., and Jensen, A.C. (1999). Seasonal changes in the level and diel pattern of activity in the European lobster *Homarus gammarus*. *Mar. Ecol. Prog. Ser.* **186**, 255-264.

Soeda, H., Yoza, K., Shimamura, T., and Hasegawa, E. (1987). On the swimming behavior of chum salmon in early migratory season off the coast of Hokkaido, Okhotsu Sea. *Nippon Suisan Gakkaishi* **53**, 1827-1833.

Solomon , D.J., and Storeton-West, T.J. (1983). Radio tracking of migratory salmonids in rivers; development of an effective system. *Fish. Res. Tech. Rep., MAFF Direct. Fish. Res.* (75), 11 pp.

Solomon, D.J., and Potter, E.C.E. (1988). First results with a new estuarine fish tracking system. *J. Fish Biol.* **33**, Supplement A, 127-132.

Standora, E.A., and Nelson, D.R. (1977). A telemetric study of the behavior of free-swimming Pacific angel sharks, *Squatina californica*. *Bull. So. Calif. Acad. Sci.* **76**, 193-201.

Starr, R.M., Heine, J.N., and Johnson, K.A. (2000). Techniques for tagging and tracking deepwater rockfishes. *N. Amer. J. Fish. Mgmt.* **20**, 597-609.

Stasko, A.B. (1975). Progress of migrating Atlantic salmon (*Salmo salar*) along an estuary, observed by ultrasonic tracking. *J. Fish Biol.* **7**, 329-338.

Stasko, A.B., and Polar, S.M. (1973). Hydrophone and bow-mount for tracking fish by ultrasonic telemetry. *J. Fish. Res. Bd Can.* **30**, 119-121.

Stasko, A.B., and Rommel, S.A. (1974). Swimming depth of adult American eels (*Anguilla rostrata*) in a saltwater bay as determined by ultrasonic tracking. *J. Fish. Res. Bd Can.* **31**, 1148-1150.

Stasko, A.B., and Horrall, R.M. (1976). Method of counting tailbeats of free-swimming fish by ultrasonic telemetry techniques. *J. Fish. Res. Bd Can.* **33**, 2596-2598.

Stasko, A.B., and Pincock, D.G. (1977). Review of underwater biotelemetry, with emphasis on ultrasonic techniques. *J. Fish. Res. Bd Can.* **34**, 1261-1285.

Stasko, A.B., Horrall, R.M., Hasler, A.D. (1976). Coastal movements of adult Fraser River sockeye salmon (*Oncorhynchus nerka*) observed by ultrasonic tracking. *Trans. Am. Fish. Soc.* **105**, 64-71.

Stasko, A.B., Horrall, R.M., Hasler, A.D., and Stasko, D. (1973). Coastal movements of mature Fraser River pink salmon (*Oncorhynchus gorbuscha*) as revealed by ultrasonic tracking. *J. Fish. Res. Bd Can.* **30**, 1309-1316.

Steingrund, P (1999). Studies of vertical migration of wild Faroe Plateau cod by use of Data Storage Tags. *ICES Annual Science Conference*, CM 1999/AA:02 (8 pp. mimeo).

Stewart, B.S., Leatherwood, S., Yochem, P.K., and Heide-Jørgensen, M-P. (1989). Harbor seal tracking and telemetry by satellite. *Mar. Mamm. Sci.* **5**, 361-375.

Storeton West, T.J., Mitson, R.B., and Greer Walker, M. (1978). Fish heart rate telemetry in the open sea using sector-scanning sonar. *Biotelemetry* **5**, 149-153.

Sturlaugsson, J. (1995). Migration study on homing of Atlantic salmon (*Salmo salar* L.) in coastal waters W-Iceland. *ICES Annual Science Conference*, CM 1995/M:17, 13 pp. (mimeo).

Sturlaugsson, J., and Johannsson, M. (1996). Migratory pattern of wild sea trout (*Salmo trutta* L.) in SE-Iceland recorded by data storage tags. *ICES Annual Science Conference*, CM 1996/M:5, 23 pp. (mimeo).

Sturlaugsson, J., and Thorisson, K. (1997). Migratory pattern of homing Atlantic salmon (*Salmo salar* L.) in coastal waters W-Iceland, recorded by data storage tags. *ICES Annual Science Conference*, CM 1997/CC:09, 16 pp. (mimeo).

Sturlaugsson, J., Jonsson, I.R., and Tomasson, T. (1998). Sea migrations of anadromous arctic char (*Salvelinus alpinus*) recorded by data storage tags. *ICES Annual Science Conference*, CM 1998/N:22, (summary).

Sundström, L.F., and Gruber, S.H. (1998). Using speed-sensing transmitters to construct a bioenergetics model for subadult lemon sharks, *Negaprion brevirostris* (Poey), in the field. *Hydrobiologia*, 371/372. *In* Lagardère, J.-P., Bégout Anras, M.-L., and Claireaux, G. (eds.), *Advances in Invertebrates and Fish Telemetry*, Kluwer Academic Publishers, Dordrecht, The Netherlands, pp. 241-247.

Sundström, L.F., Gruber, S.H., Clermont, S.M., Correia, J.P.S., de Marignac, J.R.C., Morrissey, J.F., Lowrance, C.R., Thomassen, L., and Oliveira, M.T. (2001). Review of elasmobranch behavioral studies using ultrasonic telemetry with special reference to the lemon shark, *Negaprion brevirostris*, around Bimini Islands, Bahamas. *Envir. Biol. Fishes* **60**, 225-250.

Sureau, D., and Lagardère, J.-P. (1991). Coupling of heart rate and locomotor activity in sole, *Solea solea* (L.), and bass, *Dicentrarchus labrax* (L.), in their natural environment by using ultrasonic telemetry. *J.Fish Biol.* **38**, 399-405.

Taillade, M. (1992). Animal tracking by satellite. *In* Priede, I.G., and Swift, S.M. (eds.). *Wildlife Telemetry.* Ellis Horwood, New York, pp. 149-160.

Tanaka, H., Takagi, Y. Iwata, M., and Naito, Y. (1998). The behavior and ambient temperature of homing chum salmon monitored by a data logger. *Proc. NIPR Symp. Polar Biol.* **11**, pp. 62-73.

Tesch, F.-W. (1972). Versuche zur telemetrischen Verfolgung der Laichwanderung von Aalen (*Anguilla anguilla*) in der Nordsee. *Helgoländ. Wiss. Meeresun.* **23**, 165-183.

Tesch, F.-W. (1974). Speed and direction of silver and yellow eels, *Anguilla anguilla*, released and tracked in the open North Sea. *Ber. Deut. Wiss. Komm. Meeresforsch.* **23**, 181-197.

Tesch, F-W. (1978). Horizontal and vertical swimming of eels during the spawning migration at the edge of the continental shelf. *In* Schmidt-Koenig, K., and Keeton, W.T., (eds.). *Animal Migration, Navigation, and Homing*. Springer-Verlag, Berlin, Heidelberg, New York, pp. 378-391.

Tesch, F.-W. (1979). Tracking of silver eels (*Anguilla anguilla* L.) in different shelf areas of the North East Atlantic. *Rapp. PV Réun. Cons. Int. Explor. Mer* **174,** 104-114.

Tesch, F-W. (1989). Changes in swimming depth and direction of silver eels (*Anguilla anguilla* L.) from the continental shelf to the deep sea. *Aquat. Living Res.* **2**, 9-20.

Tesch, F-W., Westerberg, H., and Karlsson, L. (1991). Tracking studies on migrating silver eels in the central Baltic. *Meeresforschung* **33**, 183-196.

Thorsteinsson, V. (1995). Tagging experiments using conventional tags and electronic data storage tags for the observations of migration, homing and habitat choice in the Icelandic spawning stock of cod. *ICES Annual Science Conference*, CM 1995/B:19 Ref. G, 16 pages (mimeo).

Thorsteinsson, V., and Marteinsdottir, G. (1998). Size specific time and duration of spawning of cod (*Gadus morhua*) in Icelandic waters. ICES CM 1998/DD:5, 18 pp. (mimeo).

Trefethen, P.S. (1956). Sonic equipment for tracking individual fish. *Spec. scient. Rep. U.S. Fish Wildl. Serv.* (179), 11 pp.

Tremblay, M.J., Duggan, R., O'Dor, R., Curtis, C., Webber, D., and Andrade, Y. (1999). Daily movements of lobster from ultrasonic tracking [abstract]. *J. Shellfish Res.* **18,** 307.

Tricas, T.C., Taylor, L.R., and Naftel, G. (1981). Diel behavior of the tiger shark, *Galeocerdo cuvier*, at French Frigate Shoals, Hawaiian Islands. *Copeia* **1981**(4), 904-908.

Urquhart, G.G., and Smith, G.W. (1992). Recent developments of a fixed hydrophone array system for monitoring movements of aquatic animals. *In* Priede, I.G., and Swift, S.M. (eds.), *Wildlife Telemetry*, Ellis Horwood, New York, pp. 342-353.

Voegeli, F.A., and Pincock, D.G. (1981). Determination of fish swimming speed by ultrasonic telemetry. *Biotelem. Pat. Mon.* **7**, 215-220.

Voegeli, F.A., and Pincock, D.G. (1996). Overview of underwater acoustics as it applies to telemetry. *In* Baras, E., and Philippart, J.C. (eds.), *Underwater Biotelemetry*, Proceedings of the First Conference and Workshop on Fish Telemetry in Europe. University of Liège, Belgium, pp. 23-30.

Voegeli, F.A, and McKinnon, G.P. (1996). Recent developments in ultrasonic tracking systems. *In* Baras, E., and Philippart, J.C. (eds.), *Underwater Biotelemetry*, Proceedings of the First Conference and Workshop on Fish Telemetry in Europe. University of Liège, Belgium, pp. 235-241.

Voegeli, F.A., Lacroix, G.L., and Anderson, J.M. (1998). Development of miniature pingers for tracking Atlantic salmon smolts at sea. *Hydrobiologia*, 371/372. *In* Lagardère, J.-P., Bégout Anras, M.-L., and Claireaux, G. (eds.), *Advances in Invertebrates and Fish Telemetry*, Kluwer Academic Publishers, Dordrecht, The Netherlands, pp. 35-46.

Voegeli, F.A., Smale, M.J., Webber, D.M., Andrade, Y., and O'Dor, R.K. (2001). Ultrasonic telemetry, tracking and automated monitoring technology for sharks. *Envir. Biol. Fishes* **60**, 267-281.

Voglis, G.M., and Cook, J.C. (1966). Underwater applications of an advanced acoustic scanning equipment. *Ultrasonics* **4**, 1-9.

Wada, K., and Ueno, Y. (1999). Homing behavior of chum salmon determined by an archival tag. (*NPAFC Doc. 425*). Hokkaido National Fisheries Research Institute. 116 Katsurakoi, Kushiro, Hokkaido, 085-0802, Japan. 29 pp. (mimeo).

Walker, R.V., Myers, K.W., Davis, N.D., Aydin, K.Y., Friedland, K.D., Carlson, H.R., Boehlert, G.W., Urawa, S., Ueno, Y., and Anma, G. (2000). Diurnal variation in thermal environment experienced by salmonids in the North Pacific as indicated by data storage tags. *Fish. Oceanogr.* **9**, 171-186.

Walters, C., and Maguire, J-J. (1996). Lessons for stock assessment from the northern cod collapse. *Rev. Fish Biol. Fish.* **6**, 125-137.

Ward, P., Carlson, B., Weekly, M., and Brumbaugh, B. (1984). Remote telemetry of daily vertical and horizontal movement of *Nautilus* in Palau. *Nature* **309**, 248-250.

Wardle, C.S., and Kanwisher, J.W. (1974). The significance of heart rate in free swimming cod, *Gadus morhua*: some observations with ultra-sonic tags. *Mar. Behav. Physiol.* **2**, 311-324.

Watson, W.H., Vetrovs, A., and Howell, W.H. (1999). Lobster movements in an estuary. *Mar. Biol.* **134**, 64-75.

Webber, D.M., and O'Dor, R.K. (1986). Monitoring the metabolic rate and activity of free-swimming squid with telemetered jet pressure. *J. Exp. Biol.* **126**, 205-224.

Welch, D.W., and Eveson, J.P. (1999). An assessment of light-based geoposition estimates from archival tags. *Can. J. Fish. Aquat. Sci.* **56**, 1317-1327.

West, G.J., and Stevens, J.D. (2001). Archival tagging of school shark, *Galeorhinus galeus*, in Australia: initial results. *Envir. Biol. Fishes* **60**, 283-298.

Westerberg, H. (1979). Counter-current orientation in the migration of the European eel. *Rapp. PV Réun. Cons. Int. Explor. Mer* **174**, 134-143.

Westerberg, H. (1982a). Ultrasonic tracking of Atlantic salmon (*Salmo salar* L.) - I. Movements in coastal regions. *Rep. Inst. Freshwat. Res., Drottningholm* (60), 81-101.

Westerberg, H. (1982b). Ultrasonic tracking of Atlantic salmon (*Salmo salar* L.) - II. Swimming depth and temperature stratification. *Rep. Inst. Freshwat. Res., Drottningholm* (60), 102-120.

Westerberg, H., Sturlaugsson, J., Ikonen, E., and Karlsson, L. (1999a). Data storage tag study of salmon (*Salmo salar*) migration in the Baltic: behaviour and the migration route as reconstructed from SST data. *ICES Annual Science Conference*, CM 1999/AA:06, 18 pp. (mimeo).

Westerberg, H., Eveson, P., Welch, D., Karlsson, L., and Ikonen, E. (1999b). Data storage tag study of salmon (*Salmo salar*) migration in the Baltic: the performance of tags. *ICES Annual Science Conference*, CM 1999/AA:07, 13 pages (mimeo).

Wilson, R.P., Ducamp, J.-J., Rees, W.G., Culik, B.M., and Niekamp, K. (1992) Estimation of location: global coverage using light intensity. *In* Priede, I.G., and Swift, S.M. (eds.), *Wildlife Telemetry*, Ellis Horwood, New York, pp. 131-134.

Wolcott, T.G., and Hines, A.H. (1989). Ultrasonic biotelemetry of muscle activity from free-ranging marine animals: a new method for studying foraging by blue crabs (*Callinectes sapidus*). *Biol. Bull.* **176**, 50-56.

Wroblewski, J.S., Bailey, W.L., and Howse, K.A. (1994). Observations of adult Atlantic cod (*Gadus morhua*) overwintering in nearshore waters of Trinity Bay, Newfoundland. *Can. J. Fish. Aquat. Sci.* **51**, 142-150.

Wroblewski, J.S., Goddard, S.V., Smedbol, R.K., and Bailey, W.L. (1995). Movements of Atlantic cod (*Gadus morhua*) within the spring thermocline in Trinity Bay, Newfoundland. *J. Mar. Biol. Assoc. UK* **75**, 265-284.

Wroblewski, J.S., Nolan, B.G., Rose, G.A., and de Young, B. (2000). Response of individual shoaling Atlantic cod to ocean currents on the northeast Newfoundland Shelf. *Fish. Res.* **45**, 51-59.

Yamanka, K.L., and Richards, L.J. (1993). Movements of transplanted lingcod, *Ophiodon elongatus*, determined by ultrasonic telemetry. *Fish. Bull. US* **91**, 582-587.

Yamashita, H and Miyabe, N. (2000). Report of 1999 bluefin tuna archival tagging in the Mediterranean Sea conducted by Japan. ICCAT SCRS/00/26. 15 pp. (mimeo.).

Yano, A., Ogura, M., Sato, A., Sakaki, Y., Ban, M., and Nagasawa, K. (1996). Development of ultrasonic telemetry technique for investigating the magnetic sense of salmonids. *Fish. Sci.* **62**, 698-704.

Yano, A., Ogura, M., Sato, A., Sakaki, Y., Shimizu, Y., Baba, N., and Nagasawa, K. (1997). Effect of modified magnetic field on the ocean migration of maturing chum salmon, *Oncorhynchus keta*. *Mar. Biol.* **129**, 523-530.

Young, A.H., Tytler, P., Holliday, F.G.T., and MacFarlane, A. (1972). A small sonic tag for measurement of locomotor behaviour in fish. *J. Fish Biol.* **4**, 57-65.

Young, A.H., Tytler, P., and Holliday, F.G.T. (1976). New developments in ultrasonic fish tracking. *Proc. Roy. Soc. Edinb.* (B) **75**, 145-155.

Yuen, H.S.H. (1970). Behavior of skipjack tuna, *Katsuwonus pelamis*, as determined by tracking with ultrasonic devices. *J. Fish. Res. Bd Can.* **27**, 2071-2079.

Zeller, D.C. (1997). Home range and activity patterns of the coral trout *Plectropomus leopardus* (Serranidae). *Mar. Ecol. Prog. Ser.* **154**, 65-77.

Zeller, D.C. (1998a). Spawning aggregations: patterns of movement of the coral trout *Plectropomus leopardus* (Serranidae) as determined by ultrasonic telemetry. *Mar. Ecol. Prog. Ser.* **162**, 253-263.

Zeller, D.C., and Russ, G.R, (1998b). Marine reserves: patterns of adult movement of the coral trout (*Plectropomus leopardus* (Serranidae)). *Can. J. Fish. Aquat. Sci.* **55**, 917-924.

Zeller, D.C. (1999). Ultrasonic telemetry: its application to coral reef fisheries research. *Fish. Bull. US* **97**, 1058-1065.

Archival and Pop-up Satellite Tagging of Atlantic Bluefin Tuna

Barbara A. Block,[1] Heidi Dewar,[1#] Susanna B. Blackwell[1],[+] Tom Williams,[2] Eric Prince,[3] André M. Boustany,[1] Chuck Farwell,[2] Daniel J. Dau[1] and Andy Seitz[2]

[1]*Tuna Research and Conservation Center, Stanford University, Pacific Grove, California*
[2]*Monterey Bay Aquarium, Cannery Row, Monterey California*
[3]*Southeast Fisheries Center, 75 Virginia Beach Drive, Miami, Florida*
[#]*Present address: Pfleger Institute of Environmental Research, 1400 North Pacific Street, Oceanside, California*
[+]*Present address: Greeneridge Sciences, 1411 Firestone Road, Goleta, California*

Abstract: Pelagic fish have historically been a challenge to study because of their large size and highly migratory movements. Previous technological limitations have recently been overcome using archival and pop-off satellite tags, enabling studies of long-term movements, oceanographic preferences and behaviors. Archival tags record information on depth, ambient and internal temperatures, and light levels. Their major advantage lies in the extensive detail of this information and the ability to extract geolocation and oceanographic information in addition to biological data. We have deployed 279 archival tags in Atlantic bluefin tuna (*Thunnus thynnus thynnus*) in the western North Atlantic. To date, 40 of these have been reported as recaptured from both the western Atlantic and the Mediterranean Sea. Detailed records up to 3.6 years in length have been obtained demonstrating that Atlantic bluefin prefer the top 200 m of the water column and spend more than half their time in the upper 40 m. Atlantic bluefin maintain a high internal body temperature despite encountering a wide range of ambient temperatures (2-30°C). Patterns of feeding behavior have emerged providing data on how often and when fish feed at sea. Geolocation estimates for electronic tagged western Atlantic bluefin derived from archival and pop-up satellite archival tags indicate these bluefin show visitation and aggregation in New England, Carolina, the Gulf of Mexico as well as the Mediterranean. Pop-up satellite tags have been deployed on 120 west Atlantic bluefin tuna. Ninety percent of the pop-up tags scheduled to transmit have delivered data or position information on time. Both types of electronic tag data can be combined with oceanographic data to reveal a complete picture of how and where these fish forage in the pelagic realm.

J.R. Sibert and J.L. Nielsen (eds.), Electronic Tagging and Tracking in Marine Fisheries, 65–88.
© 2001 *Kluwer Academic Publishers. Printed in the Netherlands.*

1. INTRODUCTION

Atlantic bluefin are known to migrate from warm temperate waters to polar seas in short durations (Mather *et al.*, 1995). The scale of these migrations makes it a challenge to study their seasonal movements. Although acoustic tracking of bluefin has revealed significant information about their unique thermal biology, depth movements, and behavior (Carey and Lawson, 1973, Carey *et al.*, 1984, Lutcavage et al., 1999, 2000), little information on the long-term movement patterns of bluefin tuna have been obtained. Conventional tags have shown that Atlantic bluefin tuna range widely throughout the North and South Atlantic (reviewed in NRC, 1994). The successful deployment in recent years of new electronic tagging technologies to examine the migrations of oceanic vertebrates has opened the door for studies in the pelagic realm (Delong *et al.*, 1992, McConnell *et al.*, 1992, Gunn *et al.*, 1996, Metcalfe and Arnold, 1997, Block *et al.*, 1998a, b, Lutcavage *et al.*, 1999, Klimley and Holloway, 1999; Friedland *et al.*, 1999, LeBoeuf *et al.* 2000). In this paper we overview the results we have obtained from electronic tagging of Atlantic bluefin tuna (Block *et al.*, 1998a; Block *et al.*, 1998b).

Two types of electronic tags have been used in this study, archival and pop-up satellite tags. Archival tags are data logging tags that record depth, temperature and light (Klimley *et al.*, 1994, Metcalfe and Arnold, 1997). For use on fish, the archival tags are surgically implanted inside the peritoneal cavity. Pressure and internal temperature sensors are located within the body of the tag. A stalk protruding from the archival tag carries the irradiance and water temperature sensors externally. Information on the diving behavior, ambient and internal temperature preferences are obtained (Figure 1). Additional data on the thermal structure of the water column can be recorded if dive profiles are examined as depth versus temperature profiles (Figure 1A). The pop-up satellite tag technology combines archival tags with satellite technology. Pop-up satellite tags archive information while on the fish and remain continuously submerged. The tag pops off the fish at a predetermined time, surfaces and downloads data via the Argos satellites (Block *et al.*, 1998a, Lutcavage *et al.*, 1999). The most useful aspect of the pop-up satellite tag is their capability of delivering a fisheries-independent data set on the distribution of fish. Current generations of the pop-up satellite archival transmitting tags built by Wildlife Computers of Redmond, Washington log data in a user set interval (one-two minutes) in a similar fashion to an archival tag, and transmit a subset of the data on ambient temperature, depth and temperature profiles of the water column, and geolocation to the user via the ARGOS satellites (Figure 2). Both types of electronic tags provide a record of environmental variables and behavior of the animal carrying the tag. While data logging tags are relatively new to studies of fish biology (Gunn *et al.*, 1994, Metcalfe and Arnold, 1997, Block *et al.*, 1998a, 1998b, Lutcavage *et al.*, 1999), archival and satellite tags have been used extensively to study the biology of marine mammals and birds (McConnell *et al.*, 1992, Bost *et al.*, 1997, Costa, 1999, Koudil et al. 2000, LeBoeuf *et al.* 2000, Polovina *et al.* 2000).

1.1 Archival Tags

The implantable archival tags deployed in this study record a high-resolution data set from four sensors: pressure, external and internal temperature and light. From light intensity it is possible to estimate latitude and longitude (Hill, 1994). Archival tag light records allow an estimate of geoposition to be achieved on a fixed buoy with an accuracy at best of 0.5° in longitude, and 1.5-° of latitude (Hill, 1994, Klimley *et al.*, 1994, Block *et al.*, 1998b, Welch and Eveson, 1999, Hill and Braun, this volume). Importantly, during the equinox or times of prolonged diving, positions may be considerably far from the actual position. In practice, longitude is a robust estimate of position in light based geolocation however latitude must be augmented with sea surface temperature and bathymetry data to confirm individual positions. To obtain a track of a fish on an oceanic scale, data are filtered in a consistent fashion to remove outlier positions. To compare data sets between researchers working on similar species, it would be ideal to publish error estimates around positions or appendices providing the time of midnight or dawn from one position to the next. Both the implantable Northwest Marine Technology and Wildlife MK7 tags have been extensively tested in a variety of experimental conditions. Both have been found to be accurate at predicting longitude over large segments of the year. The positional accuracy of longitude alone is sufficient to track the basin-scale movements of bluefin in the Atlantic. Accurate latitude data are more challenging to generate and has a greater variation (Hill and Braun, this volume). Although the data storage capabilities of implantable archival tags are high (to date they store 2 megabytes of data), their major limitation is the need to recapture the fish to obtain the information. This requires the deployment of large numbers of archival tags on heavily exploited fish, and the coordination of a multinational recovery program. This limits the use of this electronic tag technology on most marine organisms.

1.2 Pop-Up Satellite Tags

Satellite-linked data recorders have expanded our understanding of the fine scale movements of a number of marine birds (e.g., Jouventin *et al.*, 1994),

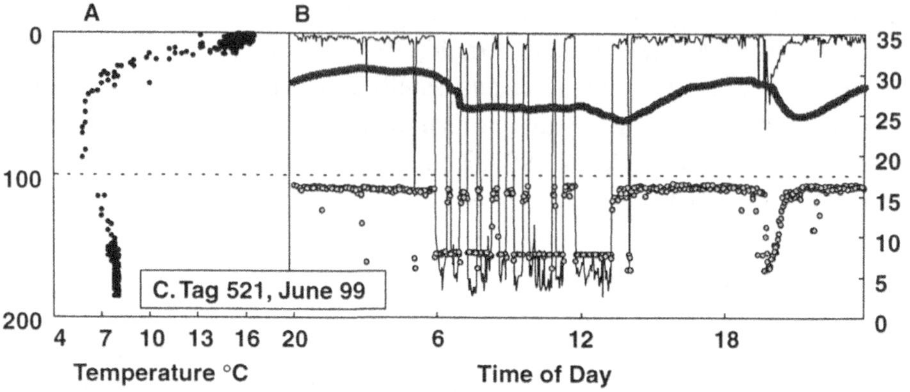

Figure 1. A day in the life of a giant bluefin recorded with an archival tag. (A) Depth versus temperature data obtained in two minute intervals for the day permits reconstructing the vertical temperature structure of the water column. This reveals a shallow mixed layer and a sharp thermocline for the bluefin in New England waters. (B) Archival tags acquire data in two-minute intervals for depth (black), internal temperature (gray) and external temperature (light gray circles). The bluefin tuna is feeding along the continental shelf off New England.

sea turtles (e.g., Renaud and Carpenter, 1994, Polovina *et al.* 2000),and marine mammals (e.g., McConnell *et al.*, 1992). Since the antenna on the satellite transmitter on the animal must be out of the water to communicate with an orbiting satellite, the technology has mainly been used on air-breathing vertebrates which surface regularly. Satellite tags have also been used successfully in basking sharks (Priede, 1984) and blue sharks (Carey, unpublished data) because these fish spend significant amounts of time basking at the surface. Only recently have satellite technologies been adapted for use on completely submerged marine vertebrates.

Pop-up satellite tags combine data storage tags with satellite transmitters (Block *et al.*, 1998a, Lutcavage *et al.*, 1999). The pop-up satellite devices communicate with the ARGOS satellites that serve to both up-link data and calculate an end-point location. The tags final position is calculated using the Doppler shift of the transmitted radio frequency in successive uplinks received during one pass. Importantly, the tags are fisheries independent and they do not require retrieval for data acquisition. The first generation of these tags built by Microwave Telemetry (Maryland), functioned as a single point tag and provided an end-point location and a small archive of temperature data (Block *et al.*, 1998a, Lutcavage *et al., 1999*). The positions of the fish between the time of release and recapture are documented in relationship to daily temperature preferences. Pop-up satellite tags have been deployed on a wide variety of fish including Atlantic and Pacific bluefin tuna, yellowfin, albacore, mola, blue and striped marlin, white, thresher and salmon sharks (Marcinek *et al.*, 2001; Block, unpublished data).

Second generation pop-up satellite archival tags (PSAT or PAT) combine pop-up satellite and archival technologies into one electronic tag (the PAT tag of Wildlife Computers, Inc. has a dual function as an external archival and pop-up satellite tag). Initially, the Wildlife Computer built pop-up archival satellite tags recorded ambient temperature data for the entire mission in thirty minute intervals. More recent generations of the tag log depth, light and ambient temperature in user set intervals (most often two-minute intervals) and provide onboard processing of longitude. Data are summarized in bins and transmitted to the user as a percentage occupancy of the depth or temperature bins set by the user. The tags transmit raw light data at sunrise and sunset, and provide with the software the capacity to calculate longitude. From the light data, sea surface temperature and bathymetry data latitude may also be calculated. The tag transmits a depth/temperature profile that provides information on the vertical structure of the water column, sea surface temperature and maximum daily depth, and the depth and temperature behavior of the animal carrying the tag summarized in user set hourly intervals from the one to two minute continuous data acquisition (Figure 2).

In contrast to implantable archival tags, few data sets are available verifying the positional accuracy of the pop-up satellite tags against traditional technologies (radio-linked satellite transmitters or GPS). Recent tests on a fixed buoy (Block, unpublished data) indicated that the accuracy of generating positional data from the archival light data set on the PAT tag was similar to implantable archival tags. Tests by the manufacturer are underway, providing a direct correlation between radio-linked satellite determinations and light based geolocation from pop-off satellite tags applied to marine mammals that surface at regular intervals (Hill, pers. com.). The major limitation of the geolocating pop-up tags, in contrast to implantable archival tags, is that only a limited fraction of the raw light data from which the longitude estimation is derived is transmitted back to the user. Importantly, if recovered the PAT archival tag is equivalent to an implantable record and provides depth, ambient temperature and light data for post-processing of position.

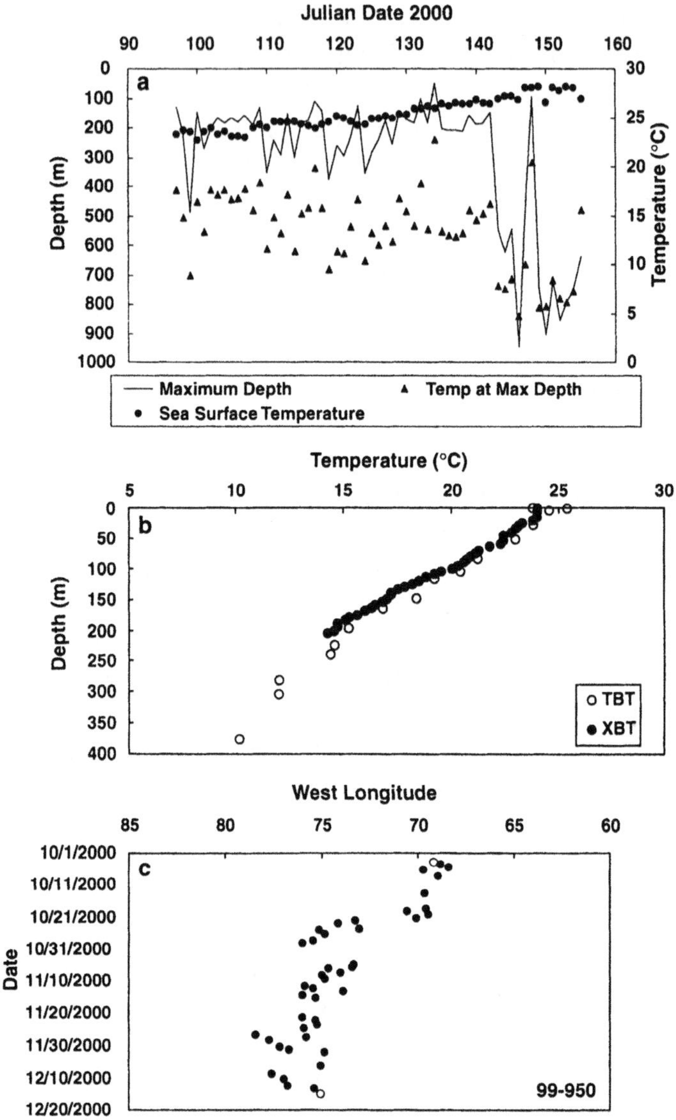

Figure 2. The types of data that are transmitted via ARGOS from archived data set on a Wildlife Computer pop-up satellite transmitting archival tag. (A) Maximum depth (solid line), surface temperature (black circles) and minimum temperature (solid triangles) from an Atlantic bluefin tuna in the Gulf of Mexico. (B) Comparison of a tuna bathythermograph and an XBT profile from a Gulf of Mexico fish. (C) An example of longitude data transmitted from an Atlantic bluefin tuna released in Nantucket, Massachusetts with a pop up position off of Cape Hatteras, North Carolina. In addition to these data, depth and temperature measurements obtained at a user set interval (usually two minutes) (see Figure 9) are transmitted back and a full archival record can be obtained if the tag is recovered.

2. ELECTRONIC TAG DEPLOYMENT

The Atlantic bluefin tuna electronic tagging program (Block *et al.*, 1998a,b) has focused tag deployments in three locations (North Carolina, New England and the Gulf of Mexico). These geographic regions represent both feeding and spawning aggregations of bluefin tuna. When possible, genetic samples are taken from tagged fish, as well as gonad and stomach samples from fish caught and landed concurrently in the fishery to assess maturity and feeding in the region. The implantable archival tags have all been deployed at Cape Hatteras and Cape Lookout, North Carolina. For the past decade bluefin fishers have been aware that the bluefin are aggregating close to shore in winter months, presumably to feed on menhaden (Boustany *et al.*, this volume). The lack of a targeted commercial fishery enhances the opportunity for tag and release fishing efforts.

The "Tag-A-Giant" program of Stanford University, the Monterey Bay Aquarium and the National Marine Fisheries Service (hereafter referred to as TAG) conducted a pilot study in 1996 to develop methods for catching large bluefin (n=10) and bringing them aboard a sport fishing vessel for surgery, external tagging and release (Block *et al.*, 1998b). Tagging efforts in 1997 and 1999 led to tagging of an increased number of bluefin (269) and fish of larger year classes (Table 1). Archival tagging necessitates shipboard surgery, which requires a platform close to the water. Charter fishing vessels of the North Carolina sport fishing fleet provide an ideal surgery deck as many have ample space and "tuna doors" close to the water. Procedures have been described previously and involve pulling the bluefin onto a wet vinyl mat, keeping their eyes covered with a wet cloth, and irrigating their gills continuously with seawater. By bringing the fish on board, surgery can take place and accurate measurements of length and girth can be obtained.

Table 1. Archival tag summary for bluefin tuna released off North Carolina in 1997 and 1999

	1997 Hatteras	1999 Morehead City
N	160	109
Mean Curved Length (cm)	192	209
Mean Mass (kg)	122	159
Maximum (kg)	234	265
% Recovered	17.5%	14%

3. POST-TAGGING SURVIVORSHIP OF BLUEFIN CAUGHT ON SPORTFISHING GEAR

A number of factors indicate that the techniques developed for implantion of archival tags results in high survivorship. The recapture rate of archival tagged Atlantic bluefin in this program is approximately 15.5%. This recovery rate of the tagged fish is significantly higher than that of conventionally tagged fish released from the region (Tables 1 and 2). This is most likely due to the enhanced reward ($1000) provided for return of the tags. To examine short-term survivorship of bluefin caught on heavy tackle with circle and J-hooks, twenty-two fish (Figure 3) have been electronically tagged (acoustic and pop-up satellite tags) after shipboard handling and in some instances surgery (Block *et al.*, 1998a). For the short-term pop-up satellite tagged bluefin (3-14 days), only a single tag failed to report. The short-term results with pop-up satellite tags indicate a high reliability of the electronic tag technology built by all manufacturers of the instruments. Several acoustically tracked Atlantic bluefin (Boustany *et al.*, this volume) were also tracked in the same region, to examine recovery from handling and surgery. In one bluefin tuna an acoustic tag was implanted surgically to provide some basis for comparison with the archival-tagged fish surgery and handling procedures. Although the sample sizes are small, all of the acoustically tracked fish demonstrated survivorship after one or two days of post-release activity.

Table 2. Release and recapture of conventional tagged Atlantic bluefin tuna 1996-1997

Releases	Hatteras Releases	Hatteras Recaptures N (percent of Total)	New England Recaptures
1997	2,493	85 (3.41%)	46 (1.85%)
1996	2,875	39 (1.36%)	25 (0.87%)

The activities of fish in the water column were documented by the single-point pop-up satellite tags that recorded the water temperature hourly, or at thirty minute intervals in the first generation tags built by Microwave Telemetry and Wildlife Computers respectively (Block *et al.*, 1998a). In second generation tags built by Wildlife Computers, both temperature and pressure were recorded in one to two minute intervals which improved the ability to discern if the fish survived. These records (not shown) indicate the fish were actively moving through the water column during the short-term experiments described above. No tags prematurely released, however one short-term PSAT tag landed on the beach after release (bluefin are often within 2-7 nm of the shore), was recovered and provided a full archival record of the fish during the period the tag was attached. This PSAT archival record matched well with implantable archival data from the region which together indicate a shallow pressure distribution (0-40 m), which corresponds to the shelf depth in the Carolinas. The ability to recover a full archival record from a PSAT tag is a benefit of this type of pop-up tag. The instruments in effect become external archival tags and are highly visible compared to internally implanted tags. To date in the Atlantic six PSAT tags have been recovered and all have provided a full archival record.

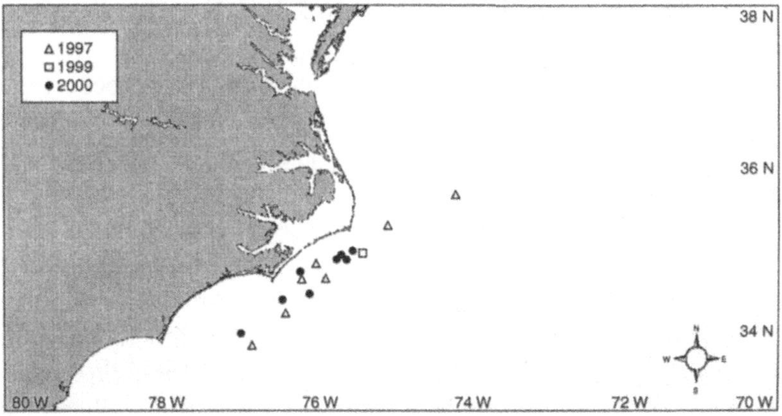

Figure 3. Short term positions of sixteen pop-up satellite tags off the North Carolina Coast. Ninety-five percent of the short-term tags reported positions indicative of high survivorship from the procedures associated with tag and release from hook and line. Both J-hooks and circle hooks were used and no significant difference in survivorship was noted. The tags reveal a preference for the warm waters off the Carolina coast in the winter time and aggregation in the region after deployments from 1-14 days.

Deployments of longer duration off North Carolina, in which pop-up satellite tags were deployed for 30 to 180 days (n=47), indicated a high survivorship for pop-up satellite tagged fish as 92% of the tags successfully transmitted position, temperature or pressure data. Five individual bluefin were double-tagged with both the archival and pop-up technology to examine the effects of surgery (Block *et al.*, 1998a). These fish, in all cases, underwent handling identical to the archival tagged fish and therefore give a preliminary estimation of survival rates for archival tagged fish. A single archival tagged fish whose pop-up tag did not surface was presumed dead, although other explanations (i.e. mechanical failure) are possible. In more recent deployments (1999-2000), temperature and pressure data in two minute intervals has been logged on the long-term tags. In addition to the survivorship data, this provides the opportunity to discern whether the tags remained on the fish up until the point of release, or if the tags were prematurely released over the long duration the tags were on the fish. Important factors that improve survivorship likely include the use of heavy tackle (60 kg line and 182 kg leaders) leading to shortened fight times, the use of circle hooks on the majority of the fish captured (removed from the fish in all cases), low impact handling techniques and short surgery or pop-up tagging durations. The survivorship data are not applicable to any other gear type used in recreational or commercial bluefin tuna fishing.

4. ARCHIVAL TAG RECOVERIES

Forty-three archival tagged Atlantic bluefin tuna have been reported as recaptured in the western North Atlantic, eastern North Atlantic, central Atlantic, Gulf of Mexico and Mediterranean Sea as of October 2000. The time at liberty for recaptured archival tagged fish has ranged between 160 and 1310 days. Thirty one of these tags have been returned intact to U.S. researchers. The majority of the recaptures are in the western Atlantic (33) and ten have been confirmed (11 reported) in the east Atlantic or Mediterranean Sea. Several tags have been returned from central Atlantic longline fisheries.

A comparison of archival and conventional tagged bluefin in the Carolinas indicates similarity as well as differences in the geographical pattern of returns. The primary site of archival tag returns is over the northeast continental shelf of the U.S. and Canada where the commercial fishery lands most of the western quota. Forty-five percent of the archival tagged bluefin tuna recaptured in the West have come from the same region near the Great South Channel, east of Cape Cod, Massachusetts. In contrast, conventional tag recapture results from releases in the Carolina fishery between 1994 and 1999 indicate that about 60% of recoveries occur in North Carolina during the same year, or the year following release. To date, not a single archival tagged bluefin has been recaptured in North Carolina. This may be due to the major reduction in fishing effort in 1998 due to adverse weather. The El Niño event resulted in colder than usual waters close to the Carolina shore, thereby changing the dynamics of the North Carolina fishery. In addition, the overall intensity of the recreational fishery appears to be declining as fewer fish are now being conventionally tagged and reported to the NMFS (Prince, pers. com.). However in 2000 a commercial fishery was opened in the region to allocate remaining quota and to date no tags have been recovered.

Archival tagged fish have been recaptured throughout the eastern Atlantic and Mediterranean sea by fishers of seven nations; however, the recovery of the surgically implanted tag has at times been problematic. This has become a significant issue as the program has evolved and emphasizes the need for continuous outreach and education in an archival tag program. Such efforts must be continued throughout the duration of the program and should be more than a one-time event. The percentage of tags which were successfully removed from an archival tagged fish and returned was 100% in 1997 (n=2), 72% in 1998 (n=9), 55% in 1999 (n=11) and 90% in 2000-2001 (21). The large variation in return rate may be due to high initial education efforts upon release of the tags in 1997 and reduced communications about the 1999 release. A redoubling of education efforts in 2000 may be associated with the elevation of returns. Personal contact with fishers and the distribution of educational materials about the results have been helpful for improving interest among fishers to return the tags. Archival tags have been discarded with the viscera at sea, or shipped intact in the fish to Japan despite the green and white conventional tags on the fish indicating the presence of an internal tag. In one case the sensor stalk was cut from the archival tag and returned along with the conventional tags from Italy. The tag was left in the fish. Continuous education about the archival tag program and recovery methodologies is critical to successful recapture of the fish and tags. When recovering tags, fishers in the western North Atlantic bluefin tuna fishery are instructed to obtain scientific assistance, which is readily available along the Canadian and U.S. coasts . Building awareness in all observer programs is also critical (as illustrated in the case of a Canadian observer aboard a Japanese longliner within Canada's EEZ, who recorded catching an archival tagged bluefin and returned the conventional tags, but did not retrieve the archival tag).

A number of factors result in problems with tag recovery: 1) the tag gets covered with mesentery making it a challenge to find for the untrained eye; 2) the tag works its way into the body cavity, making it a puzzle to discern where it is; and 3) some tags were placed in an extraperitoneal position in the ventral muscle layer and not in the body cavity, complicating the recovery efforts by having two locations (only 10% of the tags are in this position). Such tags are often, but not always, marked with a third conventional tag, attached directly to the instrument ventrally. The ideal solution is to allow only trained scientists to remove tags. However, this is rarely possible as most fish are gutted and cleaned at sea.

4.1 Types of Archival Tags

Archival tags implanted in Atlantic bluefin tuna were obtained from two manufacturers, Wildlife Computers of Redmond, Washington and Northwest Marine Technology of Shaw Island, Washington. Two generations of the Wildlife Computer archival tags have been deployed. The Wildlife Computers first generation MK7 archival tag was initially designed for implantation in the Atlantic bluefin tuna tagging program. The first generation tag has a flattened rectangular shape and a 13 cm flexible external stalk. The tag color was fluorescent green, increasing the ability of fishers to find the tag among the viscera. This tag (used in 63% of the 1997 deployments) was programmed to record pressure, ambient and internal temperature, and light intensity every two minutes, yielding detailed archival records for up to 1.4 years (Figure 2). In an attempt to obtain longer records, approximately 20% of the tags were cycled on and off over a period of days.

All MK7 tags recovered to date have had data sets logged in memory and not a single tag has failed to record data. However, the sensor stalk in the first generation MK7 tags had durability problems at the point of entry into the tag and all tags failed to record light and external temperature at some point during the first year of the mission. After the sensor stalk failed, pressure and internal temperature data were often recorded for one or two years, providing the opportunity to use dawn and dusk diving algorithms to generate longitudes. The stalk was strengthened in second generation MK7 archival tags and has shown improvement in data recovery with geolocation records based on light levels of up to 1.6 years being obtained thus far in giant bluefin. However, stalk failures are still a problem in this second generation of the instrument most likely due to the increased stiffness of the stalk. Giant bluefin tuna have broken the stalk during the spawning season, perhaps due to interaction with a second fish, and have broken the stalk during feeding dives where the depth pattern suggests they are on a shelf (possibly they are breaking the stalk on the bottom). The more robust external stalk is, however, easier to locate according to fishers who have found both types of tags. First generation Wildlife Computers tags were analyzed using custom software generated by the laboratory that permits depth correction and rapid data display using Matlab (Math Works Inc.) routines. Software permits rapid display of data and routines for extraction from the data records of sea surface temperature, minimum and maximum external temperature, internal temperature and depth, ascent and decent rate as well oscillatory periods associated with possible breeding behaviors. Second generation tags can be examined with geolocation software supplied by the manufacturer or examined with custom software. The major advantage of the Wildlife Computer tags is the use of non-volatile memory which allows data to be recovered even if an electronic failure occurs (stalk damage or water intrusion).

Northwest Marine Technology (NMT, V1.1) provides archival tags that record pressure, ambient temperature, internal temperature and ambient light for up to ten years. The major advantage of this tag is the capability of obtaining on-board geolocation estimates and rapid display of the entire data record via histogram profiles for pressure and temperature data sets. Theoretically, the tag has the capacity to store the position information in the "day log" for a decade. The disadvantage of the tag is that in the first generation tag the memory was small and the detailed two minute archival depth, temperature and light records were stored for only 2-3 months. During the remaining portion of the mission, the tag logs a daily surface water temperature measurement which is very useful for ground-truthing latitude. Depth, external temperature and internal temperature measurements are recorded into histogram bins for this portion of tag deployment. Two additional temperature measurements are recorded daily for the lifetime of the mission at depths preset by the user that can be used to roughly estimate the position of the thermocline or water depth.

Deployments of NMT tags in captive bluefin, as well as on a free-floating drifter at sea, indicate that the algorithms for on-board longitude calculations are accurate, especially when a filter for outliers is used (Block et al., 1998b, Musyl *et al.*, 2001). Archival tags were glued on to the pelage of an elephant seal in 1997 during a translocation experiment in Monterey Bay to examine geolocation against a radio-linked satellite recorder (Block, unpublished data). This experiment revealed that a major limitation of the first generation NMT tag was a less sensitive light sensor than manufacturers anticipated (routinely the light is not recorded between 100-150 m depending upon the water mass). This resulted in periods in the archival records of both elephant seals and bluefin tuna where the tag does not record informative light data. Bluefin in early spring occasionally occupy depths greater than 150 m in the water column for several days or weeks. Problems in detecting light also emerged at dusk and dawn during offshore periods when Atlantic bluefin would regularly dive to great depths, presumably after squid. Despite this limitation, the tags recorded excellent light curves and accurate geopositions during periods when bluefin were on the continental shelf and had limited diving activities or periods during the year when the fish displayed more surface oriented behavior.

NMT tags have had a series of electronic failures which include pressure sensor failure during the mission (n=3) due most likely to over-pressurization by bluefin diving (to depths>1000 m), water intrusion on the electronics (n=2), and battery failure (n=1). NMT tags of the first generation did not utilize non-volatile memory, reducing the opportunity to recover data after electronic failures. Loss of the pressure sensor interrupted acquisition of geolocation and sea surface temperature data at the point in the record this occurred. Despite these problems it is important to note that the NMT tags performed flawlessly on several occasions in giant bluefin, and provided the longest records to date of migrations (1300 days) with complete geolocation and sea surface temperaure records. The NMT tag has proven to be highly suitable for small Pacific bluefin that remain closer to the surface than Atlantic bluefin tuna (Kittigawa *et al.*, 2000, Marcinek *et al.*, 2001).

The NMT tag is packaged in a cigar-shaped stainless steel cylinder. One problem with the slender shape of the tag is its capacity to migrate into the body cavity. This occurred on four separate occasions during the 1997 deployments. This interfered with geolocation estimates and external temperature measurements, but did not affect depth or internal temperature data. This internalization of the tag did not happen in all cases and a 1996 NMT tag remained in the proper position for 1330 days. To prevent the movement into the body cavity from occurring in 1999 deployments, a surgical attachment procedure was developed to limit movements of the tags. A stainless steel loop was soldered to the housing of the tag by the manufacturer and the monofilament of a conventional tag was knotted to this loop. After the archival tag was inserted into the body cavity and sutured in place, the conventional tag head was placed into the skin and muscle near this ventral region, providing an anchor which prevented movement of the tag. Three tags attached with this new technique have been retrieved and all remained in the proper place for the duration of the deployment.

4.2 Data Recovery

Figure 1 shows a day in the life of a bluefin tuna obtained from an archival tag. Archival tag records provide extensive detail of bluefin tuna vertical movements, thermal biology and ambient temperature preferences (Figure 1). In addition, oceanographic information is easily obtained. Many stereotypical behavior patterns have emerged in the extensive data records collected from the returned archival tags. These include distinctive onshelf and offshelf behaviors, descents and ascents in the water column at speeds of 1-6.6 m s,$^{-1}$ sunrise and sunset dives (particularly in the early spring and summer months), and long durations spent at the surface, presumably to bask in the warmer waters. Diurnal vertical migrations suggestive of deep scattering layer or squid feeding (Carey and Robison, 1981) are frequent in the records during offshore periods in deeper waters. On shelf behavior is often characterized by periods of oscillatory diving to shallow shelf depths (40-200 m) in the morning and basking at the surface in the afternoon. In addition to providing detailed information on depth movements and temperature preferences, the records contain information on geoposition of the fish, oceanography and foraging ecology.

Figure 4 illustrates the depth preferences of two Atlantic bluefin tunas over the course of one year. Bluefin can be found throughout the entire water column to over 1000 m but spend most of their time in the top 50 m. While individual daily records indicate fish can dive quickly to great depths (first generation tags were limited to 800 m; second generation tags are limited to 1000 m and individual bluefin occasionally exceed this depth for long durations), such dives are seasonal, mainly occurring during early spring and summer when the bluefin have migrated off the continental shelf. Deep dives also show up when surface waters are warmest. In the Gulf of Mexico where some western tagged Atlantic bluefin have migrated into to spawn, dives to over 1000 m are quite common. It is not clear at this point if these are feeding dives or thermoregulatory dives associated with a cooling off in the warm Gulf waters.

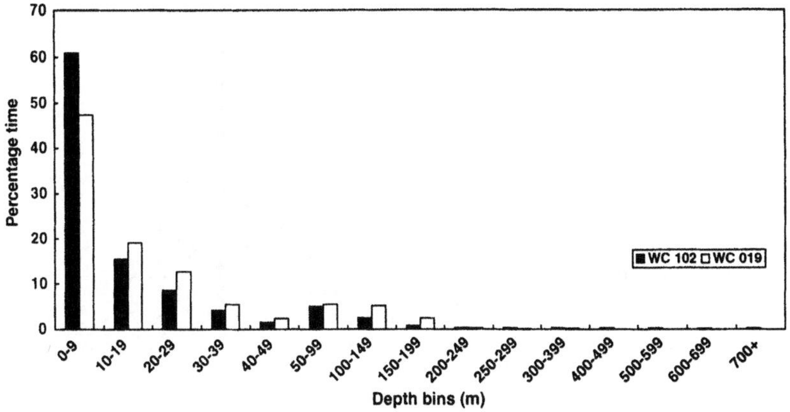

Figure 4. Depth preference histograms for 2 archival tagged bluefin recovered in the western North Atlantic. The data are collected at two-minute intervals and binned as shown. Record length is 330 days for 97-102 and 344 days for 97-019. Bluefin tuna of these year classes (see text) prefer the top 20-40 m of the water column and stay above 200 m most of the time. Deep dives (>300 m) are relatively infrequent events and occur most often in springtime while the bluefin are off the continental shelf.

Figure 5 shows the external temperatures experienced by two Atlantic bluefins over the course of a year and illustrate the wide range of temperatures experienced by these fish. To date, the lowest ambient temperature recorded on an archival tag is 2.8°C and the warmest temperature has been 30°C. Internal temperatures are generally warmer than ambient waters (18°C to 32°C) and show less variation. Positioning of the archival tag varies within each individual and results in significant variation in thermal record experienced by each fish. The ability to detect feeding events by monitoring peritoneal temperature is possible in giant bluefin and provides direct evidence of feeding activity. When these endothermic fish ingest prey items in northern waters, the ectothermic prey is generally colder than the internal temperature of a bluefin tuna, and the tag therefore records a depression of the fish's internal temperature. Feeding events can be discerned in this way as long as the fish remains in cold water. Long stretches without observable internal temperature variation often indicate migratory periods; however, we have noted that one cannot easily discern in large bluefin (>150 kg) feeding events in warmer waters which makes it challenge to discern migration versus feeding activity behaviors.

In addition to returning information on bluefin biology, archival tags also return information on the oceanography surrounding the tagged fish. The movements of the bluefin through the water column are often rapid, and the tag's relatively fast data acquisition rate allows one to use the diving fish as an autonomous probe for measuring depth and temperature profiles. We call such profiles "tuna bathythermographs" or TBTs. By examining the depth

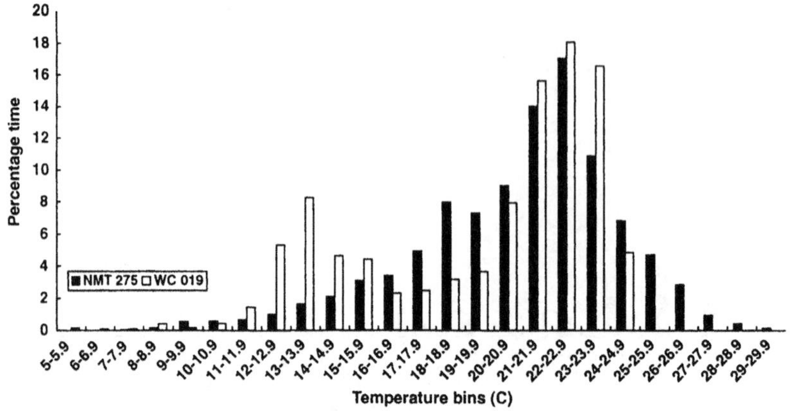

Figure 5. External temperature histograms for 2 archival tagged bluefin recovered in the western North Atlantic. The data are collected at two-minute intervals and binned. Record length is 77 days for NMT 275 and 344 days for 97-019. Both fish showed a preference for 21-24 degree waters.

and temperature data daily, we can reconstruct profiles of the water column with information on the extent of the mixed layer,the position of the thermocline or the movement of the bluefin in relationship to inversions or oceanographic fronts (Figure 1a and 6). Oceanographic data from the tags can be combined with remote sensing to examine what environmental conditions the fish prefers relative to the environments available. This is important for interpreting what environmental conditions are associated with foraging ecology.

To summarize, the behavioral data indicate Atlantic bluefin spend the majority of their time in surface waters, often above the thermocline, but can dive to depths over 1000 m. They most often dive at sunrise and sunset, presumably to feed on organisms in the scattering layer. Feeding events, based on the internal temperature information, can be observed daily in almost all records, with seasonal absences during migratory periods. Finally, archival tags provide extensive information on the water column and the movements of the bluefin in relationship to the local oceanography. The behaviors recorded with archival tags on Atlantic bluefin tuna are similar to those described for Pacific and Southern bluefin tuna (Marcinek *et al.*, 2001; Gunn and Block, 2001).

4.3 Seasonal Movements Observed with Archival Tags

Perhaps the most important contribution of the archival tag to fisheries management is the ability to estimate position using light level curves. This provides a view of large-scale migrations at a daily, weekly and seasonal interval. The quality of these records depends upon several factors which include the amount of time bluefin tuna are diving, when these dives occur and the

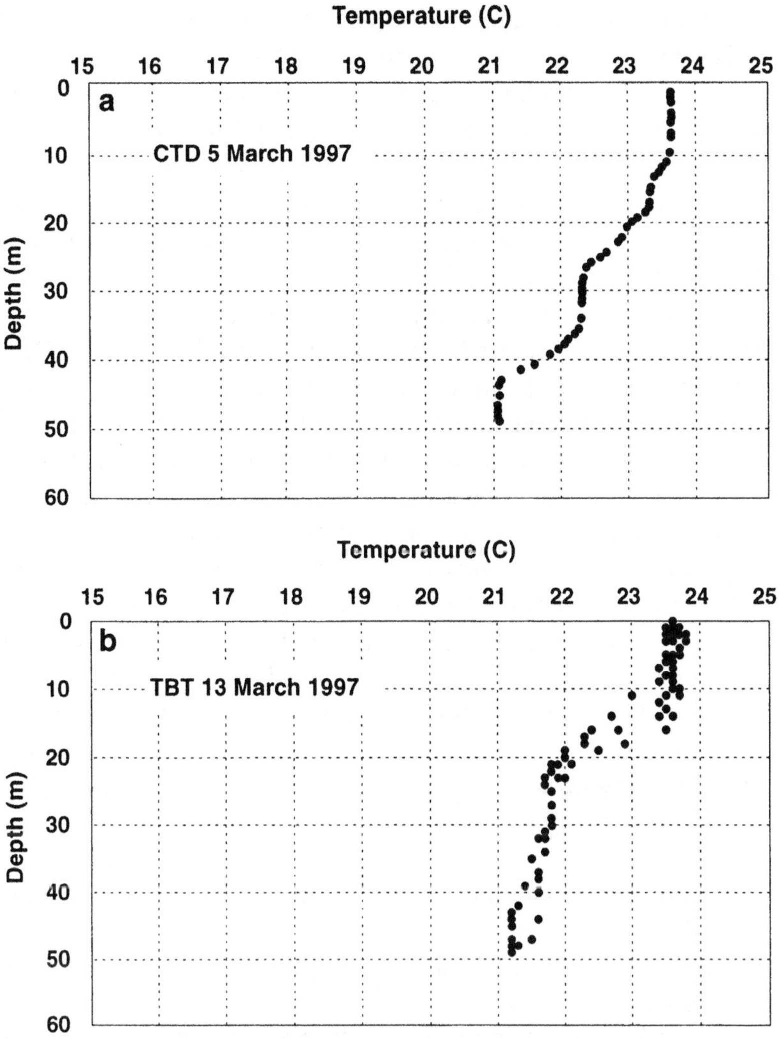

Figure 6. Archival tags reveal oceanographic features of the water column. a) Temperature profile of the water column off North Carolina, taken with a Seabird conductivity/temperature and depth probe (CTD). b) Similar profile compiled from the data collected by an archival tagged tuna in the same area.

time of year. From our experience, geolocation by light works reasonably well in West Atlantic bluefin because much of their time is spent on the continental shelf, which limits their diving ability and therefore yields better light level measurements. Fish of the primary ages tagged (7-11) appear to be in feeding aggregations on the continental shelf in the Carolinas in the winter (75W), and in New England in the summer and early fall (68-70W). Once the bluefin tuna move offshore into frontal edges of Gulf Stream waters, deeper diving patterns interrupt the light curves and make depth corrections of light levels difficult and in many cases impossible.

The pattern of bluefin tuna movements varies depending upon the individual. The migrations of two bluefin tuna (NMT 153,Wildlife Computers 97-019) as viewed through longitude tracks are shown in Figures 7 and 8. In the NMT 153 track, the data are obtained directly from the day log and a two degree per day filter has been used on the data set to eliminate outliers. Acoustic tagging records suggest that speeds of 2-3 km per hour are common for bluefin tuna and thus the possible distances covered in a 24 h period most likely will not exceed this filter.

Bluefin tuna 153 spent three years (1996-1999) along a similar track in the west Atlantic. For the Wildlife Computer tag (Figure 8), because of light stalk failure, we have combined light-based estimates of geolocation with estimates generated using sunrise and sunset dives. By using the time of the sunrise and sunset dives as an indicator event of true sunrise and sunset, we can get a reasonable longitude estimate. The accuracy of the methodology was checked by calculating longitudes based on sunrise sunset dives when the stalk worked and comparing them to longitudes obtained light (Figure 8 inset). Figure 8 (WC-019) shows points estimated with both techniques (circles and triangles) on similar days and a high degree of overlap. From both tags and the general picture that emerges is that the bluefin remained in the vicinity of the tagging location (North Carolina) for about 40 days, then progressed offshore to the east in the springtime. The offshore movement was shown to be along the path of the Gulf Stream with simultaneously tagged PSATs on bluefin (Block et al., 1998a). The bluefin migrate back onto the continental shelf in the summertime (68-70 W). Latitude estimates for the summer positions as well as sea surface temperature indicate that the positions are associated with New England waters. The highest return of all archival tags takes place in the summer and coincides with the commercial fishery's peak effort in the western Atlantic (June-September) off of New England. One year after tagging, the bluefin tuna positions are at longitudes and in sea surface temperatures that indicate a return to North Carolina. Examination of the bathymetry reveals a signature of the Carolina coastal shelf in which extremely shallow maximum depths (40-60 m) and warm water temperatures (18-23°C) are characteristic of this period. This movement pattern is suggestive of a western loop and represents a residency period of one to three years (019 vs. 153) within the western Atlantic for young adult bluefin tuna. This particular pattern of migratory movements has appeared in a considerable number of archival tags recovered in the west. Bluefin tagged in the west Atlantic have also shown the ability to cross the ocean basin and move into the Mediterranean Sea (Block *et al.,* 2001). Thus, tag recaptures and geolocation data from archival tagged western Atlantic bluefin tuna link these Carolina released bluefin tuna to New England and Canadian waters, the Gulf of Mexico, the Central and East Atlantic and the Mediterranean Sea.

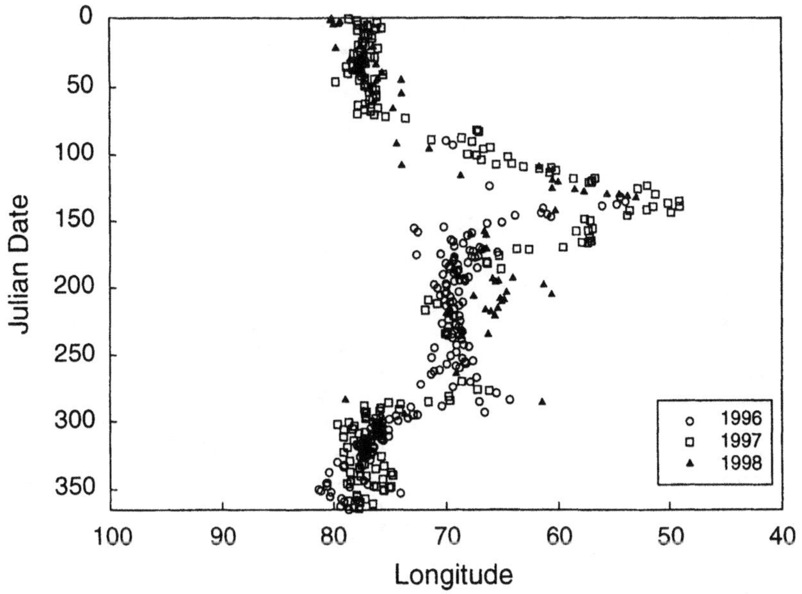

Figure 7. Three years (1996-1999) in the life of a bluefin tuna (NMT 153) showing three seasons of western residency with no visitation to the Gulf of Mexico. The general pattern is Carolina waters in the winter, offshore in the spring, New England in the summer and back to Carolina in the fall. Bathymetry and sea surface water temperature are used to augment the latitude data set (not shown).

4.4 Movements of Bluefin from Pop-Up Satellite Tags

The migratory paths of bluefin released from 1997 to 2000, based on the seasonal position of the pop-up satellite tag single end points show a distribution similar to the one described above for the two archival tags (Block *et al.*, 1998a; Lutcavage *et al.*, 1999). The pattern suggests aggregation in North Carolina along the continental shelf in winter (Figure 3), in the mid-Atlantic or offshore along the Gulf Stream in spring and in New England in the fall. This overlap with the archival records is not surprising given many of these fish were tagged concomitantly with the archival fish (Block *et al.*, 1998a). Pop-up tag endpoint positions in the winter remain close to the continental shelf of the US in the vicinity of North Carolina and the mid-Atlantic shelf. Pop-up satellite tagged bluefin had end-point positions along the edge of the Gulf Stream to the north and east of North Carolina or in New England in early summer when released from a start point in Carolina. Bluefin were often located on the frontal edges of the Gulf Stream as well as on warm core rings (reported in Block *et al.*, 1998a). Importantly, this pattern was repeated temporally in 1999 tagged bluefin released in the Carolinas (Block *et al.*, 2001),).

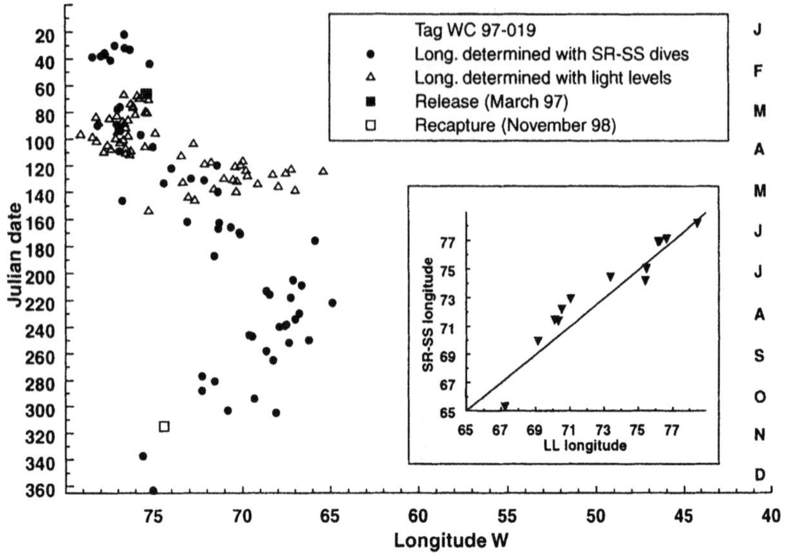

Figure 8. Seasonal longitude estimates for an archival tagged bluefin tuna (WC-019). Filled circles are estimates obtained using sunrise and sunset dives (SR-SS) and filled triangles are estimates obtained with light levels. The insert shows the agreement between longitude estimates using the two above techniques on the same days.

4.4.1 Pop-up satellite tagging from purse seines

Atlantic bluefin have been released with pop-up satellite tags using other methodologies and in several locations by the TAG Program. Large bluefin were tagged and released in a single school from a purse seine in New England in September of 1998 (9) and on hook and line in 2000 using techniques similar to the Carolina tagged fish (14). Pop-up satellite tags from the purse seine released New England fish surfaced in the first five months along the U.S. continental shelf as the bluefin appeared to be migrating southward in winter in concordance with archival data from implantable tags (Block *et al.*, 2001). In January, a large bluefin tuna from the purse seine release was just south of the Carolina release site for the 1999 archival tags. In 2000 releases several bluefin released in New England were in the Carolinas in winter and spring upon tag release. Several tags surfaced in springtime in the west Atlantic close to the boundary at 45° between the two stocks and one fish crossed the stock boundary. This result is similar to Lutcavage *et al.* (1999) although linkage of New England fish to Carolina has not been emphasized in this study. New England fish released in 1998 displayed fidelity back to the release location as a pop-up tag timed to release12 months after deployment surfaced in the vicinity of New England. The New England data set, although small, supported the positional information from the implantable archival results. Pop-up end points most often reported from the western

Atlantic however springtime releases were located in the central Atlantic. Only two fish were timed to pop-off at the proposed breeding period (March) but neither visited a known spawning ground. However the pop-off satellite tag date might have been premature (early) for assessing spawning in these western Atlantic bluefin tuna. All of the tags (100%) released from New England reported to a satellite, indicating that this technique (release from a purse seine) was highly successful. Five tags released over the initial 0.5-5 month period had complete archival temperature records and did indicate the tags had remained on the fish for the length of the deployments. The limitation of the data set from New England is that with a 6-12 month deployment and the use of single point tags, a limited archive of temperature was recorded and we could not be sure what any fish was doing before or after the point of release. Fish that released along the mid-Atlantic could have arrived at this position due to premature release. Importantly, the first generation tag did not record a temperature archive up to the last point the tag was on the fish. Thus it was difficult to confirm attachment to the fish up until the point of pop-off. These fish could have gone on to spawn in any location later in the season and to some extent illustrate the limitation of this approach. However, when viewed in light of the evolution of the technology, they demonstrated that 180-365 day deployments were possible.

4.4.2 Pop-up satellite tagging of spawning bluefin using longline gear

Fifteen giant bluefin tuna were tagged in the Gulf of Mexico in 1999 and 2000 with PSAT tags. To date this research effort on western bluefin is the most difficult. The spawning area targeted for tagging efforts is 180 nm offshore and access to bluefin is best in the winter and spring months when rough weather prevails. The best results occur with short longline sets of less than three hours which utilize approximately 120 hooks. Longer sets result in mortality of bluefin in the warm (22-30°C) Gulf waters. Efforts to sample all fish after a mortality for gonad condition as well as to obtain genetic and stomach content samples provides significant information on the spawning cycle and behavior in the Gulf. Thus far, the smallest bluefin measured in the Gulf of Mexico during tagging operations was 217 cm (curved length). Out of five bluefin (3 females) sampled over two seasons (mortalities), none were in spawning condition in March (n=3), but all females were histologically close to spawning (yolked oocytes) in late April (2). Bluefin tagged in the Gulf of Mexico with pop-up satellite tags timed for 60-365 days provide the first direct observations for understanding the preferred ambient temperatures presumably for breeding, the location of breeding activity, and residency period within a breeding area. All of these tags have full archival records on depth and temperatures and reveal a wide range of depth movements (0-1000 m) and a preference for warm waters while in the Gulf of Mexico (Figure 9). The data provide an excellent comparison with the archival data which is obtained at two minute intervals. The Gulf data reveal a warmer temperature preference (predominantly 22-26°C) at this time of year.

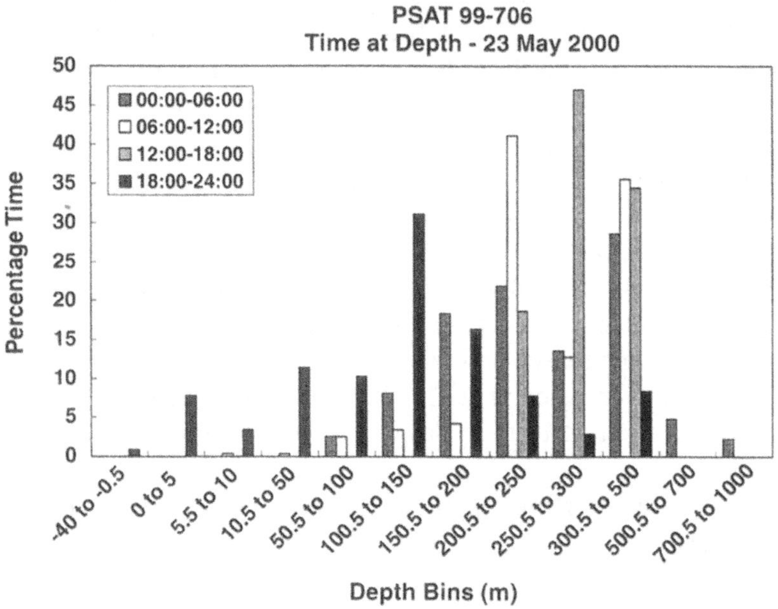

Figure 9. A day in the life of a bluefin from depth and temperature data archived from a pop-up satellite tag of a bluefin tuna on the spawning grounds in the gulf of Mexico.

5. SUMMARY

Electronic tags are providing exciting new information on the biology of Atlantic bluefin tuna. Both types of electronic tags have advantages and disadvantages but used together they are capable of providing enormous insight into the movements and natural history of the fish. The synthesis of electronic tagging data and remote sensing will provide the first view of the habitat utilization of bluefin tuna as well as other pelagic fishes over long temporal periods. The new information provided by electronic tagging data sets should improve the ability of managers to implement recovery plans for the Atlantic bluefin tuna.

ACKNOWLEDGEMENTS

The research was supported by the National Marine Fisheries Service, National Science Foundation, the Packard, Pew and MacArthur and Monterey Bay Aquarium Foundations and the TAG A Giant Campaign. We thank Andreas Walli and Shana Beemer for assistance on data and figure analysis. We thank the captains and crews of the many vessels that aided the TAG research program, the fishers and processors that recovered the archival tags, and the assistance of Sue Vermillion, Richard Novak and Doug Fudge is gratefully acknowledged.

REFERENCES

Block, B.A., Dewar, H., Farwell, C., and Prince, E.D. (1998a) A new satellite technology for tracking the movements of Atlantic bluefin tuna. *P. Natl. Acad. Sci. USA* **95**, 16, 9384-9389.

Block, B.A., Dewar, H., Williams, T., Prince, E.D., Farwell, C., and Fudge, D. (1998b) Archival Tagging of Atlantic bluefin tuna (*Thunnus thynnus thynnus*). *Mar. Tech. Soc. J.* **32**, 37-43.

Block, B.A., Dewar, H., Blackwell, S.B., Williams, T., Prince, E.D., Farwell, C. J., Boustany, A., Teo, S., Seitz, A., Walli, A., and Fudge, D. (2001) Electronic tags reveal movements, depth preferences and thermal biology of Atlantic bluefin tuna (*Thunnus thynnus thynnus*). *Science. In press.*

Bost, C.A., Georges, J.Y., Guinet, C., Cherel, Y., Putz, K., Charrassin, J.B., Handrich, Y., Zorn, T., Lage, J., Le Maho, Y. (1997) Foraging habitat and food intake of satellite tracked king penguins during the austral summer at Crozet Archipelago. *Mar. Ecol. Prog. Ser.*, **150**, 1-3, 21-33.

Carey, F.G., and Lawson, K.D. (1973) Temperature regulation in free-swimming bluefin tuna. *Comp. Biochem. Physiol.* **44A**, 375-392.

Carey, F. G., and Robison, B.H. (1981) Daily patterns in the activities of swordfish, *Xiphias gladius*, observed by acoustic telemetry. *Fish. Bull.* **79**, 277-292.

Carey, F.G., Kanwisher, J.W., and Stevens, E.D. (1984) Bluefin tuna warm their viscera during digestion. *J. Exp. Biol.* **109**, 1-20.

Costa. D.P. (1999) The role of physiology in the behavior of diving mammals: Insights from animals in nature. Proceedings, 25th Annual Meeting, European Underwater and Baromedical Society, EUBS 99, Haifa and Eilat, Israel, pp. 233-239.

Delong, R. L., Stewart, B.S., and Hill, R.D. (1992) Documenting migrations of northern elephant seals using day length. *Mar. Mammal Sci.* **8**, 155-159.

Friedland, K.D., Walker, R.V., David, N.D., Myers, K.W., Boehlert, G.W., Urawa, S., Ueno, Y. (1999) Influence of open-ocean orientation chum salmon during their return migration based on data from data storage tags. ICES Council Meeting Papers, AA 08, 29 pp.

Gunn, J., Polacheck, T., Davis, T., Sherlock, M., and Nicolau, M. (1994) The development and use of archival tags to study the movement, behavior and physiology of the southern bluefin tuna, *Thunnus maccoyii*. IATTC, 45th Annual Tuna Conference. National Marine Fisheries Service, La Jolla, CA. 85 pp.

Gunn, J., Polacheck, T., Davis, T., and Wotherspoon, S. (1996) Update on CSIRO's studies of southern bluefin tuna using archival tags. 47th Annual Tuna Conference. National Marine Fisheries Service, La Jolla, CA. 56 pp.

Gunn, J., and Block, B.A. (2001) Acoustic, archival and pop-up satellite tagging of tunas. *In* Block, B.A., and Stevens, E.D. (eds.) Tunas: Physiology, Ecology and Evolution. *Fish Physiology* Vol 19. Academic Press (in press).

Hill, R.D. (1994) Theory of geolocation by light levels. *In* LeBoeuf, B.J. (ed.) *Elephant Seals*. California Press, pp. 227-236.

Jouventin, P., Capedeville, D., Cuenot-Chaillet, F., Boiteau, C. (1994) Exploitation of pelagic resources by a non-flying seabird: satellite tracking of the king penguin throughout the breeding cycle. *Mar. Ecol. Progr. Ser.* **106**, 11-19.

Klimley, A.P., Prince, E.D., Brill, R.W., and Holland, K. (1994) Archival tags: Present and future archival tag working group. NOAA Technical Memorandum, NMFS-SEFSC-357. National Marine Fisheries Service, Southeast Fisheries Science Center, 30 pp.

Klimley, A.P., and Holloway, C. (1999) Homing synchronicity and schooling fidelity by yellowfin tuna. *Mar. Biol.* **133**, 308-317.

Kitagawa, T., Nakata, H, Kimura, S., Itoh, T., Tsuji, S., and A. Nitta. (2000) Effect of ambient temperature on the vertical distribution and movement of Pacific bluefin tuna *Thunnus thynnus orientalis*. *Mar. Ecol. Prog. Ser.* **206**, 251-260.

Koudil, M., Charrassin, J., Le Maho, Y., Bost, C. (2000) Seabirds as monitors of upper-ocean thermal structure. King Penguins at the Antarctic polar front, east of Kerguelen sector. *CR Acad. Sci. III-Vie* **323**, 377-384.

Le Boeuf, B .J., Crocker, D.E., Costa, D.P., Blackwell, S.B., Webb, P.M., and Houser, D.S. (2000) Foraging ecology of northern elephant seals. *Ecol. Monogr.* **70**, 353-382.

Lutcavage, M.E., Brill, R.W., Skomal, G.B., Chase, B.C., and Howey, P.W. (1999) Results of pop-up satellite tagging of spawning class fish in the Gulf of Maine: Do North Atlantic bluefin tuna spawn in the mid-Atlantic? *Can. J. Fish. Aquat. Sci.* **56**, 173-177.

Lutcavage, M. E., Brill, R. W., Skomal, G. B., Chase, B. C., Goldstein, J.L., and Tutein, J. (2000) Tracking of North Atlantic bluefin tuna (*Thunnus thynnus*) in the northwestern Atlantic using ultrasonic telemetry. *Mar. Biol.* **137**, 347-358.

Marcinek, D.J., Blackwell, S.B., Dewar, H., Freund, E., Farwell, C., Dau, D., Seitz, A., and Block, B.A. (2001) Depth movements and muscle temperatures of Pacific bluefin tuna measured with ultrasonic telemetry. *Mar. Biol.* **138**, 869-885.

Mather, F. J., Mason, J. M., and Jones, A. C. (1995) Life history and fisheries of Atlantic bluefin tuna. NOAA Technical Memorandum, NMFS–SFSC. 370, 165 pp.

McConnell, B.J., Chambers, C., and Fedak, M.A. (1992) Foraging ecology of southern elephant seals in relation to the bathymetry and productivity of the Southern Ocean. *Antarct. Sci.* **4**, 393-398.

Metcalfe, J.D., and Arnold, G.P. (1997) Tracking fish with electronic tags. *Nature* **387**, 665-666.

NRC (1994) An assessment of Atlantic bluefin tuna. National Academy Press, Washington, D.C. 148 pp.

Polovina, J.J., Kobayashi, D.R., Parker, D.M., Seki, M.P., and Balazs, G. H. (2000) Turtles on the edge: movement of loggerhead turtles (*Caretta caretta*) along oceanic fronts, spanning longline fishing grounds in the central North Pacific, 1997–1998. *Fish. Oceanogr.* **9**, 1, 71-82.

Priede, I.G. (1984) A basking shark (*Cetorhinus maximus*) tracked by satellite together with simultaneous remote sensing. *Fish. Res.* **2**, 201-216.

Renaud, M.L., and Carpenter, J.A. (1994) Movements and submergence patterns of loggerhead turtles (*Caretta caretta*) in the Gulf of Mexico determined through satellite telemetry. *Bull. Mar. Sci.* **55**, 1-15.

Welch, D.W., and Eveson, J.P. (1999) An assessment of light-based geoposition estimates from archival tags. *Can. J. Fish. Aquat. Sci.* **56**, 1317-1327.

Movements and Temperature Preferences of Atlantic Bluefin Tuna (*Thunnus thynnus*) off North Carolina: A Comparison of Acoustic, Archival and Pop-Up Satellite Tags

Andre M. Boustany, David J. Marcinek,[1] John Keen, Heidi Dewar,[2] and Barbara A. Block
Tuna Research and Conservation Center, Stanford University, Pacific Grove, CA
[1]Present address: Department of Radiology, University of Washington Medical Center, Seattle, WA
[2]Present address: Pfleger Institute of Environmental Research, Oceanside, CA

Abstract: Electronic tagging technologies are providing new methodologies for tracking open ocean fish. In this paper, we compare the behavior and environmental preferences of Atlantic bluefin tuna tracked with acoustic, archival and pop-up satellite tags off the coast of Cape Hatteras, North Carolina. Four bluefin tuna were acoustically tracked for periods ranging from 2-27 hours. Data from the acoustic tracks were compared to results from archival and pop-up satellite tagging studies that were being conducted concurrently. The 2 minute sampling rate of the archival and pop-up satellite archival tags was compared to the 1–2 second sampling rate from the acoustic tags and it was determined that the lower resolution was sufficient to accurately describe behavior of bluefin tuna. Data from all types of tags show the same patterns in bluefin tuna behavior throughout the course of this study. All three techniques indicate that the fish remained for a significant period of time on the North Carolina continental shelf in waters that ranged from 19–24°C, near the strong thermal gradient formed by the Gulf Stream and the colder waters of the Labrador Current. It is hypothesized that a large aggregation of spawning prey fish is what attracts bluefin tuna to this area in the winter months.

1. INTRODUCTION

Atlantic bluefin tuna (*Thunnus thynnus*, Scombridae) range throughout the northern Atlantic Ocean from the subtropical Mediterranean Sea and Gulf of

J.R. Sibert and J.L. Nielsen (eds.), Electronic Tagging and Tracking in Marine Fisheries, 89–108.
© 2001 *Kluwer Academic Publishers. Printed in the Netherlands.*

Mexico to polar latitudes (Mather et al., 1995; NRC 1994). Their large range and highly vagile lifestyle has made direct observation of movement patterns difficult. Traditionally, the most common methods for investigating the movement patterns of highly migratory pelagic fishes have been analysis of catch statistics (Clay, 1990; Mather et al., 1995), conventional tagging studies (Jones et al., 1997; Sibert et al., 1999) and acoustic tracking (Carey and Lawson, 1973; Carey and Robison, 1981; Holland et al., 1990a; Block et al., 1992a, 1997; Marsac and Cayre, 1998; Brill et al., 1999; Dagorn et al., 2000). The use of catch statistics and conventional tagging studies are problematic as a result of the difficulty in correcting for anthropogenic effects such as spatial and temporal shifts in fishing effort.

Until recently, the best method for studying large pelagic fish in their environment has been through acoustic telemetry. Acoustic tracking allows for the observation of fine scale behavioral patterns as well as simultaneous profiles of oceanographic parameters, permitting a better understanding of how the environment influences fish behavior and distribution (Carey and Lawson, 1973; Carey and Olson, 1982; Carey et al., 1984; Block et al., 1992a, 1997; Cayre and Marsac, 1993; Josse et al., 1998; Brill et al., 1999).

Acoustic tracking provides a fine resolution of data over short time scales, but owing to the high cost of the tracking vessel and human effort required to track fish, tracks are most often less than five days (Gunn and Block, in press). New electronic tag technologies, such as archival and pop-up satellite tags, have addressed these limitations by allowing tags to collect data over long time scales independent of tracking vessels (Gunn et al., 1994; Itoh et al., 1997; Metcalfe and Arnold, 1997; Block et al., 1998a, 1998b, this volume; Lutcavage et al., 1999). As these tags are designed to collect data for durations of several years, the sampling rate is reduced compared to acoustic tags (usually a 2 minute interval vs. every second).

These new tagging technologies hold great promise in examining unanswered questions about pelagic species, yet few studies have been able to compare the data they provide—in essence ground truth the technology to the acoustic standard (Marcinek et al., 2001). Acoustic, archival and pop-up satellite tags all collect similar types of data at varying sampling rates. What effect these different data sets have on our ability to make inferences about the behavior of pelagic fish needs study before data from the different tags can be used interchangeably. In order to answer these questions, it is necessary to use these distinct tagging technologies concurrently.

The large catch and release fishery off the coast of North Carolina has provided scientists with a "natural laboratory" to address these questions. Acoustic, archival and pop-up satellite tags have all been used to study bluefin tuna off the coast of North Carolina (Skomal and Chase, 1997; Block et al., 1998a,b, this volume). Data returned from these tags were used to examine the behavior and environmental preference of bluefin tuna in the Carolinas, as well as survivorship of these fish after catch and release. Information returned from the three electronic tag types is compared to examine the concordance of data between these different tagging technologies.

2. MATERIALS AND METHODS

Four Atlantic bluefin tuna (185–206 cm curved length) were acoustically tracked in March of 1997 for periods ranging from 2 to 27 hours. During the same period, 37 single point pop-up satellite tags and 160 archival tags were also deployed. Acoustic, archival and pop-up satellite tagged fish were all handled in the same manner to better assess the effects of handling stress on survivability (Block et al., 1998a,b). Tuna were caught on rod and reel from sport-fishing vessels using heavy line (90 kg line, 180 kg leader) and circle hooks to minimize fight time and trauma to the fish. All fish were brought on board the fishing vessels through "tuna doors" located near the water line at the rear of the boats. Large padded mats were used to cushion the fish while researchers aerated the gills with a seawater hose. Acoustic tags were affixed to the fish by means of two dart heads constructed of medical grade nylon. Dart heads were located at the front and the rear of the acoustic tag body to minimize movement of the tag and abrasion of the fish's skin. Darts were soaked in betadine and placed in the dorsal musculature eight centimeters below the second dorsal fin. One of the acoustic tagged fish (9704) was also surgically implanted with an internal temperature tag in a similar manner as the archival tagged bluefin tuna (Block et al., 1998b). After tagging, fish were released through the tuna door and tracking of acoustic fish began aboard an 11-meter research vessel.

The acoustic tracking methodologies employed in this study have previously been described in detail (Block et al., 1997). Swimming depth of the acoustic tagged fish was transmitted by a 34 or 50 kHz V22EX pressure transmitter tag (V22 with extra power, Vemco Ltd., Armdale, Nova Scotia). The signal was received using a towed Vemco VR-41 hydrophone array, with four element spaced 90 degrees apart. Depth and (in the case of the one fish that was internally tagged) temperature readings were recorded using a Vemco VR-60 receiver. Direction of the fish relative to the boat was determined by a Vemco VR-28 receiver by examining the difference in signal strength among the four array elements. Absolute position of the research vessel was obtained through the ship's onboard GPS unit. GPS position was updated every 4–8 seconds and this information was incorporated into the same data file as depth of the fish. Speed of the vessel was estimated by calculating the distance traveled between successive GPS points. Position information was later used to construct tracks of the fish as well as obtain estimates of swimming speed.

A Seabird portable CTD (conductivity, temperature, depth) was used to measure changes in temperature, dissolved oxygen and pH with depth. XBTs (expendable bathythermograph, Sippican) were also used to profile temperature with depth when the speed of the tracking vessel was too fast to successfully deploy the CTD. Water column profiles were taken at 3-4 hour intervals during a track. Sea surface temperatures (SSTs) were also recorded at these times through the use of a handheld digital thermometer, which was calibrated against a Bureau of Standards mercury thermometer. Fish movements were also compared to large scale SSTs by overlaying tracks on advanced very high resolution radiometry (AVHRR) satellite images provided by NOAA. Water depth was measured every ten minutes using the ship's on board depth finder.

3. RESULTS

3.1 Acoustically Tracked Fish

Acoustic tracking occurred March 4-21, 1997 for periods ranging from 2 to 27 hours (Table 1). All tracks were continuous except 9703, which was followed for 2 hours on March 13 before the track was terminated. This fish was relocated 8.5 days later while fishing in a school this fish had joined. It was then followed for an additional 10 hours. The second portion of this track is treated separately (9703b) because of the long period of time between segments of the track. All tracks were terminated due to deteriorating weather conditions.

Table 1. Summary of track data for Atlantic bluefin tuna (*Thunnus thynnus*). BL = body lengths

Track no.	Starting date	Curved length (cm)	Est. mass (kg)	Length of track	Starting lat	Starting lon	Avg speed (km/hr)	BL/sec
9701	3/4/97	185	110	27 hours	35.085'	75.340'	3.14	0.87
9702	3/8/97	187	114	4.75 hours	36.095'	75.342'	3.04	0.84
9703a	3/13/97	206	150	2 hours	35.032'	75.445'	4.89	1.22
9704	3/17/97	205	148	16 hours	35.060'	75.357'	3.97	1.00
9703b	3/21/97	206	150	10 hours	36.075'	75.293'	2.52	0.67

3.1.1 Vertical Movements

Depth records are shown overlaid on isotherms for tracks 9701, 9703b and 9704 (Figure 1). Upon release, all fish immediately swam to the bottom and stayed at depth for at least one hour before making any excursions to the surface. Three fish remained in the shallow water over the continental shelf (< 90 m) with only fish 9704 swimming offshore into water deeper than 100 m for a portion of the track. This fish dove to a depth of 220 m for approximately an hour, spending the remainder of the track above 100 m. Swimming patterns in shallow water are characterized by frequent oscillatory

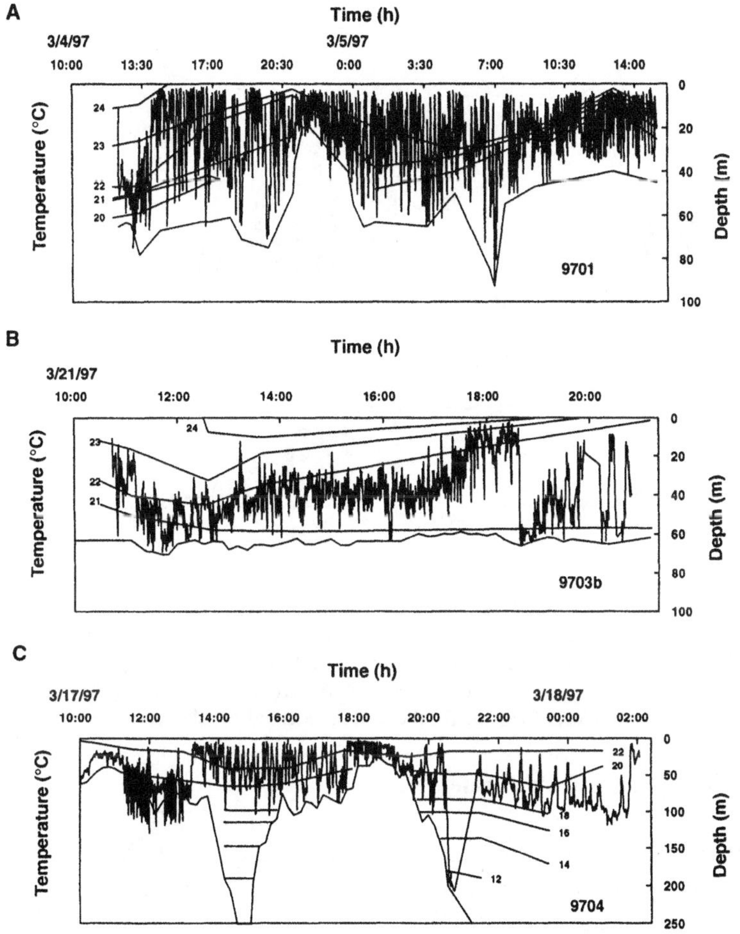

Figure 1. Depth records of Atlantic bluefin tuna plotted over isotherms and estimates of bottom topography. (a) Track 9701. (b) Track 9703b. (c) Track 9704. Note: axes are not at the same scale.

dives throughout the entire water column, with no apparent preference for a single depth. Histograms of time spent at depth indicate equal occurrence at all depths between 10 and 50 m (Figure 2a). Most fish showed no significant difference in preferred depth between night and day. The exception to this was with fish 9704, which had a greater average depth during the night than during the day (Kolmogorov-Smirnov test $p<0.05$). This is most likely due to the fact that a greater proportion of the nighttime tracking was done in the deeper offshore waters and does not necessarily signify any diel pattern. Fine-scale examination of the depth record indicates that bluefin tuna exhibit the oscillatory diving pattern stereotypical of other tunas that have been tracked. These short vertical excursions lasted 2-5 minutes and covered a range of 10-20 m.

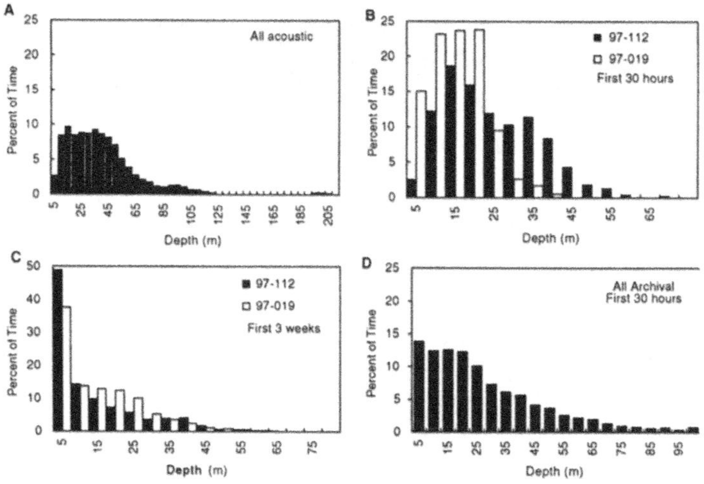

Figure 2. Percent of time spent at depth (a) All acoustic tagged fish (b) Archival tagged fish WC 97-112 and 97-019 for the first 30 hours after release. (c) Archival tagged fish WC 97-112 and 97-019 for the first 3 weeks after release. (d) All archival tagged fish (n = 12) which remained off the coast of North Carolina for the first 30 hours after release.

3.1.2 Temperature Preferences

Acoustic tracking occurred in the shallow waters over the continental shelf, narrowing the environmental range to which the fish were exposed. In winter, a cyclonic eddy from the Gulf Stream overlies the continental shelf between Cape Hatteras and Cape Lookout, North Carolina. This provides an environment with warm temperatures throughout the water column. The warm water and shallow depth resulted in most fish experiencing a narrow temperature range (19–24°C) with only one fish (9704) experiencing water colder than 19°C. This fish encountered water below 12°C when it swam off the continental shelf into deeper water. Dissolved oxygen concentrations also showed little variability over the course of the tracks, remaining between 4.5–6 ml O_2/L H_2O throughout the water column (Figure 3).

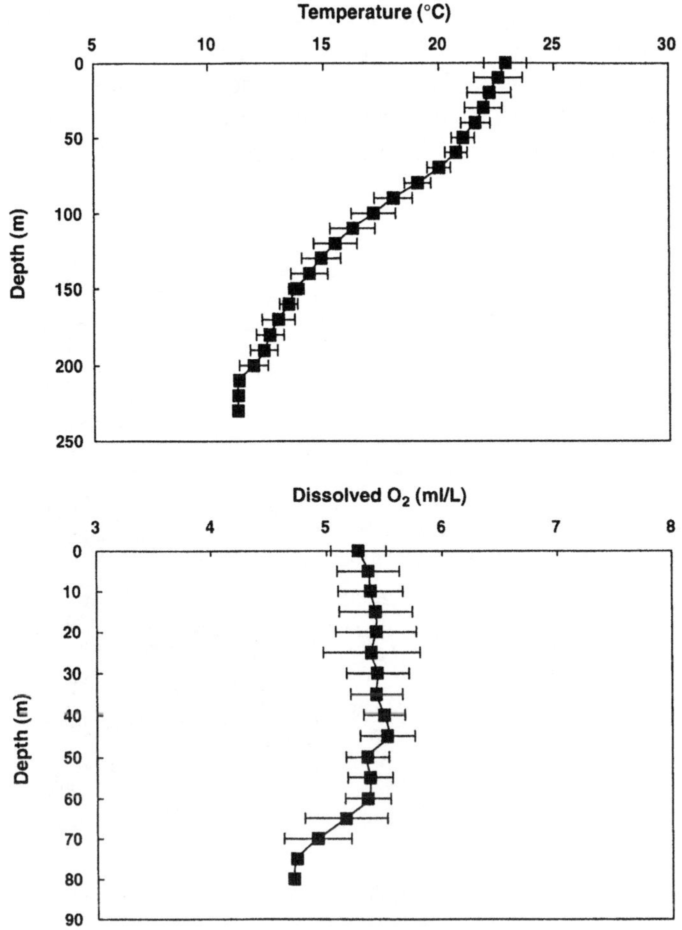

Figure 3. Average (± 1 SD) water profiles of a) temperature with depth and b) dissolved oxygen with depth. These measurements include all CTD (temperature and oxygen) and XBT (temperature) casts.

3.1.3 Horizontal Movements

All bluefin tuna showed similar movement patterns in relation to thermal fronts, closely following the strong frontal boundary formed by the Gulf Stream coming in contact with the much cooler nearshore waters. Fish stayed on the warm side of this front, in surface waters that ranged from 21 to 24°C (Figure 4). Tracks 9701, 9702 and 9703 all proceeded to the southwest (Figures 4, 5a, 5b). Both portions of track 9703 showed the same pattern of movement, following the thermal front and traveling to the southwest. When this fish was relocated it was not far from the area where the initial tagging took place, 8 days earlier. The only fish to show different movement patterns was 9704. This fish was tagged in the same areas as the others but moved to the northeast instead of the southwest (Figure 5c) This course took the fish off the continental shelf. Like the other fish, 9704 also remained on the warm side of the front in temperatures greater than 21°at the surface.

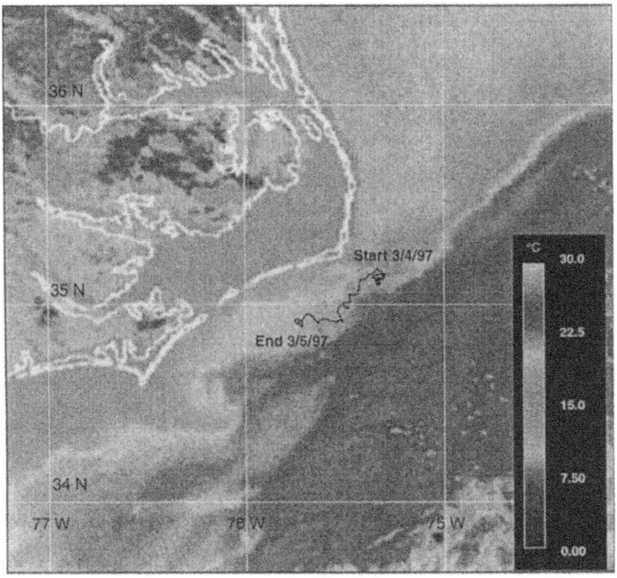

Figure 4. Acoustic track 9701 with AVHRR sea surface temperature for 3/5/97.

3.2 Archival Tagged Fish

Concomitant with acoustic tracking in 1997, 160 archival tags were implanted in medium and giant Atlantic bluefin tuna (Figure 5a), (Block et al., 1998b).To date, 27 of those archival tagged bluefin tuna have been recaptured and data recovered from 21 of the tags. Most of these fish remained for periods of a few days to several months within the Carolina coastal province. The data from two archival tagged bluefin tuna (WC 97-019 and WC 97-112, Wildlife Computers, Redmond, WA) were examined in detail and compared to those from the acoustically tracked fish. These fish (193 and 210 cm CFL respectively) were tagged on March 7 and 21 in the same areas as the acoustic tagged fish. The first three weeks of data from the archival tagged fish were used to compare to the acoustic tagged fish. Geolocation estimates based on light level indicate these fish remained in the North Carolina area for this time period and, therefore, allow for a direct comparison between acoustic and archival data (Figure 6).

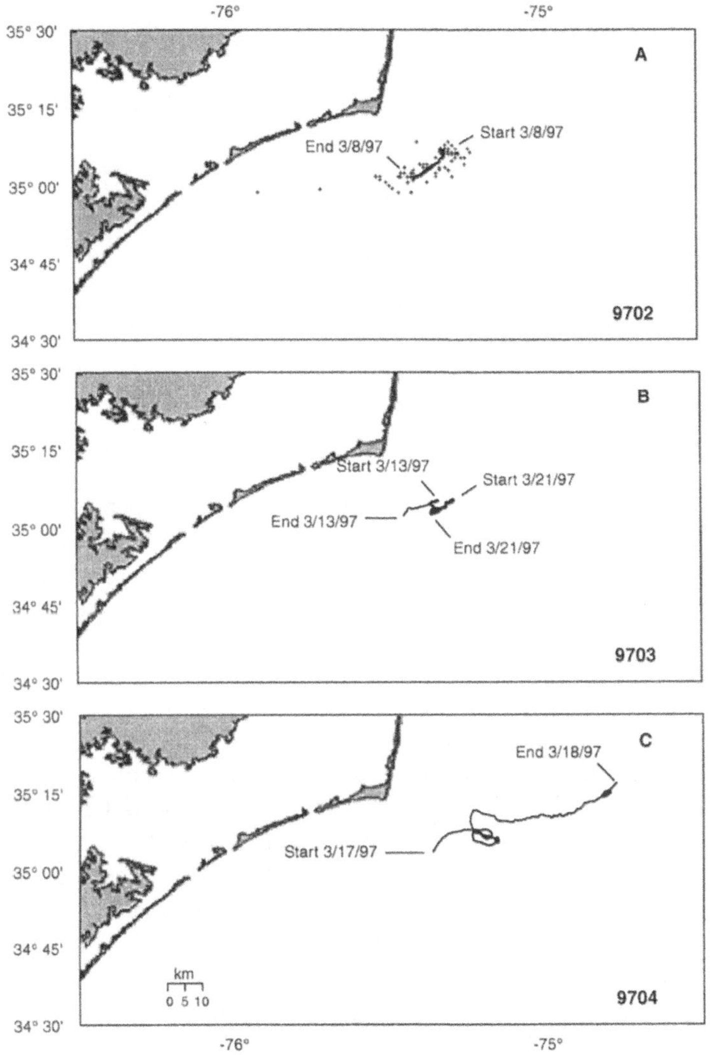

Figure 5. (a) Acoustic track 9702 with 1997 archival tag release positions. (b) Acoustic track 9703a and 9703b. (c) Acoustic track 9704.

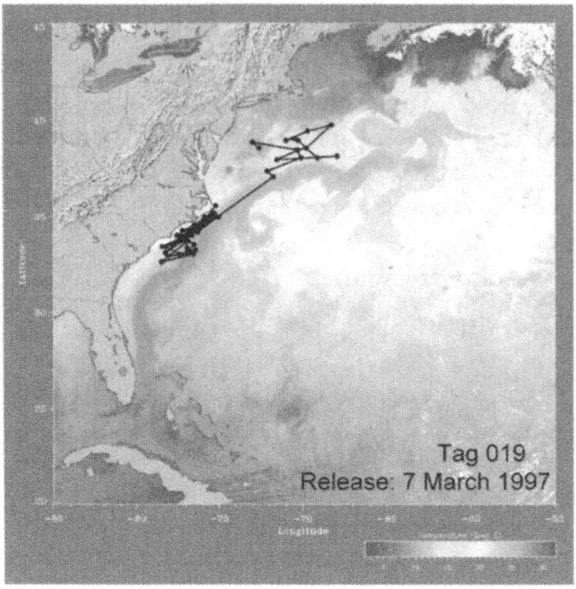

Figure 6. Geolocation estimates based on light level for archival tagged fish 97-019 showing 5 weeks of residency off the coast of North Carolina. Large black circle is the recapture location on November 14, 1998.

Ambient temperature preference, percent of time spent at depth, and overall diving pattern are compared for these archival and acoustic tagged fish. The external temperature range experienced by the archival tagged fish over the first three weeks was 16–25° C. These fish spent a majority of their time in waters between 19 and 24° C (mean = 22.6 ± 1.3). This is similar to the range of water temperatures in which the acoustic tagged fish were found. Internal temperatures of the archival tagged fish ranged from 23 to 27°C over the same time period (25.3 ± 0.51).

Percent of time spent at depth shows more variation in the archival tagged fish than in the acoustic tagged fish. During the first 30 hours of the archival tag data, these fish spend most of their time at depths between 10 and 40 m (Figure 2b). This is similar to the depth preference observed for all acoustic tagged fish over the same time period (Figure 2a). However, as the archival tracks progress, these fish begin to spend a greater portion of their time near the surface. Throughout the three weeks examined, archival tagged fish spent 43% of their cumulative time in the top 5 m (Figure 2c). This is in contrast to the much lower occupancy in the surface layer by the acoustic tagged fish. This shift in depth preference as the tracks progress is a common occurrence and is seen in all archival tagged fish (n= 12) which remained off the coast of North Carolina (Figure 2d). During the first 30 hours, archival tagged fish spent less than 14% of the time in the top 5 m of the water column while after this period, the occupancy above 5 m was over 40%.

Diving patterns of bluefin tuna are compared between the two types of tags. The archival tags were programmed to take measurements at 2-minute intervals. This is in contrast to the 1-2 Hz sampling rate for the acoustic tags. To test what effect this decreased sampling rate has on perceived diving pattern, one point was taken every two minutes from acoustic track 9701 and compared to the same track under the higher sampling rate (Figure 7a,b). Although resolution decreases with the 2-minute sampling rate, the observed behavior remains consistent over the course of a day as diving pattern is very similar in the two plots. We see a similar diving pattern in archival tagged fish 97-112 while it was still off the coast of North Carolina (Figure 7c).

3.3 Pop-Up Satellite Tagged Fish

Nine short-term (3-14 day) pop-up satellite tags were also deployed over the same time period as the archival and acoustic tags in 1997 (Block et al., 1998a). Tags were deployed on bluefin tuna between February 24 and March 17 and popped up between March 2 and March 20, 1997. These first generation Microwave Telemetry (Columbia, MD) tags recorded one water temperature measurement every hour and transmitted this information to an Argos satellite after the tag surfaced. All tags released on time and successfully transmitted temperature data to the satellite. This indicated 100% survivorship from the catch, handling and tagging experience. End point location and ambient temperature data from the pop-up tags are similar to the data from the acoustic tracks. All tags surfaced near Cape Hatteras and Cape Lookout, in the warm waters of the Gulf Stream close to the western frontal zone. Temperature data from the pop-up tags suggest that these fish spent most of their time on the warm side of the front as the mean water temperature for the nine short term pop-up tagged fish was 21.3 ± 1.1°C (Block et al., 1998a). This falls within the range of temperatures experienced by the acoustic and archival tagged fish.

End point locations in relation to time of tagging also show behavior similar to those of the acoustic tagged fish. Of the seven tags that reported end point position, the four that released between March 2 and 13 (8832, 8833, 8834, 8836) popped up to the southwest of the tagging location, moving in the same direction as the three acoustic fish followed over the same time period. The remaining pop-up tags (8969, 8971, 8972) were attached on March 17, at the same time as acoustic tag 9704. Two of these fish showed movement similar to this acoustic tagged fish, travelling to the northeast of Cape Hatteras. One of these pop-up tagged fish, 8971, also

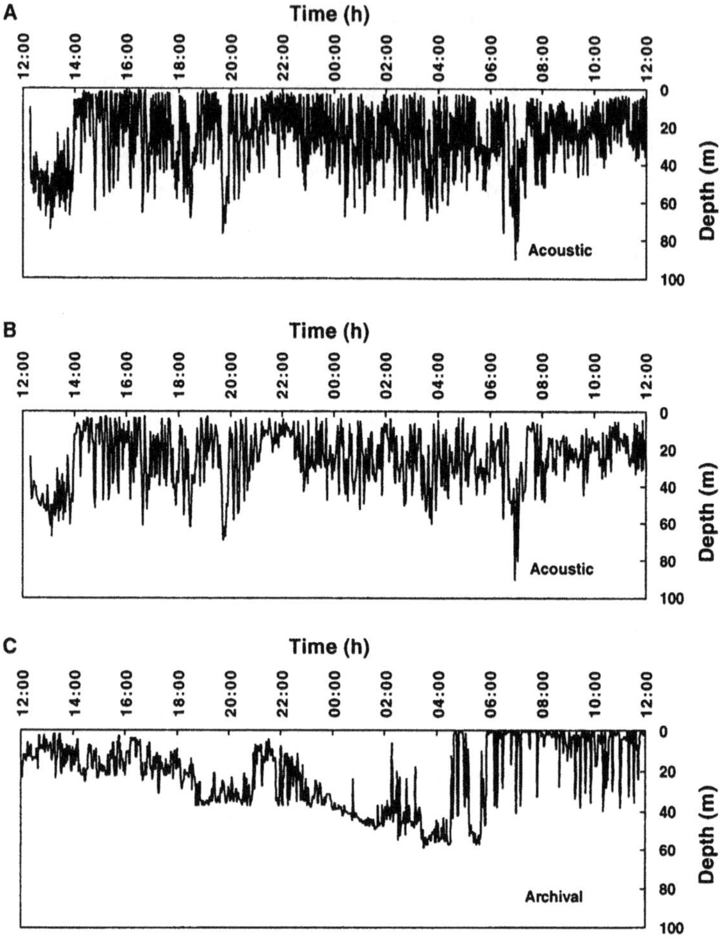

Figure 7. Comparison of acoustic and archival sampling rates. (a) Acoustic track 9701, 5 second sampling rate. (b) Acoustic track 9701, 2 minute sampling rate. (c) Archival tagged fish WC 97-112 on March 22-23, 2 minute sampling rate. Note change in diving behavior at 0600 on March 23. This was approximately 40 hours after the tagging event.

ended up offshore of the continental shelf, where it encountered water as cold as 14°C (Figure 8). Since the end point location of this fish was well within the warm waters of the Gulf Stream, it is suggested that this fish experienced these colder waters at depth, like acoustic tagged fish 9704.

In addition to these data, additional archival and pop-up satellite archival tags have been deployed from 1999-2001. The pop-up satellite tags currently in use by this research program record depth, ambient temperature and light level at one to two minute intervals, similar to archival tags (Wildlife

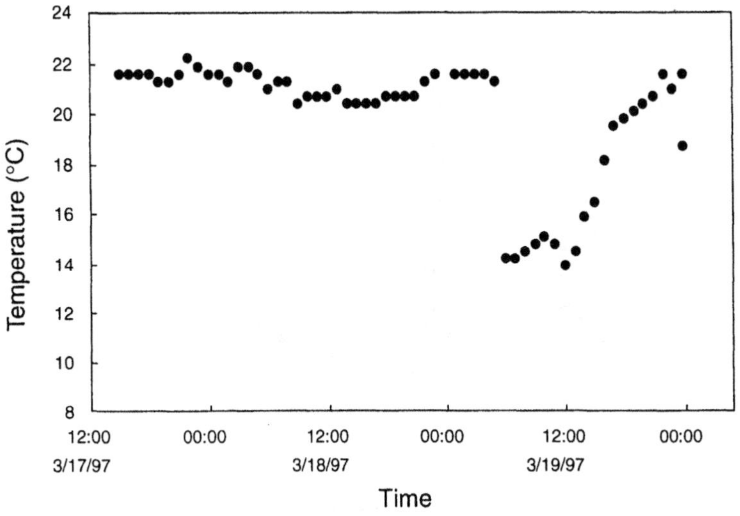

Figure 8. Temperature record of pop-up satellite tagged fish 8971. This fish was tagged on March 17.

Computers, PAT tags). When these data are transmitted to the satellites, they are summarized in user defined intervals to decrease the bytes transmitted. In 1999-2000, 30 of these pop-up archival tags were deployed near Cape Lookout, North Carolina. Eleven of these tags were deployed short term (3–15 days) to test the technology, examine survivorship from J hooks used in the troll fishery and to compare the behavior of fish tagged in the Cape Hatteras area to those tagged near Cape Lookout, 100 km to the southwest. Once again, a high survivorship was documented as all but one tag transmitted data. A single tag hit the beach in the local area and was recovered providing an additional full archival record of the bluefin tuna's behavior in the region. Data from these fish show much the same depth and thermal preference as fish tagged in the previous years. These fish show a strong preference for the top 10 m of the water column and the narrow temperature range between 18 and 24°C (Figure 9). As in 1997, all pop-up tags surfaced on the warm side of the thermal front, in SSTs above 21°C.

4. DISCUSSION

Acoustically tracked fish all showed the same general behavior patterns. After the tagging event, fish descended to 40 to 80 m and stayed there for at least an hour. This behavior, most likely a recovery phase due to the stress ing of the capture and tagging event, has previously been observed in bluefin

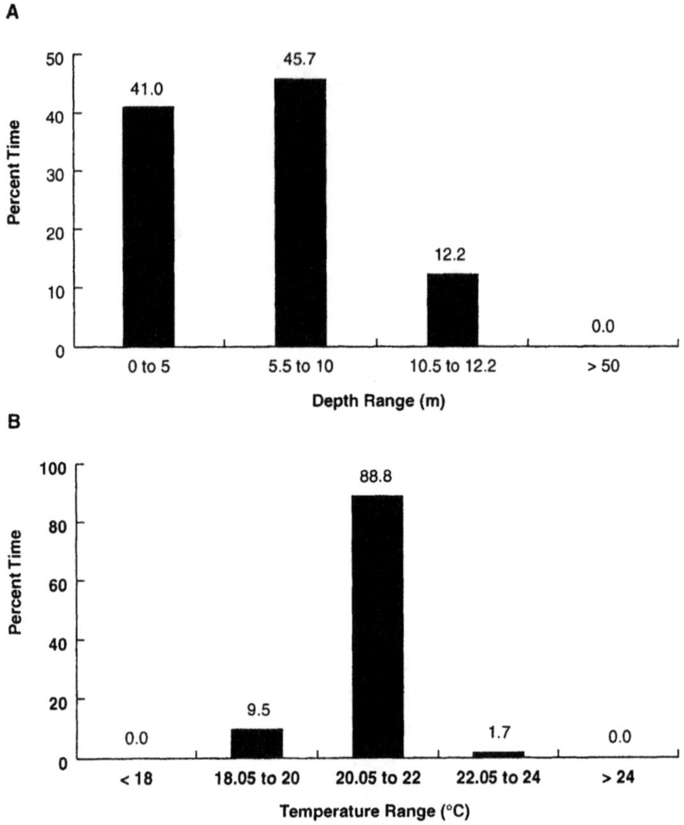

Figure 9. Daily a) depth and b) temperature preference for pop-up satellite archival tagged fish 99-550. This fish was tagged December 12, 1999. Data shown are for December 14, 1999.

tuna (Carey and Lawson, 1973), yellowfin tuna (Holland et al., 1990a) and marlin (Holland et al., 1990b; Block et al., 1992b; Brill et al., 1993). During this recovery period bluefin tuna also increased swimming speed, a response seen previously in marlin (Block et al., 1992b). Acoustic tagged fish maintained higher than average speeds for the first few hours after release compared to the remainder of the track. This is most apparent in examining the two portions of track 9703. This fish maintained speeds of 1.22 body lengths per second over the first portion of the track compared to 0.63 body lengths per second over the second portion of the track. This increase in speed may be to increase water flow over the gills and, therefore, oxygen uptake. This would be especially important immediately after a fight on hook and line. Previous acoustic tracking of bluefin tuna in this area found that fish were physiologically stressed after being caught on sportfishing gear (Skomal and Chase, 1997).

Movement patterns in relation to the environment were also similar for all acoustic tagged fish. Most of the tracking was done in the narrow range of water temperatures between 19 and 24°C. Less than 2% of the acoustic tracking was done in water temperatures below 19°C. Examination of the bathymetry in the region where tracking occurred shows that the shallow water limits temperature variation with depth. However, there exist large variations in temperature horizontally. As this area is near a strong thermal front, differences in SST up to 15°C can be observed over the course of 1 nautical mile. The fact that all these fish restricted themselves to such a limited range in SST's (21–25°C) is interesting considering the bluefin tuna's endothermy (Carey and Teal, 1969; Carey and Lawson, 1973; Linthicum and Carey, 1973). Data from archival tags have shown that bluefin tuna have the ability to spend days at a time in waters below 12°C (Block et al., this volume; Gunn and Block, in press). Although physiologically possible, fish tracked in 1997 spent a minimal amount of time in the colder waters of the nearby Labrador Current or below the thermocline in the deep waters off the continental shelf.

4.1 Comparability of Data

4.1.1 Temperature Preference

Data returned from archival and short-term pop-up satellite tags show the same general trends as those obtained from acoustic tags. Ambient temperatures experienced by the pop-up and archival tagged bluefin tuna closely match those from the acoustic tagged fish. Fish tracked with all three tag types (acoustic, archival and pop-up satellite tags) showed a strong preference for the temperature range between 19 and 25°C. One acoustic tagged fish, one first generation pop-up tagged fish and one pop-up satellite archival tagged fish experienced waters below 17°C, but even these fish spent the majority of time in waters above 18°C. Cold waters could be experienced either offshore below the thermocline or inshore. Acoustic tagged fish encountered warm temperatures as they closely followed the thermal front formed by the Gulf Stream. This frontal zone, between the cool coastal waters and the warm Gulf Stream, consists of eddies with SSTs between 21 and 25°C. An examination of the archival and pop-up data suggest that these fish remained within this region as well. Archival tagged fish remained predominately in areas where the SST was between 21 and 24°C. All pop-up tags surfaced near the edge of the Gulf Stream, in SSTs similar to those of the acoustic and archival tagged fish. This and the close correlation between overall temperatures experienced between the different tags suggest that the pop-up fish also remained in the mixing zone between the Gulf Stream and the Labrador Current.

4.1.2 Depth Preferences

Depth preferences were similar in all three tag types with a pressure sensor. Acoustic, archival and pop-up satellite archival tagged fish all spent a significant time between the surface and 50 m, with only acoustic tagged fish 9704 traveling off the continental shelf into waters deeper than 100 m. The narrow temperature range experienced by the first generation pop-up tagged fish (Block et al., 1998a) suggests that these fish also spent a majority of time in the shallow waters of the continental shelf. The only pop-up tagged fish to encounter colder waters, 8971, showed temperatures similar to the acoustic tagged fish, 9704, that swam into deeper waters. Both of these fish were tagged on the same day after being caught from the same school and it is possible they remained in the same school. Both fish swam offshore beyond the continental shelf where the acoustic fish dove to a depth of 225 m, encountering water temperatures below 12°C. It was within a day that the pop-up tagged fish recorded water temperatures below 14°C.

The only difference in data reported between electronic tagging technologies emerged in the comparison between percent of time at depth for the acoustic versus the archival tagged fish. While they were off the coast of North Carolina, the archival tagged fish spent a large portion of their time (>40%) in the top 10 m of the water column, an area where acoustic fish spent less than 11% of their time (Figures 2a and 2c). This may be due to the fish assuming modified behavior patterns for the first day after being tagged due to stress from the tagging event. As none of the acoustic tracks was over 30 hours long, these fish may have been avoiding the surface for the entire time they were tracked. However, even when acoustic tagged fish 9703 was relocated 8.5 days after being tagged, it remained below the surface, spending less than 10% of its time in the top 10 m. It is unlikely that it was still exhibiting modified behavior due to the tagging event at this time. Archival tagged fish also stayed away from the surface for the first 30 hours after release (Figures 2b and 2d). This shift in diving pattern and depth preference can be seen in the archival track of 97-112 approximately 40 hours after the tagging event (Figure 7c).

While all the tagging data from 1997 show much congruence, it remained unresolved whether these patterns would persist from one year to the next. Short term pop-up satellite archival tagging conducted in the winter of 1999–2000 addressed this question and also examined the behavior of fish tagged in the Cape Lookout area, about 100 km to the Southwest of Cape Hatteras.

The depth and temperature preferences of these fish, as well as the pop-off locations of these tags suggest that fish tagged in this area are part of the same body of fish that appear off Cape Hatteras. Depth and temperature preferences were the same as those for fish tagged two years previous, indicating that all fish electronically tracked, regardless of year, were utilizing the environment in a similar way. Pop-off end point locations for the pop-up satellite archival tags also showed complete overlap with pop-off locations from 1997 short term pop-off tags (Block et al., this volume).

4.2 Bluefin Tuna Behavior off the Coast of North Carolina

The recreational fishery for bluefin tuna in North Carolina first appeared in the early 1990s. Fish in this area range primarily between 150 and 215 cm curved fork length. Fisheries landings data suggest that bluefin tuna appear off the coast of North Carolina in late November and December. By mid April, most fish have left the area (Jones et al., 1997; E. Prince, unpublished data). It is suspected that most of the fish that reside off North Carolina in the winter make their way north and northeast in the spring to latitudes above 40° where they remain throughout the summer months (Mather et al., 1995; Block et al., 1998a). Many of the fish in this study could be confirmed to have stayed in this area for durations of 14-90 days, suggesting that these fish are not merely migrating through the area. This corroborates other studies that have examined the tag and recapture of bluefin tuna released in this area using conventional tags (Jones et al., 1997).

The question of what attracts bluefin tuna to the coast of North Carolina over the winter months can be examined by looking at the behavior of these fish in relation to the environment they encounter. The acoustic tracks show that these fish spend a large majority of their time very close to the strong thermal gradient formed by the warm waters of the Gulf Stream to the east and the colder waters of the Labrador Current to the west. Several authors have examined the relationship between thermal gradients and location of bluefin tuna (Maul et al., 1984; Reddy et al., 1995) and have found a strong preference of tuna for areas with strong thermal edges. This is most likely the result of "edge effects" where two different water masses moving against each other cause an increase in upwelling and vertical mixing that brings colder, nutrient rich water to the surface where photosynthesis can take place (Tranter et al., 1983). The occurrence of strong fronts near river mouths and estuaries, as in North Carolina, also enhances rates of horizontal mixing, further increasing primary production (Govoni, 1993). While the near-shore waters in this area experience a peak in productivity in the summer, the waters offshore (>12 km) of North Carolina, where bluefin tuna occur, experience a peak in primary productivity and zooplankton in winter (Weinstein, 1981).

This peak in offshore productivity coincides with a strong winter spawn of 17 species of fish which release eggs on the mid to outer continental shelf of North and South Carolina (Nelson et al., 1977). Larvae from this spawn take advantage of the abundance of zooplankton, mainly copepods, until they are transported to the numerous bays and estuaries in North Carolina. Young fish mature in these estuaries before returning to the ocean the following fall. Three species, Atlantic menhaden (*Brevoortia tyrannus*), spot (*Leiostomus xanthurus*) and Atlantic croaker (*Micropogonias undulatus*) make up about 75% of these fish in number (Warlen and Burke, 1990). Spawning for these three fish occurs between November and March, peaking in December through January with 76% of menhaden spawning in these two months (Warlen and Burke, 1990; Warlen, 1993). Preferred spawning temperatures for *B. tyrannus* and *L. xanthurus* range from 18 to 24°C, a zone of water approximately 20 km wide that occurs over the middle and outer continental shelf at this time of year (Checkley et al., 1988; Govoni, 1993; Warlen, 1993). This area, mid to outer continental shelf, and temperature range also coincide with the presence of bluefin tuna during the winters off of Cape Hatteras. The co-occurrence both spatially and temporally of bluefin tuna and these large spawning schools suggests that bluefin tuna come to the waters off North Carolina in the winter months to feed on this abundant prey source.

REFERENCES

Block, B., Booth, D., and Carey, F.G. (1992a) Direct measurement of swimming speeds and depth of blue marlin. *J. Exp. Biol.* **166**, 267-284.

Block, B.A., Booth, D.T., and Carey, F.G. (1992b) Depth and temperature of the blue marlin (*Makaira nigricans*) observed by acoustic telemetry. *Mar. Biol.* **114**(2), 175-183.

Block, B.A., Keen, J.E., Castillo, B., Dewar, H., Freund, E.V., Marcinek, D.J., Brill, R.W., and Farwell, C. (1997) Environmental preferences of yellowfin tuna (*Thunnus albacares*) at the northern extent of its range. *Mar. Biol.* **130**(1), 119-132.

Block, B.A., Dewar, H., Farwell, C., and Prince, E.D. (1998a) A new satellite technology for tracking the movements of Atlantic bluefin tuna. *Proc. Nat. Acad. Sci.* **95**(16), 9384-9389.

Block, B.A., Dewar, H., Williams, T., Prince, E., Farwell, C., and Fudge, D. (1998b) Archival tagging of Atlantic bluefin tuna (*Thunnus thynnus thynnus*). *Mar. Tech. Soc. J.* **32**, 37–46.

Brill, R.W., Holts, D.B., Chang, R.K.C., Sullivan, S., Dewar, H., and Carey, F.G. (1993) Vertical and horizontal movements of striped marlin (*Tetrapturus audax*) near the Hawaiian Islands, determined by ultrasonic telemetry, with simultaneous measurement of oceanic currents. *Mar. Biol.* **117**(4), 567-574.

Brill, R.W., Block, B.A., Boggs, C.H., Bigelow, K.A., Freund, E.V., and Marcinek, D.J. (1999) Horizontal movements, depth distribution and the physical environment of adult yellowfin tuna (*Thunnus albacares*) near the Hawaiian Islands recorded using ultrasonic telemetry, with implications for their physiological ecology. *Mar. Biol.* **133**, 395-408.

Carey, F.G., and Teal, J.M. (1969) Regulation of body temperature by the bluefin tuna. *Comp. Biochem. Physiol.* **28A**, 205-213.

Carey, F.G., and Lawson, K.D. (1973) Temperature regulation in free swimming bluefin tuna. *Comp. Biochem. Physiol.* **44A**, 375-392.

Carey, F.G., and Olson, R.J. (1982) Sonic tracking experiments with tunas. International Commission for the Conservation of Atlantic Tunas, Madrid, [Collect. Vol. Sci. Pap.], XVIII(2), pp. 458-66.

Carey, F.G., and Robison, B.H. (1981) Daily patterns in the activities of the swordfish *Xiphias gladius*, observed by acoustic telemetry. *Fish. B-NOAA* **79**, 227-92.

Carey, F.G., Kanwisher, J.W., and Stevens, E.D. (1984) Bluefin tuna *Thunnus thynnus* warm their viscera during digestion. *J. Exp. Biol.* **109**, 1-20.

Cayre, P., and Marsac, F. (1993) Modelling the yellowfin tuna (*Thunnus albacares*) vertical distribution using sonic tagging results and local environmental parameters. *Aquat. Living Resour.* **6**, 1-14.

Checkley, D.M., Raman, S., Maillet, G.L., and Mason, K.M. (1988) Winter storm effects on the spawning and larval drift of a pelagic fish. *Nature* **335**, 346-348.

Clay, D. (1990) Atlantic bluefin tuna (*Thunnus thynnus thynnus*): a review. World Bluefin Meeting, LaJolla, California, USA.

Dagorn, L., Bach, P., and Josse, E. (2000) Movement patterns of large bigeye tuna (Thunnus obesus) in the open ocean, determined using ultrasonic telemetry. *Mar. Biol.* **136**(2), 361-371.

Govoni, J. (1993) Flux of larval fishes across frontal boundaries examples from the Mississippi River plume front and the western Gulf Stream front in winter. *Bull. Mar. Sci.* **53**, 538-566.

Gunn, J., Polacheck, T., Davis, T., Sherlock, M., and Nicolau, M. (1994) The development and use of archival tags to study the movement, behavior and physiology of the southern bluefin tuna (*Thunnus maccoyii*). Proceedings of the 45th Annual Tuna Conference, 85 pp.

Gunn, J., and Block, B.A. (in press) Advances in acoustic, archival and pop-up satellite tagging. *In* Block, B.A., and Stevens, E.D. (eds.) *Tuna Ecology, Physiology and Evolution*, Academic Press, San Diego.

Holland, K.N., Brill, R.W., and Chang, R.K.C. (1990a) Horizontal and vertical movements of yellowfin and bigeye tuna associated with fish aggregating devices. *Fish. B-NOAA* **88**, 493-508.

Holland, K.N., Brill, R.W., and Chang R.K.C. (1990b) Horizontal and vertical movements of Pacific blue marlin captured and released using sportfishing gear. *Fish. B-NOAA* **88**, 397-402.

Itoh, T., Tsuji, S., and Nitta, A. (1997) Experiments of archival tags on the pacific bluefin tuna juveniles. Proceedings of the 47th Annual Tuna Conference, 37 pp.

Jones, C.D., Oritz, M., Judge, M.T., and Prince, E.D. (1997) Overview of the cooperative tagging center release and recapture activities for highly migratory species: 1954- present, International Commission for the Conservation of Atlantic Tunas, Madrid, [Collect. Vol. Sci. Pap. ICCAT], SCRS/97/70, 23 pp.

Josse, E., Bach, P., and Dagorn, L. (1998) Simultaneous observations of tuna movements and their prey by sonic tracking and acoustic surveys. *Hydrobiologia* **371-372**, 61-69.

Linthicum, S.D., and Carey, F.G. (1972) Regulation of brain and eye temperatures by the bluefin tuna. *Comp. Biochem. Physiol.* **43A**, 425-433.

Lutcavage, M.E., Brill, R.W., Skomal, G.B., Chase, B.C., and Howey, P.W. (1999) Results of pop-up satellite tagging on spawning size class fish in the Gulf of Maine. Do North Atlantic bluefin spawn in the mid-Atlantic. *Can. J. Fish. Aquat. Sci.* **56**, 173-177.

Marcinek, D.J., Blackwell, S.B., Dewar, H., Freund, E.V., Farwell, C., Dau, D., Seitz, A.C., and Block, B.A. (2001) Depth and muscle temperature of Pacific bluefin tuna examined with acoustic and pop-up satellite archival tags. *Mar. Biol.* (in press).

Marsac, F., and Cayre, P. (1998) Telemetry applied to behaviour analysis of yellowfin tuna (*Thunnus albacares*, Bonnaterre, 1788) movements in a network of fish aggregating devices. *Hydrobiologia* 371-372, 155-171.

Mather, F.J., Mason, J.M., and Jones, A.C. (1995) Life history and fisheries of Atlantic bluefin tuna. NOAA Technical Memorandum, NMFS–SFSC. 370. 165 pp.

Maul, G.A., Williams, F., Roffer, M., and Sousa, F.M. (1984) Remotely sensed oceanographic patterns and variability of bluefin tuna (*Thunnus thynnus thynnus*) catch in the Gulf of Mexico. *Oceanol. Acta.* **7**, 469-480.

Metcalfe, J.D., and Arnold, G.P. (1997) Tracking fish with electronic tags. *Nature* **387**, 665–666.

National Research Council (1994) *An Assessment of Atlantic Bluefin Tuna*. National Academy Press, Washington, DC. 148 pp.

Nelson, W.R., Ingham, M.C., and Schaaf, W.E. (1977) Larval transport and year class strength of Atlantic menhaden (*Brevoortia tyrannus*). *Fish B-NOAA* **75**, 23-41.

Reddy, R., Lyne, V., Gray, R., Easton, A., and Clarke, S. (1995) An application of satellite-derived sea surface temperatures to southern bluefin tuna and albacore off Tasmania, Australia. *Sci. Mar.* **59**, 445-454.

Sibert, J.R., Hampton, J., Fournier, D.A.and Bills, P.J. (1999) An advection-diffusion-reaction model for the estimation of fish movement parameters from tagging data, with application to skipjack tuna *Katsuwonus pelamis*. *Can. J. Fish. Aquat. Sci.* **56**, 925-938.

Skomal, G.B., and Chase, B.C. (1997) Preliminary results on the physiological effects of catch and release on bluefin tuna (*Thunnus thynnus*) caught off Cape Hatteras, North Carolina. International Commission for the Conservation of Atlantic Tunas [Collect. Vol. Sci. Pap. ICCAT], vol. 46, no. 2, pp. 314-320.

Tranter, D.J., Leech, G.S., and Airey, D. (1983) Edge enrichment in an ocean eddy. *Aust. J. Mar. Fresh. Res.* **34**, 665-680.

Warlen, S.M., and Burke, J.S. (1990) Immigration of larvae of fall-winter spawning marine fishes into a North Carolina estuary. *Estuaries* **13**, 453-461.

Warlen, S.M. (1993) Spawning time, growth and recruitment of larval spot *Leiostomus xanthurus* into a North Carolina estuary. *Fish B-NOAA* **91**, 8-22.

Weinstein, M.P. (1981) Plankton productivity and the distribution of fishes on the southeastern U.S. continental shelf. *Science* **214**, 351.

The Relationship Between Food Intake and Visceral Warming in Southern Bluefin Tuna (*Thunnus maccoyii*)

Can we predict from archival tag data how much a tuna has eaten?

John Gunn[1], Jason Hartog[1] and Kirsten Rough[2]
[1]*CSIRO Marine Research, P.O. Box 1538, Hobart, Tasmania, Australia, 7000, gunn@marine.csiro.au*
[2]*Tuna Boat Owner's Association, P.O. Box 1146, Port Lincoln, South Australia, Australia, 5606*

Key words: visceral warming, archival tags, southern bluefin tuna

Abstract: Using archival tags implanted in the body cavity of juvenile southern bluefin tuna (SBT) being held in fattening cages, we examine the effect of varying ration (of clupeoid bait fish) on the patterns and extent of warming in the viscera. Our experiments were run in summer and winter to examine the effect of ambient water temperature on the relationship between ration and visceral warming. In both seasons, we found strong relationships between the amount of food eaten and the extent of visceral warming. For clupeoid prey and the size of fish we worked with, the relationships make it possible to predict ration and intake from various measures of visceral warming calculated from archival tag data. The time taken for the viscera to reach maximum heat increment seems the easiest and most robust measure for prediction. Our three year old SBT maintained a constant differential of 2-3°C between their basal visceral temperature and ambient water temperature throughout a 6-7°C seasonal change in water temperature. For a given ration, the extent of warming of the viscera during digestion relative to the basal visceral temperature was significantly higher in winter than in summer, indicating either improved heat conservation and/or higher aerobic metabolic rates in winter. Despite the greater warming, viscera were still significantly cooler during digestion in winter than in summer. Even though visceral temperatures were cooler in winter than in summer, rates of digestion were very similar in the two seasons. Lipid content of the food varied between seasons and may account for a slowing of digestion in summer. However, it is unlikely that this accounts for the apparent lack of a thermal effect on digestion rates. We hypothesise that this is facilitated through positive thermal compensation during winter, through the activation of cold adapted isozymes in the digestive system.

J.R. Sibert and J.L. Nielsen (eds.), Electronic Tagging and Tracking in Marine Fisheries, 109–130.
© 2001 *Kluwer Academic Publishers. Printed in the Netherlands.*

1. INTRODUCTION

Tunas are known to digest food much faster than other piscivorous fishes (Magnuson, 1978; Olsen and Boggs, 1986; Brill 1996), a characteristic that allows them to feed frequently when food resources are abundant. The primary mechanism for fast digestion appears to be elevated visceral temperatures, facilitated by the presence of a visceral rete (Collette, 1978). In Atlantic bluefin tuna (*Thunnus thynnus*), elevated temperatures promote increased activity of the digestive enzymes trypsin and chymotrypsin in the caecum (Stevens and McLeese, 1984).

Carey *et al.* (1984) found that temperatures in the viscera (in particular the stomach and caecum) of large Atlantic bluefin tuna were significantly elevated during and after the digestion of food. Their experiments showed the basal temperature of the viscera was 3-6°C above ambient water temperature, and that following a meal the visceral temperature rose to 10-15°C above ambient. The visceral warming is thought to be due to a combination of factors. First, and probably most important, is the heat produced by specific dynamic action (i.e., the result of metabolic heat production during digestion). Second is the heat released during hydrolysis of food, and a third, minor source of heat derives from the mechanical energy used to process a meal in the stomach. Using ultrasonic tags placed in the stomach, Carey *et al.* (1984) described the pattern and duration of heat production in the viscera, and found that the general pattern of visceral heating was similar regardless of the size of meal ingested. In cases where tuna were fed a large meal and not fed the next day, it took 36-48 h for the viscera to return to basal levels.

In 1994, we began a series of experiments on wild juvenile southern bluefin tuna (SBT) (*Thunnus maccoyii)* in the Great Australian Bight using archival tags (Gunn et al., 1994). We surgically implanted the tags, which incorporate a temperature sensor, into the body cavity immediately under the caecum and stomach. The data from these tags showed regular patterns of warming and cooling of the viscera, very similar to those described by Carey *et al.* (1984). To confirm that the temperature variation was due to feeding, we also implanted archival tags into fish that were being kept for a few months in fattening cages (Clarke *et al.*, 1997). In these cages, fish were fed at regular intervals, and by matching the recorded feeding times with patterns of visceral warming we were able to confirm that temperature elevations provide an accurate record of when feeding events occur (Gunn *et al.*, 1994).

The archival tags we had placed into wild fish were programmed to collect data on visceral temperature every four minutes for up to five years, so we speculated that these data may provide long-term records of feeding activity in the fish we were studying. However, because the visceral temperature also changes in response to changes in ambient water temperature (Carey *et al.* 1984), we needed to be able to separate the effects of variations in visceral temperature due to changes in ambient water temperature from those due to feeding. To do this, we developed a thermodynamics model for SBT, which provided an objective method for separating these effects (S. Wotherspoon and J. Gunn, unpubl. data) .

Once we could identify and remove the influence of environmental temperatures from those related to feeding, we conducted an experiment to determine whether a relationship existed between the amount of food a fish ate and the extent and nature of visceral heating. This paper reports on the results of this experiment, which was run once in summer and again in winter to determine whether the relationship was influenced by ambient temperature.

2. MATERIALS AND METHODS

2.1 Tagging Method and Experimental Design

To measure variation in visceral temperature, we surgically implanted Vemco TX temperature loggers (archival tags) into the body cavity of the tuna. The archival tags were placed so that they lay along the belly wall ventral to the stomach and the ceaca. As far as possible, they were implanted in the same position in every fish. The surgery involved making an 8-10 cm long, 2 cm deep incision into the belly wall, approximately 12 cm anterior to the vent, with a sharp paring knife. The incision did not cut through the peritoneum, which was perforated by the surgeon's finger. The tag, which had been sterilised in alcohol and coated in Betadine before implantation, was inserted carefully through the hole in the peritoneum made by the finger, with the thermistor pointing towards the head. A 3 ml dose of Amoxil, a broad spectrum antibiotic, was injected into the opening left by the tag. The wound was then sealed using one or two stitches with a Vicryl suture. The experiment was conducted once in the Austral summer and again in winter.

Tags were programmed to record visceral temperature every minute over a period of 1 month in summer. The fish for the winter experiment had been used in a separate feeding experiment prior to our work and were programmed to record visceral temperature every 5 minutes. We used fish that had been tagged in summer to avoid stress-related problems commonly associated with surgery during the colder months.

In summer, we used 12 cages (20 m in diameter and 15 m deep), each stocked with 2 fish. The fish we used were three years old and had been caught by purse seine in the Great Australian Bight three weeks before the experiment began. At the beginning of the experiment, each fish was fitted with an archival tag, weighed and measured (Table 1). The fish used were similar in size (91-95 cm fork length and 15.2-18.5 kg) and condition (weight/length^3 : 19.5-22.2). One fish in each cage was also tagged with an external spaghetti tag, to allow individuals to be recognised during the feeding operations.

Table 1. Size and condition of fish at the start of the summer and winter ration experiments.

	Summer	Winter
Fork Length (cm)		
Range	91 – 95	100 – 118
(Mean)	(93)	(105)
Body Weight (kg)		
Range	15.2 – 18.5	21.4 – 36.9
(Mean)	(17.0)	(28.7)
Condition (W/L^3)		
Range	19.5 – 22.2	21.2 – 24.9
(Mean)	(20.9)	(23.3)

Fish were then allowed to recover from the stress of handling for 7 days before we began feeding them varied rations for 18 days. Rations of 1, 3, 5, 7, 9, and 11% of body weight were offered, with the order randomised for each cage.

We used 6 cages in winter, each initially stocked with 1-2 fish. As they already carried archival tags implanted at the beginning of summer, fish were weighed and measured prior to commencement of the experiment (Table 1). The winter fish varied considerably in size (100-118 cm fork length, 21.4-36.9 kg). All the fish had been similar in size when captured in the wild, and this variation resulted from differences in growth rates during their six months in captivity. Fish were allowed to recover from the stress of handling for 7 days. They were then fed rations of 1, 2, 3, 4, 5, and 7% body weight over an 18 day period, with the order of the rations for each cage randomised. The maximum ration used in winter was smaller than we used in summer because fish in the cages generally feed less actively when water temperatures are lower than 15°C (CSIRO unpubl. data).

At each feeding event, the number of pilchards eaten by each individual fish was recorded. Pilchards had been air-thawed and weighed, so it was possible to calculate the exact weight of each meal and use this to calculate the ration (meal weight/body weight). Thus, if an individual fish chose not to eat the ration prescribed in the experimental design, we were still able to relate the visceral heating to a known food intake and ration.

Following completion of the each ration experiment, fish were recaptured, weighed, measured and the archival tags removed. Data collected from the tags were then downloaded into the CSIRO ARCHTAG software package for analysis.

2.2 Data Analysis

Figure 1 shows the form of the data collected by the archival tags. CSIRO ARCHTAG analysis software allowed us to estimate a fish's basal visceral temperature over a specified period (in our case this was generally 48 hours, including the 12 hours before and after the period of visceral warming related to a feeding event). In both the summer and winter experiments, ambient water and basal visceral temperatures remained relatively stable throughout the 18-day periods over which rations were varied. Thus, basal temperature was relatively easy to calculate as a linear function.

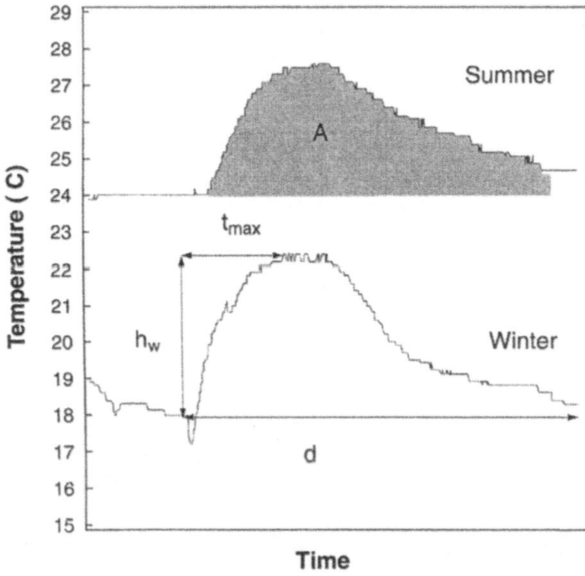

Figure 1. Visceral warming in summer and winter. A is the area of the region bounded by the temperature curve and the fishes' basal temperature; h_w is the maximum heat increment; d is the measure of duration of the feed, t_{max} is the time taken until the maximum heat increment is reached.

After determining the basal temperature, the ARCHTAG analyst identified the times at which the warming began and ended (by clicking a mouse on the data display). These points are defined as the time at which temperature first increases above, or decreases below, basal at the start of the feeding event and the time at which visceral temperature returns to basal when the heating event due to feeding is completed. The occasions where temperatures decreased at the start of the feed were rare (<5%) and occurred as a result of the food, which had been thawed to air temperature prior to feeding, remaining below ambient water temperature when feeding occurred[1]. Given the start and end points, the software computes four measures of visceral warming for each feeding event:

1. Feed measure is defined by the trapezoidal integration of the area underneath the temperature curve and above basal body temperature ("A" in Figure 1).

The area under a curve is usually obtained by calculating the integral of the mathematical function that describes the curve. Our temperature curves are difficult to model by an explicit function, so we use a simple numerical method—trapezoidal integration—to determine the integral.

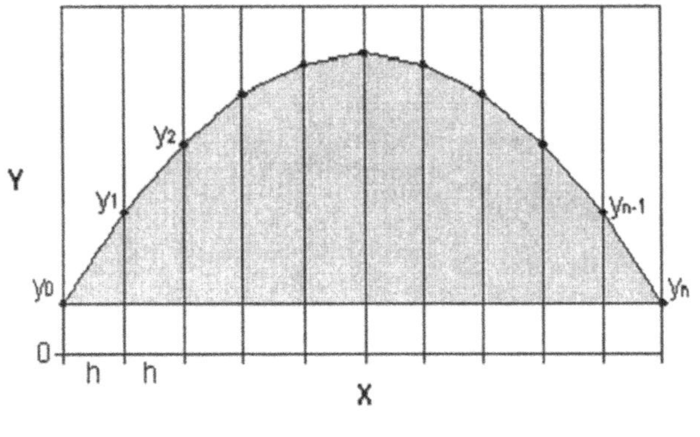

Figure 2. Diagrammatic representation of the visceral warming curve created by joining adjacent points by a straight line. The area calculated by trapezoidal integration is shaded.

The trapezoidal method involves dividing the time between the start and the end of the feeding event into equal intervals. For convenience, we use the sampling rate of the archival tags—1 minute in summer, 5 minutes in winter. We then sum the areas of the each trapezoid created by drawing vertical lines up every interval along the time axis, and drawing lines which join each adjacent temperature point (Figure 2). There is a small amount of error involved in this estimation, caused by the approximation of a curve by a straight line between the two temperature points. In our case, the error is negligible because the visceral temperature increase and decrease are both

[1] Similar cooling following a meal is observed in archival tag records for wild fish (CSIRO, unpublished data), when food consumed in cold water briefly cools the viscera.

slow, relative to the sampling interval, and smooth. The mathematical formula used is

$$Area = h/2(y_0 + 2(\sum_{j=1}^{n-1} y_j) + y_n) - y_0(x_n - x_0)$$

where y_j= visceral temperatures (°C) at each sampled time step, y_b = the temperature at the beginning of the feeding event (i.e., basal temperature for this experiment), h = sampling interval. The area estimate produces a measure of total visceral warming, which we call a "feed measure."

2. Feed duration is defined as the time interval between the start and finish of visceral warming due to feeding ("d" in Figure 1).

3. Maximum heat increment is defined as the maximum increase in temperature above basal temperature ("h" in Figure 1).

4. Time max is is defined as time taken from the onset of visceral warming to the point at which "h" is reached ("t" in Figure 1).

Using these four measures, we analysed the relationships between visceral warming and feed intake, and visceral warming and ration (feed weight (g)/ body weight (g)) using standard ANCOVA, ANOVA and univariate statistical techniques. The SYSTAT statistical software package was used for all analyses.

On 27 occasions (out of a total of 188) in the summer experiment it was obvious from the archival tag data that the experimental fish had eaten at times when they were not fed as part of the ration experiment. These feeding events appeared as minor secondary peaks in the visceral warming pattern. During the summer, small pelagic fishes often recruit into the tuna cages and we assume from our data that occasionally they become prey of the tuna. We removed from our analyses all cases where these secondary feeding events occurred. Secondary feeding events were not observed in the winter experiment.

3. RESULTS

3.1 Environmental Conditions and Food Energy Levels

The water temperatures within the cages were on average 6-7°C warmer during the summer experiment than during the winter experiment (Figure 3). The basal visceral temperatures were also 6-7°C higher in summer than in the winter. The differential between ambient water and basal visceral temperatures remained more or less constant at 2°C, regardless of ambient water temperature.

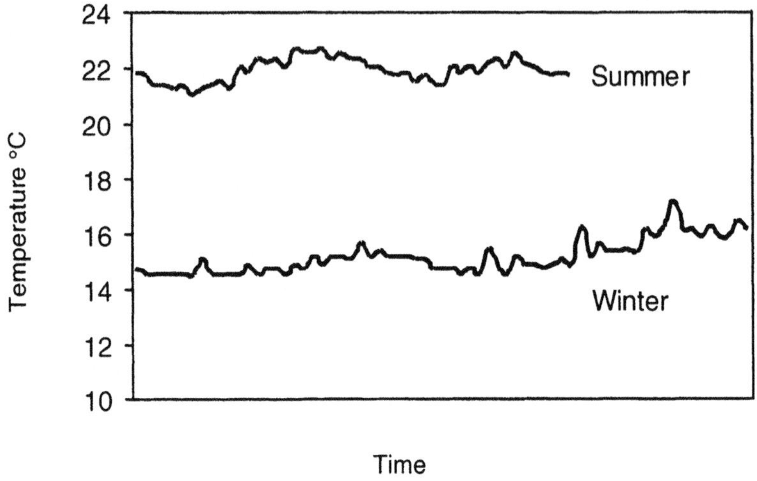

Figure 3. Variation in water temperatures in the cages during the summer and winter experiments.

There were differences in the protein and lipid content (based on wet weight proximate analyses) and the energy levels of the food used in the two experiments. During summer the pilchards were 17.9% crude protein, 7% total lipid and negligible carbohydrate, while in winter they were 19.7% crude protein, 3.2% total lipid and again negligible carbohydrate. The impact of these seasonal differences in fat and protein levels on the amount of heat produced during hydrolysis was calculated using the metabolizable energy equivalents of dietary lipid, protein, and carbohydrate reported by Brett and Groves (1979). When adjusted for digestibility and non-faecal excretory losses, these were 8.0, 4.23, and 1.6 kcal/g for lipid, protein and carbohydrate, respectively. Using a calorific density of 1.0 kcal/g for the wet weight proportions of each dietary constituent (Cummins and Wuycheck 1971), 1577 kcal (1.31 kcal/g) would have been available following a 1200 g meal in summer. In winter, 1306 kcal (1.09 kcal/g) would have resulted from the same size meal. Thus, the difference in the calories released and available for metabolism was approximately 20% higher in summer than in winter.

3.2 Relationship Between Feed Intake and Ration

Since the size (length and weight) of individuals varied significantly between the summer and winter experiments (Table 1), and within the winter experiment there were two size classes of fish, the relationships between ration and food intake differed significantly between the experiments (Figure 4).

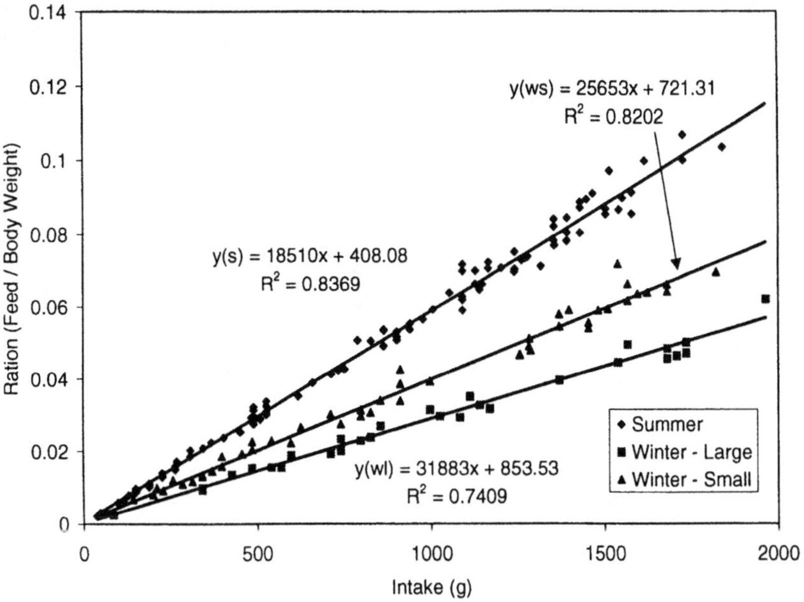

Figure 4. The relationship between food intake and ration for the three size classes of fish used in the ration experiments (15.5-18 kg fish in summer, 21-27 kg fish in winter, and 31.7-36.9 kg fish in winter).

3.3 Relationship Between Visceral Warming Parameters: Intake and Ration

In both summer and winter, the relationships between ration and all of the visceral warming measures were linear within individuals (Figures 5a and b). The same was true for the relationships between food intake and warming measures. The relationships among all of the variables examined are shown in Figure 6. With the exception of the maximum heat increment ("height" in Figure 6), there were strong linear relationships between all of the variables in both summer and winter.

3.3.1 Feed measure : food intake

The relationship between feed measure and food intake differed between seasons (Fig 7); during summer the feed measure at a given level of food intake was much lower than it was in winter. The slopes of the linear relationships for individuals did not differ significantly (F=1.427, df=16, p=0.111). When the data were pooled for all fish, a linear model provided the best fit (r^2 = 0.90). During winter, tests for homogeneity of slopes revealed two groups, one of seven fish and another of three (F = 1,179, df = 6,

$p = 0.331$, and $F = 0.836$, df = 2, $p = 0.447$, respectively). The first group consisted mainly of fish less than 27 kg, while the other consisted of large fish, >31 kg. Despite the difference in slopes among individuals, when data for all fish were pooled, both linear ($r^2 = 0.83$) and second order polynomial ($r^2 = 0.86$) models provided good fits.

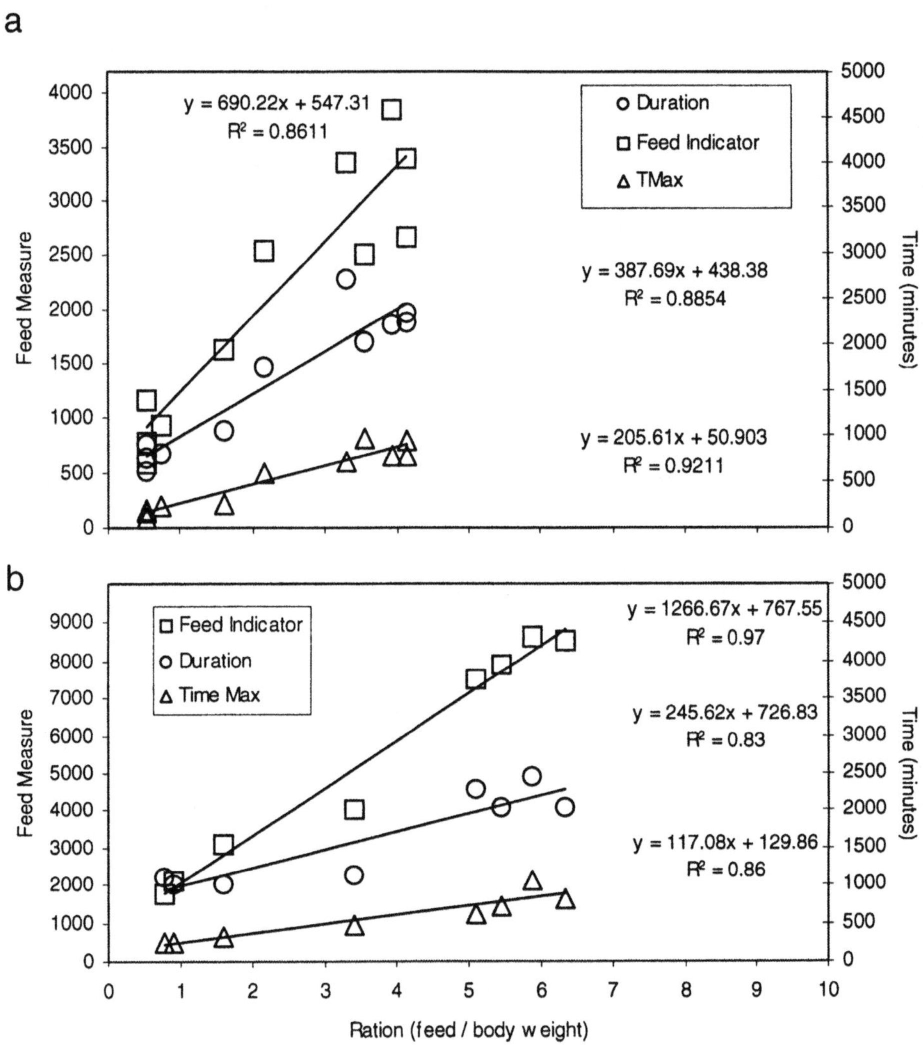

Figure 5. Relationship between ration and 3 of the 4 measures of visceral warming in summer (top) and winter (bottom). Feed measure is scaled on the left side. Duration and TMax have their scale on the right side.

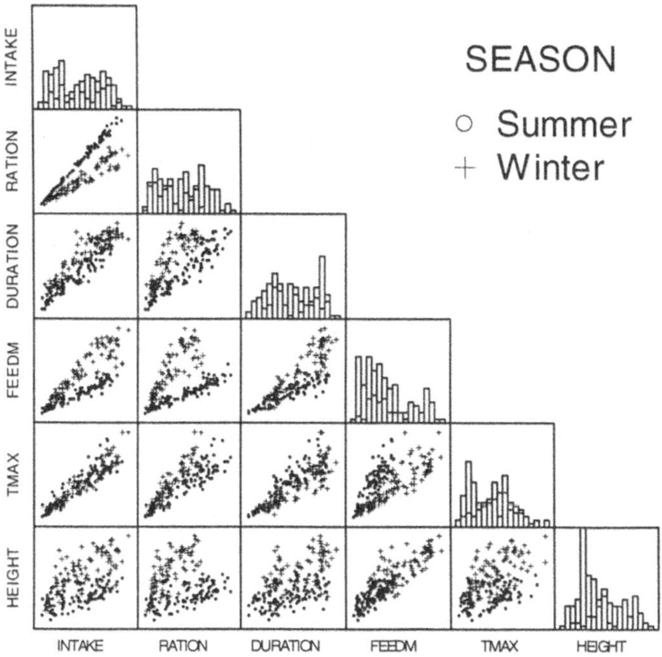

Figure 6. Pairwise comparisons of the feed measures and food intake using data pooled across all fish in summer (o) and winter (+)

3.3.2 Feed measure : ration

Throughout the range of rations used, feed measure was significantly higher in summer than in winter (Figure 7). In summer, the relationship was linear for all individuals (Figure 4); the slopes of individual relationships did not differ significantly ($F = 0.991$, df = 16, $p = 0.475$); and a linear model provided the best fit to the pooled data ($r^2 = 0.84$) (Figure 7). In winter, the relationship was linear for the majority of individuals (Figures 4 and 8). However, there was significant heterogeneity in slopes among individuals. Least squares mean plots revealed two groups, one of seven fish and another of three. Six of seven members in the first group weighed less than 27 kg, whereas all of the fish in the second group were large and weighed more than 31 kg. Within each group, the slopes of linear models did not differ significantly ($F = 1.781$, df = 6, $p = 0.123$, and $F = 2.308$, df = 2, $p = 0.132$, respectively).

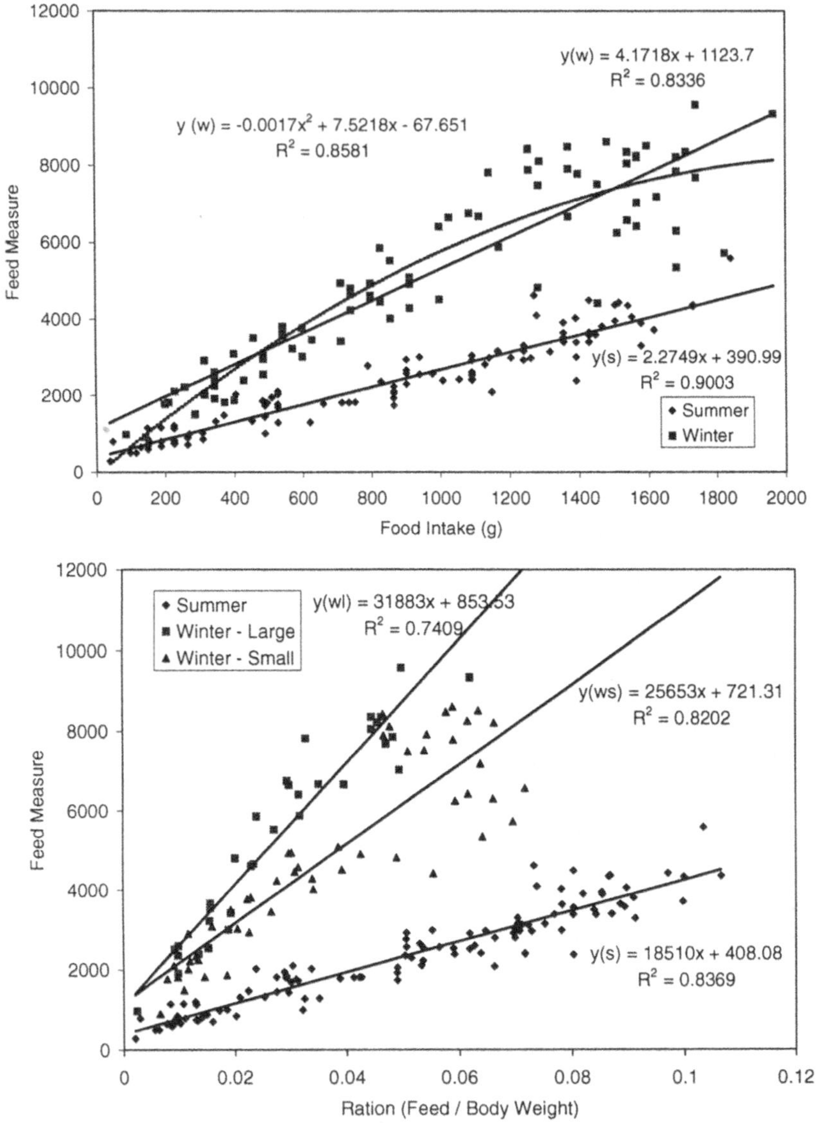

Figure 7. Relationships between feed measure and food intake (top) and feed measure and ration (bottom) for summer and winter. For winter ration, where the response varies with size, large (>31 kg) and small (<27 kg) fish are shown separately

3.3.3 Feed duration : food intake

Feed duration and food intake were closely related in all individuals in both the summer and winter experiments. The relationships were linear over most of the range of rations used and, within experiments, the slopes of linear models did not differ significantly among fish ($F = 1.124$, df = 16, $p = 0.329$ for summer; $F = 0.972$, df = 9, $p = 0.470$ for winter). The relationships for pooled data (Figure 9, top) were linear in both summer ($r^2 = 0.85$) and winter ($r^2 = 0.81$) and the slopes of the seasonal models did not differ significantly ($F = 2.95$, df = 1, $p = 0.09$), although the p value is marginal. The difference between the two relationships is primarily in the intercept, which is much higher in the winter than in summer.

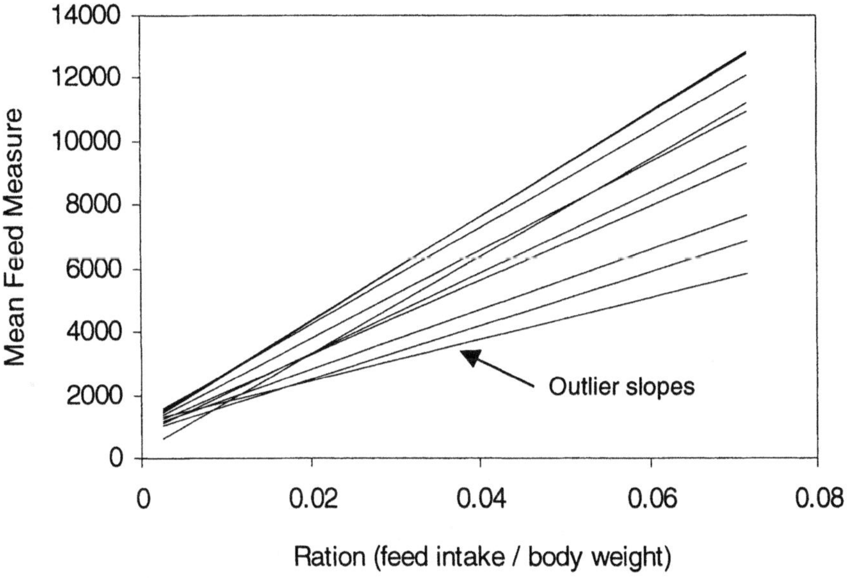

Figure 8. Individual relationships between feed measure and ration in the winter experiments showing the variation among slopes and highlighting the three outliers.

3.3.4 Feed duration : ration

The relationship between the duration of visceral warming and ration also varied between summer and winter with the slope of relationship being significantly steeper in winter than in the summer. In summer, feed duration and ration were linearly related in all individuals (Figure 5); the slopes of the relationships differed significantly among individuals ($F = 2.430$, df = 16, $p = 0.005$); but despite this, a linear model provided a good fit to data pooled across individuals ($r^2 = 0.84$) (Figure 9, bottom). In winter, the relationship was linear for individuals (Figure 5) and the slopes of the relationships did not differ significantly among individuals ($F = 1.429$, df = 9, $p = 0.195$). However, as was the case in the feed measure:ration relationship, there was a clear separation among the least squares means for the duration : ration relationships between large fish >31 kg and smaller fish weighing less than 27 kg. The duration of the feeding event was longer in larger fish (Figure 9, bottom).

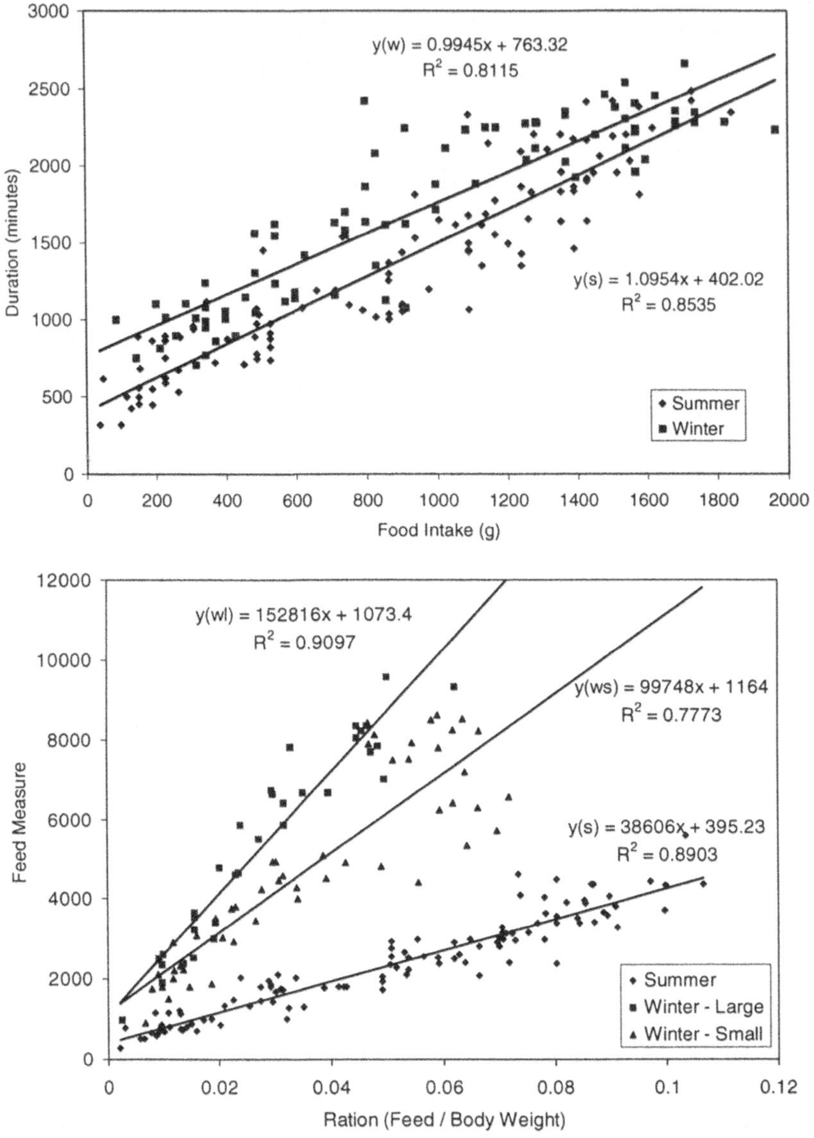

Figure 9. (top) Summer and winter relationships between duration of the feeding event and food intake. Both linear and 2nd order polynomial regressions are shown. (bottom) Summer and winter relationships between duration and ration. For the winter ration data, in which the response varies with size, large (>31 kg) and small (<27 kg) fish are shown separately.

3.3.5 Maximum heat increment

The maximum heat increment for all rations was significantly higher in winter (mean increment = 5.28°C, SE = 0.15) than in the summer (mean = 3.14, SE = 0.08). The relationship between maximum heat increment and ration was linear but weak: $r^2 = 0.362$ and 0.514 in summer and winter, respectively.

3.3.6 Time from feeding to maximum heat increment : food intake

In both summer and winter there was a strong relationship within and among individuals, between the time it took the viscera to reach their maximum heat increment and food intake (Figure 10, top). The slopes of the relationships did not differ significantly among individuals in either experiment (summer $F = 0.945$, df = 22, $p = 0.538$; winter $F = 1.128$, df = 9, $p = 0.356$), or between experiments ($F = 0.087$, df = 1, $p = 0.769$). The intercept of the winter relationship was slightly higher than in the summer, but this was not statistically significant (*t*-test - $p = 0.175$).

3.3.7 Time from feeding to maximum heat increment : ration

In both summer and winter, there was a strong linear relationship within individuals between the time it took the viscera to reach the maximum heat increment and ration (Figure 5). In summer, the slopes of the relationships did not differ significantly among individuals ($F = 0.948$, df = 16, $p = 0.522$), and there was a strong linear relationship when data for all fish were pooled ($r^2 = 0.82$) (Figure 10, bottom). In winter, there was a significant difference in the slopes for the nine fish ($F = 2.172$, df = 9, $p = 0.036$). Least mean squares plots again indicated two groups—one consisting of the fish weighing >31 kg and another of fish weighing <27 kg. The slopes for the relationship for these groups did not differ significantly ($F = 1.642$, df = 3, $p = 0.172$ and $F = 1.024$, df = 3, $p = 0.399$ for large and small fish, respectively).

Over the range of rations we used, the large fish took longer to reach their maximum visceral temperature than the small fish. Fish in the summer experiment took significantly less time to reach their maximum visceral temperature at a given ration than either size group in winter.

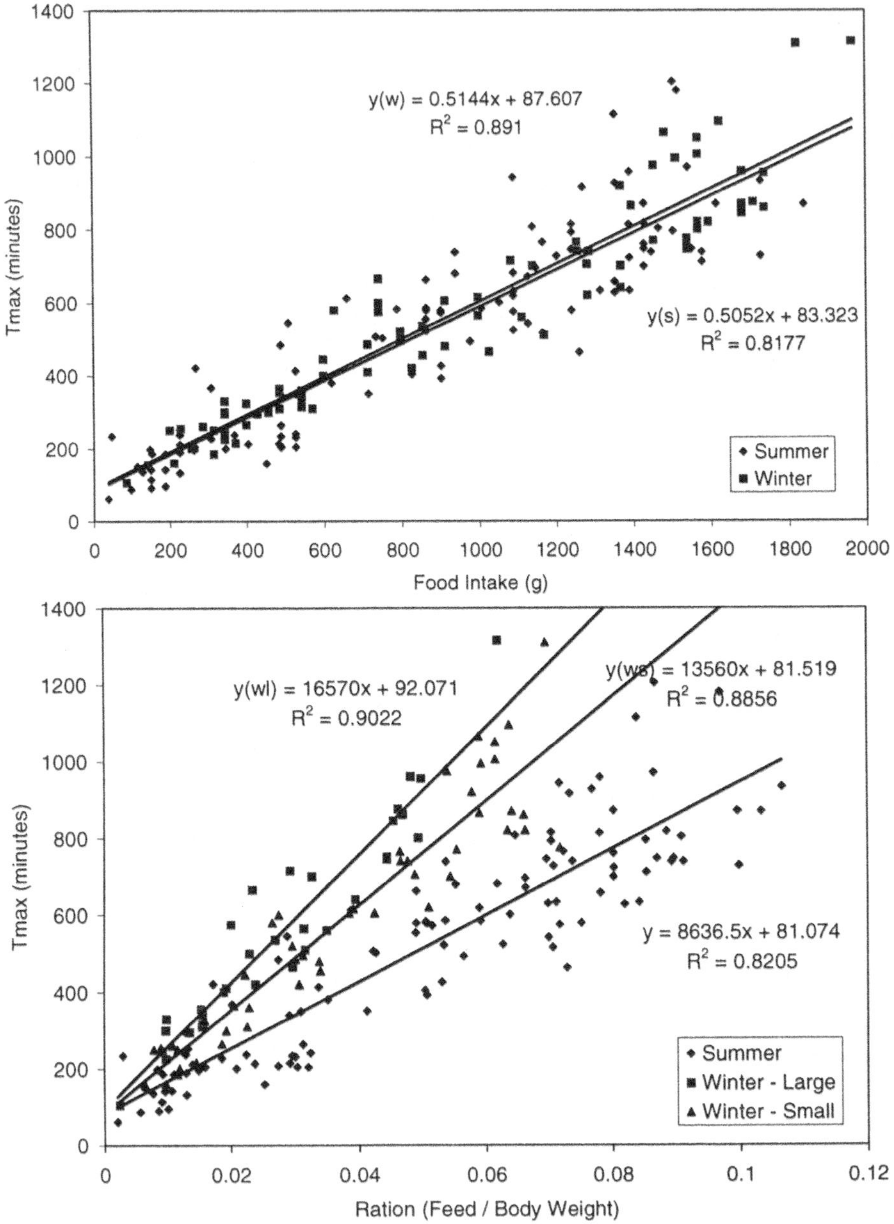

Figure 10. (top) Summer and winter relationships between time to maximum heat increment and food intake, and (bottom) summer and winter relationships between time to maximum heat increment and ration. For winter data (bottom), in which the response varies with size, large (>31 kg) and small (<27 kg) fish are shown separately.

4. DISCUSSION

Our experiments have shown that the viscera of SBT warm significantly when fish ingest rations ranging from 1-12% of body weight. The visceral warming follows a similar pattern for all rations and involves two distinct stages. The first, which starts immediately after ingestion, involves a rapid increase in visceral temperature over a period of up to 20 hours. The first stage ends when the heat increment reaches its maximum. The second stage then starts and involves a slow cooling of the viscera over a period of up to 40 hours.

The pattern of warming we observed in SBT is very similar to that described for giant bluefin tuna by Carey *et al.*, (1984). These authors concluded that the increase in visceral temperature resulted from an increase in heat production and an increase in the efficiency of heat conservation. The heat production results from heat released in the hydrolytic breakdown of food and from an increase in metabolic rate. The heat conservation derives from two factors: the activation of the visceral rete, which is well developed in the bluefin tunas (Fudge and Stevens, 1996, a translation of Eschrict & Muller, 1835), and the fact that the viscera of bluefin are well insulated by a very thick belly wall (in a 20-kg SBT the belly is 25-30 mm thick) on one side and by an air bladder on the other side.

It appears that the first stage of visceral warming corresponds to the period when there is active breakdown within the stomach. Carey *et al.* (1984) found contractions in the stomach reached their maxima at around the same time as the viscera reached their maximum temperature; they hypothesised that these contractions serve to empty the remaining fluid from the stomach into the intestine. The second stage follows the emptying of the stomach and includes a period when hydrolysis continues in the caecum, followed by the slow dissipation of heat away from the viscera after the completion of digestion.

We found significant relationships between the weight of clupeoid food eaten by our fish and all of the measurements we made to describe the warming events associated with digestion. The time it took for the viscera to reach maximum heat, which we assume following the work of Carey *et al.* (1984) to be the time to reach gastric evacuation (GET); the height of the heat increment above the basal visceral temperature; the duration of the heating event; and the area under the curve of the thermal response (which we called feed measure) all increased in a response to increased food intake and ration. In most cases a linear response was exhibited, and from these relationships it would be possible to estimate, with measurable error, the size of a meal from any of the feed measures we saw in the archival tag record. It seem likely that these relationships will vary with prey type, because the size of prey items and their composition have been shown to affect the GET and gastric evacuation rates of tunas (Olson and Boggs, 1986).

By running our experiments in summer and winter, we had hoped to be able to determine whether visceral warming patterns varied seasonally. Unfortunately, because we were working within constraints imposed by commercial fish farming operations, it was not possible to control for factors such as fish size and food composition between the two experiments. Thus, the fish we worked with in summer were smaller than those in the winter, lipid concentrations in the food varied significantly, and predation by seals in winter meant our sample sizes were small. To some extent these factors mitigate against making strong conclusions regarding seasonality at this stage. Nevertheless, we did find some striking differences and similarities in the response to varying food intake in summer and winter.

Ambient water temperatures were 6-7°C higher in summer than in winter. Despite this significant difference, basal visceral temperatures remained approximately 2°C above ambient regardless of ambient temperature. Thus, it appears that SBT in the size range we examined (15.5-39.6 kg) are capable of raising and maintaining their temperature above ambient but are either incapable of, of choose not to, elevate their visceral temperatures (and presumably also their muscle temperatures) by more than a differential of 2°C.

With ambient and basal visceral temperatures 6-7°C warmer, one might expect digestion to be significantly faster in summer than in winter. In a series of experiments using tissues harvested from giant *Thunnus thynnus*, Stevens and McLeese (1984) examined the effect of temperature on the activity of trypsin and chymotripsin from the stomach and estimated the resultant rate of digestion. They found that a warmer "tummy" means faster digestion of food and developed a model for the effect of temperature on digestion rates. If we use their model to estimate digestion rates in SBT at different temperatures, the visceral temperature differential between summer and winter of 6-7°C should result in digestion being 30% faster in summer than in winter.

We found that the relationships between meal size (grams of pilchards) and "Tmax" (i.e., GET) were almost identical in the summer and winter experiments—the slopes of the relationships and their intercepts did not differ significantly. Thus, in summer the GET for a 1000 g meal was 588 minutes, compared to 598 minutes in winter—a differential of less than 2%.

This is significantly different from what the Stevens and McLeese (1984) model would have predicted. So how and why did our SBT digest food as quickly in winter as they did in summer?

One possible explanation lies in the differences in lipid content of the food used in the summer and the winter. The pilchards used in summer were 110% higher in lipids than those used in winter (7.0%:3.2% wet weight), and the calories released and available for metabolism in summer were approximately 20% higher in summer. Lipid concentrations of greater than 15% dry weight have been shown to have an inhibitory effect on gastric motility in rainbow trout (Windell, 1978). Based on a water content estimate of 75%, the dry weight lipid composition of the pilchards we used in summer and winter would have been 28% and 12.8%, respectively. If the findings for trout can be extrapolated to SBT, one might expect the higher lipid concentrations of the food in summer to result in longer GETs, and contribute to the similarities in GET between winter and summer. Whether this difference is enough to account for the difference between our observations and the Stevens and McLeese model cannot be answered at this stage.

On the other hand, the higher calorific content of the summer food should also produce more energy and result in higher heat output during digestion in summer than in winter. If more heat was produced as a result of digestion in summer, one might expect higher rates of digestion in summer. However, we see in the heat increment data that the opposite is the case, with summer heat increments being significantly lower than those in winter.

The larger size of fish in the winter may also have an effect on digestion rates. Body size (and the correlated size of the stomach) has been positively correlated with gastric motility in fishes (Windell, 1978; Fange and Grove, 1979). Olson and Boggs (1986), however, found no significant effect of body size on gastric evacuation rates for yellowfin tuna between 9.6 and 95.4 kg. Therefore, it seems unlikely that size is important over the relatively small size range we used in our experiments. Body size could also have an effect through increased body wall thickness and resulting thermal insulation of the viscera. However, given the relatively small seasonal differences in mass and condition, this seems an unlikely explanation for all of the observed increased efficiency in winter digestion seen in our experiments.

If the effect of lipid concentration does not account for all of the difference between modelled and observed GETs, it suggests that compensatory acclimation mechanisms are providing increased efficiency in digestion at low ambient temperatures during winter (Prosser, 1973). We hypothesise that two mechanisms may be working in tandem.

First, our data indicate that the extent of visceral warming—indicated by feed measure and the maximum heat increment—is greater for a given food intake in winter than in summer. For a 1000-g meal in winter, the mean feed measure was twice as large as it was for the same size meal in summer. The maximum heat increment was also 6.3°C in winter compared to 3.7°C in summer. As noted above, seasonal differences in the thickness of the belly wall and air bladder (i.e., the thermal insulation) are unlikely to have resulted in increases in insulation sufficient to make this kind of difference. The calorific value of the food in summer was also higher than in winter. Thus, we hypothesise that the first mechanism for facilitating increased digestion efficiency in winter is a seasonal change in the heat conservation (through changes in the activity of the visceral rete) and/or the aerobic metabolism.

Despite the increased production and/or conservation of heat in winter, the maximum visceral temperatures in winter were 3-4°C lower than in summer. To achieve similar digestion rates at such different temperatures, SBT must also exhibit some positive thermal compensation through increased enzyme activity at lower temperatures (Prosser, 1973). Fudge *et al.* (1997) demonstrated the capacity for positive thermal compensation in Atlantic bluefin tuna. They found five of seven enzymes present in the visceral rete of this species showed higher activity in colder regions of the rete than in warmer regions. They suggested that thermal compensation was most likely achieved through differences in enzyme concentrations rather than the differential distribution of isozyme variants. The experiments on Atlantic bluefin by Fudge *et al.* (1997) and Stevens and McLeese (1984) were based on tissues removed from individuals rather than in situ measurements on animals subjected to a range of different temperatures. However, the selective synthesis of isozymes that function best in particular temperature ranges has been demonstrated in fish going through seasonal changes in temperature (Hochachka and Lewis, 1971, Prosser, 1973). The genetic links between Atlantic bluefin and SBT are very close. Thus, it seems likely that, if the similarities in GET between summer and winter in SBT are the result of enhanced efficiency of digestion at low temperatures, this is achieved through selective synthesis of cold adapted isozymes during the winter.

Like all of the bluefin tunas, SBT make large annual migrations. As juveniles, they spend the summer and autumn in warm coastal waters and the winters in colder oceanic waters of the Indian and Pacific Oceans. As adults they migrate between temperate latitude feeding grounds, where surface water temperatures are less than 10°C, to spawning grounds in the tropics, where surface temperatures exceed 30°C (Caton, 1991). The capacity for seasonal acclimation that we found in juveniles would be a critical adaptation, facilitating efficient digestion of food over the very broad range of thermal habitats experienced each year by migrating juveniles and adults. However, because the thermal habitat of adults is much broader than that of juveniles, it may be that the enzyme acclimation changes with ontogenetic stage.

In conclusion, we have found strong relationships between the amount of food eaten and visceral warming in juvenile SBT. On the basis of our results, we can predict with measurable uncertainty the weight of pilchards eaten by an SBT from various measures of visceral warming calculated from archival tag data. The time taken for the viscera to reach maximum heat increment seems the easiest and most robust measure for prediction. We hope to extend our experiments to include a broad range of prey types and to examine the influence of digestibility and composition (Olson and Boggs, 1986) on the relationships we described for clupeoids. It would also be useful to conduct the experiments on a range of different age groups of fish within a season, as it was difficult with the small sample sizes available during winter to quantitatively examine the effect of size on the parameters we measured.

Our data also indicate that SBT are able to maintain similar rates of digestion at different temperatures. We suggest that this is facilitated by increased metabolic rates and/or changes in the activity of the visceral rete during winter to facilitate increased warming of the viscera. These mechanisms alone would not account for the rate of digestion in winter being the same as that in summer. We hypothesise that there is also positive thermal compensation during winter in SBT through the activation of cold adapted isozymes in the digestive system. It would be interesting to examine the concentrations of enzymes in the viscera of SBT at different times of the seasonal temperature cycle, to determine whether different enzymes are synthesized in different seasons, or whether the concentrations of a common suite of enzymes change in response to temperature variation. Before the relationships we have discovered are used in a more general context for predicting food intake, it is essential that the effects of varying food type and lipid content are determined.

ACKNOWLEDGEMENTS

The project was supported by funding from the Fisheries Research and Development Corporation Grant 97/363 under its Southern Bluefin Tuna Aquaculture Subprogram. We thank Brett Glencross of the South Australian Research and Development Institute who coordinated the logistic support for the project and the farm crew who drove boats and provided logistical support. We are especially grateful to Megan Laurie of the Tuna Boat Owners Association who provided invaluable technical support for the project, often under difficult conditions; Toby Patterson (CSIRO) who provided invaluable support with preparation of the figures and analyses; Naomi Clear (CSIRO) who conducted the surgery for the second experiment; and Ann Cowling (Australian National University Statistical Consulting Unit) who provided statistical advice. The genesis of the project was a very productive collaboration with Simon Wotherspoon of the University of Tasmania. Tim Davis (CSIRO) and Bob Olson of the Inter-American Tropical Tuna Commission provided valuable comments in reviews of the manuscript.

REFERENCES

Brett, J.R., and Groves, D. (1979) Physiological energetics. *In* Hoar, W.S., Randall, D.J. and Brett, J.R. (eds.), *Fish Physiology*, Vol. VIII: *Bioenergetics and Growth*, Academic Press, New York, NY, pp. 279-352.

Brill, R.W. (1996) Selective advantages conferred by high performance physiology of tunas, billfishes, and dolphin fish. *Comp. Biochem. Physiol.* **113A**, 3-15.

Carey, F. G., Teal, J.M., and Stevens, E. D. (1984) Bluefin tuna warm their viscera during digestion. *J. Exp. Biol.* **109,** 1-20.

Caton, A.E. (1991) Review of aspects of southern bluefin tuna biology, population and fisheries. *In* Deriso, R.B., and Bayliff, W.H., (eds.). World meeting on stock assessment of bluefin tunas: strength and weaknesses, Inter-American Tropical Tuna Commission, Special Report No. 7. La Jolla California, 357 pp.

Clarke, S., Smart, A., van Barneveld, R., and Carter, C. (1997) The development and optimisation of manufactured feeds for farmed southern bluefin tuna. *Austasia Aquac.*, **11**(3), 59-62.

Collette, B.B. (1978) Adaptations and systematics of the mackerels and tunas. *In* Sharp, G.D., and Dizon, A.E. (eds.),*The Physiological Ecology of Tunas*, Academic Press, New York, San Francisco, London, pp. 7-39.

Cummins, K.W., and Wuycheck, J.C. (1971) Caloric equivalents for investigations in ecological energetics. *Int. Assoc. Theor. Appl. Limnol. Comm.*, **18**, 1-159.

Fange, R., and Grove, D. (1979) Digestion. *In* Hoar, W.S., Randal, W.S., and Brett, J.R. (eds.) *Fish Physiology*, Vol. VIII. *Bioenergetics and Growth.* Academic Press, New York, NY, pp. 161-260.

Fudge, D.S., and Stevens, E.D. (1996) The visceral retia mirabilia of tuna and sharks: an annotated translation and discussion of the Eschricht & Muller 1835 paper and related papers. *Guelph Ichthyol. Rev.* **4**, 1-5.

Fudge, D.S., Stevens, E.D., and Ballantyne, J.S. (1997) Enzyme adaptation along a heterothermic tissue: the visceral retia mirable of the bluefin tuna. *Am. J. Physiol.* **272**, R1834-R1840.

Gunn, J.S., Polacheck, T., Davis, T.L.O., Sherlock, M., and Betlehem, A. (1994) The development and use of archival tags for studying the migration, behaviour and physiology of Southern Bluefin Tuna, with an assessment of the potential for transfer of the technology to groundfish research. Proc. I.C.E.S. Mini Symposium on Fish Migration, 21, 23 pp.

Hochachka, P.W. and Lewis, J.K. (1971) Effects of temperature and pH on fish LDH. *Comp. Biochem. Physiol.* **39B**, 925-934.

Magnuson, J.J. (1978) Digestion of food consumption by skipjack tuna (*Katsuwonus pelamis*). *Trans. Am. Fish. Soc.* **98**, 379-392.

Olsen, R.J. and Boggs, C.H. (1986) Apex predation by yellowfin tuna (*Thunnus albacares*): independent estimates from gastric evacuation and stomach contents, bioenergetics, and caesium concentrations. *Can. J. Fish. Aquat. Sci.* **43**, 1760-1775.

Prosser, C.L. (1973) Temperature, Chapter 9. *In* Prosser, C.L. (ed.) *Comparative Animal Physiology*, 3rd ed., W. B. Saunders, Philadelphia, pp. 362-428.

Stevens, E.D. and McLeese , J.M. (1984) Why bluefin tuna have warm tummies: temperature effects on trypsin and chymotripsin. *Am. J. Physiol.* **246,** R487-R494.

Windell, J.T. (1978) Digestion and the daily ration of fishes. *In* Gerking, I. (ed). *Ecology of Freshwater Fish Production.* Wiley and Sons, New York. pp. 159-184.

Aggregating Behavior of Yellowfin and Bigeye Tuna Tagged With Coded Ultrasonic Transmitters Around FADs in Okinawa, Japan

Itaru Ohta, Shinichiro Kakuma, and Kiyoaki Kanashiro
Okinawa Prefectural Fisheries Experiment Station, Okinawa, Japan

Key words: aggregating behavior, FADs, *Thunnus albacares*, *Thunnes obesus*, coded ultrasonic transmitter

Abstract: Eight yellowfin tuna (*Thunnus albacares,* 48-99 cm FL) and three bigeye tuna (*T. obesus,* 56-64 cm FL) were tagged with coded ultrasonic transmitters inserted into the abdominal cavity of the fish. We released the tuna at two huge FADs (Nirai-1 and Nirai-8) off the south of Okinawa Island, Japan between 18 August and 15 October 1999. The automated acoustic receivers were attached to both the Nirai to detect signals from each coded transmitter. Submersible data loggers were also attached to monitor the environmental factors such as water temperature, salinity, and current. Eight of the 11 tuna stayed around the FADs continuously for 5 to 24 days. There were two diel patterns in frequency of detected signals per hour. The frequency of detection reflects swimming range of tuna around FADs. Six tuna were swimming nearer to the FADs in the nighttime than in the daytime. On the other hand, two other yellowfin tuna were swimming nearer to the FADs in the daytime and swimming farther from the FAD in the nighttime. There was a negative correlation between body size of the tuna and the mean of detected signals per hour. This would mean that the larger tuna was swimming farther from the FAD. There was also a negative tendency between the current velocity and the mean of detected signals. Two yellowfin tuna moved between Nirai-1 and Nirai-8. In both cases, the tuna moved against the current.

1. INTRODUCTION

Fish Aggregating Devices (FADs) have enhanced tuna fisheries in tropical areas. The FAD fisheries are important, occupying 25% of coastal fisheries production in Okinawa. Yellowfin tuna (*Thunnus albacares,*

J.R. Sibert and J.L. Nielsen (eds.), Electronic Tagging and Tracking in Marine Fisheries, 131–145.
© 2001 *Kluwer Academic Publishers. Printed in the Netherlands.*

Scombridae) and bigeye tuna (*T. obesus,* Scombridae) are major species and dominate over 70% of production from the FAD fisheries. In spite of this, we have very limited information about migration and aggregating behavior of these tuna. The horizontal and vertical movements of these tuna around FADs were well documented using telemetry methods with acoustic devices (Yonemori, 1984; Holland *et al.*, 1990; Cayre, 1991; Bach *et al.*, 1998; Brill *et al.*, 1999). Although these methods can provide the details of short term swimming behavior of each individual, they are not appropriate for long term monitoring of aggregating behavior. Klimley and Holloway (1999) conducted the long term monitoring of yellowfin tuna around FADs in Hawaiian waters using coded ultrasonic transmitters and automated acoustic receivers. This 13 month study clarified aggregating behaviors, such as duration of stay around FADs, homing synchronicity, and schooling habit.

We used the same coded transmitters and automated acoustic receivers to monitor tuna behavior in Okinawan waters. Furthermore, we simultaneously monitored environmental factors such as temperature, salinity, and current with the submersible data loggers attached to the FADs. The main objectives of our study are to understand 1) the duration of the stay of yellowfin and bigeye tuna around FADs; 2) movement among FADs; 3) the relationship between aggregating behavior and environmental factors around FADs.

2. MATERIALS AND METHODS

We released 8 yellowfin and 3 bigeye tuna tagged with coded ultrasonic transmitters between 18 August and 16 October 1999. All tuna were tagged at two huge FADs called "Nirai." The Nirai is a steel raft of 7 m in diameter, moored with chains and reinforced wires (see Kakuma, 2001a). Nirai-1 and Nirai-8 are located about 20 km offshore, depth about 1300 m, south of Okinawa Island (25-59.5'N, 127-59.34'E; 25-53.61'N, 127-41.01'E, respectively) (Fig. 1). These FADs are about 33 km apart from each other. We attached the automated acoustic receivers to both Nirai to detect signals from the transmitters, and the submersible data loggers to record ambient water temperature, salinity, and current.

With the 50-m research vessel "Tonan-maru" of the Okinawa Prefectural Fisheries Experimental Station, we fished the tuna mainly by hand line and trolling and hauled up the fish on the deck by a hand net. After measuring the weight and the length, we made a ventral incision, following procedure with Klimley and Holloway (1999), 2 cm long and 1 cm deep, 4 cm anterior and 2 cm dorsal of vent on the left side of the fish. The peritoneal membrane was punctured with a finger, and the transmitter coated with honey was inserted through the opening with injection of oxytetracycline (50 mg per 1 kg in body weight). Then this cut was closed with the surgical bond (3M Co., Vet bond). Through the operation, the tuna was laid down on the wet sponge seat, its eye covered by a wet towel, and seawater was pumped into the mouth. Three yellowfin tuna were tagged at Nirai-1, and 5 yellowfin and 3 bigeye tuna were tagged at Nirai-8 between 18 August and 15 October 1999 (Table 1).

The coded ultrasonic transmitters (VEMCO Ltd., V16-4H-R256) were cylindrical, 16 mm in diameter, 65 mm long. The theoretical operating life of the transmitters is 21 months. Each tag was distinguishable by a unique pulse interval. The automated acoustic receivers (VEMCO Ltd., VR1) and the submersible data loggers (AANDERAA Ltd., RCM9) were attached to the Nirai with stainless steel chains at 4 m depth. The RCM9 can record ambient water temperature, salinity, and current (velocity and direction) using the Doppler effect.

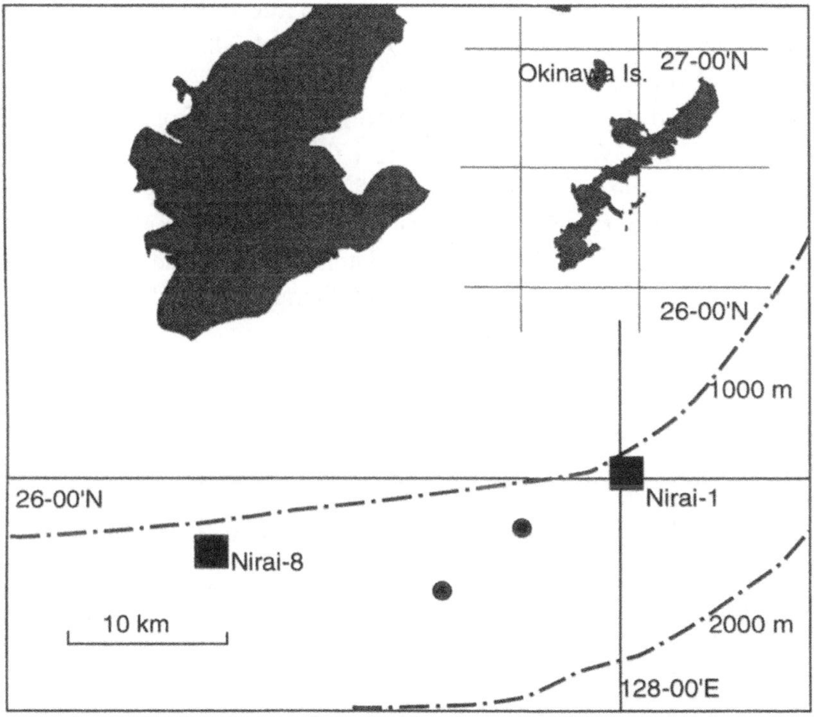

Figure 1. Locations of FADs in southern part of Okinawa Island. Solid squares indicate the huge FADs, Nirai, as monitoring stations. Solid circles indicate the small FADs of fisheries cooperative.

Table 1. Release and stay information of 11 tuna tagged with coded ultrasonic transmitters.

no.	release				Nirai-1				Nirai-8			
					duration of stay		interval of detection		duration of stay		interval of detection	
	FL (cm)	BW (kg)	date	site	hour	day	median (s)	max. (h)	hour	day	median (s)	max .(h)
YF#2	59.8	3.7	13 Oct.	N8	-	-	-	-	553	24	67	3.5
YF#3	47.5	2.8	18 Aug.	N1	12	1	62	2.0	92	5	65	0.6
YF#4	95.0	14.0	30 Sep.	N1	357	16	63	8.9	-	-	-	-
YF#5	51.5	2.4	30 Sep.	N1	323	15	62	2.0	-	-	-	-
YF#6	99.0	16.0	14 Oct.	N8	-	-	-	-	6	1	120	0.5
BE#20	56.0	5.3	14 Oct.	N8	-	-	-	-	350	16	68	4.9
YF#21	61.0	4.8	15 Oct.	N8	-	-	-	-	218	10	68	1.7
BE#22	64.0	5.1	15 Oct.	N8	-	-	-	-	3	1	140	0.2
BE#23	64.0	5.2	15 Oct.	N8	-	-	-	-	8	1	70	0.5
YF#24	66.0	5.4	15 Oct.	N8	-	-	-	-	142	7	71	7.2
YF#25	84.0	11.0	15 Oct.	N8	15	2	69	2.6	329	15	69	2.6

To determine the signal-detection range of the receiver, we lowered 5 transmitters to the depths of 10, 50, 100, 200, and 300 m, respectively. As the vessel and transmitters drifted away from the receiver attached to the FAD, we recorded the time and the horizontal distance from the FAD using a radar on the boat. We repeated this test three times, because the signal-detection range varied with the depth of transmitters (Fig. 2). Maximum straight distance from each transmitter to the receiver ranged from 603 to 787 m with mean 680 m. However, we determined that the signal detection range of the receiver was a column 550 m in diameter and 300 m in depth.

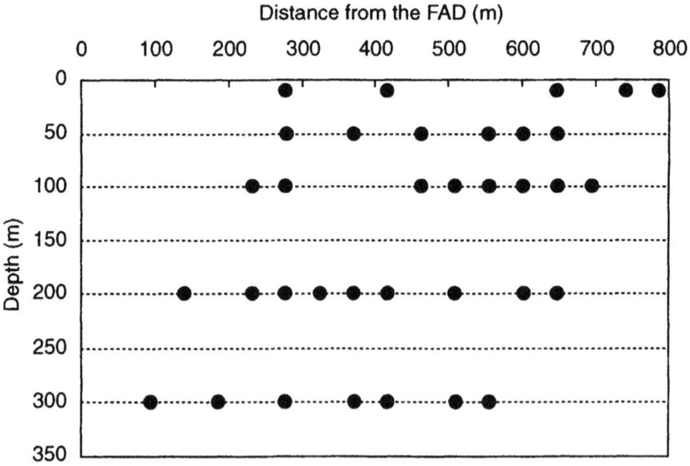

Figure 2. The results of test to determine the signal detection range of the automated receiver. Each plot indicates signal detection.

3. RESULTS

3.1 Duration of Stay Around FADs

We obtained the signals from 11 tuna tagged with the transmitters at the two monitoring stations, including two yellowfin tuna that moved between Nirai-1 and Nirai-8. These eleven tuna visited the FADs soon after release. They stayed at the FADs for various successive periods. Four tuna stayed at Nirai-1 for 12 to 357 h (median = 169 h or 7.0 days), and 9 tuna stayed at Nirai-8 for 3 to 553 h (median = 189 h or 7.8 days).

Three yellowfin tuna were tagged at Nirai-1 (Fig. 3; Table 1): YF #3 stayed for 12 hours and then left; after 17 days, the tuna visited at Nirai-8, and stayed there for 5 days. YF#4 stayed for 16 days until recaptured by a fisherman. YF#5 stayed for 15 days.

Five yellowfin and three bigeye tuna were tagged at Nirai-8 (Fig. 3; Table 1). YF#2 stayed for 24 days, and this record was the longest in this study. BE#20 and YF#21 stayed for 16 days and 10 days, respectively. YF#24 stayed for 7 days until recaptured by a fisherman. The cut on the belly of the tuna for the transmitter was closed almost completely. YF#25 stayed for 15 days; ten days after the tuna left Nirai-8, the tuna visited Nirai-1 and stayed there for 2 days. YF#6, BE#22, and BE#23 stayed for only 6, 3, and 8 hours respectively.

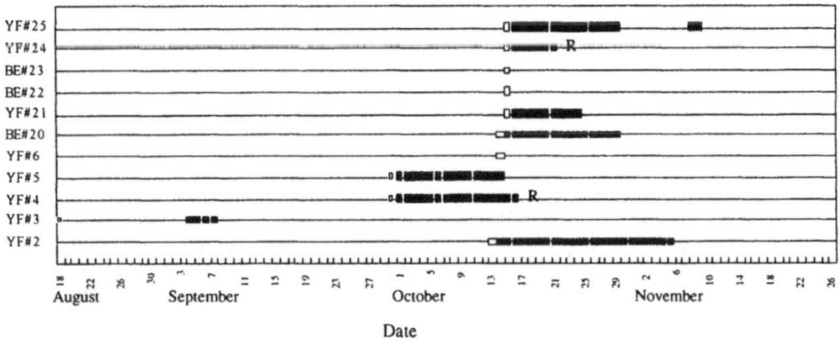

Figure 3. The chronology of daily records of 11 tuna tagged with coded ultrasonic transmitters from 18 August to 26 November 1999. Squares and circles indicate Nirai-1 and Nirai-8, respectively. Open symbols mean tagging day. R is the day that the fish was recaptured by fishermen.

3.2 Frequency of Detected Signals

Frequency of detected signals per hour was analyzed for 8 tuna that stayed for more than 5 days. Two diel patterns were detected in all 8 signal records. The time series data of detected signals for YF#3 and YF#4 are shown in Figure 4. Signal detection of YF #3 increased in nighttime (mean: 47.5 times/h) and decreased in daytime (mean: 26.2 times/h). Contrary to this, signal detection of YF#4 increased in daytime (mean: 45.1 times/h) and decreased in nighttime (mean: 30.6 times/h), though diel difference of detection was obscure for the first 4 days of its stay. The time series data of detection for 8 tuna are summarized in Figure 5. There were two patterns in fluctuation of detection for 8 tuna. Frequency of detection for 5 yellowfin (YF#2, #3, #5, #21, and #25) and 1 bigeye (BE#20) tuna were significantly higher in nighttime than in daytime; on the contrary, the remaining 2 yellowfin tuna (YF#4 and #24) were detected significantly more often in day time than in nighttime (Table 2).

The median of interval of detected signals for 8 tuna ranged from 62 to 71 seconds through their stay (Table 1). However, sometimes these tuna were beyond the signal detection range for a few hours. The maximum interval of the detection ranged from 1.7 to 7.2 hours (Table 1). Figure 6 shows frequency of periods where the signal was not detected for 1 hour. For 6 of 8 tuna, YF#2, #3, #5, #21, #25, BE#20, no detection periods often occurred from 18:00 to 20:00, especially from 18:00 to 19:00 (around sunset). YF#4 and YF#24 were absent from the FADs in the nighttime, and YF#4 was often absent for a few hours.

3.3 Relationship Between Body Size of Tuna and Detected Signals

The relationship between body size of the tuna and the mean of detected signals per hour is shown in Figure 7. The means of detected signals were significantly (unpaired t-test: $p < 0.05$) greater at Nirai-1 (38.4 to 44.9 times/h) than at Nirai-8 (from 22.6 to 35.9 times/h). For the 6 tuna that stayed at Nirai-8, there was a negative correlation between the mean of detected signals and fork length of tuna (correlation coefficient $r = 0.914$; $p < 0.05$). At Nirai-1, the mean of detected signals for YF#5 was greater than that for YF#4 (unpaired t-test; $p<0.001$).

3.4 Current at FADs

Velocity and direction of the current at Nirai-1 and Nirai-8 from 15 August to 15 November 1999 is shown in Figure 8. Although the current at both FADs was similar, velocity of the current differed during the stay of the tuna considerably. From 3 to 7 September when YF#3 stayed at Nirai-8, the mean velocity of the current was 11.2 cm/s (SD: 7.7). From 30 September to 15 October when YF#4 and YF#5 stayed at Nirai-1, the mean velocity was

17.7 cm/s (SD: 12.5). From 13 October to 5 November when 5 tuna (YF#2, #21, #24, #25, BE#20) stayed at Nirai-8, the mean velocity was 44.2 cm/s (SD: 19.0). The current velocity significantly differed among each period (MANOVA: $p < 0.001$).

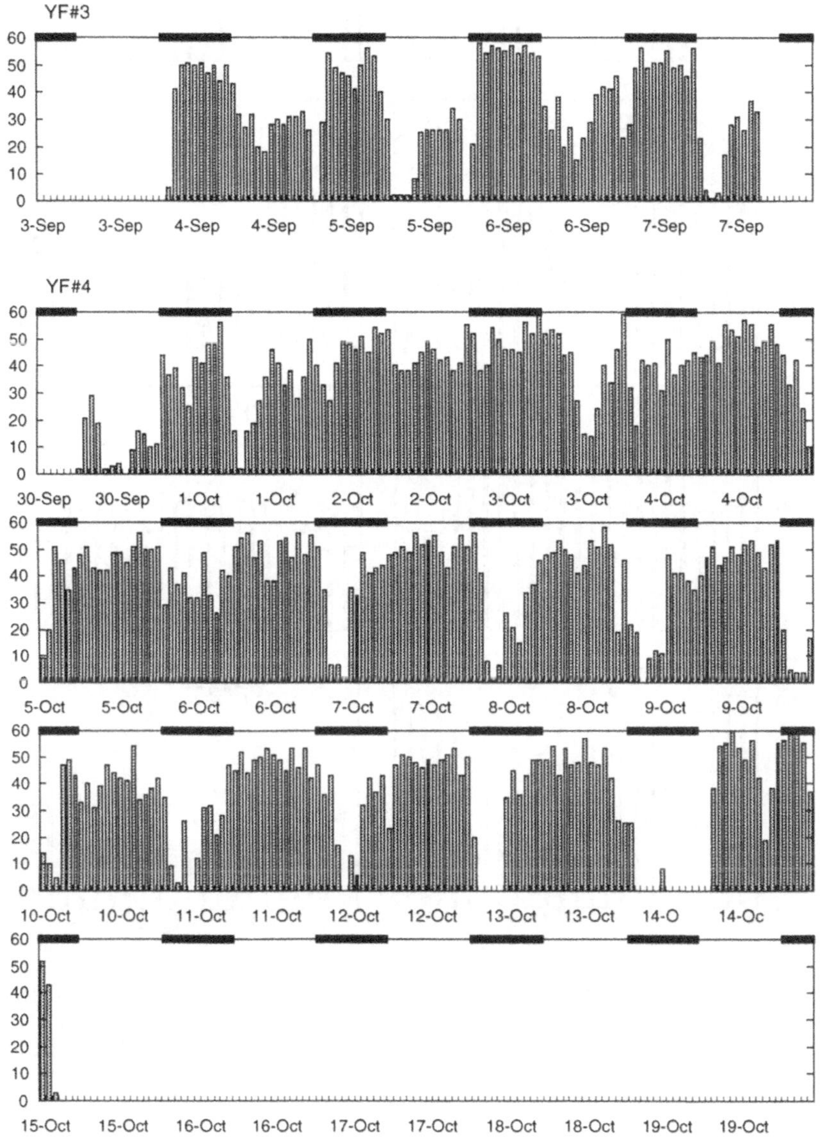

Figure 4. Time series of frequency of detected signals per hour for yellowfin tuna #3 stayed at Nirai-8 from 3 to 7 September, and YF#4 stayed at Nirai-1 from 30 September to 15 October until recaptured on 16 October. Above horizontal black bars indicate the nighttime.

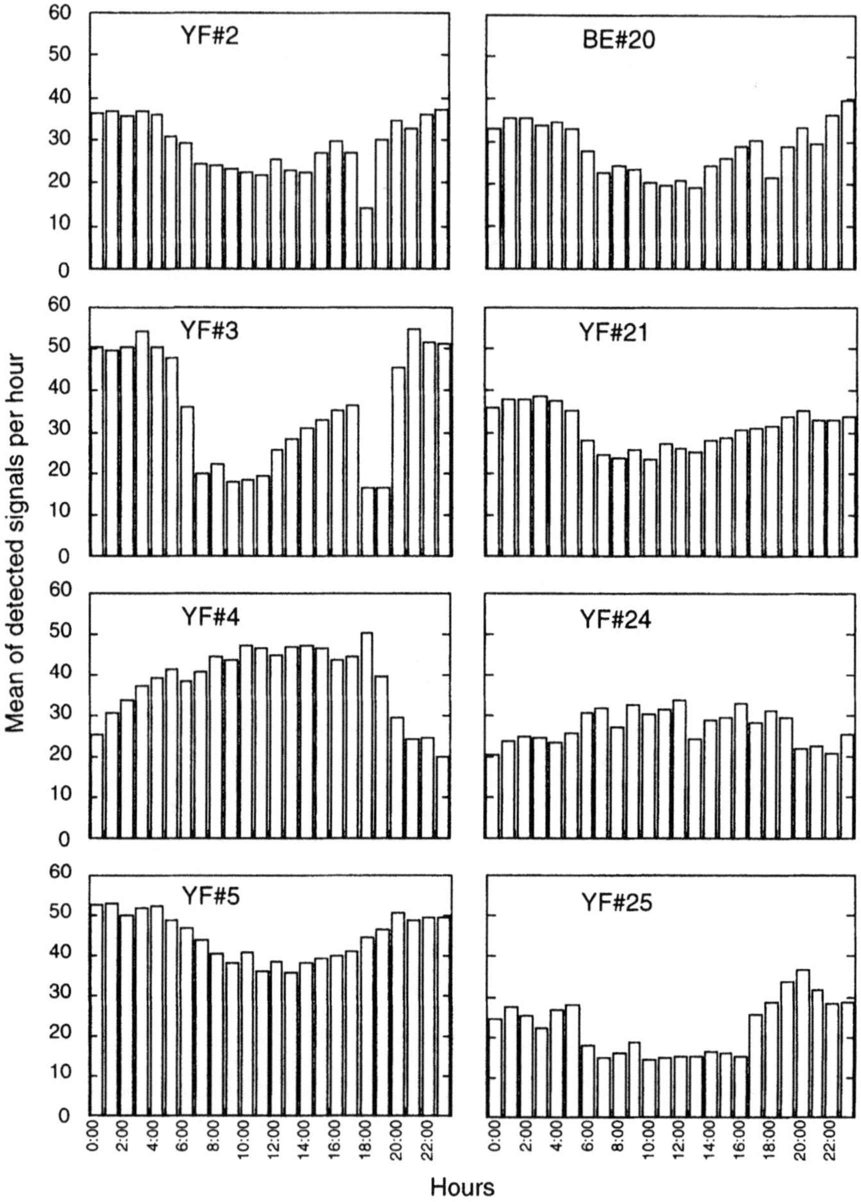

Figure 5. Frequency of the mean of detected signals per hour for 8 tuna. Above horizontal black bars indicate the nighttime.

Table 2. Statistical results of daily comparisons in detected signals between day and night by paired t-test

	mean ± SD/h							
ind.#	day		night		N	t	df	P value
YF#2	24.51	5.62	34.70	7.77	22	-5.012	21	0.0001***
YF#3	26.15	6.56	47.45	4.26	3	-4.869	2	0.0397*
YF#4	45.14	6.42	30.63	9.29	14	3.989	13	0.0015**
YF#5	39.29	8.13	50.38	4.48	13	-4.410	12	0.0009***
BE#20	23.47	7.16	33.87	5.75	14	-4.695	13	0.0004***
YF#21	26.72	3.08	35.53	3.59	8	-5.594	7	0.0008***
YF#24	30.16	4.34	23.95	6.20	5	2.902	4	0.0440*

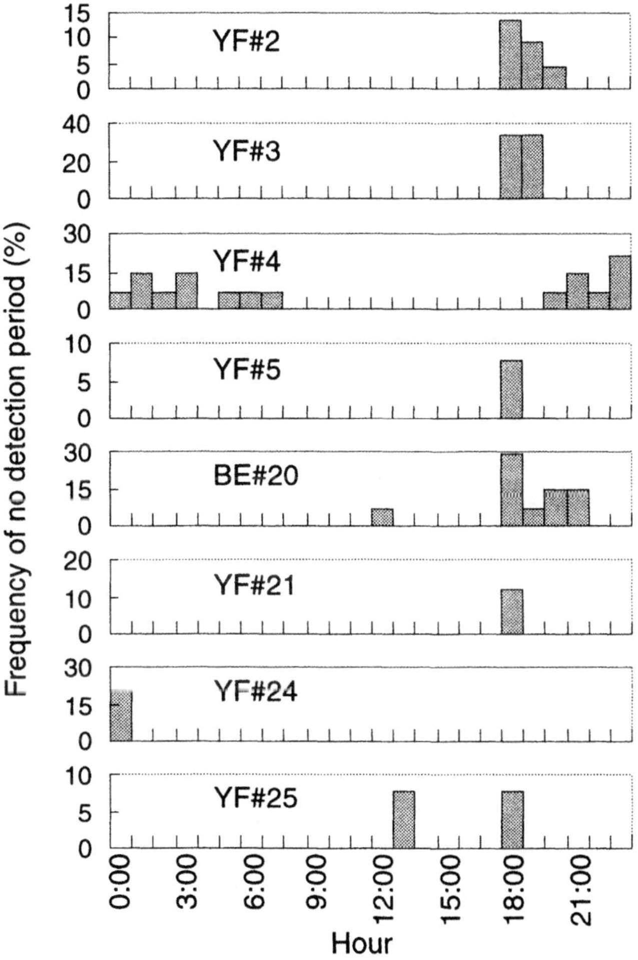

Figure 6. Frequency of the period that signal was not detected during 1 hour for 8 tuna. Above horizontal black bars indicate the nighttime

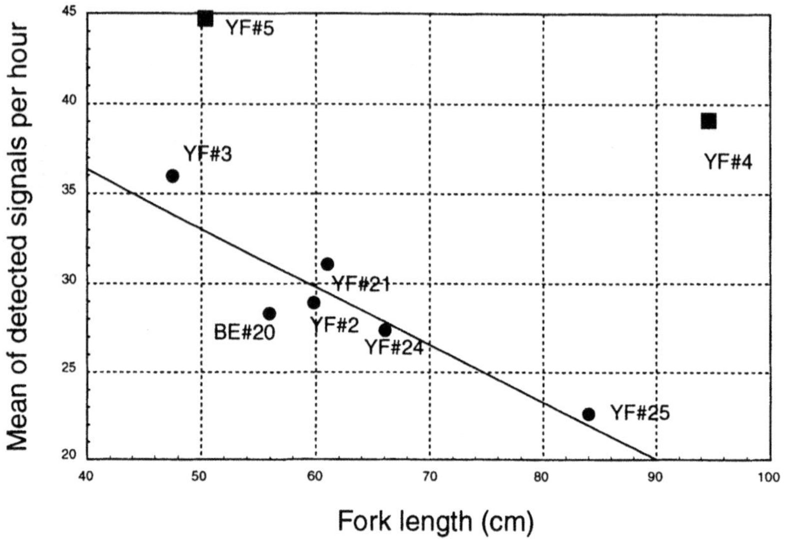

Figure 7. Relationship between body size of tuna and the mean of detected signals per hour at each FADs. Black circles and squares indicate at Nirai-8 and Nirai-1, respectively. A regression line is for relationship at Nirai-8 (correlation coefficient r = 0.914; $p < 0.05$) excluded the data at Nirai-1.

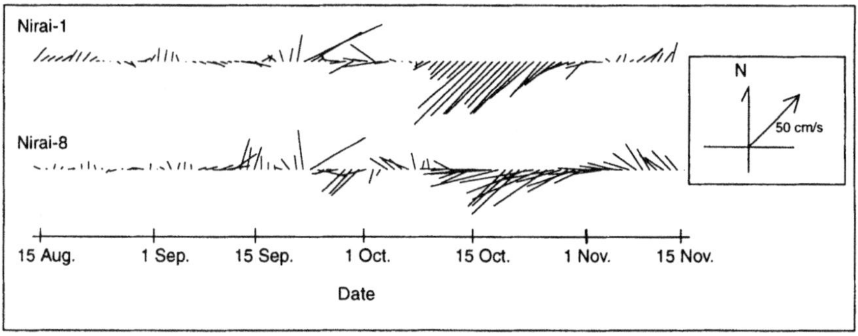

Figure 8. The daily mean of direction and velocity of the current at each FADs from 15 August to 15 November.

Two yellowfin tuna moved between Nirai-1 and Nirai-8. From 18 August to 3 September when YF#3 moved westward from Nirai-1 to Nirai-8, the mean direction and velocity of the current at Nirai-1 were NE/E; 9.7 cm/s (SD: 7.8), and were NW; 12.8 cm/s (SD: 7.7) at Nirai-8. From 29 October to 8 November when YF#25 moved eastward from Nirai-8 to Nirai-1, the mean direction and velocity of the current at Nirai-8 were W/S; 32.7 cm/s (SD: 14.3), and they were WSW; 15.4 cm/s (SD: 12.1) and WSW at Nirai-1. Both fish movements were against the current.

4. DISCUSSION

Two yellowfin tuna were recaptured by fishermen. This means that they resumed normal feeding behaviour after the transmitters were inserted. The ventral cut of the tuna was closed almost completely after 7 days, and we did not find any abnormality of the internal organs.

In this study, 8 tuna stayed around FADs for successive periods, and most of them stayed continuously. They were rarely absent from the FAD or beyond the detectable range for more than a few hours. This was different from the previous work in Hawaiian waters (Klimley and Holloway, 1999); in their study, tagged tuna repeatedly visited the FADs a few times a day for a short period. Furthermore some tuna visited the FAD intermittently during 9 months. In our study, 8 tuna stayed at FADs for 5 to 24 days continuously. According to conventional tagging studies of tuna around FADs from 1999 to 2000 in Okinawa (Ohta *et al.*, unpublished data), 35% of 138 recaptured yellowfin tuna were caught within 10 nmi from the release point, ranging from 2 to 108 days after their release. Forty percent of them were recaptured after more than 30 days. Therefore, some of the yellowfin tuna in Okinawan waters tended to stay around the FADs for over a month.

On the other hand, some yellowfin tuna left the FADs after a short stay. They might leave the FADs because of their seasonal migration. Kakuma (2001b) reported the seasonal transition of weight composition of yellowfin tuna caught around FADs in southwestern Okinawa Island from 1990 to 1998. There were two distinct weight groups, the lighter weight group (2 kg BW in May) and the heavier weight group (15 kg BW in May) considered to represent the cohorts. The weights of these groups increased monthly, presumably as individual tuna grew. Although the catches of both weight groups considerably decreased in winter, the lighter weight group seemed to mature and appear again in next spring as the heavier weight group.

Yellowfin tuna often moved among neighboring FADs (Holland *et al.*, 1990; Cayre, 1991; Bach *et al.*, 1998; Brill *et al.*, 1999; Klimley and Holloway, 1999). Holland *et al.* (1990) suggested that the fish could learn the FAD positions and navigate precisely between them even though the FADs are at least 10 nmi apart. Two tuna moved between Nirai-1 and Nirai-8 in this study. Considering the small numbers of tagged tuna and the receivers, some of the tuna would often move among other FADs, including small FADs of fisheries cooperatives. Therefore, the tuna that left the monitoring FADs could have moved to other FADs.

There were two diel patterns in frequency of detected signals per hour. The frequency of detection per hour would reflect aggregating behavior of tagged tuna around FADs. Because the transmitter emitted a signal every 60 seconds on average, maximum detectable signals per hour should be 60.

Therefore the frequency would indicate the duration that the tagged tuna was within the signal detection range of the receiver. This, then probably indicates the changes in swimming range of the tuna around FADs.

Previous works by telemetry tracking demonstrated diel patterns of vertical and horizontal movements of yellowfin and bigeye tuna (Yonemori, 1982; Holland *et al.*, 1990; Cayre, 1991; Block *et al.*, 1997; Bach *et al.*, 1998; Josse *et al.*, 1998; Brill *et al.*, 1999). According to these studies, for both species, especially bigeye tuna, swimming depth was remarkably shallower in nighttime than in daytime. Brill *et al.* (1999) suggested that vertical movement of yellowfin tuna would be limited by the change in water temperature from surface to deep water rather than an absolute water temperature. Monthly vertical distribution of water temperature at a fixed point near the Nirai-8 from July to November 1999 is shown in Figure 9. The surface water temperature ranged from 29.26 to 25.94° C, and at 300 m depth from 17.01 to 15.78° C. Additionally, Holland *et al.* (1990) reported that FADs tended to bring tuna closer to the surface than they would normally be in offshore areas of the ocean. We confirmed that the signal detection range of the receiver was, at least, 550 m in diameter and 300 m in depth. Considering vertical distribution of water temperature in this study area and the signal detection range, the horizontal movement rather than the vertical movement would cause the fluctuation of signal detection. Therefore, when a smaller number of signals was detected, the tuna would be frequently swimming farther horizontally from the FAD, beyond the signal detection range. The majority of the tuna were swimming closer to the FADs in the nighttime than in the daytime with short time excursion around the sunset. However, two yellowfin tuna were swimming nearer to the FADs in the daytime than in the nighttime. They were often beyond the signal detection range of the receivers for a few hours. There were only the two fish that were recaptured in this study.

In a daytime investigation by trolling around FADs, over 70% of juvenile yellowfin tuna were distributed within 500 m from FADs in Okinawa (Kawasaki, 1984) and 83% of yellowfin tuna were distributed within 200 m from the FAD in Vanuatu (Cillaurren, 1994). Holland *et al.* (1990) reported that yellowfin and bigeye tuna tended to remain tightly associated with FADs or the reef dropoff during the day, moved away at night, and returned the next morning. These facts may correspond with the aggregating behavior of the two yellowfin tuna closer to the FADs in the daytime.

We found a negative correlation between body size of tuna and the mean detected signals per hour at Nirai-8, and perhaps at Nirai-1. This may mean that the larger tuna swam farther from the FAD than the smaller ones. Then we noticed that the mean velocity of the current at Nirai-8 was significantly greater than at Nirai-1 during the stay of the tuna, and the mean detection at

Nirai-8 was less than at Nirai-1. This fact may indicate that the greater the current, the farther the tuna swim from a FAD.

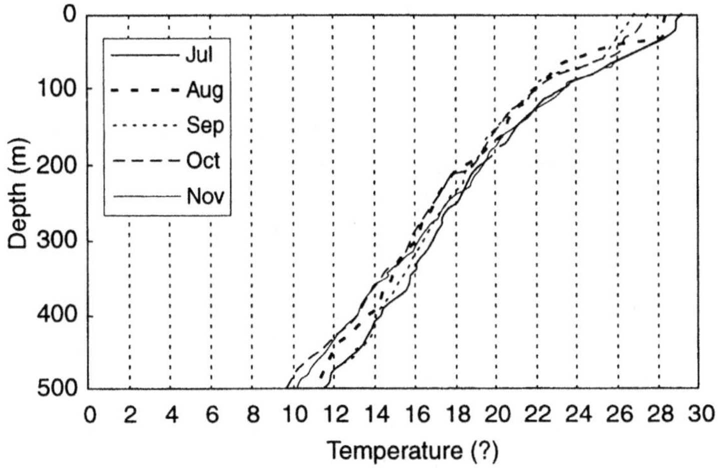

Figure 9. Monthly vertical distribution of water temperature at the fixed point in southern Okinawa near the Nirai-8 from July to November, 1999.

Many fishermen in Okinawa give attention to the current when they fish around FADs. They almost always try to fish in the upcurrent area of the FADs. Holland *et al.* (1990) reported that one of their tracked yellowfin tuna patrolled almost exclusively in the upcurrent area of the FAD in the daytime. In our daytime observations with the echo sounder, a high density of fish appeared almost exclusively in the upcurrent side of the FAD. These facts indicate that tuna have a habit of aggregating in the upcurrent area of the FADs and are able to perceive the current direction at least around the FADs. This could explain why the two yellowfin tuna moved between Nirai-1 and Nirai-8 against the current.

To better understand aggregating behavior of tuna around the FADs, it will be necessary to compare regional difference of biotic factors (i.e., species, body size of tuna, food organisms, predators, human activity) and abiotic factors (i.e., current, water temperature, dissolved oxygen, topography, density of FADs) in each area.

5. SUMMARY

Eight yellowfin tuna (*Thunnus albacares,* 48-99 cm FL) and three bigeye tuna (*T. obesus,* 56-64 cm FL) were tagged with a coded ultrasonic transmitter inserted into abdominal cavity of the fish. We released the tuna at two huge FADs (Nirai-1 and Nirai-8) south of Okinawa Island, Japan between 18

August and 15 October 1999. Automated acoustic receivers were installed at both FADs to detect signals from the coded transmitters. The submersible data loggers were also attached to monitor the environmental factors such as water temperature, salinity, and current.

Eight of the 11 tuna stayed around the FADs continuously for 5 to 24 days. There were two diel patterns in frequency of detected signals per hour. The frequency of detection reflects swimming range of tuna around FADs. Six tuna were swimming nearer to the FADs in the nighttime than in the daytime. On the other hand, 2 other yellowfin tuna were swimming nearer to the FADs in the daytime and swimming farther from the FAD in the nighttime.

There was a negative correlation between body size of the tuna and the mean of detected signals per hour. This would mean that the larger tuna were swimming farther from the FAD. There was also a negative tendency between the current velocity and the mean of detected signals. Two yellowfin tuna moved between Nirai-1 and Nirai-8. Both fish movements were against the current.

ACKNOWLEDGEMENTS

We sincerely thank K. Holland, J. Sibert, and D. Itano for advice in our study. We thank A. P. Klimley, who reviewed the manuscript. We thank all the fishermen who supplied information about fishing methods and recapture of the tagged fish. This study was supported by the government of Okinawa prefecture.

REFERENCES

Bach, P., Dagorn, L., Josse, E., Bard, F.-X, Abbes, R., Bertrand, A., and Misselis, C. (1998) Experimental research and fish aggregating devices (FADs) in French Polynesia. *SPC FAD Information Bull.* **3**, 3-18.

Block, B.A., Keen, J.E., Castillo, B., Dewar, H., Freund, E. V., Marcinek, D. J., Brill, R. W., and Farwell, C. (1999) Environmental preferences of yellowfin tuna (*Thunnus albacares*) at the northern extent of its range. *Mar. Biol.* **130**, 119-132.

Brill, R.W., Block, B.A., Boggs, C.H., Bigelow, K.A., Freund, E.V., and Marcinek, D.J. (1999) Horizontal movements and depth distribution of large adult yellowfin tuna (*Thunnus albacares*) near the Hawaiian Islands, recorded using ultrasonic telemetry: implications for the physiological ecology of pelagic fishes. *Mar. Biol.* **133**, 395-408.

Cayre, P. (1991) Behaviour of yellowfin (*Thunnus albacares*) and skipjack tuna (*Katsuwonus pelamis*) around FADs as determined by sonic tagging. Proceeding of symposium on artificial reefs and fish aggregating devices as tools for the management and enhancement of marine fishery resources. 41-60.

Cillaurren, E. (1994). Daily fluctuations in the presence of *Thunnus albacares* and *Katsuwonus pelamis* around fish aggregating devices anchored in Vanuatu, Oceania. *Bull. Mar. Sci.* **55**(2-3), 581-591.

Holland, K.N., Brill, R.W., and Chang, R.K.C. (1990) Horizontal and vertical movements of yellowfin and bigeye tuna associated with fish aggregating devices. *Fish. Bull.* **88**(3), 493-507.

Josse, E., Bach, P., and Dagorn, L. (1998) Simultaneous observations of tuna movements and their prey by sonic tracking and acoustic surveys. *Hydrobiologia* 371/372, 61-69.

Kakuma, S. (2001a) North West Pacific regional synthesis on moored FADs. Proceeding of tuna fishing and fish aggregating devices (FADs) at Caraibe Martinique 1999 (in press).

Kakuma, S. (2001b) Current, catch and weight composition of yellowfin tuna with FADs off Okinawa Island, Japan. Proceeding of tuna fishing and fish aggregating devices (FADs) at Caraibe Martinique 1999 (in press).

Kawasaki, K. (1984) Effects of FADs in Okinawa Island. Annual Report of Okinawa Prefectural Fisheries Experimental Station. 1-6. (in Japanese)

Klimley, A.P. and Holloway, C.F. (1999) School fidelity and homing synchronicity of yellowfin tuna, *Thunnus albacares*. *Mar. Biol.* **133**, 307-317.

Yonemori, T. (1984) Swimming behavior of tuna with ultrasonic telemetry. *Newsletter Far Sea Fisheries Research Laboratory.* 44, 1-5. (in Japanese)

Using Radio-Acoustic Positioning and Telemetry (RAPT) to Define and Assess Marine Protected Areas (MPAs)

Ron K. O'Dor,[1] Jill P. Aitken,[1] Russ C. Babcock,[2] Stephania K. Bolden,[3] Satoquo Seino,[4] Dirk C. Zeller,[5] and George D. Jackson[6]

[1]*Biology Department, Dalhousie University, Halifax, Nova Scotia, Canada B3H 4J1*
[2]*Leigh Laboratory, University of Auckland, PO Box 349, Warkworth, New Zealand,*
[3]*National Marine Fisheries Service, Miami, 75 Virginia Beach Drive, FL 33149, USA,*
[4]*Department of Earth Sciences and Astronomy, College of Arts and Sciences, University of Tokyo, Komaba, Meguro, Tokyo 153-8902, Japan,*
[5]*Fisheries Centre, 2204 Main Mall, University of British Columbia, Vancouver, B.C., Canada V6T 1Z4.*
[6]*Institute of Antarctic and Southern Ocean Studies, University of Tasmania, GPO Box 252 – 77, Hobart, Tasmania 7001, Australia*

Key words: MPA, Marine Protected Area, pinger, RAPT, Radio-acoustic positioning and telemetry

Abstract: Radio-acoustic positioning telemetry (RAPT) continuously triangulates positions of up to eight acoustic transmitters by relaying sound arrival times to a shore-based computer via autonomous, moored radio-sonobuoys. As transmitters on animals or people telemeter information via RAPT, it records their position with an accuracy of as little as one meter in as much as a cubic kilometer of water from a range of up to ten kilometres—scales appropriate to many marine protected areas (MPAs). RAPT studies in Lizard Island, Australia; Exuma Cay Park, Bahamas; Moriye Bay, Japan, and Leigh Laboratory Reserve, New Zealand, four significant MPAs, are reviewed showing the power of the technology for understanding and managing many aspects of a range of biological reserve types. Most studies to date have focused on monitoring fish (*Lutjanus carponotatus*, Lutjanidae; *Plectropomus leopardus*, Serranidae; *Epinephelus striatus*, Serranidae; *Pagrus auratus*, Sparidae) home ranges and migrations, but breeding and feeding behaviours of larger invertebrates such as horseshoe crabs (*Tachypleus tridentatus*, Limulidae) and squids (*Sepioteuthis lessonianna* and *Loligo vulgaris*, Loliginidae) have also been examined. RAPT systems can also record environmental changes in

J.R. Sibert and J.L. Nielsen (eds.), Electronic Tagging and Tracking in Marine Fisheries, 147–166.
© 2001 *Kluwer Academic Publishers. Printed in the Netherlands.*

real-time directly to the computer on shore for such variables as temperature, salinity, tide height, current direction and current speed from fixed-position transmitters; these are easily compared to animal activities. Tracking tagged divers or ROVs while they are videotaping provides an accurate and reproducible way of documenting long-term changes in the sessile organisms that make up benthic habitats; only the buoy moorings need remain in the water between surveys. Finally, the location and condition of diving tourists can be monitored, to protect both them and the environment!

1. INTRODUCTION

The term marine protected area (MPA) is intentionally imprecise to encompass various levels of control over marine environments, from complete no-entry zones to no-take zones to local gear or species restrictions. Any MPA can serve multiple purposes, including these: (1) protecting habitats of threatened species, (2) providing breeding reserves for commercial species, (3) sustaining or increasing yields to surrounding fished areas due to net adult spillover, (4) elevating public awareness of marine life and (5) enhancing and accommodating ecotourism. When the Dalhousie group (http://is.dal.ca/~raptnet) began developing commercial radio-acoustic positioning and telemetry (RAPT) systems with VEMCO Ltd. (Shad Bay, Nova Scotia, Canada) in 1985, RAPT was seen as a tool for studying the behaviour, physiology and energetics of marine animals. However, as the system has evolved, it has become clear that RAPT has great potential for monitoring critical animal movements in MPAs (Holland et al., 1996), recording environmental parameters and expediting the description and mapping of marine habitats (O'Dor et al., 1998).

In the RAPT system, moored buoys with self-contained power use tuned hydrophones to record arrival times of acoustic 'pings' from frequency-matched transmitters implanted in free-ranging animals. The received data are then relayed to a shore-based computer via radio. The intervals between pings can encode telemetered information on parameters of animals or habitat. At the same time, the different arrival times of the pings at the hydrophones are used to continuously triangulate the position of the transmitter and thereby provide a record of the animal's location. In this way, RAPT technology can provide detailed information about physiology, movements and habitat usage over long periods of time.

This type of high-resolution, long-term movement data can be used to assess the utility of MPAs. When placed in appropriate configurations, the RAPT system can be used to acquire quantitative information on the movements of animals across MPA boundaries. This type of information is important in assessing the role of MPAs in protecting animals within the reserve and in evaluating whether the MPA enhances populations in adjacent areas. The high resolution (a few meters) of the positions allows observation of mating and schooling behaviour (Sauer et al., 1997). In reserves where breeding occurs, it can follow behavioural interactions between individuals with a resolution of meters, insuring that reserve areas fully contain such critical life-history events.

With respect to habitat, RAPT can also be used to track divers moving within the range of the array. The triangulated positions of the divers can be used to integrate visual observations or videotaped records of habitat type into an accurate map of the study site. When repeated over time such maps prove a basis for comparisons that indicate the impact of environmental change.

In terms of supporting ecotourism and public awareness of marine life, a base station computer monitor allows the public on land to view real-time actions of animals regardless of weather or visibility without interfering with animals and without even getting their feet wet. This interface could take full advantage of modern animation techniques. For more dynamic tourists, the system gives precise coordinates of the animals they want to see. Furthermore, tagged tourists can even be monitored, ensuring their welfare and compliance with regulations while boating or diving.

There have now been several applications of RAPT for defining, assessing or monitoring marine reserves, although none to date have fully exploited its potential. This article reviews these applications, including some in progress, and points to trends for future applications as well as to specific improvements in technology needed to enhance them.

2. WHAT IS RAPT?

RAPT arose from early experiments with hardwired hydrophone systems used to triangulate loud whales (Watkins and Schevill, 1972) or acoustically tagged animals (Hawkins et al., 1974; Maynard and Webber, 1987; Wartzok et al., 1992), because wired systems were difficult to deploy and move. RAPT technology, pioneered by McKibben and Nelson (1981) and commercialized by VEMCO Ltd. (O'Dor et al., 1989; O'Dor and Webber, 1991; Bjordal et al., 1993), uses sonobuoys that replace wire links with radio to the base-station computer. RAPT is mobile, works in 3-D and intermittently tracks multiple targets with an accuracy of a few metres (O'Dor et al., 1998). The area monitored can be extended over several kilometres with multiple buoys (Figure 1).

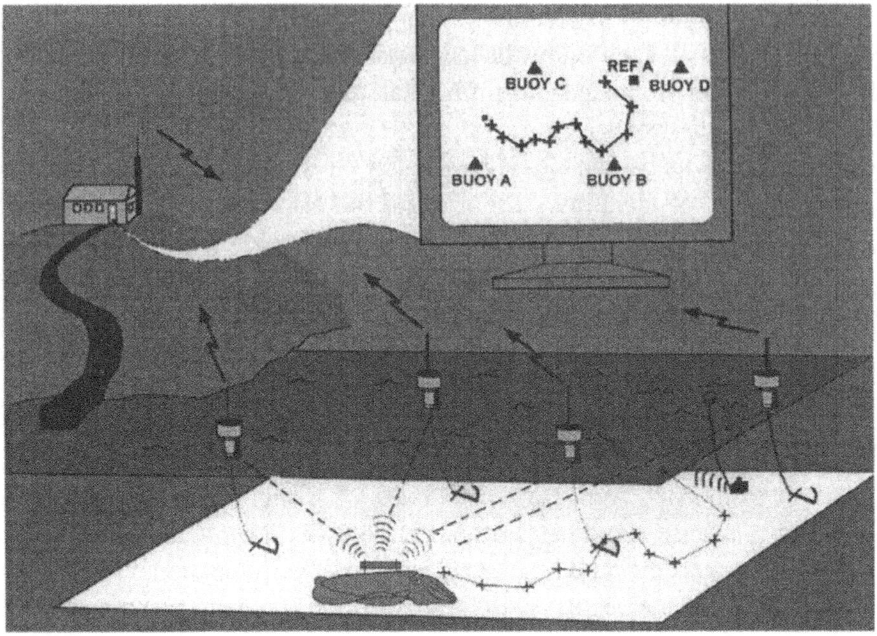

Figure 1. Schematic diagram of a four buoy 3D RAPT system with a reference pinger.

The RAPT technology is relatively new, but there are about fifteen systems in use around the world in a wide range of projects, as well as several hardwired systems using similar software (O'Dor et al., 1998). Analytical approaches to RAPT are developing rapidly, and users exchange ideas through http://is.dal.ca/~raptnet. Among the challenges is the fact that position accuracy declines rapidly outside the buoy array particularly in "shadow zones" behind buoys (Wartzok et al., 1992) and degrades with complex bottom morphology or extreme weather conditions (Smith et al., 1998).

Simply presenting the vast quantities of four-dimensional data collected can be a challenge. Figure 2 is an example from the Leigh study (Section 4.1.3) of the commercially available VRAP software (VEMCO) display of the tracks of multiple tags in real-time on the base station screen. These are stored in D-files and can be viewed individually as simple animations in playback mode, but further post-processing is required for statistical analysis. VRAP also stores R-files containing all raw arrival time data, which can be brought into matrices in Matlab (O'Dor et al., 1998) for more complete 3-D, or even 4-D, display and analysis. Positional data in 2-D can be analysed in GIS as illustrated in the Exuma Cay example (Section 4.1.2) in Figure 3. Figure 4 overlays two separate studies at Lizard Island (Section 4.1.1) on a vectorised map. The tracks can also be presented as animations because temporal patterns can play an important part in documenting spatial behaviours. Commercial Windows-based software is under development and will provide many of these features in the real-time program.

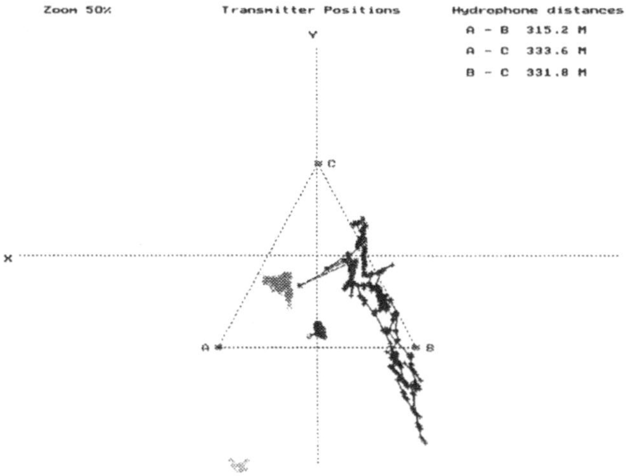

Figure 2. VRAP playback output from five snappers (*Pagrus auratus)* tagged with V16 pingers. The array was located at the position indicated in Figure 6 with distances as indicated. The period covered in the figure was approximately mid-day to 1600 h. Four of the fish are displaying site-attached behaviour while the fifth (adjacent to buoy B) has made two excursions well outside the array.

Commercial hardware presently available is limited in its capacity to study animal interactions because of its inability to continuously and simultaneously monitor multiple transmitters. It must switch between frequencies and it does not collect acoustic data during radio transmission. Systems with multiple, overlapping buoy triangles ("cells"), as described below, allow error estimates from multiple positions. Transponder systems under development also reduce errors. Digital signal processing for the buoys is under development to allow systems to track multiple, simultaneous targets. However, there is a trade-off between power consumption and maintenance requirements.

Continuous monitoring of rare or sentinel species in marine reserves or disturbed areas is feasible with surgically implanted tags, which have minimal impact on animal health or behaviour (Swanberg and Geist, 1997; Bolden and Ewing, 1999; Zeller, 1997, 1998, 1999). Conservation of rare species requires knowledge of their ranges throughout life, which can best be established by continuous tracking.

Public interest in conservation is heightened by RAPT displays on shore-based computers. This provides real-time information about animal activities even when weather or other factors limit direct observation; several public preservation centres are interested in providing such indirect "viewing" to their visitors, reducing disappointment when animals are not visible and decreasing direct visitor pressure. Continuous monitoring may also reveal behaviour changes giving early warning of environmental degradation or recovery.

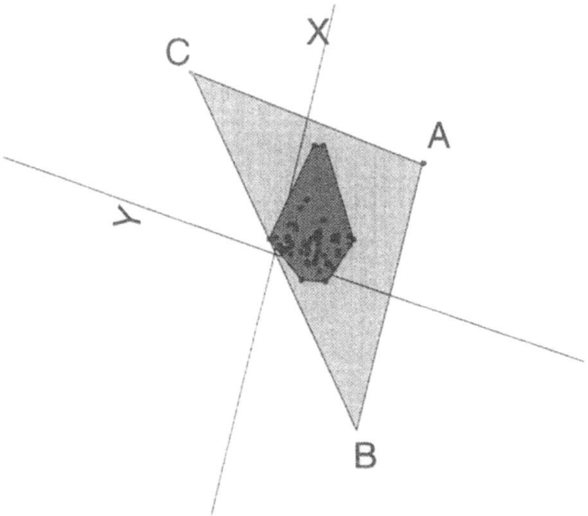

Figure 3. The RAPT deployment at Exuma Cay, Bahamas. A, B and C indicate the 3 RAPT telemetry buoys located about 400 m apart. X and Y denote the axes of the VRAP software Cartesian coordinate system. Each dot denotes a position fix. The minimum convex polygon for fish #16 during a 24 h period between 08 July – 09 July, 1998, is darker and encompasses an area of 2810 m^2 as determined by the Animal Movement Analyst written as an Arc View® extension by Hooge and Eichenlaub (1997).

Bolden (this volume) describes the first direct application of RAPT to evaluate fish movements within an MPA. Many techniques have been applied to monitor changes in fish abundance within MPAs (Willis et al. 2000), but RAPT is now providing unique continuous records of individuals that substantiate estimates made from statistical approaches (Millar and Willis, 1999). Earlier acoustic tracking studies on patterns of fish movement in relation to MPAs (Zeller, 1997; Zeller and Russ, 1998) have also been extended to include RAPT to provide the temporal and spatial resolution needed to monitor movements across MPA boundaries. Seino et al. (1999) used RAPT to monitor the movements of endangered horseshoe crabs near their breeding and nursery grounds, resulting in preservation of critical mudflats (Seino and Uda, 1999) and, hopefully, the eventual establishment of an MPA.

3. WHY IN MPAs?

An early project with a wired system, tracking *Octopus vulgaris* in 1985 in the Cerbere Reserve near the Laboratoire Arago in Banyuls-sur-Mer, France (McKenna, 1987; Mather and O'Dor, 1991), yielded more information about illegal fishing than about octopuses. Although the system was on the fenced and guarded grounds of a hospital, inside a no-take marine reserve, an old Catalan fisher (no doubt following a practice established long before the reserve in 1965), regularly removed our tagged octopuses by jabbing near-shore dens with a stick from land. His pre-dawn routine apparently avoided both hospital and reserve guards, but our positioning system faithfully recorded the removals.

In addition to this unique capacity to aid in enforcement of MPA regulations through continuous monitoring, RAPT has tremendous potential to test theories about how MPAs affect populations of a wide spectrum of marine organisms. Potential effects of MPAs on target species include these: (1) increased abundance, (2) increased mean individual size and age, (3) increased reproductive output, (4) enhanced recruitment inside and outside refugia, (5) maintenance of genetic diversity of stocks, and (6) enhanced fishery yields in adjacent fishing grounds. When properly established and managed, MPAs should yield results not easily achieved by more traditional management practices.

Design requirements for MPAs should include (1) dispersion/larval export potential, (2) sources and sinks of larvae, (3) protection of larger individuals, (4) movement of adults into nearby fishing grounds, (5) potential fishery yields, and (6) the range of effectiveness. Although the many long-term advantages of MPAs are attractive to environmentalists and managers, support for area closures by commercial fishers usually depends on clear and predictable evidence of cross-boundary movements out of reserves. Such movements, including the net emigration of adult biomass to outlying areas is believed to be the mechanism leading to increased fishery yields outside MPA boundaries (Alcala and Russ, 1990; Russ and Alcala, 1996, 1999).

RAPT is a powerful tool for quantifying cross-boundary movements, or "leakage" on a species-by-species and location-by-location basis. Leakage can enhance local fisheries only if the refugia are located within reach of the fishing vessels. Rates of transfer between MPAs and areas open to fishing will be species dependent. The amount of emigration from an MPA is thought to be dependent partially on the function of perimeter:area ratio (Rowley, 1994). The habitat bordering an MPA will also influence the amount of emigration as habitat-specific species may be reluctant to cross foreign habitats.

Daily activity patterns, spawning migrations and natural population densities need to be considered when assessing potential leakage (e.g., Zeller 1997, 1998, Zeller and Russ 1998, 2000). Density-dependent competition for food and shelter within MPAs can drive some individuals into adjacent harvest zones (Dugan and Davis, 1993). Animals that have limited movements and size distributions that have been lowered by fishing pressure will most likely show dramatic responses to MPAs. Animals with long life spans for which the maximum historical size is larger than is common today will also fare well.

Quantitative evidence for target species replenishing areas outside MPA is scarce (Dugan and Davis, 1993; Holland et al. 1996), although individual records, such as Bolden's (2000) recapture of a tagged Nassau grouper over 200 km from the RAPT site at a presumed spawning aggregation, make it clear that it does occur. Clear evidence for "quick" depletion once refugia are opened to harvesting is available (Russ and Alcala, 1999).

4. EXAMPLES

4.1 Animal Monitoring

4.1.1 Lizard Island, Australia

Figure 4 shows a typical layout of a RAPT system deployed in Watson's Bay in front of the Lizard Island Lodge as used in studies of the reef squid *Sepioteuthis lessoniana.* Also indicated on the figure is the 3-buoy RAPT layout used by Zeller and Russ, with some typical tracks of a coral reef snapper, the stripey *Lutjanus carponotatus* and a coral trout *Plectropomus leopardus* moving in relation to the MPA boundary that extends out from Chinaman's Ridge. This study will provide greater spatial and temporal resolution than previous manual telemetry and tagging studies of *P. leopardus* associated with MPAs at Lizard Island (Zeller, 1997; Zeller and Russ, 1998, 2000). The larger range of movements of *P. leopardus* compared to *L. carponotatus* is evident from the continuous 19 h tracks illustrated in Figure 4. These species also exhibit different behaviour patterns. *P. leopardus* is essentially day-active (see also Zeller 1997) while *L. carponotatus* is active both day and night, although this would be better illustrated through animation (O'Dor et al., 1998). The coral trout in this example was 57.2 cm fork length (FL) when a V16 transmitter was implanted into its body cavity in April, 1997. It was tracked intermittently over the next 2 years until recapture after this study in May, 1999, at 60.2 cm FL. The stripeys tracked at Lizard Island had V8 transmitters inserted orally into the stomach under full anaesthesia and retained their transmitters a maximum of 10 days, but regurgitated transmitters could be recovered on SCUBA using a handheld tracking receiver (VUR96, VEMCO) and replaced. All transmitter attachments were carried out using the method described by Zeller (1999).

The squid track shows a marked contrast to the fish behaviours. It was difficult to retain squid in the array. Four 1-m^2 floats were attached corner-to-corner to buoy B to create a squid aggregating device (SAD) based on observations of aggregations under boats (Adamo and Weichelt, 1999; O'Dor et al. 2001). Squid would orient to the SAD for several hours after release. In this example, the squid then moved across the array during slack tide and held position near buoy A until it finally departed from the bay (and the array) on the ebbing tide.

This diamond shaped array of buoys includes a fixed position reference transmitter, which also telemetered temperature with a differential pressure transducer on a rotating vane (Figure 5), which transmitted current direction and velocity. This array will serve as an example of the various features of RAPT monitoring systems described below. The dimensions of the array were chosen to allow all four buoys to receive acoustic pings from VEMCO V8 transmitters (8 mm diameter by 30 mm), thereby allowing positioning by the four buoy array. These transmitters could be carried by animals as small as 40 g. Ping arrival times were recorded at each buoy and relayed by radio to the base station (BS) located in the Lizard Island Lodge dive locker where positions where calculated with prototype four-buoy software (VRAPB 4.06, VEMCO). Larger V16 transmitters could be detected and positions calculated anywhere in the Bay, but real-time positioning of V8s was restricted to the central array by their low power output.

Although this approach reduced the number of real-time positions recorded from the smaller transmitters, the advantage of the four-buoy software is that it allowed continuous monitoring of the entire bay, an area approximately 1 km^2. The four-buoy array is made up of two equilateral triangles (the optimum configuration, Wartzok et al., 1992) and the software records all "hits" by acoustic signals on each buoy hydrophone. In post-analysis (O'Dor et al., 1998) these data can be resolved as positions and tracks using clustered hits resulting from a single ping on any three buoys. Four clustered hits can be positioned in multiple triangles to provide estimates of positional accuracy.

It is clear just from these limited data that rates of cross-boundary movement must be assessed on a species-by-species basis. The behaviours of the two fishes are quite distinct and those of the squid even more so. The fishes displayed distinct home ranges, with species-specific sizes and diurnal patterns, but they rarely left the study area covered by the RAPT system (Figure 4).

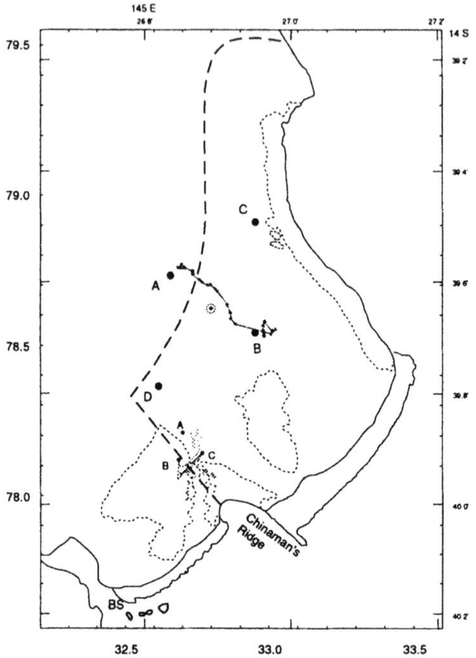

Figure 4. Two RAPT systems deployed in Watson's Bay, Lizard Island, Australia, for different studies. The map gives both kilometre grid and latitude and longitude references, and the heavy dashed line encloses a no-take MPA zone. The large four-buoy diamond was linked to the base-station (BS) at the Lizard Island Lodge and produced the illustrated 3 h track of a squid *Sepioteuthis lessonianna* in February, 2000. The smaller triangle recorded territories over 24 h for a stripey *Lutjanus carponotatus* (smaller home range) and a coral trout *Plectropomus leopardus* (divided home range), as indicated by points, and the illustrated track for a diver conducting a survey for passively tagged fish in April, 1999.

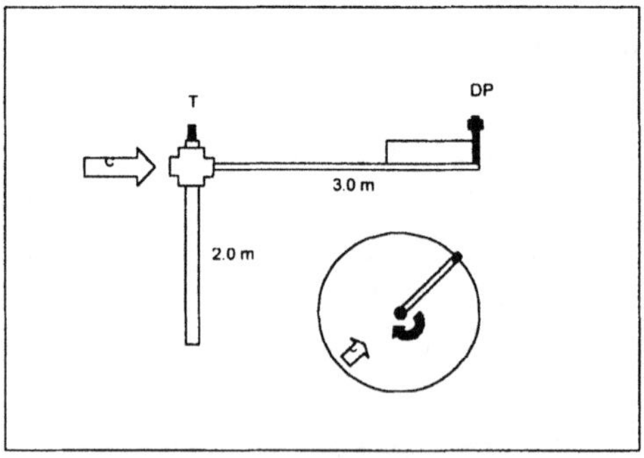

Figure 5. The RAPT-based current meter at the centre of the RAPT diamond in Figure 4. The centre post holds a fixed, reference temperature transmitter (T) that anchors a rotating vaned boom carrying a pitot-type differential pressure transducer/transmitter (DP), which measures current velocity with no moving parts. The inset below shows the meter from above. Current direction (indicated by the C-arrows) rotates the boom (dark arrow), and its angle is indicated by the DP transmitter's position on the base-station screen.

This particular coral trout clearly occupied territory on two separate patch reefs, one inside the reserve and the other outside. Caribbean reef squid have been described as occupying territories and having highly structured schools with complex social organisation (Moynihan and Rodaniche, 1982). However, RAPT confirmed earlier studies of the Lizard Island squid showing that some habitats always had squid present in them, but, in fact, individual squid were highly mobile and seldom reappeared in the same site (Adamo and Weichelt, 1999; O'Dor et al., 2001). Within a few days even acoustically tagged squid as small as 8 cm mantle length (ML) were no longer detectable around the Island, let alone within the array.

While the potential for handling-induced mortality could not be assessed in this case, such evaluations should form part of any major study. For example, previous studies on coral reef fish have indicated a high mortality rate for injured or stressed animals, requiring species and habitat specific considerations (Zeller, 1999). Given such considerations, one of the outstanding strengths of RAPT for MPAs is its ability to rapidly categorise species in terms of their mobility.

4.1.2 Exuma Cay Land and Sea Park

Three RAPT buoys were utilized to determine the home range of a Nassau grouper (*Epinephelus striatus,* Serranidae) in the Exuma Cay Land and Sea Park, central Bahamas. While the park was initially chosen as a study site owing to the presence of many large Nassau groupers, the absence of fishers also insured that the tagged individuals remained at large during the tracking period.

The buoys were positioned in a 400 m equilateral triangle over individual patch reefs, providing total acoustic coverage. Individual frequency V16 transmitters were successfully implanted into the coelomic cavity of twenty-four grouper ranging from 39.1 to 72.1 cm TL. Because V16 transmitters have greater power output (i.e., higher decibels) than the V8 transmitters in the Lizard Island study routine signal transmission over distances of greater than 400 m was possible. Nassau grouper that ventured outside the buoy array were tracked for about an additional 50 meters before their signal transmission was lost. The distance between the RAPT buoys and the base-station radio receiver/computer is conservatively rated by the manufacturer at 2 km. During this study, the base-station receiver was positioned about 1.8 km from the buoy array, but field-testing indicated that radio transmissions were satisfactory to nearly 4 km.

The RAPT buoy system is currently capable of tracking transmitters on ten individual frequencies. In this study, between four and eight individual V16 transmitters were simultaneously tracked for a period of three weeks.

The transmitters were also equipped with a pressure sensor to telemeter depth (Z), which allows plotting fish positions in three dimensions (i.e., X, Y, and Z) and improves X, Y positions during vertical movements.

The continuous monitoring available with RAPT provides not only detailed real-time movement patterns of animals; it can also confirm vitality. Acoustic tracks for 87.5% of the total 24 tagged groupers were generated for the entire three-week study period. To avoid any behavioral effects they might introduce, SCUBA divers did not enter the tracking area during the tracking period and could not verify vitality using visual sighting of individuals during the tracking period. Suspicions regarding a sudden abnormally stable position of a single grouper near day three of tracking were confirmed when divers later found the transmitter free on the bottom.

Two fish with implanted transmitters were unexpectedly recovered after the tracking study. A fisher caught a tagged *E. striatus* 185 days post-tracking at a presumed spawning aggregation approximately 220 km (in a straight ESE route) from the release point (Bolden, 2000). Another *E. striatus* tracked in 1997 was recaptured in 1998 on its original patch reef; the grouper had a well-healed incision with an encapsulated transmitter.

All 21 Nassau grouper tracked in this study remained either within the buoy array itself or very close to the periphery. Data reveal a high degree of site fidelity over a long period (weeks) with stable core areas and well-defined, overlapping home range areas. A minimum convex polygon for fish #16 (41.7 cm TL) created by the Animal Movement Analyst ArcView extension (Hooge and Eichenlaub 1997, www.absc.usgs.gov/glba/gistools/index.htm) is presented in Figure 3. Total area coverage for a 24 h tracking period of 08 July–09 July 1998 is 2810 m^2.

4.1.3 Leigh Marine Reserve

Currently there are 16 no-take MPAs (Marine Reserves) administered by the New Zealand Department of Conservation. The oldest of these, established in 1975, is the Cape Rodney to Okakari Point (CROP) Marine Reserve, at Leigh, in which a RAPT tracking study has recently been initiated. The reserve is 549 ha in size and encompasses approximately 5 km of coastline, out to 800 m from the low tide mark (Figure 6). The dominant predatory fish on this coast is the sparid snapper *Pagrus auratus*, which also forms the basis of important commercial and recreational fisheries. A study has been established to provide information on the extent of movements undertaken by *P. auratus*. The data produced will be used to plan marine reserve boundaries and optimise their size. Although improved conservation of snapper within reserves is the primary goal of this work, the data can also be used to examine cross-boundary movements that contribute to the role of marine reserves as a fisheries management tool.

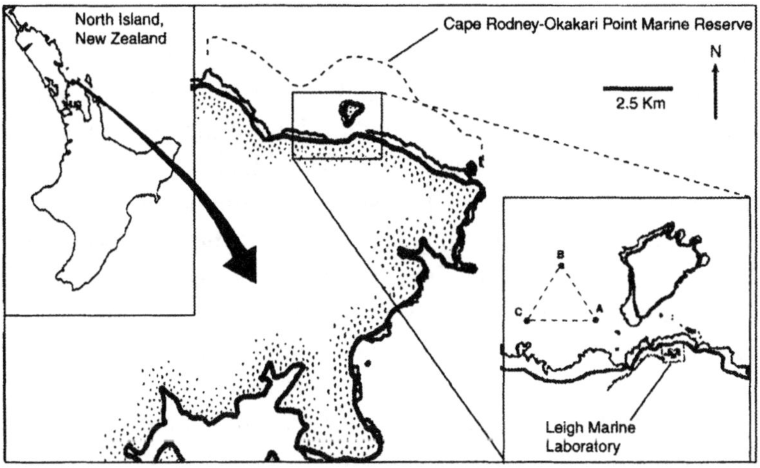

Figure 6. The location of the CROP study area as denoted by the array of buoys opposite the Leigh Marine Laboratory.

Fishers and fisheries scientists have long thought that snapper undertake seasonal migrations of many kilometers (Crossland, 1976), possibly related to water temperature or the formation of spawning aggregations (Paul, 1967), but there is little evidence to support this. Long-term studies of snapper tagged in the CROP marine reserve using visible elastomer tags (Willis and Babcock, 1998) indicated that some proportion (perhaps up to 50%) of snapper may not move more than a few hundred meters (Willis, Parsons and Babcock, unpublished data). Juvenile snapper abundances on inshore reefs and other habitats have been reported to peak over the summer (Kingett and Choat, 1981; Francis, 1995). It is possible that some snapper of all sizes follow lenses of warm water, which in marine reserves may take them across boundaries.

Angling surveys and baited underwater video (BUV) have shown that snapper abundance peaks in the reserve's centre and declines towards the edges (Millar and Willis, 1999; Willis et al., in press). Possible explanations for this pattern include increased probability of capture due to a lack of surveillance towards the reserve edges encouraging poaching, or home ranges of snapper that overlap the reserve edge. Either way, high levels of fishing on the edge of the CROP reserve could explain the gradual decrease in abundance away from the reserve center.

An array of three VEMCO sonobuoys has been set up in close proximity to the Leigh Marine Laboratory in an area averaging about 12 m in depth and including reef and sand habitats. Five snapper caught here had V16 acoustic pingers surgically inserted following similar protocols to those described by Zeller (1999). Fish appeared to overcome the trauma of capture and surgery, and quickly returned to normal behavior. Smaller V8 pingers were also being used. These tags were encased in a squid (*Notodarus sloanii*) hood and readily ingested by snapper when fed by divers. This allowed the fish to be tagged without any handling and yielding around two days of movement information before the tag was passed. Fish positions are recorded every minute, around the clock.

Data so far indicate that all fish exhibited a high degree of site attachment, spending the vast majority of their time in areas of no greater than 400 m radius. For some fish, this range is can be as small as 50 m (Figure 2). Less frequently, fish make excursions of 400–600 m radius and some fish have been observed to move outside tracking range for a period of a few hours.

The data acquired using the RAPT system provide a clear explanation of how snapper populations have been able to recover so markedly inside the CROP reserve. That is, typical home range size is encompassed by the dimensions of the reserve. The scale of movement also corresponds reasonably well with the observed edge effects, where abundance is lower at the reserve margins. This could come about through the capture of fish whose home ranges overlap reserve boundaries, as well as the capture of fish during less frequent, though longer-range, movements. The movement data will be useful for individual-based models of fish population dynamics in reserves, as well as improving our understanding of fish habitat utilization. The latter project also utilizes the RAPT system through video-based mapping conducted by tagged divers.

4.1.4 Moriye Bay

A RAPT array set up to monitor horseshoe crab breeding and overwintering activities in Moriye Bay, Oita Prefecture, in northeastern Kyushu Island, Japan, has promoted co-operation between Kitsuki City, River Department of Oita Prefecture, the local Fishery Coop and Sumiyoshihama Resort Park. RAPT's potential for protecting habitats of threatened species through elevating public awareness is clear. Studies of spawning horseshoe crabs in a conservation area, proposed by a citizen's group near the privately owned park, have helped to conserve one of the few remaining breeding populations in Japan (Seino et al. 1995, 1999; O'Dor and Seino 1997).

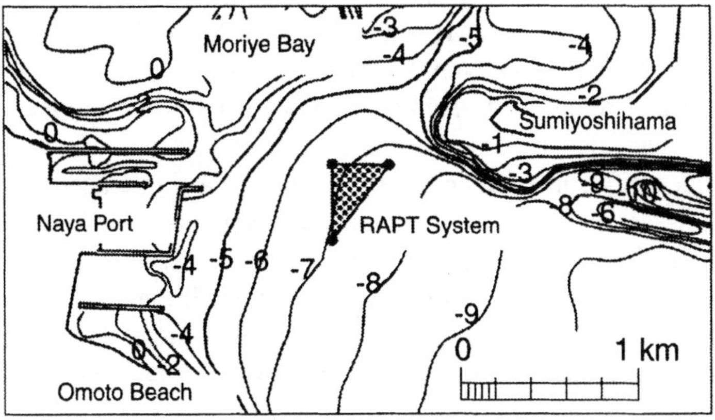

Figure 7. The Moriye Bay RAPT system is positioned at the entrance to the Yasaka and Takayama River mudflats, relaying data to a base station in Sumiyoshihama Park a. The RAPT systems position between the Sumiyoshihama sand spit and the Naya Port allows it to intercept all possible paths to and from spawning areas without blocking the channel to the fishing harbour at the top of the figure.

As indicated in Figure 7, the triangular buoy array monitors crab movements to spawning sites on the mudflats and estuaries from their assembly areas in deeper waters. Movements are recorded at the base-station in an interpretation centre in the park. These real-time data are used by volunteers to locate and observe spawners. The RAPT system was maintained in continuous operation for over four months and demonstrated the return of the crabs to the mudflats outside their breeding season. This had important consequences in a landmark decision to develop a mitigation plan for proposed flood control modifications to the Yasaka and Takayama Rivers that would have negatively impacted the critical mudflat breeding and nursery areas. This is a remarkable environmental transition for Japanese civil engineering—the first alteration in a plan made 35 years ago. This biotelemetry study with broad support from a wide variety of social groups has become an example of the use of new technology for coastal environmental monitoring (Seino et al., 1999). In 1997, the Japan National River Act was revised, and revisions continued to include the Coast Act in 1999. Future projects with the cooperation of local fishing fleets will extend the area of protection with linked reserves between the inshore nursery and offshore adult feeding areas.

4.2 Environmental Monitoring

Animals can change behaviour in response to the environment, so it is important to correlate behaviours with environmental change to distinguish biologically driven events from physically driven ones. How can we determine if a fish leaves an MPA as a result of local biology or larger scale El Nino or global warming event? Normally, RAPT systems need to include bottom-mounted reference transmitters to correct for buoy movement; these transmitters can also record environmental variables. Temperature and depth changes reflecting tides can be recorded on a single acoustic frequency and in the same files that record animal movements. In the Lizard Island study previously described, the temperature reference transmitter was mounted on a central 2 m pole. A 3-m vaned boom with a differential pressure transducer operating as a pitot tube was attached to the pole and served as a telemetering current meter (Figure 5) (Aitken et al., 2001; Webber et al. in press; Voegeli et al., 2001). The differential pressure transducer monitored velocity and current direction, which was indicated on screen as the boom rotated. Two other variables that can be routinely monitored with RAPT are salinity (important in estuaries) and light intensity, which is a key variable driving diel activity.

4.3 People Monitoring

Like animals, human divers can be monitored with RAPT, allowing remote supervision of diver safety and effectiveness. Objects and sites can be marked. Divers can potentially carry larger telemetry packages allowing transponder contact via the buoys. Transponders provide two-way communication allowing the base station to signal for return or even to advise divers on directional information. A RAPT system in an MPA which combines conservation with public diver activity can provide information on animal positions for easier location by visitors and can also monitor visitor location for safety and control. As pointed out in the introduction, RAPT can record patterns of animal disappearance from an MPA that can indicate illegal removals, but transmitters on boats and dive gear would also allow direct monitoring of visitor compliance with regulations without requiring personnel for diver supervision. Differential pressure transducers have been adapted for telemetering respiration rates and amplitude from SCUBA regulators (Aitken et al., 2001) or fin-beat frequency (Webber et al. in press), so that the same tags that ensure compliance can also ensure safety.

Human tracking also has a number of scientific applications. In the Lizard Island study, Russ and Zeller used standardized underwater visual census (UVC) surveys completed by divers carrying ultrasonic transmitters to observe fish carrying both standard external tags and acoustic transmitters. The fish positions were verified on shore by cross-referencing RAPT and diver recorded time-bases. The solid line track shown in Figure 4 was fit to the median positions using an iterative approximating spline to provide a continuous track. This track was then splined again in the time dimension to interpolate a time for any position (O'Dor et al., 1998). Thus, using standard UVC surveys in conjunction with RAPT-based monitoring of target species and divers permits assessment and adjustment of UVC for both sightability and changes in behavior due to the presence of the diver. Optimization of this technique will require use of the digital signal processing system that is currently under development.

A similar technique could rapidly survey fish with visual tags using underwater photography or video with diver position verifying fish positions on shore by cross-referencing RAPT and image time-bases. This could provide a record of locations, associations and habitat for large numbers of animals without expensive acoustic transmitters or the need for complex surgical implantation.

4.4 Habitat Monitoring

The use of RAPT to locate divers making underwater video recordings outlined above also provides the basis for rapid, non-destructive, repeatable surveys of MPA habitats. Traditional survey techniques require fixed, visible markers and even rope transect lines, which require expensive dive time to set up and may be visually unacceptable to visitors. In contrast, it requires only a few minutes to attach RAPT buoys to permanent mooring blocks set hundreds of meters apart. Subsequently, video images shot by a diver carrying a transmitter can be located with 1–2 m accuracy. Random surveys of habitat in an area can be completed rapidly as RAPT allows reconstruction of positions on shore. With a little more effort, complete bottom coverage is possible even to the extent of reconstructing a detailed map of benthic biota by grabbing individual frames and transferring them to a GIS referenced CD-ROM. Animal tracks can then be superimposed on bottom type and changes in biota with time can be documented. Many existing survey and census techniques depend on repeated observations along the same routes based on visual references in real time. These might require modification to take advantage of RAPT, but acoustic information can be substituted for visual information using diver-held directional receivers. The acoustic techniques have the advantage of working in the dark or other low-visibility situations. The Canadian Defence and Civil Institute of Environmental Medicine (DCIEM) is currently developing certifiable survey techniques using diver-held receivers to orient to surface controlled pingers inside RAPT arrays (Eaton and Webber, 2000).

The next generation of RAPT buoys will include digital signal processing, allowing continuous monitoring of multiple frequencies and transponder contact to targets. This technology can actually guide divers or potentially even ROVs through the array. With this feature, bottom surveys to monitor changes in kelp beds or coral reefs could be almost automatic, adding another dimension to RAPT applications in MPAs.

5. CONCLUSIONS

RAPT is becoming an excellent tool to aid in the design of and assessing the effects of MPAs. Results from numerous studies utilizing RAPT, currently in preparation, will provide required quantitative data to test models regarding the design of MPAs and their potential effects on animal movement and distribution. Advances in both the RAPT system and commercially available movement software will enhance the applicability of RAPT and assist in data analyses.

The next generation of RAPT will incorporate digital signal processing (DSP). As a result the RAPT system will be able to monitor multiple frequencies and pulse rate variations which will result in significantly increasing the number of trackable transmitters. Transponder functions can be used to guide divers and potentially even remotely operated vehicles (ROVs) through the buoy array. This function will automate bottom surveys to monitor, for example, changes in kelp beds or coral reefs by comparing changes between time-based images crossed referenced with RAPT positions. RAPT will also be adding the ability to incorporate locations from the Global Positioning System (GPS) into the data records.

New and improved software packages incorporating GPS and GIS will allow RAPT data to be manipulated and analyzed more easily. Coupling movement and environmental factors in fine-scale detail will ultimately allow MPA design and applicability to be quantitatively examined.

REFERENCES

Adamo, S.A., and Weichelt, K.J. (1999) Field observations of the schooling of the oval squid, *Sepioteuthis lessonianna* (Lesson, 1830) *J. Moll. Stud.* **65**, 377-380.

Alcala, A.C., and Russ, G.R. (1990) A direct test of the effects of protective management on abundance and yield of tropical marine resources. *J. Cons. Int. Explor. Mer.* **46**, 40-47.

Aitken, J.P., Webber, D.M., O'Dor, R.K., and Voegeli, F.A. (2001) *In* Acoustic telemetry of integrated differential pressures: calibration and applications. Proc. 3rd Conf. Fish Telemetry Europe, Norwich, England, pp. 21-27.

Bjordal, A., Engas, A., Soldal, A.V., and Ovredal, J.T. (1993) A new radio link telemetry system: experiences from tracking fish and crustaceans. ICES CM 1993/B:23, 6pp.

Bolden S.K. (2000) Long-distance movement of a Nassau grouper (*Epinephelus striatus*) to a spawning aggregation in the central Bahamas. Fish. Bull. U.S. **98**, 642-645.

Bolden S.K., and R.Y. Ewing (1999) Surgical implantation of ultrasonic transmitters in adult Nassau grouper (*Epinephelus striatus*). *In* Proc. Internat. Assoc. Aquatic Animal Med. **30**, 53-55.

Crossland, J. (1976) Snapper tagging in north-east New Zealand, 1974: analysis of methods, return rates and movements. *N.Z. J. Mar. Freshwater Res.* **10**, 675-696.

Dugan J.E., and Davis G.E. (1993) Applications of marine refugia to coastal fisheries management. *Can. J. Fish. Aquat. Sci.* **50**, 2029–2042.

Eaton, D., and D.M. Webber (2000) Acoustic jackstays. *In* Proceedings of the ABCANZ 3 Diving Conference, Auckland, N.Z., September, 2000.

Francis, M.P. (1995) Spatial and seasonal variation in the abundance of juvenile snapper (*Pagrus auratus*) in the north-western Hauraki Gulf. *N.Z. J. Mar. Freshwater Res.* **29**, 565-579.

Hawkins, A.D., MacLennan, D.N., Urquhart, G.G., and Robb, C. (1974) Tracking cod *Gadus morhua* L. in a Scottish sea loch. *J. Fish. Biol.* **6**, 225-236.

Holland, K.N., Lowe, C.G., and Wetherbee, B.M. (1996) movements and dispersal patterns of blue trevally (*Caranx melampygus*) in a fisheries conservation zone. *Fish. Res.* **25**, 279-292.

Hooge, P.N., and Eichenlaub B. (1997) Animal movement extension to Arcview. Ver 1.1 Alaska Biological Science Center, U.S. Geological Survey, Anchorage, AK, USA.

Kingett, P.D., and Choat, J.H. (1981) Analysis of density and distribution patterns in *Chrysophrys auratus* (Pisces:Sparidae) within a reef environment: an experimental approach. *Mar. Ecol. Prog. Ser.* **5**, 283-290.

Mather, J., and O'Dor, R.K. (1991) Foraging strategies and predation risk shape the natural history of juvenile *Octopus vulgaris*. *Bull. Mar. Sci.* **49**, 256-269.

Maynard, D.R., and Webber, D.M. (1987) Monitoring the movements of snow crab (*Chionoecetes opilio*) with ultrasonic telemetry. *In* Proceedings Oceans '87. **3**, 962-966.

McKenna, S. (1987) Energy budgets of *Octopus vulgaris* based on a telemetry value for activity in nature. B.Sc. Honours thesis, Biology Department, Dalhousie University, Halifax, Canada.

McKibben, J.N., and Nelson, D.R. (1981) A portable real-time, x-y plotting system for ultrasonic tracking of fish. *In* Long, F.M. (ed.) Proceedings of the 3rd International Conference on Wildlife Biotelemetry, University of Wyoming, Laramie, pp. 105-115.

Millar, R.B., and Willis, T.J. (1999) Estimating the relative density of snapper in and around a marine reserve using a log-linear mixed effects model. *Aust. N. Z. J. Statistics* **41**, 383-394.

Moynihan, M., and Rodaniche, A.F. (1982) The behaviour and natural history of the Caribbean reef squid *Sepioteuthis sepioidea*. *Adv. Ethol.* **25**, 1-151.

O'Dor, R.K., and Seino, S. (1997) Monitoring marine reserves with radio-acoustic positioning and telemetry (RAPT) systems. University of British Columbia Fisheries Centre Res. Rpts. **5**(1): 14-15.

O'Dor, R.K., and Webber, D.M. (1991) Ultrasonic telemetry. *The Lobster Newsletter* **4**(1), 1-2.

O'Dor, R.K., Webber, D.M., and Voegeli, F.M. (1989) A multiple buoy acoustic/radio telemetry system for automated positioning and telemetry of physical and physiological data. *In* Uchiyama, A., and Amlaner, C.J. (ed.) Proc. 10th International Symp. Biotelemetry. Tokyo, Waseda Univ. Press, pp. 444-452.

O'Dor, R.K., Andrade, Y., Webber, D.M., Sauer, W.H.H., Roberts, M.J., Smale, M.J., and Voegeli, F.M. (1998) Applications and performance of Radio-Acoustic Positioning and Telemetry (RAPT) systems. *Hydrobiologia* **371/372**, 1-8.

O'Dor, R.K., Aitken, J.P., Andrade, Y., Finn, J. and Jackson, G.D. (2001) Currents as environmental constraints on the behavior, energetics and distribution of squid and cuttlefish. *Bull. Mar. Sci.* (in press)

Paul, L.J. (1967) An evaluation of tagging experiments on the New Zealand snapper, *Chrysophrys auratus* (Foster), during the period 1952-1963. *N.Z. J. Mar. Freshwater Res.* **1**, 455-463.

Rowley R.J. (1994) Marine reserves and fisheries management. *Aquat. Conser.* **4**, 233–254.

Russ, G.R., and Alcala, A.C. (1996) Do marine reserves export adult fish biomass? Evidence from Apo Island, central Philippines. *Mar. Ecol. Prog. Ser.* **132**, 1-9.

Russ, G.R., and Alcala, A.C. (1999) Management histories of Sumilon and Apo Marine Reserves, Philippines, and their influence on national marine resource policy. *Coral Reefs* **18**(4), 307-325.

Sauer, W.H.H., Roberts, M.J., Lipinski, M.R., Smale, M.J., Hanlon, R.T., Webber, D.M., and O'Dor, R.K. (1997) Choreography of the squid's 'nuptial dance'. *Biol. Bull.* **192**, 203-207.

Seino, S., O'Dor, R.K., Hamada, T., Tsuchiya, Y., Tsuchiya, M., Nishihara, S., and Kawahara, M. (1995) Horseshoe crab *Tachypleus tridentatus* spawning patterns in Japan from acoustic telemetry. *In* Cristalli, C., Amlaner, C.J., Jr., and Neuman, M.R. (eds.) Proc. 13th Internat. Symp. Biotelemetry. Fayetteville: U. Arkansas Press, pp. 205-208.

Seino, S., Tsuchiya, Y., O'Dor, R.K., Uda, T., Nishihara, M., Kugimiya, K., and Watnabe, N. (1999) Behavioral monitoring of horseshoe crabs by radio-acoustic biotelemetry. Proc. Coastal Engineering , *Jap. Soc. Civil Eng.* **46**, 1296-1300 (In Japanese).

Seino, S., and Uda, T. (1999) *The Horseshoe Crab Mud Flats.* Oita, Japan: Oita Prefecture, 60 pp. (In Japanese).

Smith, G.W., Urquhart, G.G., MacLennan, D.N., and Sarno, B. (1998) A comparison of theoretical estimates of the errors associated with ultrasonic tracking using a fixed hydrophone array and field measurements. *Hydrobiologia* **371/372**, 9-17.

Swanberg, T.R., and Geist, D.R. (1997) Effects of intraperitoneal transmitters on the social interaction of rainbow trout. *N. Am. J. Fish. Managment* **17**, 178-181.

Voegeli, F.A., Webber, D.M., Smale, M.J., Andrade, Y. and O'Dor, R.K. (2001) Ultrasonic telemetry, tracking and automated monitoring technology for sharks. *Environ. Biol. Fish.* **60**, 267-281.

Wartzok, D., Sayegh, S., Stone, H., Barchak, J., and Barnes, W. (1992) Acoustic tracking system for monitoring under-ice movements of polar seals. *J. Acoust. Soc. Am.* **92**, 682-687.

Watkins, W.A., and Schevill, W.E. (1972) Sound source location with a three-dimensional hydrophone array. *Deep-Sea Res.* **19**, 691-706.

Webber, D.M., Voegeli, F.A., and Smale, M.J. (in press). The development of differential pressure techniques to measure swimming speed in fish. *In* Eiler, J. (ed.) Proceedings of the 15th International Symposium on Biotelemetry, Juneau, Alaska, June,1999.

Willis, T.J., and Babcock, R.C. (1998) Retention and *in situ* detectability of visible implant fluorescent elastomer (VIFE) tags in *Pagrus auratus* (sparidae). *N.Z. J. Mar. Freshwater Res.* **32**, 247-254.

Willis, T.J., Millar, R.B., and Babcock, R.C. (2001) Detection of spatial variability in relative density of fish: comparison of visual census, angling, and baited underwater video. *Mar. Ecol, Prog. Ser.* 198: 249-260.

Zeller, D.C. (1997) Home range and activity patterns of the coral reef trout *Plectropomus leopardus* (Serranidae). *Mar. Ecol. Prog. Ser.* **154**, 65-77.

Zeller, D.C. (1998) Spawning aggregations: patterns of movement of the coral trout *Plectropomus leopardus* (Serranidae) as determined by ultrasonic telemetry. *Mar. Ecol. Prog. Ser.* **162**, 253-263.

Zeller, D.C. (1999) Ultrasonic telemetry: its application to coral reef fisheries research. *Fish. Bull.* **97**, 1058-1065.

Zeller, D.C., and Russ, G.R. (1998) Marine reserves: patterns of adult movement of the coral trout (*Plectropomus leopardus* (Serranidae)). *Can. J. Fish. Aquat. Sci.* **55**, 917-924.

Zeller, D.C., and Russ, G.R. (2000) Population estimates and size structure of *Plectropomus leopardus* (Pisces: Epinephelinae): Mark-Release-Resighting. *Mar. Freshwater Res.* **51**, 221-228.

Using Ultrasonic Telemetry to Determine Home Range of a Coral-Reef Fish

Stephania K. Bolden
NOAA/NMFS, 75 Virginia Beach Drive, Miami, Florida 33149

Key words: home range, telemetry, Nassau grouper, Bahamas

Abstract: Recent advances in telemetry allow the tracking of larger numbers of individuals over longer periods of time to reveal information about home range, spawning distribution, movement patterns, and habitat preference. Identifying and defining specific research objectives and parameters prior to initiating a telemetry study is critical. This paper outlines a series of topics that, when answered systematically, will assist in identifying a telemetry system appropriate to investigate and determine the home range of a coral-reef fish. An investigation of the home range of Nassau grouper (*Epinephelus striatus*) in the Bahamas is presented as a case study with points considered important to the success of the study. The materials and methods section directly applies elements of the case study. Results include a summary of the twenty two adult (>40 cm TL) Nassau grouper tracked by a radio-linked acoustic positioning buoy system. Mean number of position fixes was 3420 ± 1795 (SE) per fish ranging from 1032-7493. Data were analyzed using the Animal Movement Analysis Extension for Arcview®. Total cumulative home range increased less than 2% daily on tracking days 14-18 and less than 1% daily on tracking days 18-21. The mean home range area as determined by 95% minimum convex polygon (MCP) method for the Nassau grouper was 18305 m^2 ± 5806 (SE); the smallest home range was 1565 m^2 and the largest was 131805 m^2.

1. INTRODUCTION

Nelson (1990) stated "the potential of ultrasonic telemetry as a tool for the study of marine fishery resources is just beginning to be realized." Ten years later the application of telemetry technology to study marine fishes is a novel field approach that continues to expand rapidly. Recent telemetry advances allow the tracking of larger numbers of individuals over longer peri-

J.R. Sibert and J.L. Nielsen (eds.), Electronic Tagging and Tracking in Marine Fisheries, 167–188.
© 2001 *Kluwer Academic Publishers. Printed in the Netherlands.*

ods of time to reveal information about home range, spawning distribution, movement patterns, and habitat preference. Ongoing technological advances in telemetry such as digital signal processing, transmitters with underwater data download capabilities, and telemetric systems that incorporate locations from the global positioning system (GPS) into the data records promise further improvement.

Identifying and defining specific research objectives and parameters prior to initiating a telemetry study is critical. Recognizing constraints regarding the study animal, its habitat, available resources, and study site are essential prior to purchasing telemetry equipment. This paper outlines a series of topics that, when answered systematically, will assist in identifying a telemetry system appropriate to investigate and determine the home range of a coral-reef fish. Those topics are as follows:

I. Identify the research question and data needs (spatial and temporal)
II. Consider behavior and ecology of the target species with particular regards to habitat and movement ecology
III. Select study site
IV. Choose telemetry systems, transmitter, and method of transmitter attachment

Following the above outline, I used my research on the determination of the home range of Nassau grouper (*Epinepelus striatus*) in the Bahamas to demonstrate points considered important to the success of that study. The materials and methods section directly applies elements of that project to the topics listed above, and it is followed by some representative results, including a brief summary of computer programs available for home-range analysis.

2. GENERALITIES

2.1 Identify the Research Question and Data Needs

2.1.1 Home range

There has been considerable debate among terrestrial biologists concerning the definition of home range and methodologies to delineate the area (Harris et al., 1990; White and Garrott, 1990). The most commonly cited definition of a home range is "that area traversed by the individual in its normal activities of food gathering, mating and caring for young" (Burt, 1943). Because this definition does not account for the bipartite life history of many coral-reef fishes, Brown's (1975) definition of home range may be more appropriate. Hence, for the purpose of this study, home range is defined as an "area in which an animal normally lives, exclusive of migrations, emigrations, dispersal movements, or unusual wanderings" (Brown, 1975). A home range likely encompasses the minimum area necessary to provide key resources required by the individual; the actual shape and size of the home range is probably determined by the nature and distribution of these resources. To benefit its occupant, the home range of an individual must be large enough to contain a sufficient quantity of food resources to enhance the animal's reproductive success (Wilson, 1975). Clearly, size of individuals, (McNab, 1963), resource availability (Brown, 1975), metabolic requirements (Arnold, 1988), and a host of other variables correlate, interact, or otherwise influence the size of a home range (Mace et al., 1983).

Studies that use visual identification of the position of a reef fish over time (generally using SCUBA) are relatively common in the literature (e.g., Randall, 1962, 1963; Collette and Talbot, 1972; Luckhurst and Luckhurst, 1978; Holland et al., 1993a; Shapiro et al., 1993; Carter et al., 1994; Tulevach and Recksiek, 1994; Sluka and Sullivan, 1996). Home ranges as defined by Brown (1975) have been investigated in only a few coral-reef fish species: lemon shark (*Negaprion brevirostris,* Carcharhinidae; Morrissey and Gruber, 1993), blue trevally (*Caranx melampygus*, Carangidae; Holland et al., 1996), and coral trout (*Plectropomus leopardus,* Serranidae; Zeller, 1997).

Home-range size coupled with ecological information provide meaningful insight to space utilization, resource partitioning, habitat selection and fish behavior on the coral reef. Quantitative data determining the size, shape and patterns of movement within the home range are important in applied ecology, fisheries, and fisheries management. Home-range theory and home-range size can provide guidance to marine reserve design (e.g., Holland et al., 1993a; Zeller, 1997), and the proportion of the home range encompassed by the reserve boundaries can influence reserve effectiveness (Kramer and Chapman, 1999). Some models of marine reserve design and effectiveness have recognized that fish movements between reserve and non-reserve areas are likely to be important for the functioning of reserves (Polachek, 1990; Roberts and Polunin, 1991; DeMartini, 1993), however there are few quantitative studies examining either the home range or fine-scale movement patterns of reef fishes.

2.1.2 Data requirements

The goal of a home range study is to accurately determine an animals' areal coverage in a specified time period. The periodicity and quantity of position points required to provide data appropriate for analyses needs to be identified. Generally tracking occurs in either short-term (one to several days) or long-term periods (weeks, months, years) using manual or automated systems in intermittent or continuous tracking regimes. The interval between position points is relevant to precision and is dictated by both the speed at which the animal can travel and also the area covered in daily activities. Habitat complexity within the study area may also affect frequency of recorded position points because structural components (e.g., coral) can interfere with acoustic signal transmission.

While intermittent tracking may define the size and shape of an animal's home range, continuous tracking allows one to investigate how the animal utilizes the space within its home range. Specifically, the following elements can be investigated with more precision and significantly less effort by continuous monitoring versus intermittent tracking: diel activity patterns and habitat usage as well as crepuscular/nocturnal activity peaks; swimming speed, direction and duration; fine-scale site and position fidelity over long periods (weeks); stability and center of core activity areas; spatial persistence of home ranges through time; daily foraging migrations; onshore-offshore movements; spatial and temporal overlap of activity areas by numerous individuals; and individual differences in performance.

To determine a home range one needs to track an animal long enough to reveal the areas that the individual frequents during its normal activities. Because there is a positive correlation between number of tracking positions and size of activity space, Odum and Kuenzler (1955) introduced a technique of plotting the number of observations as a guideline to determine sample size. They modified the "species-area" curve used by ecologists by plotting the home range area on the y-coordinate and the number of points (telemetry fixes) on the x-coordinate and called it the area-observation curve. When the curve approaches an asymptote it is assumed that adequate samples have been obtained. Sampling after that point results in little or no observed increase and most subsequent data points will fall within the already-identified home range. Odum and Kuenzler (1955) considered a one percent level as a practical point to use—that is, when each additional observation produces less than one percent increase in area. However, other percentage levels have been used (e.g., Bowen, 1982; Messier and Barrette, 1982; Morrissey and Gruber, 1993).

Ultrasonic telemetry equipment is considerably more expensive than terrestrial radio telemetry, and evaluating the cost of undertaking an acoustic-tracking project in terms of time and money is an important consideration. When planning sample size in a tracking study, size class and sex of the study species should be considered in addition to the number of individuals.

Allowances may have to be made for mortality, tag loss, and dispersal from the study area.

2.2 Consider Behavior and Ecology of Target Species

Prior to tracking, direct field observations coupled with literature searches on the behavior and ecology of the targeted species will assist in choosing suitable telemetry systems and transmitters and methods of direct transmitter attachment and will identify potential limitations in data collection. Knowledge of the habitat types used by the organism will help determine if the transmitters might be snagged on structures, overhangs, crevices, or in shelter sites. Potential behavioral effects may be identified. Life history traits of age and size at maturation, and body size also need to be considered because they will affect transmitter size and attachment. Fish mobility and habitat preferences will affect the periodicity and success of acoustic transmission.

2.3 Select Study Site

Selecting the tracking area is an important consideration when planning a telemetry study as environmental variables can affect data quality and quantity. Noise interferes with telemetry system performance (Kanwisher et al., 1974; Stasko and Pincock, 1977; Pincock and Voegeli, 1992). In shallow or coastal waters, acoustic ocean noise is 5 to 9 dB higher than in the deep ocean (Pincock and Voegeli, 1992). Soniferous marine life noise (e.g., snapping shrimp) can be 20 dB or more above sea state zero noise (Pincock and Voegeli, 1992). Water flowing by a hydrophone also creates noise, as do rain and wind (Kanwisher et al., 1974; Pincock and Voegeli, 1992). Tracking systems that transfer data to receiving stations have limitations on distance between receivers and hydrophones. Additional factors such as current flow, water depth and temperature, accessibility of power supply, probable sea state, transport and security of tracking equipment, and fishing pressure should also be considered.

2.4 Choose Telemetry System, Transmitter and Transmitter Attachment

2.4.1 Telemetry

Telemetry systems available for use in fish behavior studies can be separated into radio and acoustic categories. Radio transmitters commonly use wavelengths from 40 to 200 MHz (Stasko and Pincock, 1977; Hawkins and

Urquhart, 1983; Winter, 1983). The radio signal is transmitted and received by an antenna and is used in terrestrial and freshwater studies (Stasko and Pincock, 1977; White and Garrott, 1990). Ultrasonic transmitters use wave lengths of 20-300 kHz (Winter, 1983); the frequencies most often used in fish tracking are between 40 and 80 kHz (Kanwisher et al., 1974; Winter, 1983). Absorption losses increase with frequency, therefore lower frequencies are preferred for fish tracking (Stasko and Pincock, 1977). Unlike radio waves, ultrasonic telemetry signals are transmitted through seawater and can be received by a hydrophone (Winter, 1983). Range for both radio and ultrasonic transmitters is highly variable because it is proportional to battery size, current drain, power output and local conditions. Although water depth itself does not reduce signal detection, thermal stratification can weaken the signal and physical obstacles (e.g., corals) can block it. Individual fish can be tracked by a combination of frequency pulse and frequency channel (Winter, 1983; Pincock and Voegeli, 1992). Telemetry studies in general are limited to low sample size because of the high cost of transmitters and tracking equipment.

When choosing a telemetry system, data requirements must be considered. Most home range studies require fine data resolution, continuous monitoring, minimal disruption of behavior, and relatively long tracking periods (weeks to month). Tracking systems that remotely monitor multiple animals are usually advantageous.

2.4.2 Transmitters

Transmitters function by emitting a pulse on a specified frequency. Transmitters may emit a fixed-rate pulse at varying intervals or infrequent bursts of pulses encoded with information. The choice of transmitter size involves a trade-off between detection distance and transmitter life (Winter, 1983). A larger, stronger transmitter has longer transmitter life and better distance reception, but the weight may affect fish behavior by interfering with buoyancy compensation, balance, energetics or swimming performance (Gallepp and Magnuson, 1972; McCleave and Stred, 1975; Ross and McCormick, 1981; Summerfelt and Mosier, 1984). Smaller transmitters minimize effects on fish behavior and swimming performance, but have shorter battery-life and range. Transmitters can be equipped with sensors which measure environmental (e.g., depth, water temperature) or physiological (e.g., heart rate) variables.

2.4.3 Transmitter attachment

Transmitters can be surgically implanted into the body cavity (internal), inserted into the stomach (gastric), or fixed to the external surface of the body (e.g., Peake et al., 1997; Winter, 1983; Zeller, 1999). Studies that investigated these three procedures generally indicate that external tagging can significantly alter behavior or swimming ability. The ratio of transmitter mass to body mass (i.e. transmitter ratio) can affect the suitability of a particular attachment procedure (Marty and Summerfelt, 1986). Generally, transmitter size should be between 1.5 to 6.0% of a fish's body weight (Stasko and Pincock, 1977; Ross and McCormick, 1981; Winter, 1983; Summerfelt and Mosier, 1984). Selecting the method of transmitter attachment is both species- and species-habit specific (Zeller, 1999). Gastric transmitters have been found to affect foraging ability (Armstrong and Rawlings, 1993), thereby reducing their suitability for longer-term studies (McCleave and Stred, 1975; Zeller, 1999). Surgical implantation of transmitters into the body cavity of fish through an incision into the ventral body wall is a technique well suited for long-term experiments (Stasko and Pincock, 1977; Zeller, 1999), especially for bottom-associated reef fishes (Zeller, 1999). The disadvantage of surgically implanting transmitters is postoperative trauma. The physiological effects of surgery rather than of the transmitter itself have been found to affect swimming speed of Atlantic salmon smolts tagged with internal and external tags (Peake et al.,1997); anesthetic agents induce sedation and influence recovery differently (Zeller, 1999). Transmitters have been surgically implanted successfully into the body cavity of many fish species: channel catfish (*Ictalurus punctatus,* Ictaluridae) (Ziebell, 1973); flathead catfish (*Pylodictis olivaris*, Ictaluridae) (Hart and Summerfelt, 1975); white bass (*Morone chrysops,* Percichthyidae), smallmouth bass *(Micropterus dolomieui*, Centrarchidae), spotted bass (*Micropterus punctatus,* Centrarchidae) and largemouth bass *(Micropterus salmoides,* Centrarchidae (Peterson, 1975); lemon sharks (Morrissey and Gruber, 1993); white goatfish (*Mulloides flavolineatus,* Mullidae) (Holland et al., 1993a); and coral trout (Zeller, 1997, 1999).

3. APPLICATION OF A CASE STUDY— THE NASSAU GROUPER PROJECT

3.1 Identify the Research Question and Data Needs

This study was undertaken to rigorously examine and provide the first description of Nassau grouper home range. However, prior to determining the analytic home range, the assumption of site fidelity had to be made (Spencer et al., 1990). Site fidelity exists when the observed area used by an individual is significantly smaller than the area used if an individual's movement is random (Munger, 1984, Danielson and Swihart, 1987; Spencer et al., 1990). An individual that exhibits site fidelity is judged to possess a home range (Spencer et al., 1990). Adequate sample size also needs to be determined before home range analyses are conducted. Both the number of observations and the length of time an animal is observed can influence home range size (Winter, 1977). An area-observation curve (Odum and Kuenzler, 1955) can determine if the length of the tracking period has demonstrated a stable home range size. Determining the number of Nassau grouper to include in this study was influenced by budget constraints, other reef fish home range studies and local population density. Three studies have determined home range by intermittently recording telemetric positions of 6-39 reef fish over periods from hours to months recording up to 8002 position fixes (Morrissey and Gruber, 1993, Holland et al., 1993b; Zeller, 1997). This is the first attempt at determining home range of a coral-reef fish with continuous tracking.

3.2 Behavior and Ecology of the Nassau Grouper

The Nassau grouper are large tropical western Atlantic serranids generally found near shallow high-relief coral reefs and rocky bottoms to a depth of at least 90 m (Bannnerot, 1984; Heemstra and Randall, 1993). Adults are unspecialized, top-level predators (Randall, 1965) and lead solitary lives outside of spawning periods (Smith, 1971). Only adult Nassau grouper (>40 cm TL) were investigated in this home range study to avoid potential ontogenetic shifts in home range. Because Nassau grouper have the ability to home (Bardach et al., 1958) and apparently remain within highly circumscribed areas for extended periods of time (Randall 1962, 1963; Carter, et al., 1994), transporting individuals from one reef to another for a home range study was not feasible. Hence, sample size (i.e. number of fish) was limited to the individuals present on the study reefs. Nassau grouper reproduce during the winter in transient, site-specific spawning aggregations (Colin, 1992; Domeier and Colin, 1997) with probable long-distance spawning migrations (Colin, 1992). To avoid probable winter emigration, the tracking study was conducted during the summer (June-August) to avoid potential spawning migrations. Unfortunately the home range study on the Pacific coral trout (Zeller, 1997) was not available when this study commenced; it would have offered guidance on methodology and anticipated home ranges.

3.3 Study Site

The Exuma Cay Land and Sea Park is located in the central Bahamas (24°34'N, 76°33'W). It was established as a park in 1958 and encompasses a section 35.4 km long by 12.9 km wide of the northern Exuma Cays, covering an area of 456 km^2 (Wells, 1988). The 1986 bylaws prohibited fishing in the entire area, creating a no-take (no-fishing) zone. The islands in the park are oriented NW-SE along the edge of the Exuma Sound and are less than 8 km long, separated by inlets or tidal channels which range from 200 m to >2 km wide. The islands consist of mostly oolitic limestone, formed by the precipitation of calcium carbonate (Sealey, 1992). The currents within Exuma Sound are variable and may be significantly affected by wind direction (Colin, 1992; Shenaker et al., 1993). To the south, the Caribbean Sea is subject to northeasterly trade winds throughout most of the year (Milliman, 1973). Tides are semi-diurnal, with a range of 0.5-1 m, at a rate of 1 to 2 knots (Milliman, 1973).

The patch reefs in the Exuma Cays often occur in clusters of several hundred, rising up to 10 m from the sandy sea floor in water 10 to 15 m deep (Sealey, 1992). Two nearby patch reefs were used for this study: Burgandy and Plane Reefs. Both reefs are in depths of 9-10 meters, are tidally influenced, and are surrounded by a transition area of moderate to dense seagrass *(Thalassia testudinum)* and sand flat. The reefs are approximately 0.6 km apart, separated by a small island and a sand flat. Burgandy Reef (21895 m^2) is 11% larger than Plane Reef (19471 m^2).

3.4 Telemetry System, Transmitters and Transmitter Attachment

3.4.1 Transmitters

Individual frequency acoustic transmitters were selected for this study (V16P-4H, VEMCO Ltd., Shad Bay, Nova Scotia). These cylindrical transmitters (16 mm diameter, 80 mm long, and 12 grams) are pressure-sensitive, powered by a lithium battery, and have a battery life of 60 days. The transmitters emit a continuous signal of one acoustic pulse per second and provide opportunity for continuous surveillance. The size of the transmitter is dictated by the cylindrical shape of the transducers and batteries; the transmitters are cast in resin in one end of the tube, with the depth sensor protruding on the other end. These 12 g transmitters constitute less than 1% of the total weight of a 40-cm Nassau grouper (Table 1).

Table 1. Summary of ultrasonically tagged Nassau grouper 1997 and 1998 with post-operative field observations. Post-operative survival status is divided into visual observations (either weekly or yearly) and tracking verification.

				Post-operative survival status		
				Visual		
Fish #	Total length (cm)	Weight (kg)	Tag ratio	Weeks	Year	Tracking data at 3 weeks
1[a]	58.1	5.1	0.24	5, 6		yes
2[b]	47.1*	2.7	0.45		1	yes
3[b]	61.2	5.9	0.20	5, 6	1	yes
4[b]	42.0	1.9	0.63			no
5[b]	72.1	9.7	0.12			yes
6[b]	58.8	5.1	0.24			no (transmitter failure)
7[b]	44.1	2.2	0.55		1	yes
8[b]	41.4	1.8	0.68			yes
9[c]	54.3	4.1	0.30	3, 4, 5, 6		yes
10[c]	46.2	2.5	0.48	4		yes
11[c]	59.1	5.3	0.23	4, 5		yes
12[c]	66.1	7.5	0.16	4, 5, 6		yes
13[c]	40.2	1.6	0.73	4, 5, 6		yes
14[c]	42.3	1.9	0.63	3, 4, 5		yes
15[c]	58.9	5.3	0.24	5, 6		yes
16[c]	41.7	1.8	0.68	5, 6		yes
17[c]	62.0	6.2	0.19	1, 2, 3		yes
18[c]	69.2	8.6	0.14	3		yes
19[c]	50.1*	3.2	0.37			no
20[c]	56.2	4.5	0.26			yes
21[c]	39.1	1.5	0.79	3		yes
22[c]	43.9	2.0	0.59	1, 2, 3		yes
23[c]	45.0	2.3	0.51	3		yes
24[c]	45.2	2.3	0.51			yes

Superscript following fish number refers to suture type: a = stainless steel suture; b = monofilament nylon suture; c = chromic gut suture. * indicates fish recaptured one year later with peritoneal transmitter encapsulated and released as fish #19 with new transmitter. Tag ratio = Ratio of tag mass to body mass. Weights were estimated using the power function $W = aLb$ (where W=weight, L=length, a = 2.14E-05, and b=3.03). Modified from Bolden and Ewing, 1999.

3.4.2 Transmitter attachment

Because the Nassau grouper is a coral reef dweller, internal surgical implantation of the transmitter was chosen to reduce the potential of snaring the transmitter on overhangs and crevices.

3.4.3 Fish Capture

Animals used in the tracking study were caught in traditional fish traps at various locations around the reef. Rectangular fish traps (~60 x 70 x 90 cm) were constructed of vinyl coated hexagonal mesh (2 x 3 cm) wire with a single funnel entrance and single hatch door. Fish were captured and released within 5 meters of capture location. Traps were baited with canned herring, fresh queen conch (*Strombus gigas*), canned dogfood, or a combination of the three. Traps were monitored 2-3 times daily by SCUBA divers. Fish traps were also used as recovery cages following transmitter insertion.

3.4.4 Anesthesia

MS-222 was utilized in this study as a calming agent and anaesthetic during surgical implantation of transmitters into Nassau grouper. The anesthetic has been used at concentrations levels from 10 to 1,000 mg/l in fish and has an induction time of less than 15 minutes (Argent Chemical Supplies, Redmond, WA). Optimal level of anesthesia for surgery is total loss of reflex. Loss of reflex is determined and monitored in the fish by its lack of response to touch, slow and steady rate of opercular movement, and loss of equilibrium (i.e., the fish rests on its side in the anesthesia bath) (Ross and Ross, 1984).

Prior to removal from a submerged trap, Nassau grouper captured in fish traps were given a "stun solution" of the anesthetic via squirt bottles through the trap mesh. The stun concentration (1 g/L seawater) was squirted toward the fishes' mouths to induce passage to the gills, thereby tranquilizing them. Once the fish was tranquilized (usually indicated by slow swimming or slight alteration of body orientation in the water column), the fish was removed from the trap and placed into a plastic bag for transport to the surface. In some cases we were unable to tranquilize the fish by the squirting the stun solution toward the mouth; in those cases, the entire trap was enfolded in a large plastic bag (137 x 112 x 183 cm) and the anaesthetic was squirted into the bag. There was no occurrence of swim-bladder expansion.

At the surface, fish were immersed into an MS-222 anesthetic bath (125mg/L seawater). Stage 3 of anesthesia (McFarland, 1959; Ross and Ross, 1984) was obtained within 4-6 minutes after immersion in the bath as indicated by loss of equilibrium and loss of reaction to stimulation. Fish were then removed from the anesthetic bath, measurements were made, and a standard external tag was placed into the dorsal musculature. The fish were returned to the anesthetic bath while preparations for surgery were made. For surgery, the fish were again removed from the anesthetic bath and laid on a table. The head of the fish was covered during the procedure to maintain moisture and cover to eyes. The fish did not receive any anesthetic during the surgical procedure (see below). Following surgery, the fish was temporarily held in the anesthetic bath before being placed into the transport bag, slowly returned to the ocean bottom, and placed into a secured cage (closed fish trap) for 24-36 hours.

3.4.5 Surgical implantation

An approximate 1-3 cm paramedian incision into the left lateral body wall was made with a disposable scalpel midway between the pelvic fin origin and the cloaca. This incision was made in two steps: the first cut was through the integument and muscle and the second cut through the parietal peritoneum into the coelomic cavity. Acoustic transmitters were then inserted through the incision into the coelomic cavity of the grouper to a position anterior to the anus and behind (at least 9 cm) the pelvic fin to ease tension on the incision. The incision was closed with a single row of three interrupted sutures through the peritoneum and the dermis using a curved cutting needle (FS-1) swaged with 2-0 chromic gut (Ethicon®), 2-0 black monofilament nylon (Ethicon®), or stainless steel. The antibiotic Amikacin® was injected into the dorsal musculature with a 1-cc 27 G1/2 tuberculin syringe and needle, using 5 mg/kg of fish (Carpenter et al., 1996). Mean surgery time was less than 5 minutes (n=8).

3.4.6 External tags

To identify individuals with internal transmitters, external dart tags fitted with a pennant streamer (Floy Tag & Manufacturing, Inc. Seattle, WA) were implanted into the dorsal musculature of the Nassau grouper, just below the origin of the dorsal fin between the second and third scale rows. Pennants attached to the tag were either green, yellow or white and stamped with a unique alphanumeric combination. Clear anti-fouling spray paint was applied to the pennant to reduce fouling. Tags were inserted into the dorsal musculature with a steel applicator while the fish was under MS-222 anesthesia. These dart tags allowed individuals to be easily identified underwater by divers as they could be differentiated by unique codes of color, letter and number.

3.4.7 Recovery

To reduce postoperative trauma, surgery was performed one day and the fish was released at its original point of capture 24-36 hours later. Prior to release, animals were fed de-shelled queen conch (*Strombus gigas*) and observed while in the recovery pen; normal feeding and swimming behavior was observed for each individual.

3.4.8 VRAP buoy system

A radio-linked acoustic positioning (VRAP) buoy system (VEMCO Ltd., Shad Bay, Nova Scotia) was used to track individual Nassau grouper. This buoy system (also called RAPT, see O'Dor et al., this volume) consists of three buoys (each containing a hydrophone, receiver, and two-way radio link) positioned in an equilateral triangle over the reef (see Figures 1 and 3, O'Dor et al., this volume). A base station containing a personal computer with a two-way radio, antenna, and an interface box receives the radio signal via UHF from the buoys and delivers them to a PC every time a pulse signal is detected. The position of a tagged individual (resolution = 1-2 m) is sent by a two-way radio link and a microprocessor controller (O'Dor et al., 1998). The signal arrival times at the three buoys are used to calculate the position of the tagged fish with respect to the three buoys. Tracking provides real-time positions of animals, with minimal disruption of behavior. Individuals could be identified by separate acoustic frequency.

3.4.9 Observation protocol

Fish were continuously tracked 24 hours a day for a three week period. Subsequent daily diver observations identified individual fish via external tags.

4. RESULTS AND DISCUSSION OF THE NASSAU GROUPER PROJECT

4.1 Identify the Research Question and Data Needs

A total of 526,489 positions were recorded in this home range study. Number of position points were subsequently reduced to 76,038 by extracting the mode per tracking period (15 second interval). All analyses were conducted on the modal positions to reduce deviation. Mean number of position fixes was 3420 ± 1795 (SE) per fish (n=22) ranging from 1032 to 7493.

Data were analyzed using with the Animal Movement Analysis Extension (AMAE) for Arcview®, written by Hooge and Eichenlaub (1997) and available at http://www.absc.usgs.gov/glba/gistools/index.htm. The AMAE is a collection of more than 30 functions to aid in the analysis of animal movement data with real integration into a Geographic Information System (GIS). Numerous other home range data analysis programs are available: Ranges 5, CALHOME, TELEM8, KERNELHR, and TRACKER. For a review of these programs and corresponding internet sites see Larkin and Halkin (1994) and Gallerani Lawson and Rodgers (1997). The advantages and disadvantages of the most frequently used home-range analysis methods is discussed by Harris et al.(1990).

4.1.1 Site fidelity

Site fidelity was examined by a Monte Carlo simulation of position points from each individual fish using the AMAE. Site fidelity exists when the observed area used by an individual is significantly smaller than the area used if an individuals movement was random (Munger, 1984, Danielson and Swihart, 1987; Spencer et al., 1990). One hundred simulations were run on the positions of each fish (n = 22) with the arithmetic mean being chosen as the starting point of each simulation. In all cases the observed Nassau grouper movement was more constrained than random movement paths ($p > 0.99$).

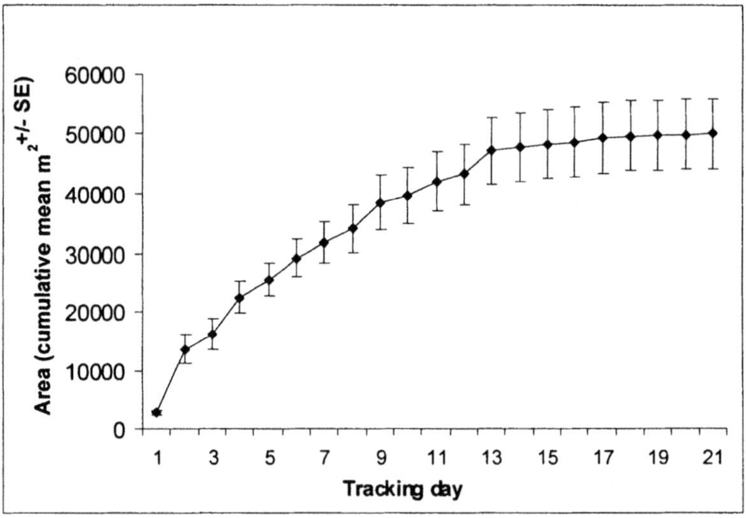

Figure 1. Area-observation curve for the Nassau grouper. Values are the mean cumulative home range area as calculated by the 95% minimum convex polygon for the Nassau grouper (n = 22) per tracking day with standard error bars.

4.1.2 Area observation curve

An area-observation curve was created for the Nassau grouper (Fig. 1) following the procedure of Odum and Kuenzler (1955). Area-observation curves are created by calculating and plotting the cumulative home range area per tracking period (i.e. tracking day). Sampling could cease when the area-observation curve asymptotes, that is when an additional tracking period (or fix) produces an increase in the cumulative home range area less than some predetermined value (Odum and Kuenzler, 1955); this value could be from 1% (Odum and Kuenzler, 1955) to 10% (Messier and Barrette, 1982; Morrissey and Gruber, 1993). For the Nassau grouper, the area-observation curve increased steeply from day 1-14 with daily increases in cumulative home range between 3 and 21% (Fig. 1). Cumulative home range area increased less than 2% daily during days 14-18, and less than 1% daily during tracking days 18-21 (Fig. 1). Although tracking could have ceased on day 18 with the small (<1%) increase in cumulative home range area, data from all 21 tracking days were included in the home range analyses.

4.1.3 Minimum convex polygon

Because analyses indicated that the Nassau grouper exhibit strong site fidelity and that adequate sampling was conducted during this investigation, the home range of the Nassau grouper was then determined by the minimum convex polygon (MCP). The MCP method is one of the earliest and most common methods of delimiting home range (Hayne, 1949). MCP is the most often used home range method in fish telemetry for both freshwater (Winter and Ross, 1981) and marine (e.g., Holland et al., 1993b; Morrissey and Gruber, 1993; Zeller, 1997) systems. MCP is a non-statistical technique that makes no a priori assumptions, which has been shown to be an accurate depiction of telemetrically monitored animal movements (Cranford, 1977).

MCP is calculated by simply connecting the extreme outermost positions of an animal's home range so that the smallest possible convex polygon encloses all other locations (Bekoff and Mech, 1984). This often represents the maximum home range size in telemetry studies. Potential sample size bias is surely the MCP's most serious deficiency, because the method is sensitive to movement near the periphery of the home range, is affected by outliers, and may include large areas never utilized by the animal (Jennrich and Turner, 1969; Schoener, 1981; Anderson, 1982; Worton, 1987). Without modification, the MCP method may result in in an overestimation of the home range area (Morrissey and Gruber, 1993). Therefore the Nassau grouper data were modified in two ways: the modal position points were used to reduce deviation, and 5% of the positions were statistically identified as outliers and omitted.

Outliers were eliminated before the MCPs were calculated as suggested by White and Garrott (1990). Outliers were identified as those positions farthest from the center of activity (COA). The COA is the area of highest utilization (Ables, 1969) and is based on the arithmetic mean point of all positions (Jennrich and Turner, 1969). COAs were calculated by the AMAE and all positions located outside of the COA 95% ellipse were considered outliers and were omitted from the analyses (Novick and Stewart, 1982; Kleeberger, 1985; White and Garott, 1990). Hence, 95% MCPs were then created for each individual fish and the home range area calculated (see Figure 3, O'dor et al., this volume). Mean number of position fixes per fish (n=22) for the 95% MCP analyses was 3364 ± 379 (SE) ranging from 1011 to 7485.

The mean home range area as determined by 95% MCP for the Nassau grouper (n=22) was 18305 m^2 ± 5806 (SE); the smallest home range was 1565 m^2 and the largest was 131805 m^2. Mean daily home range for all fish (n = 22) combined (Fig. 2) supported the conclusions of the area-observation curve as the mean home range varied without trend from day 1-13, then declined from day 14-21.

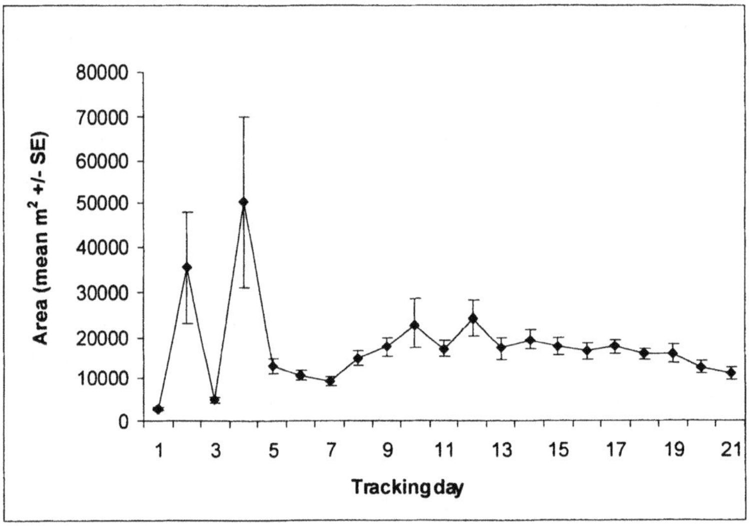

Figure 2. Mean home range area for the Nassau grouper per tracking day. Values are mean home range area for all fish (n=22) combined with standard error bars.

4.2 Behavior and Ecology of the Nassau Grouper

Twenty three adult (>40 cm TL) Nassau grouper were internally implanted with ultrasonic transmitters and released between 1997 and 1998 (Table 1). All individuals survived the surgical procedure. While all three suture types (chromic gut, nylon, and stainless steel) were successful in closing the incision with comparable healing times, the absorbable chromic gut is preferred for use in the Nassau grouper because there was less morbidity, and no residual or foreign body incited tissue damage. The disadvantages in other studies when utilizing chromic gut sutures (e.g., adverse tissue reaction and dehiscence due to premature absorption and loss of tensile strength) were not observed during this study. The disadvantage associated with the non-absorbable materials (i.e. nylon and steel) was increased tissue morbidity with skin and tissue crimping; however, there was no increase in mortality. For example, the stainless steel sutures in fish #1 (Table 1) were intact 185 days post-operative when the individual was captured by a local fisherman with the transmitter implanted (Bolden, 2000).

Visual sighting of individuals (n=6) three weeks after surgery revealed healing wounds with slight swelling, and at 4-6 weeks (n=10) the incision was barely noticeable. Three individuals sighted a year after surgery had no visible external scar, and the transmitter was still implanted (the outline was visible through the abdomen). A single individual recaptured a year later revealed a well-healed incision with an encapsulated transmitter. All but three animals were verified to be alive at three weeks post-operative by tracking data; two individuals perished, and one individual's transmitter malfunctioned and was never recorded (Table 1). Hence, post-operative survival at three weeks was 92%. Most fish were also identified by divers at the end of the study confirming vitality and demonstrating clearly that internal implantation of transmitters was highly successful.

Potential emigration from the home reef for spawning was avoided as all fish remained extremely faithful to their home reef throughout the study. Three individuals remained at their exact home reef one year later (Bolden, 2000). One individual was recaptured 185 days postoperative in a presumed spawning aggregation 256 km away (Bolden, 2000).

4.3 Study Site

Investigating the home range of a coral reef fish in a no-take marine park (i.e. The Exuma Cay Land and Sea Park) was beneficial as no fish were removed from the area during the study. Although an individual was later captured by a fisher (Bolden 2000), the individual was recaptured outside park boundaries. The relatively shallow depth (9-10 m) of the study reefs was conducive to both the surgical procedure, as no fish exhibited signs of barotrauma, and to relatively unrestricted SCUBA access. However, the complex coral reef environment did interfere with acoustic transmission as indicated by the range of position fixes per fish. While the remote location of the park introduced logistical challenges (e.g., equipment transport, utilities, lodging, etc.), it reduced security concerns and neither the animals nor the equipment were abused or molested.

4.4 Performance of the Telemetry System, Transmitter and Transmitter Attachment

The VRAP buoy system is a self-powered remote telemetry system that can continuously record position fixes via transmitters without interfering with animal behavior. It is well suited for home range studies as it allows tracking of numerous individuals over relatively long periods of time. Although deployment is challenging and the electronics are complicated, the system performs extremely well in the coral-reef environment. The VRAP system has been used to successfully track reef-associated fish and invertebrates (see also O'Dor et al., this volume) and generates both the quantity and quality of data required to determine home range.

5. CONCLUSIONS

Compiling information to systematically define parameters and determine limitations is critical to the success of a home range study. Clearly defining research questions will dictate data quality and quantity and influence selection of telemetry system and transmitter. Personal observations coupled with published reports will also influence transmitter type and also method of attachment. Population density, fishing pressure and access to utilities will influence the choice of a study site.

The rich field of analyzing movement patterns that has been developed and used by the terrestrial biologists (see White and Garrott, 1990) combined with recent telemetric advances and the utility of the AMAE promise future success in defining and interpreting the home range and fine-scale movement patterns of coral-reef fishes.

ACKNOWLEDGEMENTS

This research was conducted under a grant from NOAA Office of Protected Species, and a Research Permit from the Govt. of the Bahamas. I thank National Marine Fisheries Service for financial support. The NMFS does not recommend or endorse any proprietary product or material mentioned in this publication. Special appreciation to Bell Island Ltd. and K. James for logistical support. I am indebted to John Morrissey and Fred Voegeli for discussions and advise on biotelemetry. The critical review by Dirk Zeller greatly improved the content of this manuscript and is appreciated.

REFERENCES

Ables, E.O. (1969) Home range studies of red foxes (*Vulpes vulpes*). *J. Mammal.* **50**, 108-120.

Anderson, D.J. (1982) The home range: a new nonparametric estimation technique. *Ecology* **63**, 103-112.

Armstrong, J.D. and Rawlings, C.E. (1993) The effect of intragastric transmitters on feeding behavior of Atlantic salmon, *Salmo salar*, parr during autumn. *J. Fish. Biol.* **43**, 646-648.

Arnold, S.J. (1988) Behavior, energy and fitness. *Am. Zool.* **28**, 815-827.

Bannerot, S.P. (1984) The dynamics of exploited groupers (Serranidae): an investigation of the protogynous hermaphroditc reproductive strategy. PhD dissertation, University of Miami, Coral Gables, FL 416 pp.

Bardach, J.E., Smith, C.L. and Menzel, W. (1958) Final report Bermuda Fisheries Research Program. Bermuda Trade Development Board, Hamilton, Bermuda.

Bekoff, M. and Mech, L.D. (1984) Simulation analysis of space use: home range estimates, variability and sample size. *Behav. Res. Meth., Ins. C.* **16**, 32-37.

Bolden, S.K. (2000) Long-distance movement of a Nassau grouper (*Epinephelus striatus*) to a spawning aggregation in the central Bahamas. *Fish. Bull. U.S.* **98** (3), 642-645.

Bolden, S.K. and Ewing, R.Y (1999) Surgical implantation of ultrasonic transmitters in adult Nassau grouper (*Epinephelus striatus*). *Proc. Int. Assoc. Aquatic Animal Med.* **30**, 53-55.

Bowen, W.D. (1982) Home range and spatial organization of coyotes in Jasper National Park, Alberta. *J. Wildlife Manage.* **46**, 201-216.

Brown, J.L. (1975) *The Evolution of Behavior.* Norton, New York, 761 p.

Burt, W.H. (1943) Territoriality and home range concepts as applied to mammals. *J. Mammal.* **24**, 346-352.

Carpenter, J.W., Mashima, T.Y., and Ruiper, D.J. (1996) *Exotic Animal Formulary.* Greystone Publications, Manhattan, Kansas, 432 p.

Carter, H.J., Marrow, G.J. and Pryor, V. (1994) Aspects of the ecology and reproduction of Nassau grouper, *Epinephelus striatus*, off the coast of Belize, Central America. *Proc. Gulf. Carib. Fish. Inst.* **43**, 64-110.

Colin, P.L. (1992) Reproduction of the Nassau grouper, *Epinephelus striatus* (Pisces: Serranidae) and its relationship to environmental conditions. *Env. Biol. Fish.* **34**, 357-377.

Collette, B.B. and Talbot, F.H. (1972) Activity patterns of coral reef fishes with emphasis on nocturnal-diurnal changeover. *Bull. Nat. Hist. Mus. Los Angeles* **14**, 98-124.

Cranford, J.A. (1977) Home range and habitat utilization by *Neotoma fuscipes* as determined by radiotelemetry. *J. Mammal.* **58**, 165-172.

Danielson, B.J. and Swihart, R.K. (1987) Home range dynamics and activity pattern of *Microtus ochrogaster* and *Synaptomys cooperi* in syntopy. *J. Mammal.* **68**, 160-165.

DeMartini, E.E. (1993) Modeling the potential of fishery reserves for managing Pacific coral reef fishes. *Fish. Bull. U.S.* **91**, 414-427.

Domeier, M.L. and Colin, P.L. (1997) Tropical reef fish spawning aggregations: defined and reviewed. *Bull. Mar. Sci.* **60**, 698-726.

Gallepp, G.W. and Magnuson, J.J. (1972) Effects of negative buoyancy on the behavior of the bluegill, *Lepomis macrochirus* Rafinesque. *Trans. Am. Fish Soc.* **3**, 507-512.

Gallerani Lawson, E.J. and Rodgers, A.R. (1997) Differences in home-range size computed in commonly used software programs. *Wildlife Soc. B.* **25**, 721-729.

Harris, S., Cresswell, W.J., Forde, P.G., Trewhella,W,.J., Woollard, T. and Wray, S. (1990) Home-range analysis using radio-tracking data—a review of problems and techniques particularly as applied to the study of mammals. *Mammal Rev.* **20**, 97-123.

Hart, L.G. and Summerfelt, R.C. (1975) Surgical procedures for implanting ultrasonic transmitters into flathead catfish (*Pylodictis olivaris*). *Trans. Am. Fish. Soc.* **104**, 56-59.

Hawkins, A.D. and Urquhart, G.G. (1983) Tracking fish at sea. *In* Macdonald, A.G. and Priede, I.G. eds. *Experimental Biology at Sea.* Academic Press, New York, NY. pp. 103-166

Hayne, D.W. (1949) Calculation of size of home range. *J. Mammal.* **30**, 1-18

Heemstra, P.C. and Randall, J.E. (1993) FAO Species catalogue. Vol 16. *Groupers of the World* (Family Serranidae, Subfamily Epinephelinae). An annotated and illustrated catalogue of the grouper, rockcod, hind, coral grouper and lyretail species known to date. *FAO Fisheries Synopsis.* No. 125. Vol. 16. Rome, FAO.

Holland, K.N., Peterson, J.D., Lowe, C.G. and Wetherbee, B.M. (1993a) Movements, distribution and growth rates of the white goatfish *Mulloides flavolineatues* in a fisheries conservation zone. *Bull. Mar. Sci.* **52**, 982-992.

Holland, K.N., Wetherbee, B.M., Peterson, J.D. and Lowe, C.G. (1993b) Movements and distribution of hammerhead shark pups on their natal grounds. *Copeia* **2**, 495-502.

Holland, K.N., Lowe, C.G. and Wetherbee, B.M.. (1996) Movements and dispersal patterns of blue trevally in a fisheries conservation zone. *Fish. Res.* **25**, 279-292.

Hooge, P.N. and Eichenlaub, B. (1997) Animal Movement Extension to Arcview. Ver 1.1 Alaska Biological Science Center, U.S. Geological Survey, Anchorage.

IUCN (The World Conservation Union). (1988) *Coral Reefs of the World.* Volume 1: Atlantic and Eastern Pacific. Wells, S.M. (ed.) IUCN Conservation Monitoring Centre, Cambridge, U.K.

Jennrich, R.I. and Turner, F.B. (1969) Measurement of non-circular home range. *J. Theor. Biol.* **22**, 227-237.

Kanwisher, J., Lawson, K. and Sundnes, G. (1974) Acoustic telemetry from fish. *Fish. Bull. U.S.* **72**, 251-255.

Kleeberger, S.R. (1985) Influence of intraspecific density and cover on home rnage of a plethodontid salamander. *Oecologia* **66**, 404-410.

Kramer, D.L. and Chapman, M.R. (1999) Implications of fish home range size and relocation for marine reserve function. *Envir. Biol. Fish.* **55**, 65-79.

Larkin, R.P. and Halkin, D. (1994) A review of software packages for estimating animal home ranges. *Wildlife Soc. B.* **,** 274-287.

Luckhurst, B.E. and Luckhurst, K. (1978) Diurnal space utilization in coral reef fish communities. *Mar. Biol.* **49**, 325-332.

Mace, G.M., Harvey, P.H. and Clutton-Brock, T.H. (1983) Vertebrate home range size and metabolic requirements. *In* Swingland I. and Greenwood P.J. eds. *The ecology of animal movement.* Oxford. pp. 32-53

Marty, G.D. and Summerfelt, R.C.. (1986) Pathways and mechanisms for expulsion of surgically implanted dummy transmitters from channel catfish. *Trans. Am. Fish. Soc.* **115**, 577-589.

McCleave, J.D. and Stred, K.A. (1975) Effects of dummy telemetry transmitters on stamina of Atlantic salmon (*Salmo salar*) smolts. *Trans. Am. Fish. Soc.* **32**, 559-563.

McFarland, W.N. (1959) A study of the effects of anesthetics on the behavior and physiology of fishes. *Pub. Inst. Marine Sci.* **6**, 22-55.

McNab, B.K. (1963) Bioenergetics and the determination of home range size. *Am. Nat.* **101**, 377-385.

Messier, F. and Barrette, C. (1982) The social system of the coyote (*Canis latrans*) in a forested habitat. *Can. J. Zool.* **60**, 1743-1753.

Milliman, J.D. (1973) Caribbean coral reefs. *In* Jones O.A and Endean R. eds. *Biology and geology of Coral Reefs.* Vol. I. Academic Press, New York. pp. 1-50

Morrissey, J.F. and Gruber, S.H. (1993) Home range of juvenile lemon sharks, *Negaprion brevirostris*. *Copeia* **2**, 425-434.

Moser, M.L., Olson, A.F. and Quinn, T.P. (1990) Effects of ultrasonic transmitters on juvenile coho salmon. *Am. Fish. Soc. Symp.* **7**, 353-356.

Munger, J. (1984) Home ranges of horned lizards: circumscribed and exclusive? *Oecologia* (Berlin) **62**, 351-360.

Nelson, D.R. (1990) Telemetry studies of sharks: a review with applications in resource management. *In* Pratt, H.L., Gruber ,S.H. and Taniuchi, T. eds. Elasmobranchs as living resources: advances in the biology, ecology, systematics, and status of the fisheries. NOAA Tech. Report 90. pp. 239-256

Novick, H.J. and Stewart, G.R. (1982) Home range and habitat preference of black bears in the San Bernadino mountains of southern California. *Calif. Fish. Game* **67**, 21-35.

O'Dor R.K., Andrade, Y., Webber, D.M., Sauer, W.H.H., Roberts,M.J., Smale, M.J. and Voegeli, F.A. (1998) Applications and performance of radio-acoustic positioning and telemetry (RAPT) systems. *Hydrobiologia* **371/372**, 1-8.

Odum, E.P. and Kuebzler, E.J. (1955) Measurement of territory and home range size in birds. *Auk* **72**, 128-137.

Peake, S., McKinley, R.S., Scruton, D.A. and Moccia, R. (1997) Influence of transmitter attachment procedures on swimming performance of wild and hatchery-reared Atlantic Salmon smolts. *Trans. Am. Fish. Soc.* **126**, 707-714.

Peterson, D.C. (1975) Ultrasonic tracking of three species of black bass, *Micropterus* spp. in Center Hill Reservoir, Tennessee. M.S. Technological Univ. Cookeville, Tenn.

Pincock, D.G. and Voegeli, F.A. (1992) *Quick Course in Underwater Telemetry*. VEMCO Ltd., 3895 Shad Bay, Halifax, Nova Scotia, Canada.

Polacheck, T. (1990) Year around closed areas as a management tool. *Nat. Res. Model* **4**, 327-354.

Randall, J.E. (1962) Tagging reef fishes in the Virgin Islands. *Proc. Gulf. Carib. Fish. Inst.* **14**, 201-241.

Randall, J.E. (1963) Additional recoveries of tagged reef fishes from the Virgin Islands. *Proc. Gulf Carib. Fish. Inst.* **15**, 155-157.

Randall, J.E. (1965) Food habits of reef fish of the Nassau grouper *(Epinephelus striatus)*. *Assoc. Island Mar. Labs Carib. 6th Meeting.* Jan, 1965, pp. 13-16.

Roberts, C.M. and Polunin, N.V.C. (1991) Are reserves effective in management of reef fisheries? *Rev. Fish. Biol. Fish.* **1**, 65-91.

Ross, M.J. and McCormick, J.H. (1981). Effects of external radio transmitters on fish. *Prog. Fish-Cult.* **43**, 67-72.

Ross, L.G. and Ross, B. (1984) *Anaesthetic and sedative techniques for fish.* Institute of Aquaculture, Univ. Stirling, Scotland, 35 pp.

Schonener, T.W. (1981) An empirically based estimate of home range. *Theor. Pop. Biol.* **20**, 281-325.

Sealey, N. (1992) *Caribbean World.* Cambridge University Press,

Shapiro, D.Y., Sadovy, Y. and McGehee, M.A. (1993) Size, composition, and spatial structure of the annual spawning aggregation of the red hind *Epinephelus guttatus* (Pisces: Serranidae). *Copeia* **1993**, 399-406.

Shenker, J.M., Maddox, E.D., Wishinski, E., Pearl, A., Thorrold, S.R. and Smith, N. (1993) Onshore transport of settlement-stage Nassau grouper, *Epinephelus striatus*, and other fishes in Exuma Sound, Bahamas. *Mar. Ecol. Prog. Ser.* **98**, 31-43.

Sluka, R. and Sullivan, K.M. (1996) Daily activity patterns of groupers in the Exuma Cays Land and Sea Park, central Bahamas. *Bahamas J. Sci.* Feb., 17-22.

Smith, C.L. (1971) A revision of the american groupers: *Epinephelus* and allied genera. *Bull. Am. Mus. Nat. Hist.* **146**, 69-241.

Spencer, S.R., Cameron, G.N. and Stewart, R.K. (1990) Operationally defining home range: temporal dependence exhibited by hispid cotton rats. *Ecology* **71**, 1817-1822.

Stasko, A.B., and Pincock, D.G. (1977) Review of underwater biotelemetry, with emphasis on ultrasonic techniques. *J. Fish. Res. Board. Can.* **34**, 1261-1285.

Summerfelt, R.C. and Mosier, D. (1984) Transintestinal expulsion of surgically implanted dummy transmitters by channel catfish. *Trans. Am. Fish. Soc.* **113**, 760-766.

Tulevech, S.M. and Recksiek, C.W. (1994) Acoustic tracking of white grunt, *Haemulon plumieri*, in Puerto Rico and Florida. *Fish. Res.* **19**, 301-319.

White, G.C. and Garrott, R.A. (1990) *Analysis of Wildlife Radio-Tracking Data.* Academic Press, New York, 383 pp.

Wilson, E.O. (1975) *Sociobiology.* Belknap Press, Cambridge, MA, 366 pp.

Winter, J.D. (1977) Summer home range movements and habitat use by four largemouth bass in Mary Lake, Minnesota. *Trans. Am. Fish. Soc.* **106**, 323-330.

Winter, J.D. (1983) Underwater biotelemetry. *In* Nielsen, L.A. and Johnson, D.L. (eds.) *Fisheries Techniques.* American Fisheries Society, Bethesda, Maryland, pp. 371-395.

Winter, J.D. and Ross, M.J. (1981) Methods in analyzing fish habitat utilization from telemetry data. *In* Armantrout, N.B. (ed.) *Acquisition and Utilization of Aquatic Habitat Inventory Information.* Proceedings of a symposium held Oct. 28-30, Portland, OR, pp. 273-279.

Worton, B.J. (1987) A review of models of home range for animal movement. *Ecol. Model.* **38**, 277-298.

Zeller, D.C. (1997) Home range and activity patterns of the coral trout *Plectropomus leopardus* (Serranidae). *Mar. Ecol.Prog. Ser.* **154**, 65-77.

Zeller, D.C. (1999) Ultrasonic telemetry: its application to coral reef fisheries research. Fish. Bull. U.S. **97**, 1058-1065.

Ziebell, C.D. (1973) Ultrasonic transmitters for tracking channel catfish. *Prog. Fish-Cult.* **35**, 28-32.

A Stepwise Approach to Investigating the Movement Patterns and Habitat Utilization of Goliath Grouper, *Epinephelus itajara*, Using Conventional Tagging, Acoustic Telemetry and Satellite Tracking

Anne-Marie Eklund and Jennifer Schull

NOAA-Fisheries, Southeast Fisheries Science Center, Division of Protected Resources and Biodiversity, 75 Virginia Beach Drive, Miami, FL 33149; 305-361-4271; anne.marie.eklund@noaa.gov; jennifer.schull@noaa.gov

Key words: goliath grouper, jewfish, *Epinephelus itajara*, acoustic tagging, mark-recapture

Abstract: Goliath grouper,[1] *Epinephelus itajara*, the largest grouper in the western North Atlantic, has been protected from all harvest in U.S. waters since 1990, after years of overexploitation at its spawning aggregations. We are currently assessing this species' recovery by using a variety of tagging methods, including conventional tagging and acoustic telemetry. We have been monitoring the adult populations at offshore spawning aggregations and the juveniles at their nursery areas along mangrove shorelines. Conventional mark/recapture studies enabled us to predict juvenile goliath grouper population densities, growth rates and survival rates. Conventional tagging and recaptures of both adults and juveniles have given information on habitat use and movement patterns, while manually tracking acoustically tagged fish provided fine-scale habitat use and seasonal movements. Continuous data-logging hydrophones provided long-term data on site residency of both juveniles and adults. Future studies will include the use of satellite tracking to define large scale ontogenetic and spawning migrations to previously undescribed habitats. Each method of tagging has provided answers to key questions regarding goliath grouper population biology, but every method also has had its limitations. By starting with the most economical and simplest methods, we have built upon each study by adding complexity as it is warranted.

[1] Name change from jewfish to goliath grouper, May 2001.

J.R. Sibert and J.L. Nielsen (eds.), Electronic Tagging and Tracking in Marine Fisheries, 189–216.
© 2001 *Kluwer Academic Publishers. Printed in the Netherlands.*

1. INTRODUCTION

Tagging of marine fishes has been an important aspect of fishery biology and management for many years (Nielson, 1992; Laird and Stott, 1978; Boreman, 1996). Fish can be tagged with a variety of visual marks, as well as with tags that emit acoustic or radio transmissions. Through conventional mark-recapture studies, it is possible to estimate population abundance (Krebs, 1999; Koenig and Coleman, 1998), growth rates (Cruz-Escalona et al., 2000; Turner et al., 1990) and movement patterns (Xiao, 1996; Laird and Stott, 1978) of many different species of fish. However, getting the information from fish marked with visual tags is dependent upon their recapture. In many studies fewer than 10% of the marked fish are ever recaptured. Nevertheless, recaptures of even a few marked fish have yielded important information on stock structure and range of movements (e.g., Mather et al., 1995; Schaefer and Fable, 1994). Even in these cases, however, the information gained is only the start and endpoints of the migration, with no data on where the fish had been while at large.

In terrestrial and freshwater systems, biologists are able to tag animals with radio transmitters. Shore-based and aerial radio receivers can easily pick up the signal of individual fish and other animals, within a large range of up to 1500 m (de Morais and Raffray, 1999). While radio waves do not transmit in marine systems, there have been many advances in marine biotelemetry, including the use of acoustics in tracking marine fishes. Acoustic signals can only be heard from 100-500 m, depending on ambient noise (Arendt, in press; Priede et al., 1990, Eklund, pers. obs.), but directional hydrophones can precisely locate the position of an animal in water. Manual tracking is a labor-intensive method of tracking marine fish and invertebrates (Commiskey, 1999; Gunn et al., 1999; Wetherbee and Rechisky, 1999; Yano, 1999; Yano et al., 1999; Block et al., 1992), which has yielded much information on animals' movements and behaviors. Recent advances in biotelemetry have resulted in data-log ging capabilities on hydrophones, which can be left on site for weeks or months, collecting data on the presence/absence or exact position of fish marked with acoustic transmitters (Bolden, this volume; Arendt, in press).

New technologies have made tracking fish over large distances possible by enabling the tags to transmit their location to satellites, if the animal is on the surface or if the tag has disengaged from the animal (Lutcavage et al., 1999; Welch and Eveson, 1999). In addition, continuous data on depth, temperature and geolocation can be archived on these tags, giving the researcher a continuous log of where the animal has been (Davis and Gunn, 1992; Block et al., 1998). The archival tags have been mainly used on fish that may make transoceanic migrations, such as whale sharks (Gunn et al.,

1999) and bluefin tuna (*Thunnus thynnus*) (Block et al., 1998; Davis and Gunn, 1992).

With the advent of these new technologies comes much confusion about what tagging method is best for each animal. To the uninitiated, the terminology of biotelemetry can be daunting, as it encompasses fields of engineering, oceanography and physical science, in addition to basic biology and ecology. Often, researchers are eager to use the latest and most expensive technology, without totally understanding the advantages and disadvantages of each method. It is our purpose to describe our experiences in using a few different tagging and tracking techniques and the successes and limitations that each method provided. The particular tagging method chosen depends on the question asked and in some cases, even the most current and sophisticated biotelemetry techniques cannot provide answers that the most basic mark-recapture study can provide. Each tagging method that we describe has or will provide answers that the other methods cannot, owing to their inherent limitations.

Case Study: Tagging and Tracking Goliath Grouper

The goliath grouper, *Epinephelus itajara*, is one of the largest groupers in the world, growing to at least 210 cm TL and living at least 37 years. Prior to our research, there was remarkably little information on the species, with only one study investigating its life history (Bullock et al., 1992). Its population status in recent years has prompted much interest in investigating the goliath grouper's life history and population dynamics.

Although once common along both the Gulf of Mexico and Atlantic coasts of Florida, goliath grouper populations declined rapidly through the 1980s, shortly after spawning aggregations in the eastern Gulf of Mexico were discovered (Sadovy and Eklund, 1999). In 1990, the fishery was closed by emergency rule and all harvest of this species was, and still is, prohibited in U.S. waters (GMFMC, 1990; SAFMC, 1990). The fish is now a candidate for the U.S. Threatened and Endangered Species List, is currently on the IUCN Red List of Threatened Animals (Hudson and Mace, 1996), and is considered by the American Fisheries Society to be at risk of extinction in North America (Musick et al., 2000).

Our research objectives were to assess the recovery of this species, by describing its critical habitat during different life history stages and producing recruitment forecasting by estimating the abundance of juveniles and spawning adults over time. We determined that the goliath grouper's large size, longevity and overall hardiness made it an ideal subject for telemetry studies. There were several different tagging and tracking methods we could employ successfully to answer our research questions.

This paper will focus on the aspects of our study that make use of biotelemetry methods, coupled with conventional mark/recapture studies. We have used a step-by-step approach to our tagging work, beginning with conventional tagging experiments that were later augmented by acoustic telemetry. We have gained valuable information from both conventional and acoustic tagging, with each method providing results that facilitate and map our plans for future satellite tracking. We discuss the key findings and limitations to each methodology as they relate to our overall objectives of investigating goliath grouper life history and distribution.

We have tagged juvenile goliath grouper in their nursery habitat in order to estimate absolute abundance of the juveniles as a predictor of year class strength. Our conventional and acoustic tagging studies have yielded important information on critical nursery habitat, site fidelity, movement patterns and growth and survival rates of juveniles. We have also censused adult goliath grouper on selected spawning aggregations and have marked some of these fish with visual tags and sonic transmitters in order to describe their seasonal migration patterns and their habitat use during the spawning period.

2. STUDY AREA

2.1 Juvenile Habitat

We studied juvenile populations at four sites in the Ten Thousand Islands off southwest Florida in the Gulf of Mexico: Rabbit Key Pass, Russell Pass, Fakahatchee, and Faka Union Channel (Figure 1). This region, considered the historical center of abundance for juvenile goliath grouper in Florida (Bullock and Smith, 1991), is an essentially undeveloped maze of mangrove forest islands and tidal passes, protected by the Everglades National Park and the Rookery Bay National Estuarine Research Reserve. Local environmental characteristics include water temperatures ranging from 18 to 33°C, water depths of 2 to 10 m, salinity ranges from 30 to 33 ppt, and tidal amplitudes up to 2 m. Strong tidal currents over time have eroded undercuts along many of the mangrove islands. Our observations (using SCUBA) of these undercuts revealed them to be highly complex habitat, with many goliath grouper, gray snapper (*Lutjanus griseus*), snook (*Centropomis undecimalis*), spadefish (*Chaetodipterus faber*) and several species of portunid and xanthid crabs utilizing the same areas.

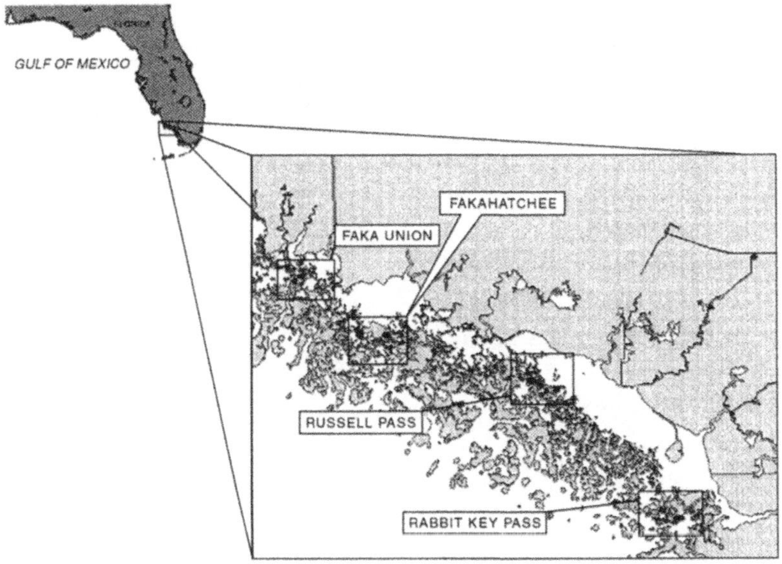

Figure 1. Four sampling areas for juvenile goliath grouper within the Ten Thousand Islands of southwest Florida, U.S.A.

2.2 Adult Habitat

Adult goliath grouper inhabit coral reefs and wrecks at depths ranging from 5 to 50 m (Sadovy and Eklund, 1999). They were once commonly found close to shore under bridges and jetties (DeMaria, 1996), but are now more often observed several kilometers offshore in the Gulf of Mexico, associated with isolated wrecks and U. S. Navy Navigation Towers at depths ranging from 30 to 50 m (Figure 2). These sites provide habitat for many reef fishes, such as grunts (haemulids), snappers (lutjanids), jacks (carangids), groupers (serranids), and wrasses (labrids), as well as for many reef invertebrates including decapod crustaceans. We concentrated our efforts thus far on these offshore wrecks and towers, because they are thought to be spawning aggregation sites and because we have been unable to locate any goliath grouper spawning aggregations on natural reef habitats.

In addition to the offshore wrecks, goliath grouper have also been observed frequently at a fishing pier at the southern end of Boca Grande, Florida (Figure 2). The pier is a popular diving and fishing location because of the concentration of tarpon (*Megalops atlanticus*), snook and snappers *(Lutjanus sp.)*, as well as goliath grouper. Located at the mouth of Charlotte Harbor, a large estuary, the pier provides habitat structure in a productive area of intense tidal flow and shallow water (0-10 m).

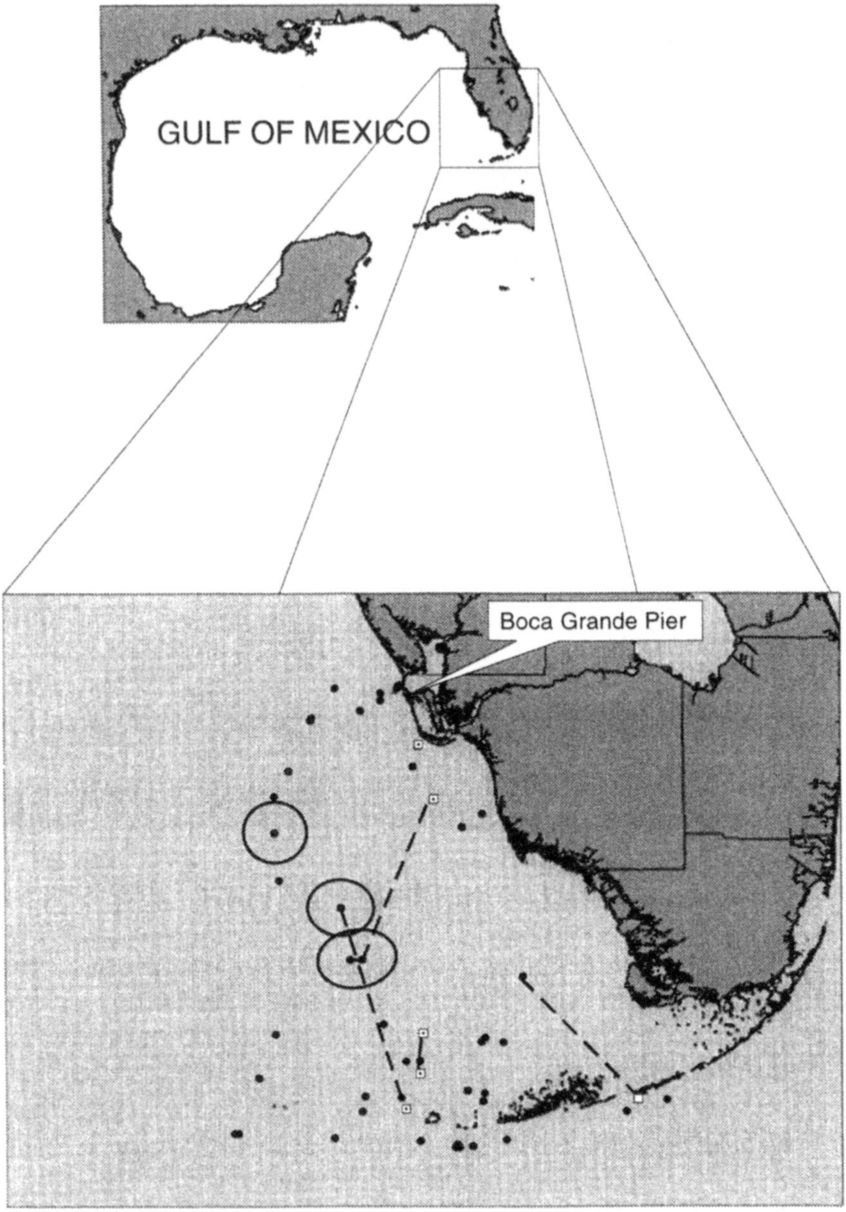

Figure 2. Locations off the coast of Florida, U.S.A. where adult goliath grouper were tagged (filled circles) and where adult goliath grouper have been observed or caught, other than at their original tagging locations (open squares). The dashed lines represent movement of goliath grouper from a tagging site to another site. The large circles depict the areas where goliath grouper are known to aggregate during the spawning season.

3. CONVENTIONAL TAGGING STUDIES

3.1 Objectives and Methodology

3.1.1 Juveniles

The primary objective of our conventional mark/recapture study was to estimate juvenile goliath grouper abundance, growth rates and survival rates. Recaptures of previously tagged individuals also allowed us to describe their preferred habitat and to document their movements, including patterns of seasonal activity and ontogenetic migrations. Together this information can be used to forecast recruitment and to promote protection of critical nursery habitat.

We initiated the juvenile mark/recapture study in June 1997. Juveniles were caught using circle hooks set on lines (20-30 lines per area) in deep undercuts along mangrove islands in four discrete areas of the Ten Thousand Islands (Figure 1). Each hook was baited with a live catfish (*Arius felis*) from which the spines had been removed, and allowed to fish for four to eight hours with the hook resting on the bottom. Captured juveniles were weighed to the nearest gram (g) and measured to the nearest mm—standard length (SL) and total length (TL). Scales and dorsal fin rays and spines were collected to age fish. Before release at the site of capture, each fish was tagged with a stainless-steel-core internal anchor tag secured on the ventral side of the body. We included on each tag, in addition to a unique tag number for identification of individual fish, the Florida Marine Research Institute's (FMRI) toll-free tagging hotline number (800-367-4461) in case anglers captured any of our tagged fish. To facilitate reporting, we distributed posters to area bait shops and marinas to inform people of the pertinent information we needed and how to get in touch with us. We recorded the tag number, location, and the length and weight of all recaptured fish. For any fish at large longer than six months, we resampled scales and dorsal fin rays and spines. All mark and recapture locations were recorded with a Global Positioning System unit (GPS) and entered into a geographical information system (GIS) mapping database (Arcview, ESRI, 380 New York St., Redlands, CA 92373) to allow characterization of juvenile habitat use, movement patterns and site fidelity.

3.1.2 Adults

We initiated tagging of adult goliath grouper in 1996 at four to five offshore aggregation sites to determine the degree of site fidelity and inter-aggregation movements. In the fall of 1998, we started tagging fish at inshore wrecks, reefs, and the Boca Grande pier to study their annual migrations to and from the spawning sites. Knowledge of the utilization of critical spawning habitat, movements, and the extent of seasonal migrations will help us describe the stock structure of this protected species.

Adult goliath grouper were tagged externally just below the dorsal fin using dart tags with large disks at the proximal end embossed with discrete tag numbers. Tagging occurred *in situ* and remotely (at distances of 3 to 4 m) by a diver using a modified spear gun. *In situ* tagging ensured that fish experienced minimal physiological stress, far less than they would have with typical onboard tagging. A disk at the proximal end of the tag contained a large number, along with the toll free Tagging Hotline number. Observations of tagged fish were made by us on our annual dives (using SCUBA), by our associate Captain Don DeMaria, by recreational divers or by commercial or recreational fishers.

3.2 Successes

3.2.1 Juveniles

The conventional tagging method proved a cost-effective means of tagging a large number of fish with a high recapture rate. We were able to estimate population abundance and growth and survival rates. We were also able to describe juvenile habitat use and their movement patterns. We tagged 148 juvenile goliath grouper (size range 205-1005 mm TL, mean = 684) in four discrete areas from June 1997 to October 1999. Our recapture rate was very high, at 37%. Twenty percent of the tagged fish were recaptured twice and several were recaptured multiple times (Figure 3). Some fish were caught 4-8 times within 35 m of their original tagging location, and 25 individuals were recaptured in the same location in which they were tagged, indicating a high degree of site fidelity. Only a few fish exhibited movements greater than 300 m: one fish in Rabbit Key Pass, one fish in Russell Pass and two fish in Faka Union (Figures 4-7).

Through the FMRI tagging hotline, we documented two fish that emigrated from the study area to offshore reefs/wrecks. One of the fish moved 31.7 km within a two-month period (Figure 8). This fish was one of the largest we had caught in the nursery habitat, at close to one meter in length. The second fish was recaptured 22.3 km and two years after it was tagged. At the time of recapture, the second fish was just over one meter in length. Since goliath grouper become reproductively mature between 1100 and 1350 mm, we suspect that this movement represents an ontogenetic migration to adult habitat. Future work using satellite telemetry may reveal more such movements (see satellite tracking section below).

3.2.2 Adults

From July 1996 through January 2000, we tagged 856 adult goliath grouper on offshore aggregation sites and inshore wrecks (Figure 2). Fifty of those fish have been re-sighted. Twelve of those tagged on offshore aggregations were re-sighted on their original tagging sites: four were re-sighted within the same spawning season (11-43 days post tagging), six were re-sighted during the subsequent spawning season, one was re-sighted in each of the following two spawning seasons, and one was observed at the same location three years after tagging. We believed that these fish might move seasonally on and off the aggregations, but we did not know the degree of site fidelity these fish might have within and among the spawning seasons. Based on the fish re-sighted at their original tagging locations, we suspected that at least some of the aggregating fish return to the same aggregation sites year after year, assuming that they leave the sites at all.

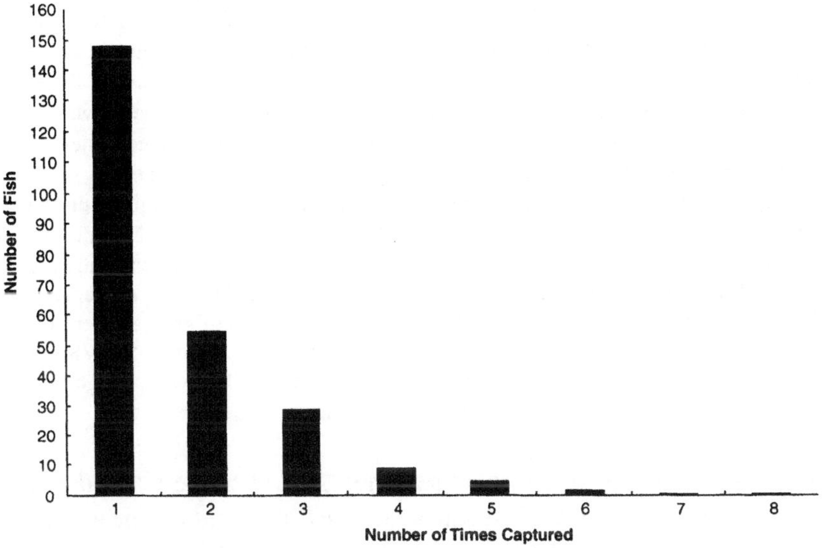

Figure 3. Number of juvenile goliath grouper tagged and recaptured in the Ten Thousand Islands of southwest Florida, U.S.A. from June 1997- October 1999.

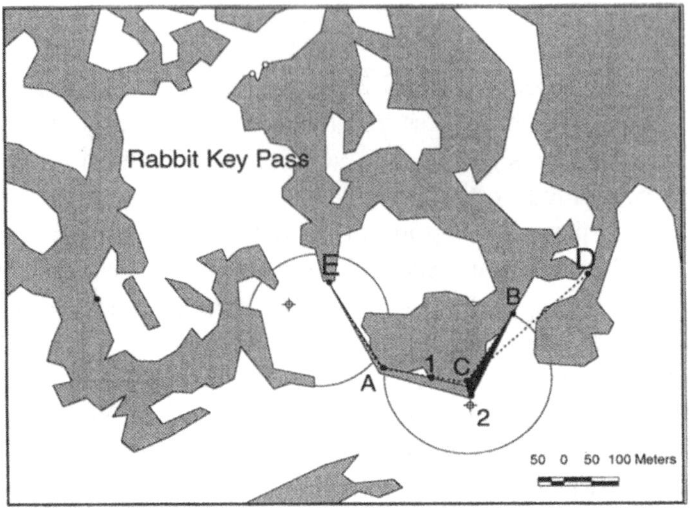

Figure 4. The Rabbit Key Pass study area for juvenile goliath grouper within the Ten Thousand Islands of southwest Florida, U.S.A. Letters represent areas where juvenile goliath grouper were tagged and recaptured and in some cases also detected acoustically. Numbers represent areas where the fish were detected acoustically, but no lines were set there to catch the fish. Letters and numbers represent areas where fish exhibited some movement to and from those sites. Dotted lines connect areas where fish were tagged to areas where they were subsequently recaptured. The two polygons describe the observed activity space of two acoustically tagged fish. The polygons were created by drawing the minimum distance along the shoreline that connects all locations where the fish were either recaptured or detected acoustically. The filled circle that does not have any letter or number designation is an area where goliath grouper were caught and subsequently recaptured, but no movement was detected to or from that area. Hollow circles are sites where goliath grouper were caught but never recaptured. The large circles represent the approximate range of the VR1 receivers, which are designated by the small circle-enclosed crosses. All fish were caught with baited circle hooks attached to set lines along mangrove shorelines.

Four fish tagged at offshore aggregation sites were re-sighted after 9-12 months, inshore either near the southwest coast of Florida or near the Florida Keys, demonstrating spawning migrations of 87-153 km (Figure 2). Two fish exhibited shorter movements among aggregation sites, one moving 16 km within 20 days and the other fish moving 22 km within four months. These small movements between offshore sites might indicate that the aggregations are not static and that individuals move among sites during the spawning season.

The remaining 32 re-sightings were of fish tagged inshore, outside the spawning period, and observed at the same site within two weeks to two months post-tagging. One fish tagged at a popular dive spot was observed at that same wreck repeatedly for 8 months.

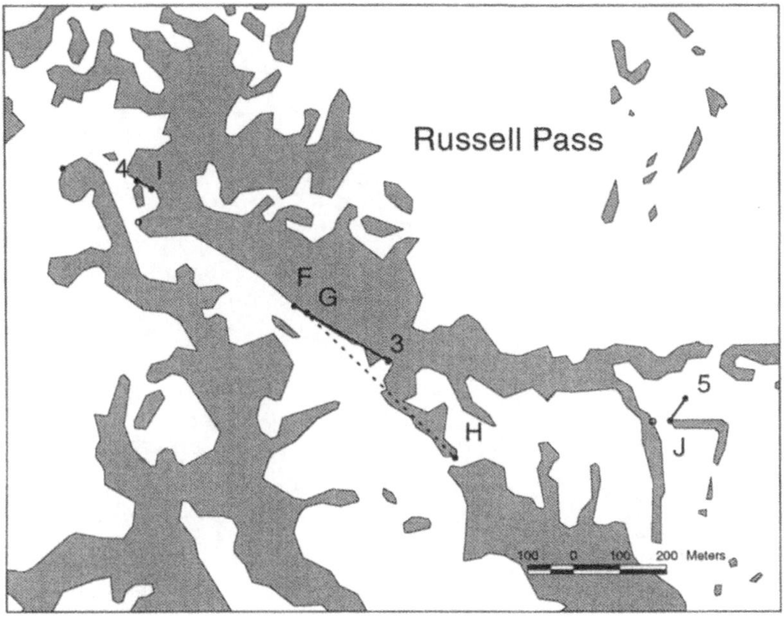

Figure 5. The Russell Pass study area for juvenile goliath grouper within the Ten Thousand Islands of southwest Florida, U.S.A. Letters represent areas where juvenile goliath grouper were tagged and recaptured, and in some cases also detected acoustically. Numbers represent areas where the fish were detected acoustically, but no lines were set there to catch the fish. Letters and numbers represent areas where fish exhibited some movement to and from those sites. The dotted line connects a site where fish were tagged to another site where the fish were subsequently recaptured. The solid lines connect sites where fish were tagged with acoustic transmitters to sites where they were later detected. The filled circle without any letter or number designation is an area where goliath grouper were caught and subsequently recaptured, but no movement was detected to or from that area. Hollow circles are sites where goliath grouper were caught but never recaptured. All fish were caught with baited circle hooks attached to set lines along mangrove shorelines.

Prior to our tagging efforts, we believed that goliath grouper might be somewhat solitary outside their summer spawning season. However, in early spring of 1999, we discovered an aggregation of goliath grouper in 5 m of water under an extended pier at Boca Grande, Florida. Since we have tagged over 200 fish in that location, we now know that these fish are much more gregarious than we had originally thought. Our misconception of goliath grouper being a solitary species (outside spawning aggregations) may be an artifact of low population numbers. Indeed, anecdotal information from historical references demonstrated that the species used to be abundant along bridges and docks all along the keys (DeMaria, 1996).

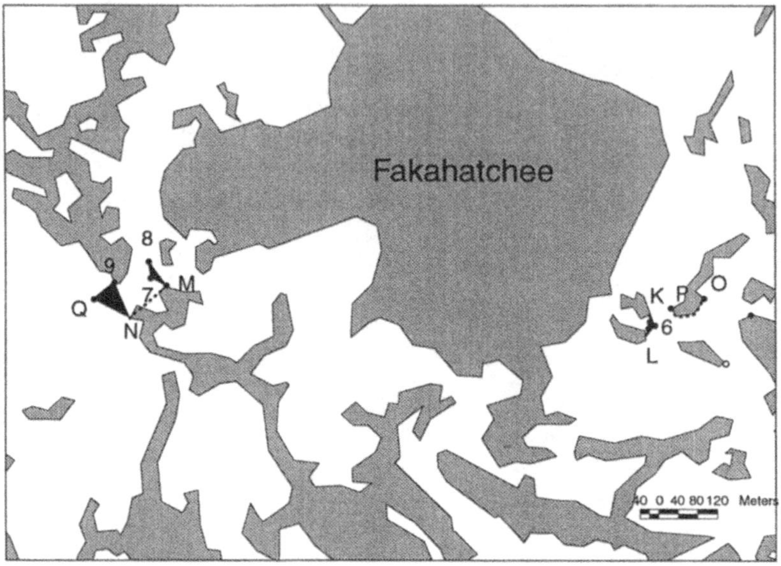

Figure 6. The Fakahatchee Island study area for juvenile goliath grouper within the Ten Thousand Islands of southwest Florida, U.S.A. Letters represent areas where juvenile goliath grouper were tagged and recaptured, and in some cases also detected acoustically. Numbers represent areas where the fish were detected acoustically, but no lines were set there to catch the fish. Letters and numbers represent areas where fish exhibited some movement to and from those sites. Dotted lines connect areas where fish were tagged to areas where they were subsequently recaptured. The three polygons describe the observed activity space of three acoustically tagged fish in the area. The polygons were created by drawing the minimum distance along the shoreline that connects all locations where the fish were either recaptured or detected acoustically. The filled circle that does not have any letter or number designation is an area where goliath grouper were caught and subsequently recaptured, but no movement was detected to or from that area. The hollow circle represents a site where goliath grouper were caught but never recaptured. All fish were caught with baited circle hooks attached to set lines along mangrove shorelines.

3.3 Limitations

There were several limitations to the mark/recapture work. First, any information on habitat and movement was limited to the tagging and recapture or re-sighting locations, with no data on the points in between capture locations. Marked fish (juveniles and adults) could only be recovered where we (or others) fished or dove, inherently biasing the descriptions of habitat. This bias is particularly true for the juvenile work, since they were captured and recaptured using baited hooks, which may attract an animal to the area. Because attraction to bait modifies typical behavior patterns, it could compromise, to some degree, descriptions of essential fish habitat. Second, we were not able to catch many juvenile fish in the winter months (Figure 9).

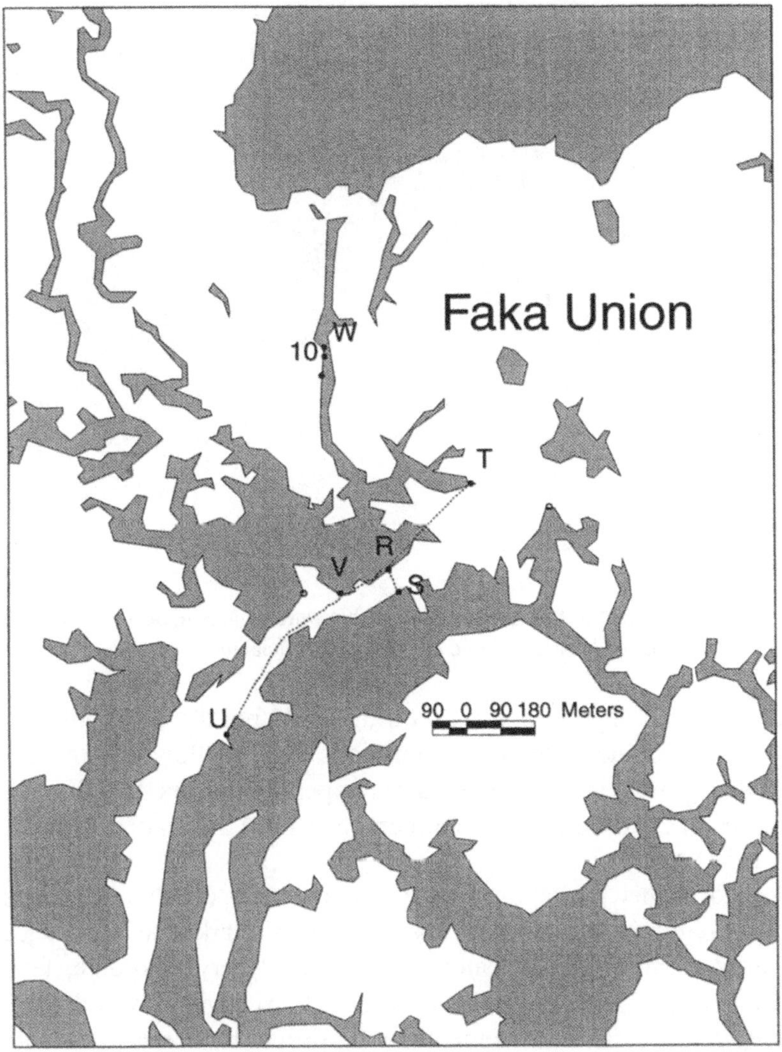

Figure 7. The Faka Union study area for juvenile goliath grouper within the Ten Thousand Islands of southwest Florida, U.S.A. Letters represent areas where juvenile goliath grouper were tagged and recaptured, and in some cases also detected acoustically. Numbers represent areas where the fish were detected acoustically, but no lines were set there to catch the fish. Letters and numbers represent areas where fish exhibited some movement to and from those sites. The dotted line connects sites where fish were tagged to other sites where the fish were subsequently recaptured. The solid line between W and 10 connects the site where a fish was tagged with an acoustic transmitter to a site where it was later detected. The filled circle without any letter or number designation is an area where goliath grouper were caught and subsequently recaptured, but no movement was detected to or from that area. Hollow circles are sites where goliath grouper were caught but never recaptured. All fish were caught with baited circle hooks attached to set lines along mangrove shorelines.

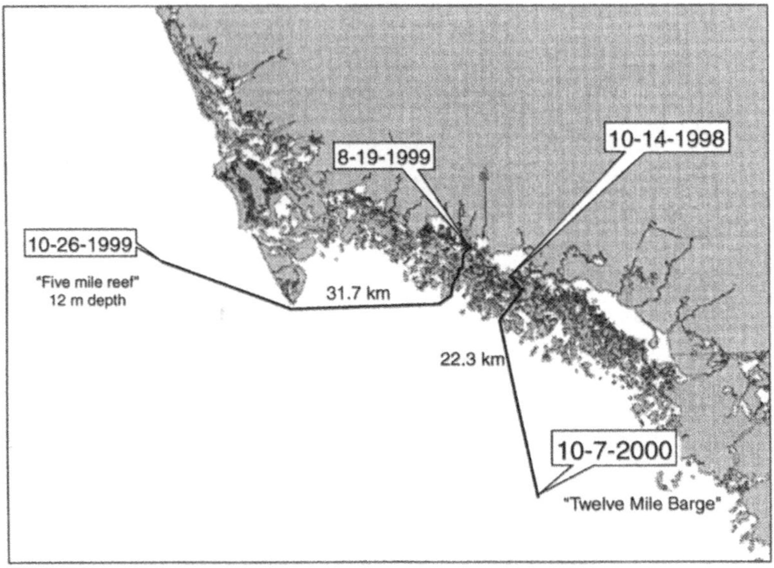

Figure 8. The movement of two goliath grouper that were tagged in the Ten Thousand Islands of southwest Florida and later recaptured on offshore artificial reefs.

Conventional tagging and recapturing could not discern whether the low catch rates were due to seasonal migrations from shallow mangrove areas to greater depths with less variable water temperature or due to reduced feeding and movement caused by reduced metabolic rates associated with cold temperatures.

We had further limitations in our tagging study of adult goliath grouper, since we were relying on visual observations of marked animals underwater or on an extremely limited catch-and-release fishery. To date, fewer than 6% of our tagged adults have been observed or caught. Even though we've been through several improvements in tag design to increase readability, the tag numbers are still difficult to read under water. Biofouling covered our early tags, and anti-fouling paint has decreased, but not eliminated, the amount of fouling on tags deployed from 1997 to the present. We also have questions about tag retention on adult goliath grouper, since we have had a few reports of tags observed on the bottom, adjacent to the Boca Grande site. The ability to detect tagged fish via acoustic telemetry can greatly enhance our work on goliath grouper movements.

4. ACOUSTIC TELEMETRY—PHASE I: DISCRETE SAMPLING

4.1 Objectives and Methodology

4.1.1 Juveniles

We were encouraged to try acoustic telemetry, based upon the results of our conventional tagging studies. The mark/recapture work revealed little to no movement of juvenile goliath grouper over two years time, suggesting that we would be able to track the fish using acoustic tags. Acoustic tagging also looked promising as a means of addressing two primary limitations we encountered during conventional tagging: (1) problems related to characterizing habitat use; and (2) problems related to seasonal migration patterns. Using hydrophones to detect tagged fish should not interfere with normal fish behavior. In addition, this method allows tracking fish year round, even during periods when catch rates using conventional methods are typically low.

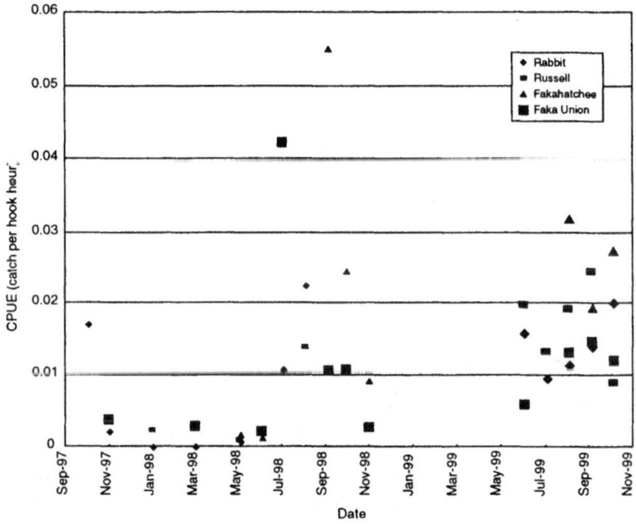

Figure 9. Seasonal catch per hook-hours of juvenile goliath grouper caught on baited circle hooks in four locations within the Ten Thousand Islands of southwest Florida, U.S.A.

We used Sonotronics' (1130 E. Pennsylvania St., Tucson, AZ 85714) transmitters that had a life expectancy of 48 months and a theoretical range through open water of one kilometer. We implanted the transmitters in the fish's peritoneal cavity through a small incision, inserted the transmitter and two internal anchor tags, then closed the incision with several surgical stitches. Using multiple anchor tags on acoustically tagged fish allowed us to immediately recognize them during conventional tagging work.

Approximately once a month, we attempted to located the acoustically tagged fish, using a portable, directional hydrophone. Each acoustic tag transmits a unique string of pulses that identifies each fish. In most cases, we could use our information on movement patterns gleaned from the conventional tagging study to give us a general idea of where to listen for the fish, if they were not near the area of last encounter. Once a fish was detected, we recorded the position with GPS. When a fish was located at the same site over several months, we confirmed that the fish was still alive by prodding the shoreline with push-poles or oars or by diving in the vicinity. This activity usually prompted the animal to move, which we could detect acoustically.

4.1.2 Adults

Because so few of the conventionally tagged adult goliath grouper were resighted and because of the problems with tag biofouling, we also tagged twelve adults with Sonotronics' ultrasonic tags. Using this method, we could detect the presence of an acoustically tagged fish without having to dive on the wrecks and observe a tagged animal. The transmitters were externally placed below the dorsal fin, in much the same way the visual tags were placed on adults. We listened for the fish with a directional hydrophone during our annual cruises to census the spawning aggregations.

4.2 Successes

4.2.1 Juveniles

Of the 14 juvenile fish tagged in the mangrove habitat, we were able to follow 9 fish successfully. Throughout the year, the majority of the fish were detected at the original tagging sites, although not all fish were detected during each sampling event. Most acoustically tagged fish remained close to the mangrove undercuts where they were originally tagged, demonstrating small activity areas of 330-6350 m^2 (Figures 4-7). Acoustic tracking revealed a second type of habitat: deep depressions of shell and rock in tidal passes, away from the shoreline (e.g., site 2 in Figure 4, site 5 in figure 5, sites 6-8 in Figure 6, site 10 in Figure 7). This habitat may be used as tidal and/or thermal refugia, although we have detected fish acoustically in the depressions year round at both high and low tide.

Fish that remained undetected for several months would sometimes reappear in the same location in which they were tagged. However, we found no seasonal pattern to their presence or absence. In fact, the majority of the fish were detected at the same sites throughout the year, giving us no evidence of seasonal migration. Although our catch rates were much lower in the colder months (Figure 9), the acoustic study provided evidence that the fish do not migrate from the area seasonally. A more likely explanation for our lower catch rates is that the colder water temperatures depress the goliath grouper metabolism and lower their movement and consumption rates. Goliath grouper are known to be susceptible to cold weather and have been reported in historical fish kills (Sadovy and Eklund, 1999).

Another aspect of goliath grouper behavior that we have learned through both our visual and acoustic tracking is that these fish are gregarious or social in nature. Several of the recaptured fish exhibited overlapping movements, and both of our acoustically tagged fish in Rabbit Key Pass have been found together at the same site for several months. Using SCUBA, we've seen as many as three goliath grouper close enough to be almost touching.

4.2.2 Adults

Since the adult goliath grouper exhibit seasonal migrations and much greater movements than the juveniles, we were less likely to detect clear patterns, especially considering our small sample size (12 fish). It does not necessarily require a large sample size, however, to gain insightful information. One fish that we tagged with a transmitter at one wreck was heard the next day on another aggregation site 6.6 km away. That information may be indicative of the dynamics of these aggregation sites. We have not detected any of our other acoustically tagged adult fish.

4.3 Limitations

We were unable to detect five (36%) of the 14 juvenile fish that were tagged acoustically. Of those five undetected fish, one died, two experienced transmitter failure, and we were unable to explain the remaining two absences. We confirmed the mortality when the transmitter, *sans* carcass, was found at the high tide mark during a spring low tide. The transmitter failures were due to problems with the magnetic switch for the battery (Sonotronics, pers. comm.). Subsequent orders of transmitters arrived without the magnetic switch, in an attempt to avoid this particular problem. Unfortunately, battery life was wasted without some means for leaving the transmitter off until time of tag placement.

We have been unable to detect four of five fish in Russell Pass since August 1999. Since all of the acoustically tagged fish in Rabbit Key and in Fakahatchee were heard throughout the fall and winter, we do not attribute their absence in Russell Pass to seasonal migration. More likely, they have moved deep in the mangrove undercuts where the mud and mangrove roots could block the signal. Another plausible explanation for the missing fish is that they move periodically between tidal refugia and their original tagging sites. These disappearances underscore one of the more problematic limitations of acoustic tracking. Namely, that non-detection of a fish does not give us enough information. Non-detection may not necessarily mean that the fish is not there, since the signal can easily be blocked in the dense undergrowth. Continuous tracking could help elucidate what had happened to these fish and to provide better information on movement and site residency patterns.

Since the adult fish were tagged on offshore wrecks that we only visit once or twice per year, our effort to detect them was limited. We learned from conventional tagging studies that at least some adults, unlike juveniles, undertake large migrations and exhibit short-term movements among aggregation sites, several kilometers apart. Our visual tagging work on the adults did not give us as much information on micro-scale movements. Fish could remain static on a small wreck site or might move off the wreck in an unknown direction. Thus, our ability to track the adults with a directional hydrophone was limited. We have only detected one of the twelve acoustically tagged fish.

Through both the visual and acoustic tagging, we have evidence of site fidelity and also of inter-aggregation movements. Using these methods, however, we could not determine whether most of the fish remain on the same site throughout the spawning period or to what extent they exhibit inter-aggregation movements. Both re-sighting of visual tags and the detection of the acoustic tags that we have used thus far merely give us a single point in time. A continuous recording of acoustic-tagged fish at each of the known aggregation sites would more adequately assess site fidelity and inter-aggregation movements of adult goliath grouper during the spawning season.

5. ACOUSTIC TELEMETRY—PHASE II: CONTINUOUS DATA LOGGING

5.1 Objectives and Methodology

Diel movements, habitat utilization and site residence times can be better described through continuous acoustic tracking either manually or with continuously recording data loggers. A hydrophone equipped with a data recorder can log presence/absence of a fish. We have much greater odds of detecting the acoustic signal through data loggers than we do with discrete monthly or annual searches using a manually operated hydrophone. Discrete sampling only gives us information about fish presence or absence at that one moment in time, but it does not provide us with any data on whether the fish was present yesterday or will be there tomorrow.

5.1.1 Juveniles

Our visual and acoustic tagging results have given us a good indication of the area of activity that the juvenile goliath grouper utilize. In Rabbit Key and Fakahatchee, we have been able to construct minimum convex polygons to describe several fish's activity space (Figures 4 and 6). In October 2000, we set up two data-logging acoustic receivers in Rabbit Key Pass at areas within those polygons (Figure 4). We used Vemco (100 Osprey Drive, Shad Bay, Nova Scotia, Canada, B3T 2C1) VR1 receivers, because they can be easily transported and can be moored on the bottom. Since they cannot be seen from the surface, we were unconcerned about theft or vandalism.

Four juvenile fish (ranging in size from 410 to 840 mm TL) were tagged internally with 16-mm acoustic transmitters that emit unique pulses every two minutes at 69 MHz, the frequency that the VR1's receive. Two fish were tagged at point E, one at point C, and one inside the creek at point D (Figure 4). One VR1 receiver was placed across from point E, moored at the bottom of a channel marker. Another receiver was placed 2 m away from point C, slightly in the channel and at the mouth of the small creek (Figure 4). We conducted range tests for each receiver and found that the ranges for the two receivers overlapped slightly. If any tagged fish were to move along the shoreline (within the already defined activity space in Figure 4), then they would be detected by one of the two receivers. Two of the tagged fish were recaptures from our conventional tagging experiment. One of the recaptures had been caught three times previously, either at point C or point D (Figure 4), and the other was originally tagged at point D. Thus, we already had some information on their activity patterns and expected them to remain in the area of the receivers' ranges. The receivers can stay submerged, collecting data for up to six months, limited by memory capacity and/or battery life.

5.1.2 Adults

Similar to the juvenile work, we set up VR1 receivers to record the presence of adult goliath grouper on three offshore spawning aggregation sites and at one inshore site (the Boca Grande pier). Since our visual and acoustic work demonstrated that some fish return to the same areas each year and that there is also some movement among the aggregation sites, continuous data loggers allowed us to monitor the dynamics of these spawning aggregations. The receivers can document the onset and extent of inshore/offshore migrations and the movement among the offshore sites. Similar work using VR1 receivers has been highly successful in describing habitat and movement patterns of tautog in Chesapeake Bay (Arendt, in press) and sharks in Sarasota Bay (Heupel and Hueter, this volume).

We used 16-mm acoustic tags in 10 adult goliath grouper at the Boca Grande Pier in June 2000 and in 20 adults on three offshore aggregations during spawning periods in August and September 2000 (Figure 2). Fish were caught on line and brought to the surface for internal tag placement. While on the surface, we measured each fish and took samples of dorsal fin rays for age and growth analysis.

5.2 Successes

5.2.1 Juveniles

After two months of data logging in the mangrove habitat, we downloaded the receivers in December, 2000. The receiver at point C (Figure 4) repeatedly (and almost daily) detected the fish tagged at that location, but it never detected either of the two fish that were tagged at point E or the fish that was tagged in the creek at point D (Figure 4). The receiver across from point E recorded the presence of both fish tagged in that area every day, but it did not detect the other fish that were tagged at points D and C. In both cases, the receiver recorded the presence of the fish intermittently, throughout each day. There were hours of time when the fish were not recorded, presumably due to the fish withdrawing farther into the mangrove undercuts, where reception of the transmitters would be limited. Further analysis will provide us with information on diel and tidal activity patterns. Over the two months of study, we found no evidence of the fish moving in our predicted activity space along the shoreline or any tagged fish moving in or out of the creek area. Perhaps, over a longer time series we will observe such movements. We predict that activity patterns will increase in the summer, when the water temperature increases.

Even with such a short time series, however, it is already apparent that we can collect data more efficiently using the data loggers than we can through our monthly checks using the manually operated hydrophone. For example, during one field day of data downloading, we collected two months of continually recorded data on three fish. On that same day, we attempted our manual tracking of nine fish previously tagged with the Sonotronics transmitters (see previous section) and found only three of them.

The information gained per man-hour of tracking was minimal compared to that gained from the data logger. The manual tracking gave us a single position for each of three fish on one single day, but the data loggers gave us presence/absence of three fish for every two minutes from October 5 to December 12.

5.2.2 Adults

Figure 10 describes the presence or absence of the ten fish that were tagged at Boca Grande Pier, from June 24, 2000 through January 23, 2001, when the data were downloaded. Overall, each fish appeared to behave independently of the others, with three fish remaining in the area almost constantly and others disappearing for days or weeks at a time, particularly during the spawning period of July-September. All of the fish disappeared in December, possibly owing to cold water temperatures. Not one of the ten fish was detected at the three aggregation sites where we had other VR1 receivers stationed.

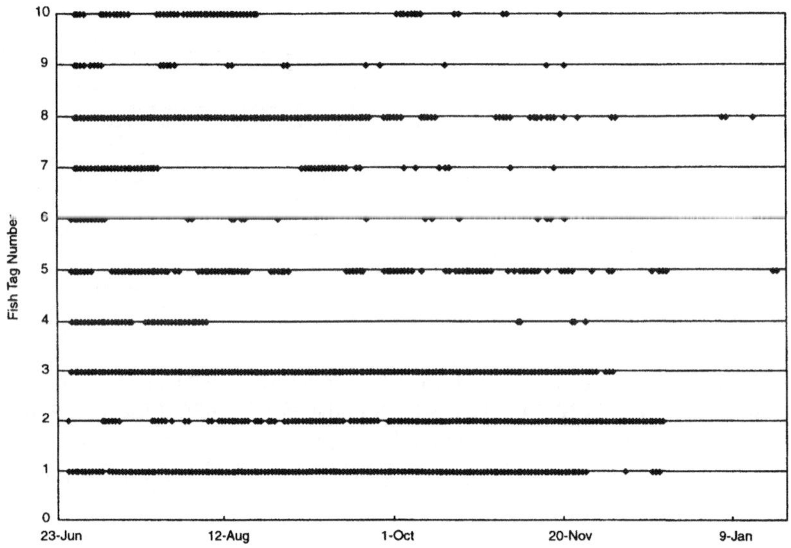

Figure 10. Detection of acoustically tagged fish by a VEMCO VR1 receiver moored at the Boca Grande Pier at the mouth of Charlotte Harbor, Florida, USA, from June 2000 through January 2001. Black diamonds indicate when each tagged fish was detected. A space with no black diamonds indicate when a fish remained undetected for more than 24 hours.

Most of the twenty adults that were tagged offshore during the spawning season remained site-faithful for much longer than we had anticipated, since we had perceived the aggregations as seasonal spawning sites. Out of 17 fish tagged on the Californian wreck in August and September, 12 were still detected on the wreck in November or early December, when receiver's memory became full (Figure 11). Four others had left the wreck in September or early October and one fish (#33) was detected only sporadically after September. Both fish tagged on a shrimp boat wreck had disappeared by the end of September, but the one fish tagged at a navigation tower remained on the tower through January 2001. The receiver at the tower also detected two of the Californian tagged fish, who visited the tower for less than a day; the Californian and the tower are 6.6 km apart. Prior to this data collection, we had anecdotal data suggesting that these aggregations were seasonal in nature. These data revealed that the majority of the fish observed during the spawning season were residents of the sites, not seasonal visitors.

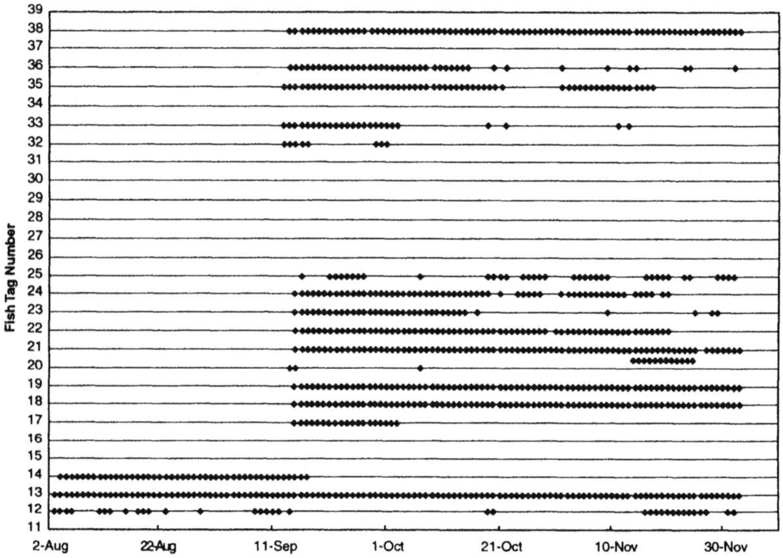

Figure 11. Detection of acoustically tagged fish by a VEMCO VR1 receiver moored at the Californian, a ship wreck in 30-40 m water depth in the eastern Gulf of Mexico, from August through December 2000. Black diamonds indicate when each tagged fish was detected. A space with no black diamonds indicate when a fish remained undetected for more than 24 hours.

5.3 Limitations

Although the data-log ging capabilities of the VR1s are extensive, they can provide no information on fish location when fish move away from receivers. We know that we lose detection capabilities for juvenile fish, when they withdraw farther into the mangrove overhangs. This loss is indistinguishable from that associated with fish leaving the area altogether. Presumably, when the juveniles grow to the adult stage they may undertake ontogenetic migrations to the reefs, where the adults reside. Acoustic tracking, either manually or with data loggers, will not give us information on any large scale and unidirectional movements of these fish.

In the adult study, finding any of the 10 fish tagged inshore on one of the three offshore sites, where receivers are moored, would be extremely good fortune. Likewise, it would be serendipitous if some of the 20 fish tagged at the summer spawning aggregations were to migrate to Boca Grande to be detected by the receiver moored on the pier. We know that these migrations are possible, based on our conventional tagging study (Figure 2); however, it is also very possible that these fish remain at the tagging sites throughout the year. If the fish do move to other unknown places both inshore and offshore, the route that they may take during these migrations is also unknown, making manual tracking difficult at best. Continuous manual tracking of these fish would be extremely labor intensive, since we do not as yet know when the onset of migration occurs, and since the migrations may well be over 50 km in length.

6. SATELLITE TRACKING

6.1 Future Objectives and Methodology

Our next phase of telemetry work involves the use of satellite "pop-off" tags. This method likely can provide answers to three key questions that thus far have alluded us using other tagging methods: (1) What is the ontogenetic migration route from the juvenile mangrove habitat to the adult offshore reef habitat? (2) What is the spatial extent of spawning migrations? (catchment area of aggregations) and (3) Where do the spawning aggregations occur on natural reef sites?

6.1.1 Juveniles

Divers have observed few young adult goliath grouper on reefs. The largest juveniles observed in mangrove habitats are about 1000 mm TL. Since there is no directed fishery for this protected species, we do not expect to receive much more information on migrations through mark/recapture. Satellite tracking with pop-off, archival tags could give us useful information on ontogenetic migrations. Although archival tags are expensive, we will only need returns from a few fish of the largest juvenile size class found in mangroves to make this project possible. Large juveniles will be equipped with archival tags, externally placed below the dorsal fin. These temperature and pressure-sensitive tags will record depth at 2% of the programmed depth range (e.g., for programmed depths ranging from 0 to 30 m, it will record depths to 0.6 m precision). The tags will be programmed to disengage from the animal one year post tagging, increasing the probability that large fish will have left the mangrove area. Once the tag reaches the surface, it will communicate with the Argos satellites, which will record its position. The tag will then be retrieved and the data downloaded, providing information on depth and temperature changes during migration and/or any temperature induced cues to migration. The depth sensor will also confirm when the tag was on the fish.

6.1.2 Adults

If fish are tagged on spawning aggregations with satellite transmitters programmed to disengage several months post-spawning, then we can discover the extent of movement after spawning. For example, it appears that some gag (*Mycteroperca microlepis*) aggregating to spawn in the northeast Gulf of Mexico may have migrated from as far away as the South Carolina coast (Coleman et al., 1999). Inferences of stock structure can be drawn from such information and would be instrumental in developing management strategies, including the design of marine protected areas.

If fish are tagged on resident sites outside the spawning season and the tags are programmed to disengage during peak spawning, we might be able to locate previously undiscovered spawning aggregation sites, perhaps including some yet unknown natural reef areas. Fish from inshore Gulf of Mexico waters, as well as those from the Florida Keys and points north along the Atlantic coast will be targeted for these satellite transmitters. By retrieving the archival tags after the satellite transmission, we will also gain data on depth changes during the spawning migration and possible temperature cues to the initiation of the migration.

6.2 Limitations

There are limits to the Argos system, particularly at the low latitude of our study area. The four Argos satellites are polar orbiters that only pass five times per day at 25 degrees North latitude, with each pass low on horizon. The position of the tag may only be resolved within several kilometers. However, the resolution of the satellite tracked positions should still be enough for us to understand the general area of ontogenetic and spawning migrations, and the archived data on the tag will give us information on the dates and depths of migration. Unlike studies conducted on transoceanic migrators, such as bluefin tuna (Block et al., 1998), we will not attempt to calculate geoposition through archived data on these tags. Block et al. (1998), among others, have been able to use data generated from light meters on the tags to calculate latitude and longitude. Since the geoposition data can only be resolved within one degree of latitude, there is no reason to collect light data for goliath grouper, which may only migrate a couple hundred kilometers at the most.

7. SUMMARY

We have presented here the evolution of our tagging studies of goliath grouper populations in southern Florida. We have used a step-wise approach in our conventional tagging and biotelemetry studies to elucidate goliath grouper life history, behavior and population dynamics. Each phase of our tagging work has led to the next phase, in ever increasing technological complexity. It is worth noting that increases in technological complexity are not necessarily accompanied by increases in information quality. For instance, some of our most important results have come from our conventional tagging studies of juvenile goliath grouper. Acoustic and satellite telemetry, however, have provided, or will provide, answers to questions that cannot be addressed adequately with visual tags.

Using conventional tags, we have learned where adult and juvenile fish are found and the extent of their movements over time. We are also able to estimate juvenile abundance, growth rates and survival rates by recapturing individuals. The acoustic work has given us a much better description of seasonal activity patterns, habitat utilization, fine-scale movements of the juveniles, and inter-aggregation movements of the adults.

Using the acoustic data to assess habitat use, we are currently testing hypotheses relating to the effects of habitat limitation, by comparing goliath grouper densities in areas with different amounts of shoreline undercuts and bathymetric heterogeneity. Without our acoustic tagging work, we would not have had enough information on habitat use to proceed. The acoustic tracking has given us a better estimation of juvenile goliath grouper home ranges, so that we could convert our abundance estimates from the mark/recapture study into density estimates.

Continuous tracking of acoustically tagged animals has facilitated our description of home ranges, daily activity patterns, and site fidelity. It is only through our acoustic data-log ging methods that we have determined the residency patterns of our adult goliath grouper, both inshore and offshore. Satellite telemetry should provide insights into ontogenetic and spawning migrations.

Each sampling method has certain limitations. Assessment of abundance and growth rates required larger sample sizes that were only economically feasible with conventional tagging methods. Smaller sample sizes, however, are sufficient to elucidate ontogenetic or spawning migrations. Therefore, the more expensive satellite tracking can be done on a smaller number of individuals, whereas the lower cost visual tags are best used for larger sample sizes.

Our stepwise approach allowed us to use the experience and information gained from each tagging method to plan subsequent experiments. We learned from our conventional tagging studies that the goliath grouper are hardy, have a high tag retention rate, and have a high degree of site fidelity. Armed with that information, we knew that an acoustic tracking experiment could be successful. The visual and acoustic tagging results were used to plan placement of the data-log ging equipment for the more intensive, continuous acoustic tracking. Discrete acoustic sampling demonstrated that juvenile goliath grouper are often, but not always in a localized area. Continuous acoustic tracking with data loggers revealed how long they remain at their home sites. Both visual and acoustic tagging helped us realize that the adults may move among aggregation sites, but continuous tracking has shown that such movements are uncommon. The data loggers also revealed that the majority of the fish at the offshore sites are residents, not seasonal visitors. Finally, results from both conventional and acoustic tagging have given us some indications of longer range ontogenetic and spawning migrations, so that satellite telemetry can be used on a small number of individuals to determine the long range movements which acoustic sampling cannot describe.

ACKNOWLEDGEMENTS

We are extremely thankful for the help and expertise of our colleagues, Christopher Koenig and Felicia Coleman, who have been an integral part of this project and without whom this research would not have been possible. We would also like to thank the many individuals that assisted us in the field, especially Don DeMaria who has single-handedly tagged all of the adult goliath grouper in our study, Larry Demere who showed us where and how to catch the juveniles, and Dr. Matt Finn who caught so many of them for us. Todd Bevis has been an important part of this project and we thank him immensely for his dedication to various aspects of the research. We would also like to thank Dave McClellan, Peter Fischel, Joe Contillo and countless others who have braved mosquitoes and lightening all in the name of science. Jim Tobias was instrumental in plotting our data onto GIS maps. NOAA-Fisheries' Office of Protected Resources, Everglades National Park, and the Curtis and Edith Munson Foundation provided funds to make this research possible.

REFERENCES

Arendt, M. (2001) Diel and seasonal activity patterns of adult tautog (*Tautoga onitis*) in lower Chesapeake Bay, inferred from ultrasonic telemetry. *Env. Biol. Fish* (in press).

Block, B.A., Dewar, H., Williams, T., Prince, E., Farwell, C., Fudge D. (1998) Archival tagging of Atlantic bluefin tuna (*Thunnus thynnus thynnus*). *Mar. Tech. Soc. J.* **32**, 37-46.

Block, B.A., Booth, D.T, and Carey, F.G. (1992) Depth and temperature of the blue marlin, *Makaira nigricans*, observed by acoustic telemetry. *Mar. Biol.* **114**, 175-183.

Boreman, J. (1996) Why tag fish? *Underwat. Nat.* **23**(2), 15-17.

Bullock, L.H., and Smith, G.B. (1991) Seabasses (Pisces: Serranidae). Memoirs of the Hour Glass Cruises, 8 (2), 243 pp.

Bullock, L.H., Murphy, M.D., Godcharles, M.F., and Mitchell, M.E. (1992) Age, growth, and reproduction of jewfish *Epinephelus itajara* in the eastern Gulf of Mexico. *Fish. Bull. U.S.*, **90**, 243-249.

Coleman, F.C., Koenig, C.C., Eklund, A.M., and Grimes, C.B. (1999) Management and conservation of temperate reef fishes in the grouper-snapper complex of the southeastern United States. *In* Musick, J.A. (ed.) Life in the Slow Lane: Ecology and Conservation of Long-Lived Marine Animals, American Fisheries Society Symposium 23, Bethesda, Maryland. pp. 233-242.

Commiskey, P.A. (1999) The use of ultrasonic tags for life history studies of red king crab in Womens Bay, Kodiak, Alaska. (*Abstract*) *Fifteenth International Symposium on Biotelemetry*, Juneau, Alaska, May 9-14, 1999.

Cruz-Escalona, V.H., Abitia-Cardenas, L.A., Campos-Davila, L., and Galvan-Magana, F. (2000) Trophic interrelations of the three most abundant fish species from Laguna San Ignacio, Baja California Sur, Mexico. *Bull. Mar. Sci.* **66**, 361-373.

Davis, T., and Gunn, J. (1992) Archival tagging provides detailed data. *Aust. Fish.* **51**, 31.

DeMaria, K. (1996) Changes in the Florida Keys marine ecosystem based upon interviews with experienced residents. The Nature Conservancy and Center for Marine Conservation Special Report, The Nature Conservancy, Key West, FL.

de Morais, L.T., and Raffray, J. (1999) Movements of *Hoplias aimara* during the filling phase of the Petit-Saut dam, French Guyana. *J. Fish Biol.* **54**, 627-635.

Gulf of Mexico Fishery Management Council (GMFMC) (1990) Amendment number 2 to the fishery management plan for the reef fish fishery of the Gulf of Mexico, 31pp.

Gunn, J.S., Stevens, J.D., Davis, L.O., and Norman, B.M. (1999) Observations on the short-term movements and behaviour of whale sharks (*Rhincodon typus*) at Ningaloo Reef, Western Australia. *Mar. Biol.* **135**, 553-559.

Hudson, E., and Mace, G. (1996) Marine fish and the IUCN red list of threatened animals. Report of the workshop held in collaboration with WWF and IUCN at the Zoological Society of London from April 29th-May 1st, 1996, Institute of Zoology, Zoological Society of London, London.

Koenig, C. C., and Coleman, F. C. (1998) Absolute abundance and survival of juvenile gag, *Mycteroperca microlepis* in seagrass beds of the northeastern Gulf of Mexico. *Trans. Am. Fish. Soc.* **127**, 44-55.

Krebs, C.J. (1999) *Ecological Methodology*. Addison-Welsey Educational Publishers, Inc., Menlo Park, California, 620 pp.

Laird, L.M., and Stott, B. (1978) Marking and tagging. *In* Bagenal, T. (ed.) *Methods for Assessment of Fish Production in Fresh Waters*, Blackwell Scientific Publications, Oxford, England, pp. 84-100.

Lutcavage, M.E., Brill, R.W., Skomal, G.B., Chase, B.C., and Howey, P.W. (1999) Results of pop-up satellite tagging of spawning size class fish in the Gulf of Maine: Do North Atlantic bluefin tuna spawn in the mid-Atlantic? *Can. J. Fish Aquat. Sci.* **56**, 173-177.

Mather, F.J., Mason, J.M., and Jones, A.C. (1995) Historical document: life history and fisheries of Atlantic bluefin tuna. NOAA Technical Memorandum, NMFS-SEFSC-370, 165 pp.

Musick, J.A., Harbin, M.M., Berkeley, S.A., Burgess, G.H., Eklund, A.M., Findley, L., Gilmore, R.G., Golden, J.T., Ha, D.S., Huntsman, G.R., McGovern, J.C., Parker, S.J., Poss, S.G., Sala, E., Schmidt, T.W., Sedberry, G.R., Weeks, H., and Wright, S.G. (2000) Marine, estuarine, and diadromous fish stocks at risk of extinction in North America (exclusive of Pacific salmonids). *Fisheries* **25**, 6-30.

Nielson, L.A. (1992) Methods of Marking Fish and Shellfish. American Fisheries Society, Special Publication 23, Bethesda, Maryland, 208 pp.

Priede, I.G., Smith, K.L., Jr., and Armstrong, J.D. (1990) Foraging behavior of abyssal grenadier fish: inferences from acoustic tagging and tracking in the North Pacific Ocean. *Deep Sea Res.* **37**, 81-101.

Sadovy, Y., and Eklund, A. (1999) Synopsis of biological data on the Nassau grouper, *Epinephelus striatus* (Bloch 1792), and the jewfish, *E. itajara* (Lichtenstein 1822). U.S. Dep. Commer., NOAA Tech. Rep. NMFS 146, and FAO Fisheries Synopsis 157, 65 pp.

Schaefer, H.C., and Fable, W.A., Jr. (1994) King mackerel, *Scomberomorus cavella,* mark-recapture studies off Florida's east coast. *Mar. Fish. Rev.* **56**, 13-23.

South Atlantic Fishery Management Council (SAFMC) 1990. Amendment Number 2. Regulatory impact review. Regulatory flexibility analysis and environmental assessment for fishery management plan for the snapper-grouper fishery of the South Atlantic region, 47 pp.

Turner, S.C., Restrepo, V.R., and Eklund, A.M. (1990) A review of the growth of Atlantic bluefin tuna, *Thunnus thynnus*. ICCAT Working Document, SCRS/90/78.

Welch, D.W., and Eveson, J.P. (1999) An assessment of light-based geoposition estimates from archival tags. *Can. J. Fish. Aquat. Sci.* **56**, 1217-1227.

Wetherbee, B.M., and Rechisky, E. (1999) Movement patterns of juvenile sandbar sharks on their nursery grounds in Delaware Bay. (Abstract) Fifteenth International Symposium on Biotelemetry, Juneau, Alaska, May 9-14, 1999.

Xiao, Y. (1996) A framework for evaluating experimental designs for estimating rates of fish movement from tag recoveries. *Can. J. Fish. Aquat. Sci.* **53**, 1272-1280.

Yano, K. (1999) Telemetry studies on the movements of the zebra shark and tawny nurse shark at the Yaeyama Islands, Okinawa, Japan. (Abstract) Fifteenth International Symposium on Biotelemetry, Juneau, Alaska, May 9-14, 1999.

Yano, K., Ochi, Y., Shimizu, H., and Kosuge, T. (1999) Diurnal swimming patterns of the diamondback squid as observed by ultrasonic telemetry. (Abstract) Fifteenth International Symposium on Biotelemetry, Juneau, Alaska, May 9-14, 1999.

Use of an Automated Acoustic Telemetry System to Passively Track Juvenile Blacktip Shark Movements

M.R. Heupel and R.E. Hueter
Center for Shark Research, Mote Marine Laboratory, Sarasota, Florida

Key words: acoustic telemetry, automated array, nursery area, *Carcharhinus limbatus*, passive tracking

Abstract: An array of 14 automated acoustic receivers were deployed within a coastal shark nursery on the central Gulf Coast of Florida during the summer of 1999. These units were used to monitor the long-term movements and behavior patterns of neonate *Carcharhinus limbatus* within the study site. Eighteen sharks were fitted with coded acoustic transmitters and passively tracked for periods of 3-159 days. The methods of equipment deployment and data collection and analysis are described. The extensive data collected from the hydrophone array were sorted and condensed using a FORTRAN program with user defined variables. Examination of the data revealed that sharks remained within a confined portion of the study site during the first three months of the study (May-July). During the months of August-November sharks began to utilize the remainder of the nursery area habitat. Active tracking of three additional sharks was used to validate results collected via the automated system and allowed comparisons between the two methods.

1. INTRODUCTION

Acoustic telemetry has been used to define the fine-scale movements of many aquatic species, including sharks. The majority of shark movement data obtained to date has been collected by active tracking efforts. Active tracking normally involves using a directional hydrophone mounted to a boat to follow a signal from an ultrasonic transmitter (e.g. Sciarotta and Nelson, 1977; Standora and Nelson, 1977; Carey and Robison, 1981; Klimley and Nelson, 1984; Carey and Scharold, 1990; Holland *et al.*, 1992; Holts and Bedford, 1993). The position of the tagged fish is regularly recorded and a

J.R. Sibert and J.L. Nielsen (eds.), Electronic Tagging and Tracking in Marine Fisheries, 217–236.
© 2001 *Kluwer Academic Publishers. Printed in the Netherlands.*

track of the animals movements is compiled (Sciarotta and Nelson, 1977; Holland *et al.*, 1992; Holts and Bedford, 1993). Because of the intensive labor and time requirements of active tracking, most studies of this kind are short-term (less than 72 hours) and include a small number of animals (generally less than ten). These studies provided important information on the habits of these species but do not provide data on longer-term movements.

In his 1990 review of shark telemetry research, Nelson reported that various researchers had tracked 200 individuals of 24 shark species. The longest reported active track was 136 hours (Carey and Scharold, 1990) and the longest intermittent tracking study was 90 days. Since this review, several technological advances have been made and longer-term telemetry studies have been conducted. Morrissey and Gruber (1993a,b) surpassed previous studies and completed 153 days of intermittent tracking of juvenile lemon sharks, *Negaprion brevirostris* (Carcharhinidae). Although there have been a few long-term intermittent tracking studies, these efforts generally contain limited information about movement patterns aside from the point in time when the animal is re-acquired.

Nelson (1990) suggested that in order to define home range, more than one 24-hour tracking session must be completed. Most current tracking studies have met and exceeded that requirement to produce information about the habits of several species (Table 1). Active tracking of scalloped hammerhead pups, *Sphyrna lewini* (Sphyrnidae), for periods of up to 72 hours revealed that this Hawaiian population had a consistent pattern of diel behavior (Holland *et al.*, 1992, 1993). Pups schooled during the day in a small core area and made wider ranging excursions to other areas at night. Studies on older *S. lewini* by Klimley and Nelson (1984) showed that sharks swam back and forth along, and remained close to, a seamount ridge during the day and left the area at night. Although sharks schooled during the day, it is unknown whether they left the seamount in groups or as solitary individuals (Klimley and Nelson, 1984).

Higher nocturnal activity levels were also found by Standora and Nelson (1977) in the Pacific angel shark, *Squatina californica* (Squatinidae). This species showed an abrupt increase in activity at dusk with maximum activity levels occurring in the middle of the night. *Squatina californica* appeared to remain in a confined area, but owing to the limited duration of tracking, home ranges could not be determined. Studies of the gray reef shark, *Carcharhinus amblyrhynchos* (Carcharhinidae) (McKibben and Nelson, 1986) and the lemon shark (Gruber *et al.*, 1988) have also reported nocturnal behavior patterns based on active tracking data.

Recent advances in acoustic receiver technology have allowed researchers to establish automated monitoring systems in sheltered areas (Klimley *et al*, 1988; Lagardère *et al.*, 1988, 1990; Bégout and Lagardère, 1995; Bégout *et al.*, 1997). These systems generally consist of bottom-mounted receivers equipped with omnidirectional hydrophones. A minimum of three hydrophones are laid out in a triangular or grid pattern to determine the location of a tagged animal. By measuring arrival times of the signal at various hydrophones, the position and movement of the fish can be determined (Lagardère *et al.*, 1988, 1990; Bégout and Lagardère, 1995; Bégout *et al.*, 1997). Information concerning fish identification and arrival/departure time is stored by the receiver and downloaded to a computer for analysis.

Table 1. Summary of the number of animals and time of tracking from acoustic telemetry studies using active and automated tracking technology.

Species	Max continuous track (hr)	Average time of continuous track	N	Intermittent tracking	Reference
Sphyrna lewini	72	N/A	3	13 days	Holland *et al.*, 1992
	72	N/A	6	12 days	Holland *et al.*, 1993
	14	N/A	13	none	Klimley and Nelson, 1984
Xiphias gladius	120	60 hours	6	none	Carey and Robison, 1981
Thunnus albacares	72	60 hours	3	none	Block *et al.*, 1997
Isurus oxyrinchus	25	22 hours	3	none	Holts and Bedford, 1993
Squatina californica	25	N/A	9	none	Standora and Nelson, 1977
Prionace glauca	136	37 hours	24	none	Carey and Scharold, 1990
	22	N/A	4	none	Tricas, 1979
	22	13 hours	14	none	Sciarotta and Nelson, 1977
Carcharhinus amblyrhynchos	50	N/A	31	2-23 days	McKibben and Nelson, 1986
Negaprion brevirostris	6	2281 positional fixes	38	153 days	Morrissey and Gruber, 1993
	101	N/A	9	1-8 days	Gruber *et al.*, 1988
Sparus aurata	none	automated receiver system	5	4-5 days	Bégout and Lagardère, 1995
Sphyrna lewini	none	automated receiver system	18	10 days	Klimley *et al.*, 1988
Carcharhinus limbatus	3816 (159 days)	automated receiver system	18	none	Heupel and Hueter (current study)

Other types of automated hydrophones do not triangulate the location of an animal, but will record when animals with transmitters are within the range of the omnidirectional hydrophone. These units provide continuous, recorded monitoring of many individuals simultaneously by recognizing an individual identification code for each animal. Klimley *et al.* (1988) used two automated acoustic monitors over a 10 day period to confirm results previously obtained by actively tracking *S. lewini* (Klimley and Nelson, 1984). Results from the acoustic monitoring system mirrored those of active tracking by showing that *S. lewini* remained on the seamount during the day and left the range of the monitors at night. The use of this technology also showed that sharks returned to their site of original capture on several days post-tagging. Results from intermittent tracking studies have produced similar findings with acoustic signals from sharks re-acquired near the location of original capture (e.g. *Carcharhinus amblyrhynchos*, McKibben and Nelson, 1986; *Negaprion brevirostris*, Gruber *et al.*, 1988; Morrissey and Gruber, 1993a,b). Automated acoustic telemetry has also been used in studies of teleost fishes (e.g. Lagardère *et al.*, 1988; Bégout and Lagardère, 1995; Bégout *et al.*, 1997).

This paper reports the implementation of an automated acoustic hydrophone array to passively track the movements of neonate blacktip sharks, *Carcharhinus limbatus* (Carcharhinidae) in a nursery area on the central Gulf of Mexico coast of Florida. Descriptions of the acoustic array and data analysis techniques are given along with preliminary results. Results from the automated system were compared to results from standard active tracking techniques.

2. MATERIALS AND METHODS

2.1 Study Species

The blacktip shark, *Carcharhinus limbatus,* is widely distributed and occurs in tropical and subtropical continental waters (Compagno, 1984; Last and Stevens, 1994). Distribution off the Atlantic coast of the Americas ranges from Massachusetts to Brazil and includes the Gulf of Mexico (Compagno, 1984). *Carcharhinus limbatus* is the most abundant medium sized shark in coastal waters of the Gulf of Mexico and the most important species in the large coastal shark fisheries, (both commercial and recreational) off the southeastern United States (Branstetter and McEachran, 1986; NMFS, 1999).

Mature *C. limbatus* are thought to produce young in alternating years after a 10-12 month gestation period, giving birth to four to seven pups (Compagno, 1984; Castro, 1996). Pups are released in inshore nursery areas from mid-April to mid-June and are thought to remain in the nursery area until early autumn. Adult and juvenile *C. limbatus* are known to migrate seasonally along the coast of the United States leaving Florida when water temperatures fall below 21°C (Hueter and Manire, 1994).

2.2 Study Site

Terra Ceia Bay, a small bay adjacent to the southeastern region of Tampa Bay, along the Gulf of Mexico coast of Florida (Figure 1), is a nursery area for neonate and juvenile *C. limbatus*. The open water area of Terra Ceia Bay is approximately 10 km^2 with a maximum water depth of approximately 4 m. The bay is fringed by mangroves and has one major entry where it joins Tampa Bay. A benthic habitat profile of Terra Ceia Bay has been compiled based on a series of 125 benthic observation stations (Heupel unpublished data). Bands of seagrass beds (*Syringodium filiforme, Thalassia testudinum, Halodule wrightii, Halophila* sp.) line the shallow near-shore areas and there is a uniform mud bottom in areas greater than 2 m deep. The extreme north and south ends of the bay have silty sediment and the area near Bird Key (inside Terra Ceia Bay) contains many corals and sponges. Terra Ceia Bay has limited water flow and a normal tidal range of 1 m. Salinities and temperatures in the bay range from 24 to 31 ‰ and from 26 to 32°C throughout the summer.

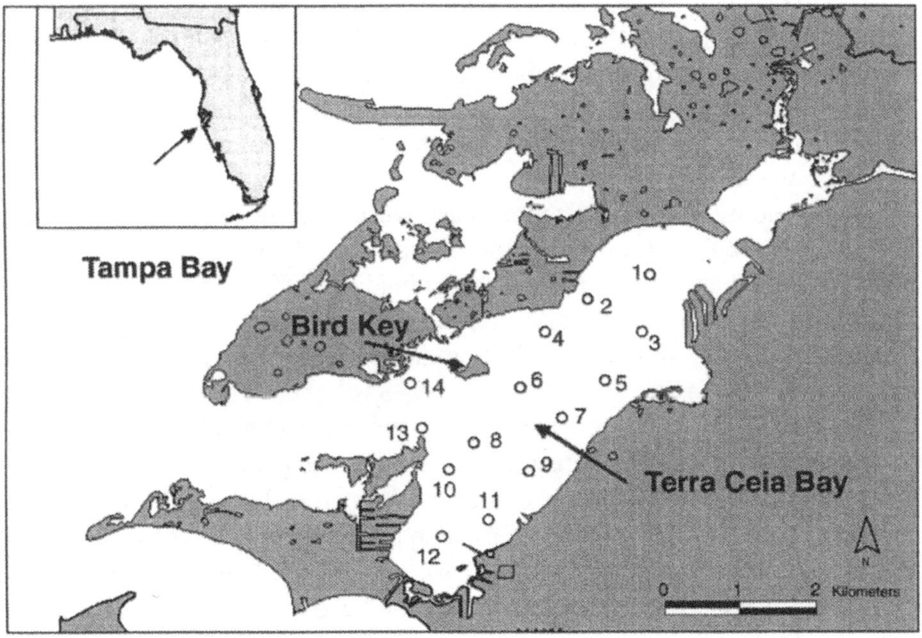

Figure 1. Map of Terra Ceia Bay showing the location and number of hydrophone stations within the study site.

2.3 Automated Acoustic Technology

To passively track the movements of sharks a series of 14 VR1 automated acoustic hydrophones (Vemco Ltd., Nova Scotia) were deployed within the study site approximately 700 m apart (Figure 1). Two of the 14 stations were placed within the mouth of Terra Ceia Bay adjacent to Tampa Bay. These two stations acted as a gate to record the movements of sharks as they exited or returned to Terra Ceia Bay. The hydrophone stations were suspended in the middle of the water column by a float and anchor system where a concrete-filled flower pot was used as an anchor and a sub-surface styrofoam float provided positive buoyancy. A ten-foot length of chain connected the concrete flower pot to a tie-down anchor that was used to prevent each hydrophone station from drifting. The use of a subsurface float made it possible to leave the equipment in a heavily utilized bay while minimizing damage or loss of equipment by keeping it out of view and below propellers. A surface float, attached to a separate tie-down anchor, was placed 10 m north of each hydrophone station to aid in relocation of the equipment. Differential GPS locations for each hydrophone station were recorded.

The hydrophone stations were deployed in late April 1999 and remained within the study site until late November 1999. Hydrophone stations recorded the identity, time and date sharks fitted with acoustic transmitters swam within range of the unit. These data were stored in flash memory until the unit was downloaded to a laptop computer via a computer interface unit (VR1PC, Vemco).

Downloading was conducted at least every two weeks to clean the units and deter biological fouling on the equipment. Downloading ports were covered with duct tape to prevent downloading difficulties, and the transducer was coated with silicone grease to slow fouling rates. Subsurface floats were cleaned to decrease fouling and increase their visibility from the boat.

Neonate *C. limbatus* were captured by rod and reel inside Terra Ceia Bay and surgically fitted with transmitters for monitoring via the automated system. Sharks were brought on board and immediately anesthetized by immersion in a 0.01 mg/l solution of tricaine methanesulfonate (MS222). After the anesthesia had taken effect, the shark was weighed and measured before being held upside down in the tank with the head and gills immersed in the seawater containing anesthetic. A 3-4 cm incision was made in the abdominal cavity and a Vemco RCODE 16 x 65 mm transmitter was inserted into the abdomen and pushed cranially. Transmitters were coated in a 70% paraffin and 30% beeswax mixture in an effort to prevent transmitter rejection (C. Lowe pers. comm.). The wound was closed with surgical sutures in the muscle and skin layers. After stitching the muscle layer, the shark was moved to a second tank containing fresh seawater and an oxygen supply. The skin layer was stitched in the fresh seawater to allow animals to recover soon after stitching was complete. Before release, all sharks were tagged with single-barb plastic dart tags (Hallprint, South Australia) for external identification. Surgical procedures generally took about 10 minutes from the time of landing to the time of release. All sharks were released in good condition.

Transmitters for the automated system were coded with a unique pulse series for each shark and pulsed at 69.0 kHz. Transmitters were programmed to transmit randomly with a minimum time between pulses of 45 seconds and a maximum of 75 seconds. Random signal transmission times prevent signals from more than one animal continuously overlapping and blocking reception by the receiver. Pause times of 45-75 seconds between transmissions provided an extended battery life of approximately 18 months for all transmitters.

Transmitters were range-tested within the study site to determine how far the signal traveled and where areas of overlap occurred between hydrophone stations. Four hydrophone stations (5, 6, 7, and 9) were tested and assumed to be representative of the other hydrophone stations in the array. Transects were run in 4 directions from each chosen hydrophone station. One transect ran directly toward the shoreline (due East), and three (North, South and West) ran towards other hydrophone stations to define the area of reception overlap. An activated transmitter was placed in the middle of the water column when the boat was 200, 300, 400, 500, 600, and 700 m from the hydrophone station. By driving directly toward another station, transects also tested the outer detection distance of the destination hydrophone.

2.4 Active Tracking

Sharks collected for active tracking efforts also were collected by rod and reel. Once landed, sharks were weighed, measured and tagged with a dart tag before attaching the transmitter. Active transmitters were 16 x 65 mm and pulsed every 2 seconds at a frequency of 50 kHz so they would not interfere with transmitters used for the automated system. Transmitters were attached to a Rototag fin tag by a thin piece of galvanized wire that corroded in seawater, allowing the transmitter to drop off after several days. Measurements and tagging were completed as quickly as possible and the shark was tracked immediately upon release. A Vemco VR60 manual receiver equipped with a directional hydrophone was used to track the movements of the shark. Location data were collected from a differential GPS every 15 minutes. The approximate direction and distance of the shark from the vessel were also recorded.

2.5 Data Analysis

Raw data collected by the hydrophone stations included the transmitter number, and time and date of detection. These data needed to be sorted and summarized to a level that was meaningful for analysis.

Software included with the VR1 units contained a data processing option called "pinger search." This function sorts the raw data by reporting the occurrences of a specific transmitter number in the stored receiver files. This function can search all data from each hydrophone station and sort the transmitter detections by time. The search function also includes an option that allows the researcher to set the time before reporting an animal absent from a station. For example, in this study, the time for reporting absent was set at five minutes. If a signal from the shark was not detected within five minutes of the previous detection the shark was considered to be absent from the receiver range. This sorting tool decreased the amount of data by reporting the hydrophone station number where the shark was present, the time and date of arrival, the time and date of departure and the number of transmitter detections that occurred during the time the shark was present. This provided detailed movement information for each shark from every hydrophone station throughout the study.

The second means of sorting the data involved inputting the raw data into Microsoft Access and sorting the data by transmitter number, date, and time. These data were then processed through a custom-written Remote Acoustic Telemetry Searcher (RATS) FORTRAN program to condense the data. Various time intervals were tested for assembling time data into discrete units. Time segments tested included 1, 2, 5, 10, 15, 20, 30, and 60 minutes. The RATS program used the appropriate time interval (e.g. 20 min) to divide the total date/time set into segments. Each segment was scanned to determine which hydrophone stations an individual shark visited during that time period. If the individual was not at any station during that period it was classed as "missing" for that time block. Processed data were examined to determine the number of hydrophone stations visited per day by individual animals and the proportion of time sharks spent at each hydrophone station during each day and also on a monthly basis. Animals that entered the study site intermittently were excluded from monthly averages because these animals were often recorded at one or two stations over a period of one to two days. These low values would have masked the results obtained from resident animals. The RATS program can be manipulated to sort the raw data according to other variables such as time of day (day/night) and tidal state (rising, falling, slack).

3. RESULTS

3.1 Range Tests

Results of range tests on three of four hydrophone stations showed a consistent reception range of at least 800 m in three out of four directions. Range in the fourth direction decreased as it approached the outer edges of the study site where decreased water depths and the presence of sea grass beds hampered signal reception. In several instances, reception range was much greater than 800 m. Signal reception as far as 1000 to 1500 m away from the transmitter location was recorded, but these results were not consistent and therefore were not considered a reliable indicator of the acoustic array reception range. Reasons for long-range detections are unknown, but because these receptions were not consistent, a working range of approximately 800 m was used in further analyses (Figure 2).

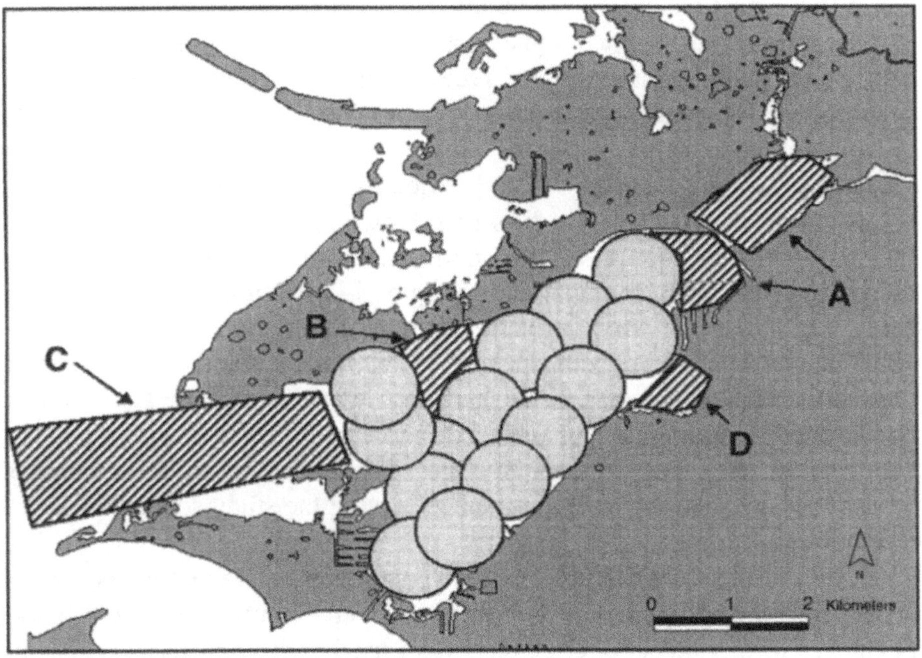

Figure 2. The approximate hydrophone range coverage of Terra Ceia Bay (circles) using a range of 800 m. Crosshatch areas A-D are "deaf" areas in the bay outside the reception range of the hydrophone stations.

Based on range tests and automated tracking results, several areas within the study site were considered to be outside the range of the hydrophone stations (Figure 2). These "deaf" areas included the northernmost portion of the bay (area A), the area north of Bird Key (area B), the mouth of the bay beyond the "gate" stations (area C), and a shallow area on the eastern side of the bay (area D). The intermittent detection of sharks at the northernmost station without detections at adjacent stations suggested that animals in this area were moving in and out of the detectable range. Attempts to manually locate sharks in the area north of station 1 were unsuccessful. Although the mouth of the study site beyond the "gate" stations was not originally part of the intended study site, it became apparent that sharks did swim into this area and returned to the main area of Terra Ceia Bay.

3.2 Passive Tracking

Eighteen neonate *C. limbatus* (11 male, 7 female) were captured and fitted with acoustic transmitters for tracking by the automated hydrophone system. Sharks were passively tracked for periods of 3 to 159 days (72-3816 hours) over a period of seven months in 1999. During this time 438,456 transmitter detections were accumulated. Two sharks appear to have died or lost their transmitters since signal detections changed from typical movement patterns to a constant signal received at one station for several days. One of these individuals was tracked for four days, the other for 37, prior to cessation of movement.

The remaining 16 sharks showed various activity patterns. Shark 1 was tracked for a period of seven days before swimming out of the study site. This individual was not re-acquired during the remainder of the study. Five sharks that left the study site after varying periods of time returned intermittently throughout the duration of the study. Approximately 25 hours spanning several days were spent "listening" for transmitter signals in Tampa Bay with a manual receiver. One shark that left the study site after 60 days was located approximately 1 km north of Terra Ceia Bay, in Tampa Bay, two weeks after it had left the study site. Six sharks lost from the range of the hydrophone stations were not tracked leaving the study site and were not re-acquired.

Before further data analysis could be completed a suitable time period for condensing the raw data was determined. RATS results from three sharks monitored for periods of over 60 days were used in this analysis. Short time segments (e.g. 1, 2, 5 min) resulted in large periods of time where sharks were recorded as missing. Long time segments (e.g. 60 min) created low missing periods, but decreased temporal resolution (Figure 3). A single factor ANOVA with a Tukey's unequal N Honest Significant Difference test showed no significant difference between 15, 20, 30 and 60 minute intervals for two of the three sharks and between 15, 20 and 30 minutes for the other.

Based on these results, a 15 minute time interval was used to maximize the temporal resolution while condensing the amount of raw data.

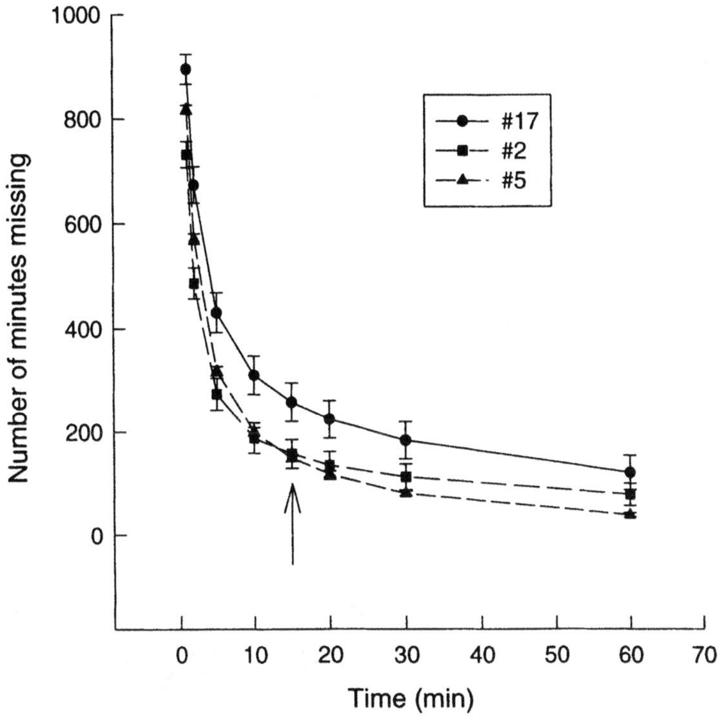

Figure 3. The amount of time sharks were determined to be "missing" from the raw data set when sorted by the RATS program at 1, 2, 5, 10, 15, 20, 30, and 60 minute intervals. Based on statistical results, a time period of 15 minutes was chosen for all database analyses.

Processed data were analyzed by the number of stations visited by sharks per day (Table 2). During the first three months of the study (May-July) sharks were generally present at 3-4 stations each day. In August, the mean number of stations visited daily increased from an average of 4.7 to 7.4 stations. The average number of stations visited remained high throughout the remainder of the study until all telemetered sharks had left the study site (November). Daily and weekly averages of visited stations showed a sharp increase in mid-August (Figure 4a,b). This change was evident in all resident sharks. From August to November resident sharks used the entire compliment of stations within the bay and at least two sharks traveled outside Terra Ceia Bay for short periods. Excursions out of the mouth of the bay were recorded on several occasions. The initial incidence of short term movement into Tampa Bay occurred in August and lasted for 11 hours. Subsequent movements outside the study area lasted for up to 13 days (days outside the study site are not included in calculations of total days tracked).

Table 2. Average number of stations used per day, by month, for individual sharks. Ranges are included in brackets. Total averages exclude animals that used the study site intermittently.

Shark no.	May	June	July	August	September	October	November
1	3 (1-6)	6 (5-7)					
2	3.8 (2-6)	5.4 (2-8)	6.2 (4-8)	7.4 (5-13)	6.9 (4-11)		
3	3 (3)	1.5 (1-3)					
4	2.5 (2-3)	2.5 (1-5)	3.2 (1-7)	1*	1*		
5	4 (3-5)	5 (3-8)	5.2 (3-7)	8.6 (5-13)	7.9 (2-11)	7.8 (4-13)	10 (10)
6	4.3 (3-5)						
7	4.5 (4-5)	3.6 (3-5)					
8		3.1 (2-4)					
9		4.3 (2-6)					
10		3.2 (1-7)	1*	1*			
11		4 (2-7)					
12		4 (1-6)					
13		4.3 (2-6)					
14		4.4 (1-7)					
15		3.6 (1-12)	1*				
16		3.8 (2-5)	5.3 (3-7)	5.8 (5-8)			4.5 (1-8)
17		4.7 (3-7)	4.1 (2-6)	7.6 (3-12)	8.1 (1-11)	8.1 (3-13)	
18			4.3 (1-10)	1*	1*	1*	1*
Total	3.6	4	4.7	7.4	7.6	8	7.3

* intermittent

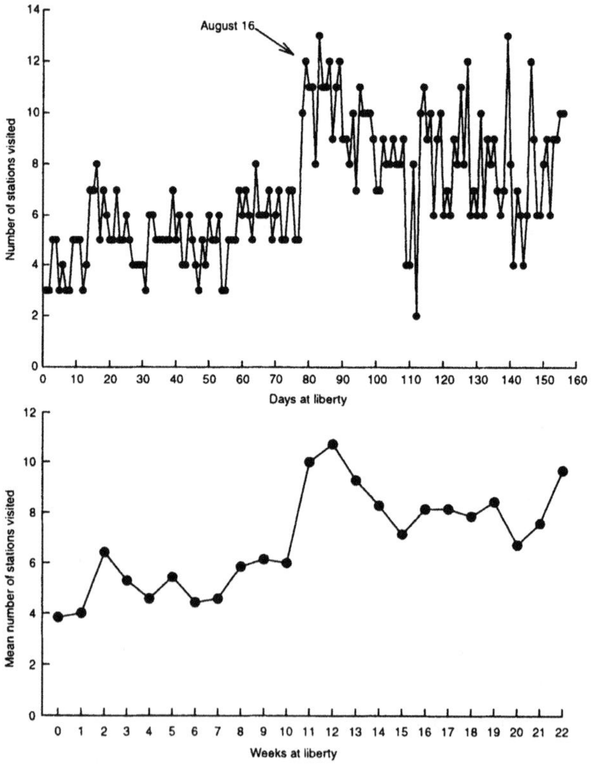

Figure 4. (top) The number of stations visited per day by shark 5 during 159 days of tracking. Increased activity began on August 16 as indicated on the graph. (bottom) The mean number of stations visited daily by shark 5 showing an increase in activity level during week 11.

Examination of the proportion of time sharks spent at each hydrophone station confirmed the use of the most northerly stations in the first three months of the study. All sharks remained in the northern portion of the bay near stations 1-3 from May-July (Table 3). The proportion of time sharks spent in the northern portion of the bay decreased over time as sharks used more of the study site and were recorded at all stations inside the bay. Although an overall summary of use of the bay indicates that sharks spent most of their time in the northern portion of the bay (Figure 5a), analysis by month revealed a distinct shift in spatial utilization of the study site during the later months of the study (Figure 5b). In August, sharks were observed to use the entire area of the bay but spent slightly more time in the middle and northern portions of the bay. In September to October, sharks were spending more time in the middle portion of the bay and little time in the northern or southern ends of the bay.

Table 3. Percentage of time sharks were recorded at each hydrophone station by month. Total column includes the average for each station throughout all months.

Station no.	May	June	July	August	September	October	November	Total
1	48.2	45.8	38.0	18.5	2.3	9.9	8.6	24.5
2	22.1	17.5	19.1	13.0	3.5	7.8	9.3	13.2
3	24.3	23.4	22.8	9.6	2.8	14.7	9.1	15.2
4	1.9	5.9	10.4	8.5	0.5	*	*	5.4
5	3.4	5.4	6.6	13.6	19.6	21.2	18.0	12.5
6	0.1	1.2	2.3	11.2	26.1	13.4	26.8	11.6
7	0	0.8	0.8	9.7	23.0	21.0	19.3	10.7
8	0	0.2	0	4.8	6.1	1.7	4.0	2.4
9	0	0	0	2.2	2.4	3.1	0.5	1.2
10	0	0	0	1.7	2.1	2.7	0.9	1.1
11	0	0	0	1.4	0.7	2.4	0.5	0.7
12	0	0	0	0.3	0.2	0.3	0	0.1
13	0	0	0	2.7	6.2	0.9	1.3	1.6
14	0	0	0	2.8	4.6	0.9	1.6	1.4

* lack of data due to equipment failure

Examination of the proportion of time sharks spent at each hydrophone station confirmed the use of the most northerly stations in the first three months of the study. All sharks remained in the northern portion of the bay near stations 1-3 from May-July (Table 3). The proportion of time sharks spent in the northern portion of the bay decreased over time as sharks used more of the study site and were recorded at all stations inside the bay. Although an overall summary of use of the bay indicates that sharks spent most of their time in the northern portion of the bay (Figure 5a), analysis by month revealed a distinct shift in spatial utilization of the study site during the later months of the study (Figure 5b). In August, sharks were observed to use the entire area of the bay but spent slightly more time in the middle and northern portions of the bay. In September to October, sharks were spending more time in the middle portion of the bay and little time in the northern or southern ends of the bay.

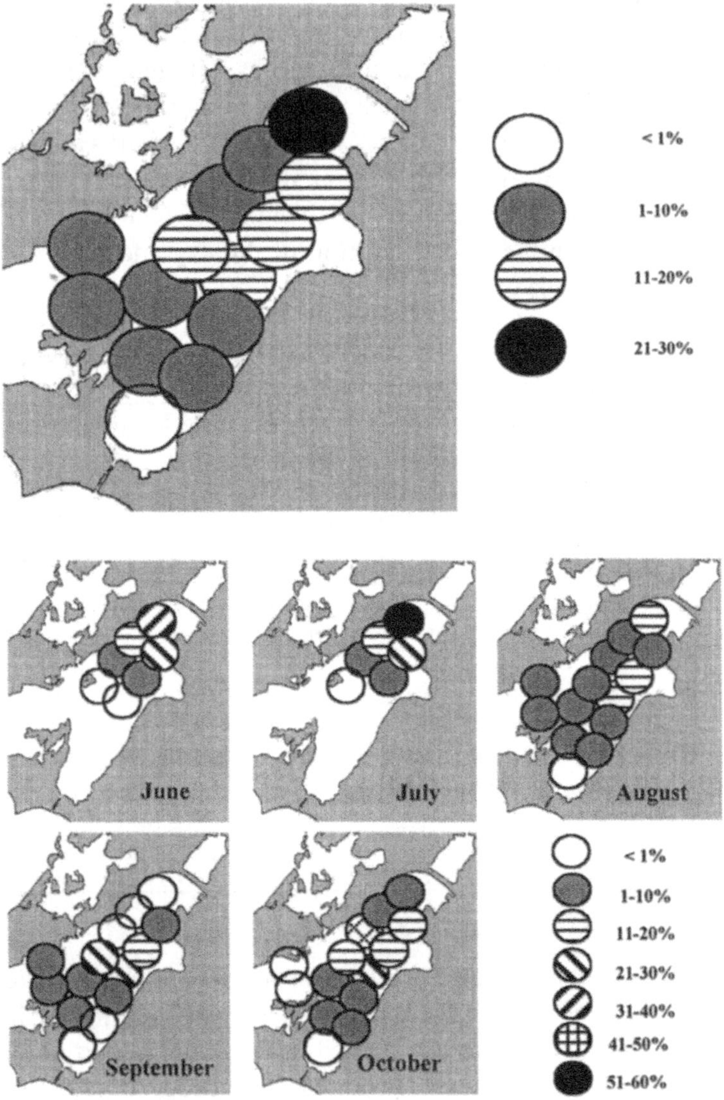

Figure 5. (top) Summary of the proportion of time shark 17 spent at hydrophone stations during the five-month period this shark was monitored. (bottom) The proportion of time spent at hydrophone stations by shark 17 displayed by month. Initial activity was confined to the northern portion of the study site until mid-August when increased movement patterns occurred. The cross-hatched circle in the month of October represents equipment failure at that location.

3.3 Active Tracking

Initial setup and maintenance of the automated monitoring system precluded time for active tracking in May or June. Three sharks, numbered 21-23 (1 female, 2 male), were actively tracked in July and early August. The female shark was caught on July 14 and tracked for a period of five hours during the day before the track was terminated because of inclement weather. Shark 21 remained within a 1 km^2 area in the northern portion of the bay near station one (Figure 6). This shark was re-acquired in the same area on July 15 and 16 and tracked for an additional two hours on July 16. The second shark was tracked for a period of 11 hours over night before the track had to be terminated because of inclement weather. Shark 22 showed a similar activity pattern to the first shark by swimming in large circles or back and forth near the location of the boat. Shark 22 also was tracked only in the northern end of the study site. The third shark was tracked for a period of 24 hours on August 12. Shark 23 remained in the northern portion of the bay for part of the track, but was also tracked moving into the middle of the bay. Although limited in duration, active tracking results showed similar movement patterns to those acquired from the automated monitoring system.

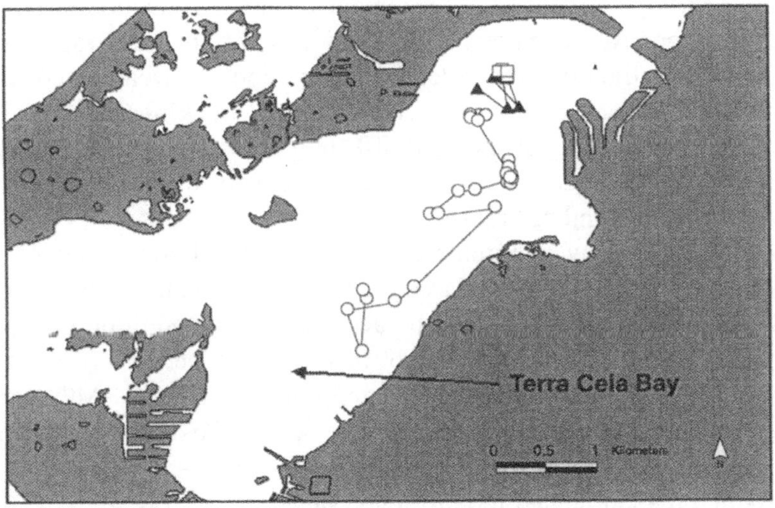

Figure 6. Map displaying the movement patterns of three sharks tracked using manual telemetry equipment. The shark represented by the open squares was tracked on July 14 and 16, The shark represented by the solid triangles was tracked on July 28, and the shark represented by the open circles was tracked on August 12. Movements of actively tracked sharks reflect the results collected by the automated monitoring system as shown in inset maps from July and August. S and E indicate the start and end of tracks.

4. DISCUSSION

Previous studies have used arrays of automated hydrophones to monitor the movements and activity levels of teleost fishes carrying transmitters (e.g. Lagardère *et al.*, 1988; Bégout and Lagardère, 1995; Bégout *et al.*, 1997). Additionally, Klimley *et al.* (1988) used two automated acoustic monitors on seamounts to detect the presence of scalloped hammerhead sharks fitted with transmitters. However, Klimley's study could only define the presence or absence of sharks and provided little information about the movements of sharks while within range of the monitors. The present study combined concepts of both approaches by deploying an array of acoustic receivers to record the presence and passively track the movements of neonate *C. limbatus* in Terra Ceia Bay.

Nelson (1990) stated that no single track to date had been long enough to demonstrate major migrations or seasonal periodicity in shark movements. He also suggested that long-term tracks are needed to uncover phenomena on a lunar or seasonal scale, such as range expansion or habitat shifts. The use of an array of automated acoustic hydrophones in this study was designed to fulfill these requirements. The collection of 159 days of continuous tracking data for one individual far exceeds what is physically possible to complete using active tracking techniques. The introduction of advanced, cost-efficient acoustic telemetry technology has provided the ability to complete long-term tracking regardless of time, tide, or weather constraints. This method avoids the labor involved in active tracking and removes the potential bias of chasing sharks and affecting their movements. Long-term monitoring also provides data beyond any recovery or reaction period that occurs after capture and handling.

The use of an array of automated acoustic hydrophones provided almost complete coverage of Terra Ceia Bay. Three of the four areas found to be outside the hydrophone array can be accessed by the addition of receivers. The area in the east of the bay cannot be monitored because of the shallow nature of this area and the presence of multiple oyster reefs. Although areas lacking hydrophone coverage existed, the areas of overlap in hydrophone coverage were well suited to hydrophone reception ranges.

Although the automated monitoring system used in this study worked effectively and provided reliable location data, the spatial resolution of those data was not amenable to analysis of fine-scale movements. In this respect, active tracking results outweigh the results obtained from an automated system. The sheer volume of data collected during this study makes analysis of fine-scale movements and location difficult. Data had to be condensed to a level that was useful and provided meaningful results. The addition of greater numbers of hydrophone stations would increase the spatial resolution by creating greater areas of overlap between stations and providing a more precise point of location. However, this approach would increase the amount of data collected, requiring further programming to be written to extract fine-scale movements. The recent availability of this technology means that analysis techniques have not been previously defined and developing these techniques requires time.

Analysis of movement patterns of neonate *C. limbatus* appear to be similar to those for young *N. brevirostris*, but different from those for young *S. lewini*. Juvenile *S. lewini* were found to remain within a core area during daylight hours and exhibit wider ranging movements after dark (Klimley and Nelson, 1984; Klimley *et al.*, 1988; Holland *et al.*, 1992,1993). Juvenile *N. brevirostris* have been defined as showing site attachment to a restricted activity space with occasional excursions to other areas (Gruber *et al.*, 1988; Morrissey and Gruber, 1993a,b). This is similar to the general patterns seen here, except that *C. limbatus* underwent a shift in spatial utilization during the course of this study. The reasons for initial preference for the northern area of the study site and the subsequent shift to the middle of the bay are currently unknown. The benthic habitat within the bay is uniform, with no one region appearing to provide better habitat for young sharks. Macpherson (1998) reported that changes in habitat use with increasing size is a common phenomenon in juvenile and adult fishes. However, based on results from growth rate studies of juvenile *C. limbatus* (e.g. Branstetter and McEachran, 1986; Killam and Parsons, 1989; Wintner and Cliff, 1996), it is unlikely that a large size change occurred and caused this shift. Changes in environmental factors such as temperature, salinity or prey density, or an ontogenetic change in behavior seem more likely to have altered the activity of these sharks.

Active tracking clearly presents the fine-scale movements of sharks in the short term and has been a successful tool in many shark studies (e.g. Sciarotta and Nelson, 1977; Standora and Nelson, 1977; Carey and Robison, 1981; Klimley and Nelson, 1984; Carey and Scharold, 1990; Holland *et al.*, 1992; Holts and Bedford, 1993). Although this technology is labor intensive, it produces results that are yet to be matched by an automated telemetry system. Active tracking during this study was limited by time and weather constraints but was consistent with data collected by the automated system. Although similar data were collected from active tracking and the automated system, results from the automated system clearly display the limitations of active tracking by providing long-term data for several animals simultaneously. Additional active tracks later in the season would presumably have provided data near the center of the bay. Results from active tracking may have been interpreted as distinct habitat use or home ranges of different individuals within the study site, rather than a spatial shift by the entire resident population as recorded by the automated system. The ability to track the movements of several animals simultaneously revealed that all of these individuals were following a similar pattern that changed over time. Tracking each shark individually over various time periods would have yielded variable, temporally discrete results for each shark and revealed little about movement patterns on a population level. Therefore, while active tracking provides fine-scale movements over a short time period, it may not completely represent what is happening in the resident population.

5. SUMMARY

The implementation of an automated acoustic telemetry system within Terra Ceia Bay, a known nursery area for *C. limbatus,* provided long-term data on the movements of sharks within the study site. Sharks were continuously tracked for periods from 3-159 days. The data collected from the hydrophone stations were sorted and condensed for analysis using a FORTRAN program that allows the user to define the time block or other variable to search the database. These analyses were used to define the number of stations visited per day by the sharks present. During the first three months of the study (May-July) sharks visited less than five stations and were confined to the northernmost portion of the study site. In the fourth month (August) sharks began to move throughout the entire study site and were detected at every hydrophone station. Analysis of the time spent at each station after the third month of the study revealed a spatial shift from the northern end to the middle of the bay. Reasons for this shift are unclear, but may be due to an environmental change.

Active tracking efforts validated the data compiled by the automated system and provided fine-scale details about the movements of sharks. The volume of data collected by the automated system and the distance between hydrophone stations decreased the spatial resolution of data, making active tracking results better for producing fine scale location data. However, the automated system provided several advantages over active tracking including decreased labor, removal of possible chasing bias involved in following an animal, the ability to track several animals simultaneously and to track in all weather conditions. Each system has advantages and disadvantages, but when combined these technologies can be used to conduct a more complete study of shark movements.

ACKNOWLEDGEMENTS

This work was supported by funding from the National Oceanic and Atmospheric Administration, National Marine Fisheries Service Grants NA87FM0588 and NA97FM0223 (to REH) and by funding from Mote Marine Laboratory. We thank Drs. C. Manire and C. Luer for their advice on surgical techniques, Dr. C. Simpfendorfer for writing the FORTRAN computer program used in this analysis, and T. Wilkie, J. Tyminski, J. Berg, S. Reinecke, E. Huber and M. Marchioni for assistance with field efforts.

REFERENCES

Bégout, M-L., and Lagardère, J-P. (1995) An acoustic study of seabream (*Sparus aurata* L.): first results on activity rhythm, effects of environmental variables and space utilization. *Hydrobiologica* **300/301**, 417-423.

Bégout Anras, M-L., Lagardère, J-P., and Lafaye, J-Y. (1997) Diel activity rhythm of seabass tracked in a natural environment: group effects on swimming patterns and amplitudes. *Can. J. Fish. Aquat. Sci.* **54**, 162-168.

Block, B.A., Keen, J.E., Castillo, B., Dewar, H., Freund, E.V., Marcinek, D.J., Brill, R.W., and Farwell, C. (1997) Environmental preferences of yellowfin tuna (*Thunnus albacares*) at the northern extent of its range. *Mar. Biol.* **130**, 119-132.

Branstetter, S., and McEachran, J.D. (1986) Age and growth of four carcharhinid sharks common to the Gulf of Mexico: A summary paper. *In* Uyeno, T., Arai, R., Taniuchi, T., and Matsuura, K. (eds.) Indo-Pacific Fish Biology: Proceedings of the Second International Conference on Indo-Pacific Fishes, Ichthyological Society of Japan, Tokyo, pp. 361-371..

Carey, F.G., and Robison, B.H. (1981) Daily patterns in the activities of swordfish, *Xiphias gladius,* observed by acoustic telemetry. *Fish. Bull.* **79**, 277-292.

Carey, F.G., and Scharold, J.V. (1990) Movements of blue sharks (*Prionace glauca*) in depth and course. *Mar. Biol.* **106**, 329-342.

Castro, J.I. (1996) Biology of the blacktip shark, *Carcharhinus limbatus*, off the southeastern United States. *Bull. Mar. Sci.* **59**, 508-522.

Compagno, L.J.V. (1984) FAO Species catalogue. Vol. 4. Sharks of the world: an annotated and illustrated catalogue of shark species known to date. Part 1. Hexanchiformes to Lamniformes. FAO Fisheries Synopsis No. 125, pp. 188-199.

Gruber, S.H., Nelson, D.R., and Morrissey, J.F. (1988) Patterns of activity and space utilization of lemon sharks, *Negaprion brevirostris*, in a shallow Bahamian lagoon. *Bull. Mar. Sci.* **43**, 61-76.

Holland, K.N., Lowe, C.G., Peterson, J.D., and Gill, A. (1992) Tracking coastal sharks with small boats: hammerhead shark pups as a case study. *Aus. J. Mar. Freshw. Res.* **43**, 61-66.

Holland, K.N., Wetherbee, B.M., Peterson, J.D., and Lowe, C.G. (1993) Movements and distribution of hammerhead shark pups on their natal grounds. *Copeia* **1993**, 495-502.

Holts, D.B., and Bedford, D.W. (1993) Horizontal and vertical movements of the shortfin mako shark, *Isurus oxyrinchus*, in the southern California Bight. *Aus. J. Mar. Freshw. Res.* **44**, 901-909.

Hueter, R.E., and Manire, C.A. (1994) Bycatch and catch-release mortality of small sharks in the Gulf Coast nursery grounds of Tampa Bay and Charlotte Harbor. Final Report, NOAA/NMFS/MARFIN Project NA17FF0378-01.

Killam, K.A., and Parsons, G.R. (1989) Age and growth of the blacktip shark, *Carcharhinus limbatus,* near Tampa Bay, Florida. *Fish. Bull.* **87**, 845-857.

Klimley, A.P., and Nelson, D.R. (1984) Diel movement patterns of the scalloped hammerhead shark (*Sphyrna lewini*) in relation to El Bajo Espiritu Santa: a refuging central-position social system. *Behav. Ecol. Sociobiol.* **15**, 45-54.

Klimley, A.P., Butler, S.B., Nelson, D.R., and Stull, T. (1988). Diel movements of the scalloped hammerhead shark, *Sphyrna lewini* Griffith and Smith, to and from a seamount in the Gulf of California. *J. Fish. Biol.* **33**, 751-761.

Lagardère, J-P., Ducamp, J-J., Frikha, L., and Spérandio, M. (1988) Ultrasonic tracking of common sole juveniles (*Solea vulgaris* Quensel, 1806) in a saltmarsh: methods and fish response to some environmental factors. *J. Appl. Ichthyol.* **4**, 87-96.

Lagardère, J-P., Ducamp, J-J., Favre, L., Mosneron Dupin, J., and Spérandio, M. (1990) A method for the quantitative evaluation of fish movements in salt ponds by acoustic telemetry. *J. Exp. Mar. Biol. Ecol.* **141**, 221-236.

Last, P.R., and Stevens, J.D. (1994) *Sharks and Rays of Australia.* CSIRO, Melbourne pp. 513.

Macpherson, E. (1998) Ontogenetic shifts in habitat use and aggregation in juvenile sparid fishes. *J. Exp. Mar. Biol. Ecol.* **220**, 127-150.

McKibben, J.N., and Nelson, D.R. (1986) Patterns of movement and grouping of gray reef sharks , *Carcharhinus amblyrhynchos*, at Enewetak, Marshall Islands. *Bull. Mar. Sci.* **38**, 89-110.

Morrissey, J.F., and Gruber, S.H. (1993a) Home range of juvenile lemon sharks, *Negaprion brevirostris. Copeia* **1993**, 425-434.

Morrissey, J.F., and Gruber, S.H. (1993b) Habitat selection by juvenile lemon sharks, *Negaprion brevirostris. Envir. Biol. Fishes* **38**, 311-319.

National Marine Fisheries Service (1999) Draft fishery management plan for Atlantic tunas, swordish, and sharks Volume 1. NMFS Highly Migratory Species Management Division, Silver Spring, MD.

Nelson, D.R. (1990) Telemetry studies of sharks: A review, with applications in resource management. *NOAA Tech Rep, NMFS* 90, 239-256.

Sciarrotta, T.C., and Nelson, D.R. (1977) Diel behavior of the blue shark, *Prionace glauca*, near Santa Catalina Island, California. *Fish. Bull.* **75**, 519-528.

Standora, E.A., and Nelson, D.R. (1977) A telemetric study of the behavior of free-swimming Pacific angel sharks, *Squatina californica. Bull. So. Cal. Acad. Sci.* **76**, 193-201.

Tricas, T.C. (1979) Relationships of the blue shark, *Prionace glauca*, and its prey species near Santa Catalina Island, California. *Fish. Bull.* **77**, 175-182.

Wintner, S.P., and Cliff, G. (1996) Age and growth determination of the blacktip shark, *Carcharhinus limbatus*, from the east coast of South Africa. *Fish. Bull.* **94**, 135-144.

Five Tags Applied to a Single Species in a Single Location: The Tiger Shark Experience

Kim N. Holland[1], Aaron Bush[1], Carl G. Meyer[1], Stephen Kajiura[1], Bradley M.Wetherbee,[2] and Christopher G. Lowe[3]
[1]*Hawai'i Institute of Marine Biology and Department of Zoology, University of Hawaii*
[2]*National Marine Fisheries Service, Northeast Fisheries Science Center, Narragansett, RI*
[3]*Dept. of Biological Sciences, California State University, Long Beach*

Key words: tiger sharks, sonic tracking, archival tags, tag-and-recapture, home range

Abstract: Five different types of tags were used to monitor the horizontal and vertical movement patterns of tiger sharks and to document site fidelity. Sonic tracking experiments gave insight into the fine-scale movement patterns of the sharks for periods of up to 50 h whereas long-lived, acoustic pingers and automated, anchored data loggers (listening posts) provided data regarding the long-term periodicity and frequency of return of tiger sharks. These electronic recaptures were compared with the rate of traditional recaptures of sharks tagged with standard dart tags and recaptured with fishing gear. In the most recent phase of research, archiving tags capable of downloading stored data via sonic modem technology (CHAT tags) were implanted in sharks. The characteristics of data derived from these tags were compared with data acquired by active sonic tracking. A single pop-up satellite tag was provided to the shark tracking project by the manufacturer. This tag was attached with a saddle fitted around the anterior edge of the dorsal fin of a shark. Again, the active sonic tracking data were used to evaluate the veracity of the data acquired form the satellite tag. Data from these various sources are beginning to paint a consistent picture of tiger shark home range size and site fidelity in the Hawaiian archipelago.

1. INTRODUCTION

In this report we present data from five different types of tag used to determine the movement patterns of tiger sharks (*Galeocerdo cuvier*) along the southern coastline of Oahu, Hawaii. Because this report is presented within

J.R. Sibert and J.L. Nielsen (eds.), Electronic Tagging and Tracking in Marine Fisheries, 237–247.
© 2001 *Kluwer Academic Publishers. Printed in the Netherlands.*

the context of a symposium on electronic tags, we intend to emphasize evaluation of the congruency of data produced by the different types of tags rather than present an extensive discussion of shark behaviour. We will discuss the contributions of the different tag types in constructing an overall picture of tiger shark movements but this research is ongoing and, consequently, the sample sizes are small for some tag types and the results are preliminary.

The impetus for this research was to acquire a better understanding of the movement patterns of tiger sharks near shorelines where there are high levels of human activity. Within this general framework, one specific goal was to evaluate whether or not fishing could cause a localized reduction in the size of a tiger shark population along a limited section of coastline. A related objective was to evaluate the possibility of catching a culprit shark after an attack (Holland et al., 1999). The feasibility of localized population reduction depends on whether the sharks are site-attached and, if so, on the typical size of the home range and the duration of site fidelity. Thus, the core question is: How often does a tiger shark return to square one? In the case of this study, square one is about 3.5 kilometers offshore of Honolulu Harbor where the sharks in this study were caught, tagged and released.

2. METHODS

Five types of tags were used in this study: 1) standard identification tags as developed by Casey and colleagues, 2) pressure sensitive acoustic tags used in active tracking, 3) long-life acoustic tags detected by submerged data loggers, 4) archiving acoustic tags that sonically download stored data to a submerged data logger, and 5) a pop-up archival tag that transmits stored data to a satellite.

2.1 Study Site and Fishing Methods

Fishing for tiger sharks was conducted along the south shore of Oahu. The entire coastline is heavily populated and the fishing area where the sharks were caught, tagged, and released is less than 5 km from the resort area of Waikiki. To capture sharks, two or three longlines were set on or near the bottom at depths of 17-75 m, about 3.5 km offshore. Each longline was fitted with 10 to 13 branch lines 10 m in length and each ending with 1.5 m of stainless steel braided leader and a 12/0 hook. These branch lines were clipped to the main line at 40-m intervals. Lines were typically baited with tuna heads provided by a local fish market, and were set around dusk and retrieved at sunrise the next morning. Soak times ranged between 8 and 13 hours.

2.2 Tag Attachment

When the mainline was retrieved, any branch lines that had caught a shark were unclipped and attached to a large float. This allowed the shark to swim freely but not escape while the rest of the gear was retrieved. The float made it possible to easily relocate the shark when the remaining gear had been retrieved and stowed. Using the branch line attached to the float, the shark was pulled alongside a 6-m skiff and a rope was placed around the caudal peduncle. This tail rope and the branch line connected to the hook in the shark's mouth allowed the shark to be restrained and rolled onto its back to induce tonic immobility (Gruber and Zlotkin, 1982; Henningsen, 1994). In this position, sharks could easily be measured, their sex determined and the tags attached or implanted. All captured sharks were tagged with identification tags but only cleanly hooked sharks >2 m total length (TL) were selected for attachment or implantation of electronic tags. Following attachment of the various types of tags, the sharks were released by removing the hook or occasionally, by cutting the leader near the hook.

Conventional external identification tags (M capsule tag, Casey and Kohler, 1992) were attached with a stainless steel barb inserted into the dorsal musculature at the base of the first dorsal fin. Contact and reward information was contained in the tag capsule.

2.3 External Acoustic Tags

External acoustic transmitters (16 mm by 75 mm, weight in water 10 g) with a nominal life-span of 3 days were attached with a stainless steel barb to the dorsal musculature at the base of the first dorsal fin (Klimley and Nelson, 1984). Only tiger sharks >2.0 m TL that were in good condition and cleanly hooked were tracked. Tracking was conducted from a 12-m vessel equipped with a directional hydrophone and acoustic receiver (Holland et al., 1983; 1990). Geographical position was determined by GPS and was automatically paired with depth data from the transmitters and stored on a computer on the tracking vessel (Holland et al., 1999).

2.4 Internal Acoustic Tags

Larger cylindrical transmitters (47 mm by 197 mm, weight in water 135 g) with a nominal life span of 8-12 months were placed intraperitoneally by inserting them through an incision in the ventral body wall while the shark was in tonic immobility alongside the skiff. The incision was sutured with nylon thread. A mixture of 3:1 paraffin to beeswax was used to thinly coat these transmitters to reduce immunological response and physical irritation.

Similarly sized transmitters of a later generation (with increased nominal life spans in excess of 18 months) were also implanted into sharks. These sharks were not actively tracked but the implanted tags could be detected by data loggers placed on the sea floor in the fishing area. That is, these transmitters were used in *passive monitoring* of whether or not the animals returned to their point of initial capture and release.

2.5 Acoustic Archival Tags

Based on our moderately high recapture rate of conventionally tagged tiger sharks and high rate of detection by the submerged data loggers of sharks implanted with long-life internal pingers, a newly developed archival acoustic tag was selected for continuation of the site-fidelity research. These tags (CHAT tags, 28.8 by 6.6 cm, nominal life-span 24 months, Vemco Ltd.) sample and store depth and water temperature data at pre-selected intervals. These data are then periodically condensed into histograms that are also of pre-selected dimensions. For this series of experiments, we selected a five-minute sample period and the data were averaged into six-hour bins. Maximum and minimum values for each six-hour period were also stored in the tag's memory.

Data stored on board the tags were sonically downloaded via acoustic modem technology to a data logger moored to the ocean floor near the normal fishing grounds where the sharks were caught and the tags implanted. Data transfer from the tag to the data logger was initiated by an exchange of recognition and identification tones between the tag and the data logger. Range tests indicated that 500 m is around the maximum effective range for data transfer. That is, data can be downloaded when the shark comes within 500 m of the data logger. The tag is programmed to download the earliest stored data first and to work forward toward the present. If contact is broken (i.e., the shark moves beyond detection range), transmission is interrupted until contact is re-established at which time data transfer recommences at the point where it was interrupted.

2.6 Data Loggers for Automated Data Collection

Data loggers were placed on the sea floor in the area where fishing and tagging were conducted. These devices continuously monitored the area in order to detect the return of sharks carrying the various types of internally implanted transmitters. The time of day and duration of contact were recorded. These data were downloaded to a computer on board the research vessel when the data loggers were retrieved by divers about every 4 to 8 weeks. Data from archival acoustic tags were recorded by data loggers modified to allow two-way (acoustic modem) communication between the tag and the data logger.

2.7 Archival Satellite Tag

One tiger shark was fitted with an archival satellite tag (Wildlife Computers Ltd.) which recorded depth and temperature information and condensed these data into histogram format for storage. This tag was externally attached to the first dorsal fin with a specially designed harness. On a preprogrammed date, an electrical current applied to a corrosible link released the tag so it could float to the surface (pop-up) and begin transmitting data to a satellite. Data were stored in pre-set depth and temperature bins and the final data were in the form of percent time spent in each bin for each four-hour time block.

3. RESULTS

3.1 Conventional Identification Tags

From May 1994 to August 1999, 133 tiger sharks were caught and tagged. Overall CPUE was 3.67 tiger sharks/100 hooks, although there were large short-term variability and some evidence of seasonal fluctuation around the mean. Seventeen individuals were recaptured (12.7%). Six of these individuals were recaptured more than once, for a total of 24 recaptures (18.0%). Average time at liberty from tagging to the first recapture was 136 days with a range of 14 to 336 days.

3.2 Active Tracking

A total of eight sharks were tracked—four with internal tags and four with tags attached to the dorsal musculature. One shark was tagged and tracked on two different occasions, resulting in a total of nine tracks ranging in duration between 7 and 50 hours (Holland et al., 1999).

Sharks tracked by acoustic telemetry showed remarkably consistent offshore movements with extended periods of straight-line swimming and movements over large distances (Figure 1). All but one of the tracks ended at

Penguin Banks (an underwater extension of Molokai'i), about 32 km away from where the tracks were started (Holland et al., 1999).

Depth data were obtained from four sharks. When in shallow water (<300 m) on the fringing reefs of Oahu and the Penguin Banks, the sharks were predominantly close to the bottom but when they moved offshore into open water they were predominantly in the mixed layer between depths of 40 and 100 m. Occasional deep movements were observed both on the outer slopes of island reefs and in open water but the open water dives were usually brief "bounce" dives. The maximum dive depth observed during active tracking was 330 m.

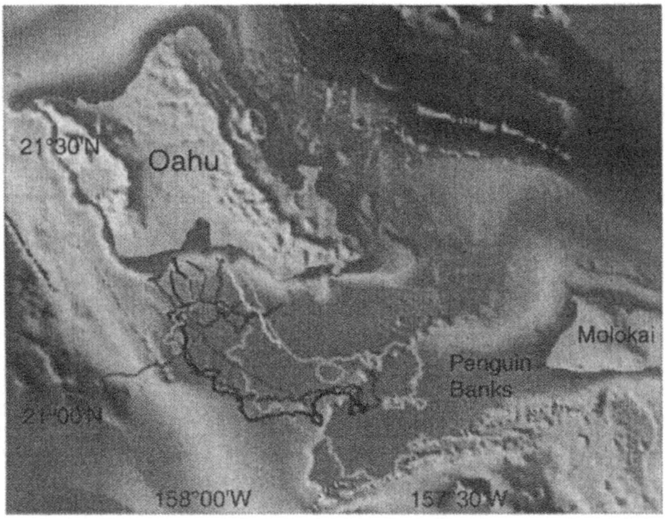

Figure 1. All nine shark tracks. All tracks started on the south shore of Oahu, and all but two ended at Penguin banks. (redrawn from Holland et al., 1999)

3.3 Internal Long-Life Pingers

Twenty sharks were implanted with long-term pingers and 10 (50%) of these were subsequently detected by data loggers in the original fishing area. Sharks were detected at all times of day, with contact durations ranging from less than 5 to 72 minutes. Figure 2 shows the duration and time of day for all detections of one shark that was a relatively frequent visitor to the site. These data are typical of all sharks that were detected in that there was no obvious rhythmicity or periodicity in the pattern of visits and they do not appear to be correlated to environmental factors such as tide or moon phase. Daytime detections were more frequent than nighttime detections. Twenty-six days elapsed between placing the tag in this shark's abdominal cavity and the first detection by the data logger.

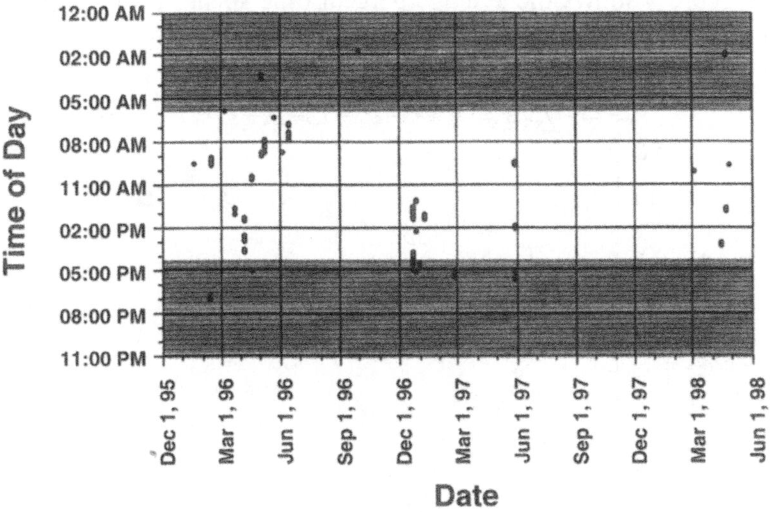

Figure 2. All detection times of a single shark with an internal pinger. Vertical height of bars indicates the duration of contact with the data logger.

3.4 Archival Acoustic Modem Tags

To date, four sharks (209-370 cm TL) have been implanted with archival acoustic modem tags, and one (25%) has been detected and successfully downloaded stored data. Time at liberty before first detection was 34 days. Three days of depth and temperature data (averaged into 6-hour bins) have been acquired to date. These data were downloaded from the shark during four separate detection events which occurred over a span of 30 days. The total contact time for these various occasions totaled 61 minutes. Depth data indicated that the shark spent most time between 30 and 80 m (Figure 3) and there were no differences between daytime and nighttime depth distributions.

3.5 Satellite Tag

An archival satellite tag was deployed on one shark (male, 180 cm TL). The tag was attached on 17 November 1999 and was programmed to release on 15 December 1999. Successful transmission of data from the tag to the satellite occurred but initial analysis of the depth data indicated that the tag came off prematurely during the night of 30 November 1999 yielding 13 days of shark movement data (temperature and depth) and 15 days of drifting at the surface. At the time of transmission (15 and 16 December 1999), the tag was located in open water approximately 140 miles southwest of

Oahu. Figure 4 shows the average percent time spent in each depth bin during the 13 days that the tag was attached to the shark.

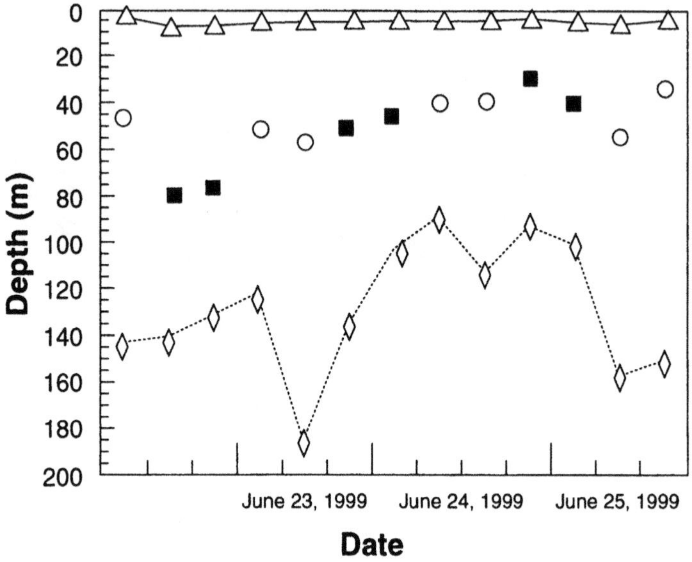

Figure 3. Data from archival acoustic tags. Depth data are recorded in 6-hour bins. The first 6-hour bin is not shown as it contained data collected before, during, and immediately after surgery. Black squares represent nighttime mean, circles daytime mean, diamonds max depth, triangles min depth.

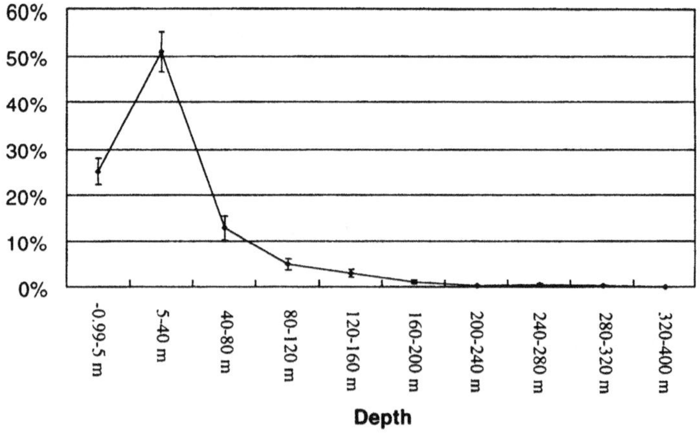

Figure 4. Depth data from archival satellite tag. Data are recorded as percent time in each depth bin over 4-hour intervals. Points are averages over all time intervals for 13 days. error bars = std error.

4. DISCUSSION

The use of different tagging methods applied to the study of tiger shark movements in a single location serves to highlight the way in which different tagging methodologies can complement each other in constructing a more complete understanding of fish behaviour. The pattern of constant motion and cross-channel movements elucidated by the active sonic tracking indicated that these sharks travel large distances on a daily basis. They did not appear to spend much time repeatedly reworking small sections of habitat and had no reticence in undertaking offshore movements. Two of the tracked sharks were subsequently recaptured at the site of release and this raised the possibility that at least some of the sharks in the area were affiliated with a home range.

Site fidelity and the possible existence of large home ranges was reinforced by the recapture of 12.7% of the sharks equipped with identification tags and by the pattern of visitation revealed by the submerged data loggers. The data logger results indicated that a large number of the sharks (50%) revisited the area of coastline where they were first captured but the visits were irregular and are often interspersed by absences of several weeks. Further, the periods of time that the sharks were within range of the data loggers were usually brief and there was no evidence of any shark spending extended periods of time in close proximity to the data logger. In summary, these results all paint a consistent picture of the behaviour of tiger sharks in this area of the ocean. They display fidelity to a home range but the size of the home range is probably very large. The sharks are constantly moving within the home range and may not revisit any particular section for periods of several weeks.

Depth profiles collected by active tracking and both types of archival tags (CHAT, pop-up) were similar, even though the collection and storage of data varied between the different methods. The similarity of the data lends confidence to the interpretation that the various tagging methods are revealing normal patterns of behaviour and that there is minimal impact on behaviour of either the external or internal tag attachment methods. Depth data from all three sources indicated sharks spent most of their time between 30 m and 100 m, although there were occasional deeper dives. It is not possible to say if the maximum depths contained within the archived data represent orientation to the bottom or blue water dives occurring during periods of open water transit.

At the outset of the phase of work using CHAT acoustic modem tags, it was recognized that data from these tags would be somewhat tangential to questions concerning shark home range site fidelity. These tags only record depth and temperature data and do not attempt to estimate geographical location. Nevertheless, depth data collected over long periods of time can indicate how much of the sharks' movements are within the typical depths used in human ocean activities. Further, this use of the CHAT and acoustic modem technologies was seen as an opportunity to demonstrate the feasibility of this technology. The preliminary results indicate that archived data can be successfully downloaded across several hundred meters from internally placed transmitters. This distance was feasible even in the noisy coastal oceanic environment. Data transfer times can probably be reduced to expedite faster transfer of the data stored in the tag.

One of the most striking aspects of the data gathered in this study is the difference in recapture rates of animals with conventional tags and those with electronic tags. The overall recapture rate of 50% for sharks with internal acoustic pingers was four times as high as for conventional tags (12.7%) and one of the four CHAT tags (25%) has been heard on four separate occasions. The probable explanation for this difference is the large disparity in sampling effort attributable to the different techniques. Whereas the fishing gear was deployed for approximately 13 hours per week, the data loggers were constantly in place and listening for tagged sharks. At least one shark was detected on the data logger on the same night that fishing gear was deployed in the immediate vicinity but which did not catch the shark. These results suggest that, for wide ranging fishes, acoustic tagging and remote monitoring techniques can be an excellent method for investigating site fidelity and residence phenomena. Similar positive results have recently been acquired in a study of tuna behavior around fish aggregating devices (Klimley and Holloway, 1999).

One underlying impetus for this research was the evaluation of the feasibility of using fishing to cause localized depletions of tiger sharks if such a reduction were deemed necessary or appropriate to improve human safety. The results from the various tagging technologies suggest that in Hawaii tiger sharks are constantly moving within very large home ranges and have a vertical distribution ranging from the surface to at least 300 m. These results indicate that localized coastal fishing would probably not result in localized population reduction unless fishing was conducted continuously for a prolonged period of time sufficient to reduce the population in the overall region of the Hawaiian Islands.

ACKNOWLEDGEMENTS

Many thanks to Tamashiro Fish Market for supplying bait for the duration of this study. Funding was provided by the division of Aquatic Resources, State of Hawaii Department of Land and Natural Resources and by the University of Hawaii Sea Grant College Program (Project R.MR-50) under an Institutional Grant from the NOAA Office of Sea Grant, U.S. Department of Commerce.

REFERENCES

Casey, J.G., and Kohler, N.E. (1992) Tagging studies on the shortfin mako shark (*Insures oxyrinchus*) in the Western North Atlantic. *Aust. J. Mar. Freshwater Res.* **43**, 45-60.

Gruber, S.H., and Zlotkin, E. (1982) Bioassay of surfactants as shark repellents. *Nav. Res. Rev.* **2**, 18-27.

Henningsen, A.D. (1994) Tonic immobility in 12 elasmobranchs: use as an aid in captive husbandry. *Zoo. Biol.* **13**, 325-332.

Holland, K.N., Wetherbee, B.M., Lowe, C.G., and Meyer, C.G. (1999) Movements of tiger sharks (*Galeocerdo cuvier*) in coastal Hawaiian waters. *Mar. Biol.* **134**, 665-653.

Holland, K.N., Brill, R.W., and Chang, R.K.C. (1990) Horizontal and vertical movements of yellowfin and bigeye tuna associated with fish aggregating devices. *Fish. Bull.* **88**, 493-507.

Holland, K.N., Brill, R.W., Ferguson, S., Chang, R.K.C., and Yost, R. (1983) A small vessel technique for tracking pelagic fish. *Mar. Fish. Rev.* **47**, 26-32

Klimley, A.P., and Holloway, C.F. (1999) School fidelity and homing synchronicity of yellowfin tuna *Thunnus albacares*. *Mar. Biol.* **133**, 307-317

Klimley, A.P., and Nelson, D.R. (1984) Diel movement patterns of the scalloped hammerhead sharks (*Sphyna lewini*) in relation to El Bajo Espiritu Santo: a refuging central-position social system. *Behav. Ecol. Sociobiol.* **15**, 45-54.

Use of Telemetry in Fisheries Management: Juvenile Sandbar Sharks in Delaware Bay

B. M. Wetherbee,[1] E. L. Rechisky,[2] H. L. Pratt,[1] and C. T. McCandless[2]
[1]*NOAA/NMFS/NEFSC, 28 Tarzwell Dr., Narragansett, RI 02882*
[2]*Fisheries, Animal and Veterinary Sciences, East Farm, Bldg. 50, University of Rhode Island, Kingston, RI 02881*

Key words: *Carcharhinus plumbeus*, telemetry, area closures

Abstract: Sandbar shark (*Carcharhinus plumbeus*) populations on the US east coast have undergone several decades of decline due to fishing pressure. One strategy for rebuilding stocks is improved survival of young sharks through enhancement of nursery areas. A key element for understanding the dynamics of shark nursery areas is increased knowledge about essential habitat for young sharks that occupy these nurseries. This study was conducted to quantify habitat use and movement patterns of juvenile sandbar sharks in Delaware Bay, which is a major east coast nursery for this species. Twenty five sandbar sharks were tracked for 2.5 to 75 h using ultrasonic telemetry. Although the sharks moved throughout the lower portion of the bay, and several sharks crossed the entire bay, individuals predominantly remained close to shore and in shallow water. The majority of sharks tracked showed movements that were heavily influenced by tidal currents, and sharks tracked on the New Jersey side of the bay had larger activity spaces, occupied deeper water, and ranged farther from shore than sharks tracked on the Delaware side. Management measures such as area/time closures would provide protection for a large number of sharks in a relatively small portion of Delaware Bay and may be improved if behavior of sharks is taken into consideration. For example, area closures on the Delaware side of the bay are likely to protect more individuals per unit area than an equal area on the New Jersey side. Our study demonstrates the utility of telemetry studies for providing information that will allow more precisely focused and effectively implemented fisheries management practices.

J.R. Sibert and J.L. Nielsen (eds.), Electronic Tagging and Tracking in Marine Fisheries, 249–262.
© 2001 *Kluwer Academic Publishers. Printed in the Netherlands.*

1. INTRODUCTION

The sandbar shark (*Carcharhinus plumbeus*, Carcharhinidae) is a cosmopolitan species and one of the most important species in US east coast commercial shark fisheries because of its abundance (Compagno, 1984), large size, quality flesh and fins (Castro et al., 1999). Declines in sandbar shark catches have been reported over the past decade and are indicative of the reduction in the east coast population (Musick et al., 1993; Castro et al., 1999).

Large numbers of young sandbar sharks are caught nearshore and in bays and estuaries, both by commercial and recreational fishermen (Camhi, 1998). Sandbar sharks are born from March to July in shallow bays and estuaries primarily between Cape Canaveral, Florida and New Jersey (Springer, 1960; Merson and Pratt, in press) and also in the Gulf of Mexico (Carlson, 1999) (Figure 1). Newborn sharks spend the first summer in these nurseries. In late September and early October, those sharks in northern nurseries such as Delaware Bay move south to warmer water off the Carolinas. The following June, the juvenile sharks move north and again spend the summer in the nursery areas (Springer, 1960; Medved and Marshall, 1981; Castro, 1993). This cycle is repeated possibly for the first 7 years of the sharks' lives, after which they remain offshore year-round (Casey et al., 1985).

Sandbar sharks are slow growing and may require as long as 30 years to reach maturity (Casey and Natanson, 1992). Therefore, recovery of populations from elevated fishing pressure may be slow, potentially requiring decades (Sminkey and Musick, 1996; Cortes, 1999). To reduce fishing pressure and to facilitate population growth, sandbar sharks were placed in the "large coastal" species category of sharks included in a fishery management plan for sharks of the Atlantic Ocean (NMFS, 1993). Bag limits and quotas for commercial and recreational fisherman stipulated in this management plan apply to federal waters farther than 4.8 km from shore, and thus are primarily of benefit to older sharks that no longer occupy nursery areas. Although a number of states have adopted fishing regulations that conform to the federal management plan, policy varies from state to state, and sandbar sharks inhabiting nursery areas are exposed to varying levels of exploitation (Camhi, 1998). The slow growth rate and the low net reproductive rate of sandbar sharks limit the rate of population growth, and models predict that populations cannot withstand even a low rate of fishing mortality on immature individuals (Sminkey and Musick, 1996; Cortes, 1999). It is not surprising then that the necessity for research, including delineation of shark nurseries, patterns of habitat use and environmental tolerances of sharks in nurseries, and the overall role of coastal/inshore habitats in supporting juvenile sharks, has recently received much emphasis (NMFS, 1998; Lent, 1999).

Figure 1. Study Site. Delaware Bay is located along the Mid-Atlantic Bight of the East Coast of the United States.

Acoustic telemetry is a technique that is emerging as an important tool in fisheries management. Telemetry studies yield information on fine-scale movements of individual sharks, which is useful for inferring habitat preferences and activity patterns (Nelson, 1990; Holland et al., 1993b). Such data is crucial for a thorough evaluation of the effects of fishing and habitat degradation on populations, and in turn for assessment of the potential success of management techniques such as area/time closures (Holland et al., 1993a; 1996). As the number of telemetry studies conducted on marine fishes increases, inclusion of data from these studies in the process of fisheries management and conservation will also increase.

Delaware Bay, one of the principal nursery grounds for sandbar sharks on the US east coast (Merson and Pratt, 2000), was chosen as site for a telemetry study to investigate movement patterns and spatial requirements of these sharks. The results of our study are presented here as an example of the utility of telemetry for providing data useful for fishery management decisions.

2. MATERIALS AND METHODS

2.1 Study Site: Delaware Bay

Delaware Bay is a large, shallow, well-mixed estuary bordering New Jersey to the north and Delaware to the south (Figure 1). Total area of the bay is approximately 2000 km^2, averaging 24 km across and 45 km at its widest point. The bay is shallow, averaging 6.3 m with extensive shoals along the New Jersey and Delaware coasts (Kraft, 1991). Currents are tidally dominated and suspended sediments cause high turbidity (Sharp, 1991) and low visibility. Bottom substrate throughout most of the bay is fine-grained muddy sand (Weil, 1976, 1977), although there are designated oyster beds within the bay. The water chemistry in the lower bay is homogeneous, and thus salinity and temperature are relatively constant throughout the study area during summer with temperatures ranging from 23 to 27°C, and salinity ranging from 26 to 30‰ (authors' unpublished data).

2.2 Acoustic Telemetry

Sandbar sharks were caught using longline or rod and reel either on the Delaware or New Jersey sides of Delaware Bay. Each shark was measured, weighed (time permitting), and an identification tag (Rototag) from the National Marine Fisheries Service Apex Predators Program was applied to the first dorsal fin (Kohler et al., 1998). Acoustic transmitters were attached to the Roto tag with thin-gauge galvanized wire and represented less than 1% of the body weight of the sharks. Field tests showed that the wire corroded and broke after approximately 1 month of exposure to seawater. Two sizes of transmitters (Vemco, Nova Scotia, Canada) were used depending on the size of the shark. Neonate sharks were fitted with a V8 model (8 mm diameter, 36 mm length, 3.5 g in water), or for larger juveniles a V16 transmitter (16 mm diameter, 59 mm length, 11 g in water) was used. Each transmitter was individually coded (60-76.8 kHz, 1000-1500 ms) and had a battery life of approximately 20 days.

Tracking was conducted from a 5.8-m motor boat equipped with a directional hydrophone, Vemco VR60 receiver, notebook computer, GPS, and depth sounder. It was estimated that the distance between the shark and the tracking boat was usually several hundred meters based on transmitter range and signal strength, and boat position was presumed to be representative of shark location. Latitude and longitude position was recorded automatically via computer at 5-min intervals and position and water depth were recorded manually at 15-min intervals. When multiple sharks carrying transmitters were detected, individuals were distinguished using Vemco VSCAN software. Minimum convex polygon analysis was used to quantify activity space, and grid-square analysis was used to examine patterns of habitat use (MacDonald et al., 1980; Winter and Ross, 1982).

2.3 CPUE

Catch per unit effort data (CPUE), expressed as sharks/hour, was obtained from gillnet and longline surveys from June through September in 1998 and 1999; 181 sets (33 gillnet and 148 longline) were dispersed among 79 stations throughout Delaware Bay.

A sinking gillnet (213 m long, 3 m deep) anchored at both ends, with stretch mesh (10.6 cm), made of #177 nylon monofilament was used. The net was generally set perpendicular to shore across the tidal current in 1-10 m of water, and sampling began approximately 20 min after setting. The net was continuously checked by pulling it across the boat while leaving the net ends anchored. Total net soak time was 3 h, although weather, catch and other conditions sometimes altered soak time.

A 50-hook bottom line was used in areas of high current, deep water, and heavy boat traffic, as well as in the same areas sampled by gillnet. Fifty gangions (each with a 4/0 longline snap, 100 cm of 1.5 cm braided nylon line, 50 cm of 3 mm stainless steel cable, and a 12/0 Mustad hook with barb depressed) were attached to a mainline that consisted of 305 m of 1.5 cm braided nylon line anchored at both ends. Fresh menhaden (*Brevoortia tyrannus*) and other local species of fish were used for bait. Soak time for longlines was 30 min.

3. RESULTS

A total of 25 sharks was tracked during June-September, 1998 and 1999; 19 tracks were started on the Delaware side of the bay and 6 on the New Jersey side (Table 1). Fourteen neonate and 11 juvenile sharks were tracked for 2.5-75 h. Sharks moved throughout much of the lower bay, with activity concentrated near shore.

3.1 Shark Activity Space

Behavior of sharks differed with the side of the bay where tracking was initiated. Activity spaces of sharks tracked on the Delaware side of the bay were generally close to shore (<3 km) over shallow water (<5 m) (Figure 2a). Sharks tracked on the New Jersey side of the bay had less restricted movements; they ranged farther from shore, spent considerable time in deeper channels (6-9 m), and had larger activity spaces (Fig. 2b). While most sharks remained in relatively shallow water on the side in which they were captured, two sharks crossed the bay, one from Delaware to New Jersey and one from New Jersey to Delaware, and two sharks moved into a deep channel (37 m) near the center of the bay; consequently these sharks had a large mean (177±119) activity space (Fig. 2c). Average activity space for sharks tracked on the Delaware side of the bay was 45±52 km^2, compared to 108±125 for those on the New Jersey side (Table 1).

Although the activity space of several sharks was extensive and the combined activity spaces of all sharks encompassed much of the lower bay, grid square analysis showed that activity was concentrated in shallow, near-shore waters, particularly between Broadkill and Primehook beaches, Delaware. Grid-square analysis also illustrated the more diffuse distribution of positional data for sharks on the New Jersey side of the bay.

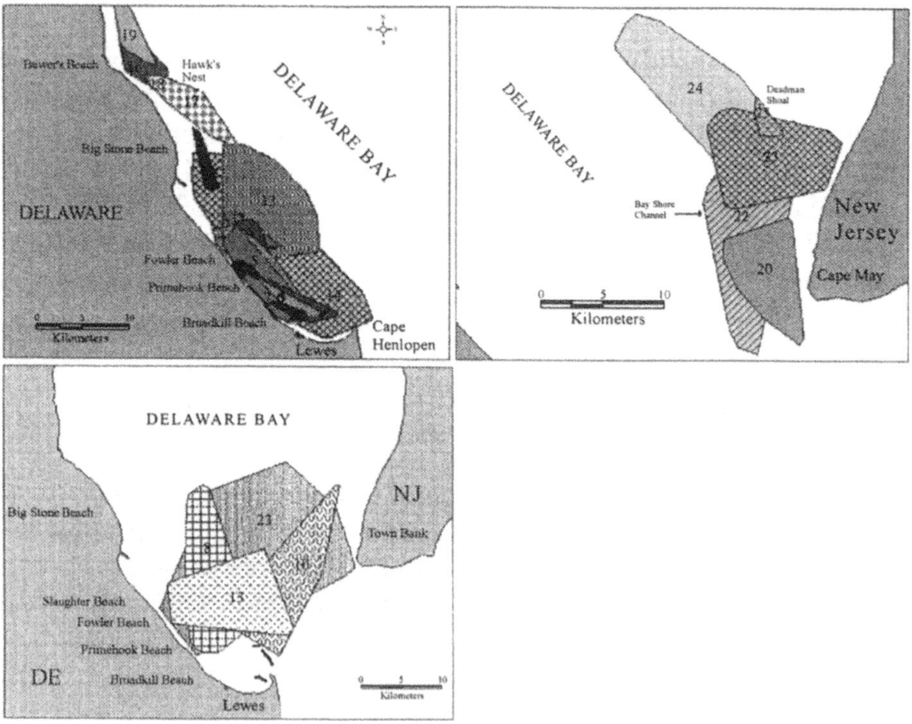

Figure 2. Activity space represented by maximum convex polygons for sandbar sharks tracked using acoustic telemetry on (upper left) the Delaware side; (upper right) the New Jersey side; and (lower left) sharks that crossed the bay or moved to a deep channel near the middle of the bay.

Table 1. Juvenile (juv) and Neonate (neo) Sandbar Sharks Tracked in Delaware Bay in 1998 and 1999.

Capture location	Stage	Sex	TL (cm)	Wt. (kg)	Date	Duration (h)	Activity* space (km^2)
DELAWARE							
Broadkill Beach	Juv.	F	74	1.2	24-Jun-98	12.3	7.3
Broadkill Beach	Neo.	F	61	1.8	29-Jul-98	6.0	2.2
Broadkill Beach	Juv.	F	131	15.0	30-Jul-98	7.3	24.0
Broadkill Beach	Juv.	M	132	NA	2-Aug-98	70.0	37.5
Broadkill Beach	Juv.	F	88	NA	13-Aug-98	53.3	29.5
Broadkill Beach	Neo.	F	61	1.7	15-Aug-98	48.3	13.5
Broadkill Beach	Neo.	F	59	NA	21-Sep-98	31.0	7.0
Broadkill Beach	Neo.	M	63	1.7	16-Sep-98	56.0	115.7
Broadkill Beach	Neo.	M	68	2.5	18-Sep-98	54.0	24.0
Broadkill Beach	Neo.	F	70	NA	24-Sep-98	24.0	121.3
Primehook Beach	Juv.	M	111	9.1	25-Jun-98	10.0	11.6
Primehook Beach	Neo.	M	68	2.3	8-Sep-99	28.5	103.4
Primehook Beach	Juv.	F	87	4.9	10-Sep-99	17.0	117.4
Primehook Beach	Neo.	M	60	1.5	11-Sep-99	52.0	170.1
E. of Mispillion River	Neo.	M	61	1.5	18-Aug-98	8.0	6.6
Bower's Beach	Neo.	M	65	NA	28-Jul-99	47.0	20.2
Hawk's Nest	Juv.	M	103	7.0	1-Aug-99	17.0	25.8
Murderkill Beach	Neo.	F	57	1.6	27-Jul-99	2.5	1.0
Kitts Hummock	Neo.	F	60	NA	30-Jul-99	32.5	14.5
							Ave(sd)=45(52)
NEW JERSEY							
Cape May	Juv.	M	124	10.8	3-Jul-99	21.0	43.0
Bay Shore Channel	Juv.	M	117	12	18-Aug-99	75.0	66.8
Bay Shore Channel	Juv.	F	115	11.3	25-Aug-99	67.0	95.2
Town Bank	Juv.	M	118	10.0	6-Jul-99	56.0	355.1
Deadman Shoal	Neo.	M	61	1.7	21-Aug-99	47.0	76.4
Deadman Shoal	Neo.	M	65	2.1	24-Aug-99	5.0	8.9
							Ave(sd)= 108(125)

*Activity space quantified by using minimum convex polygon method.

3.2 Spatial Overlap

The high degree of overlap of activity spaces for sharks tracked on consecutive days indicated a repeatable behavior, which suggests some degree of site fidelity in young sandbar sharks (Figure 3). This behavior also differed between the Delaware and New Jersey sides of the bay. For example, a shark tracked for 70 h off Delaware spent most of nearly 3 days in a small area off Fowler Beach following an initial move north after release. Overlap of activity spaces for this shark comparing consecutive days was 94.4% and 85.5% (Figure 4a). A shark tracked for 75 h on the New Jersey side of the bay had overlap values of only 11.2% and 60.8% between consecutive days (Figure 4b).

Catch data from longline and gillnet fishing supports the results obtained in our telemetry study. Sharks were caught throughout much of the bay, but shark distribution was uneven. Highest catch rates were attained in shallow, near-shore waters, particularly off Broadkill and Bigstone beaches, Delaware (Figure 5). Catch rates were lower in areas more distant from shore, and tended to be higher on the Delaware side of the bay than on the New Jersey side.

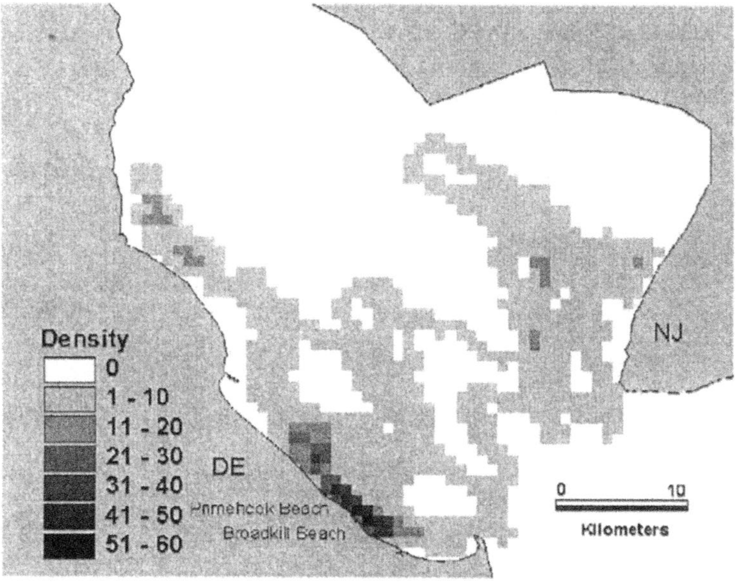

Figure 3. Grid-square presentation of habitat use of 25 sandbar sharks in Delaware Bay. Density scale represents the number of positional fixes at 15- min intervals in grids of 1 km^2.

4. DISCUSSION

Young sandbar sharks are common nearshore on both sides of Delaware Bay, and are not abundant in the deeper, middle section of the bay, possibly as a means of avoiding predation by large sharks that occur there. Based on nearly 850 h of tracking data, movement patterns exhibited by juvenile sandbar sharks in this study were heavily concentrated in a limited portion of the bay, and dependent upon the side of the bay where tracking was initiated.

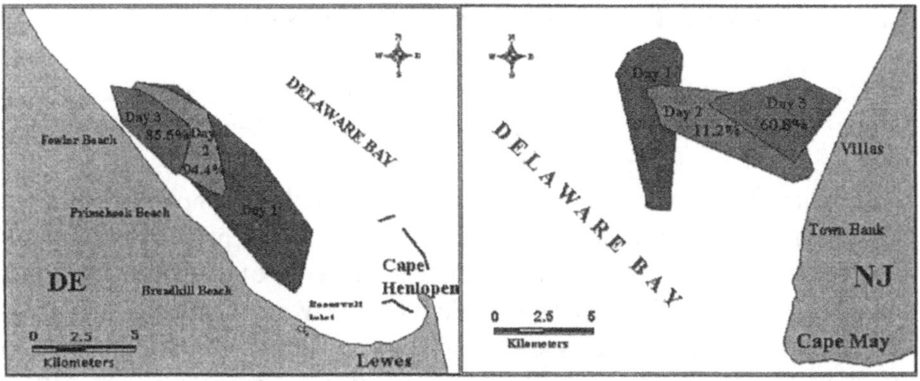

Figure 4. Daily activity space represented by maximum convex polygons for sandbar sharks tracked for (left) 70 h on the Delaware side; and (right) for 75 h on the New Jersey side of Delaware Bay. Percent overlap of activity spaces between consecutive days is shown within each convex polygon.

4.1 Shark Activity Space

Behavior patterns of young sandbar sharks in Delaware Bay appear to include repetitive movements within individuals on a daily basis and among individual sharks. Repeatable behavior within individual sharks we tracked was demonstrated by a high degree of overlap of activity spaces on consecutive days. Similar findings were reported previously for several species of shark (Holland et al., 1993b; Gruber et al., 1988; Goldman and Anderson, 1999). Repeatability of behavior among sharks in our study was demonstrated by overlap of activity spaces of different individual sharks in areas of concentrated activity. Shared activity space has also been reported for young scalloped hammerhead sharks (*Sphyrna lewini*) in a nursery area (Holland et al., 1993b). Our tracking studies were conducted over the course of two entire summers, indicating that there is a degree of site fidelity in sandbar sharks during the entire time they are residents in Delaware Bay.

4.2 Spatial Overlap

Juvenile sandbar sharks restrict the majority of their movements to a relatively small portion of Delaware Bay, and there are clearly areas where activity is concentrated, such as Broadkill and Bigstone beaches, Delaware. Telemetry data showed that multiple sharks spend considerable amounts of time in common areas during summer months, and our catch data suggest that large numbers of sharks inhabit these areas. Shared core areas of activity have been reported for other species of shark that inhabit nearshore locations or nurseries (McKibben and Nelson, 1986; Holland et al., 1993b).

4.3 Factors That May Influence Movement

There are a number of factors that may influence the distribution and habitat use of sandbar sharks in Delaware Bay. There is evidence that movements of these sharks were strongly associated with tidal currents and were generally repeated with each tidal cycle (authors' unpublished data). Medved and Marshall (1983) also found that tides influenced movements of sandbar sharks in a small Virginia bay. The presence of a shallow shelf that extends the length of the Delaware shoreline may also influence restricted movements in shallow, nearshore water on the Delaware side compared to the New Jersey side, where there is a much less extensive shallow shelf in the lower bay (Kraft, 1991). Differences in substrate may also explain the behavioral patterns observed in sharks; the New Jersey side of the bay is characterized by large oyster beds, whereas the Delaware side is predominately fine sediment with very few oysters (Price et al., 1991). Since the diet of young sandbar sharks in some locations is dominated by few prey types (Medved et al., 1985; Stillwell and Kohler, 1993), the movements of the sharks may also be related to prey distribution.

4.4 Management Strategies

Area closures have been used for management of several species of sharks and teleosts and are most successful when movement patterns or important life processes occur at a specific place at a specific time and the fishery can be directed away from these animals (Cailliet et al., 1993; Holland et al., 1996). The identified core areas occupied by large numbers of sandbar sharks during the summer in Delaware Bay meet these criteria, and area/time closure of specific locations where shark activity is concentrated has potential for protecting a large number of sharks from fishing pressure while minimizing disruption to fishing activities in the bay.

However, more precise management through area closures would include consideration of such factors as behavioral differences among locations within Delaware Bay and methods used by various fishermen to capture sandbar sharks. Based on the results of this study, closure of a small core area on the Delaware side of the bay is likely to offer protection for more sharks per unit area than an equally sized area on the New Jersey side. Also, juvenile sharks are caught in gillnets used by commercial fishermen targeting other species, as well as by rod and reel by recreational fishermen.

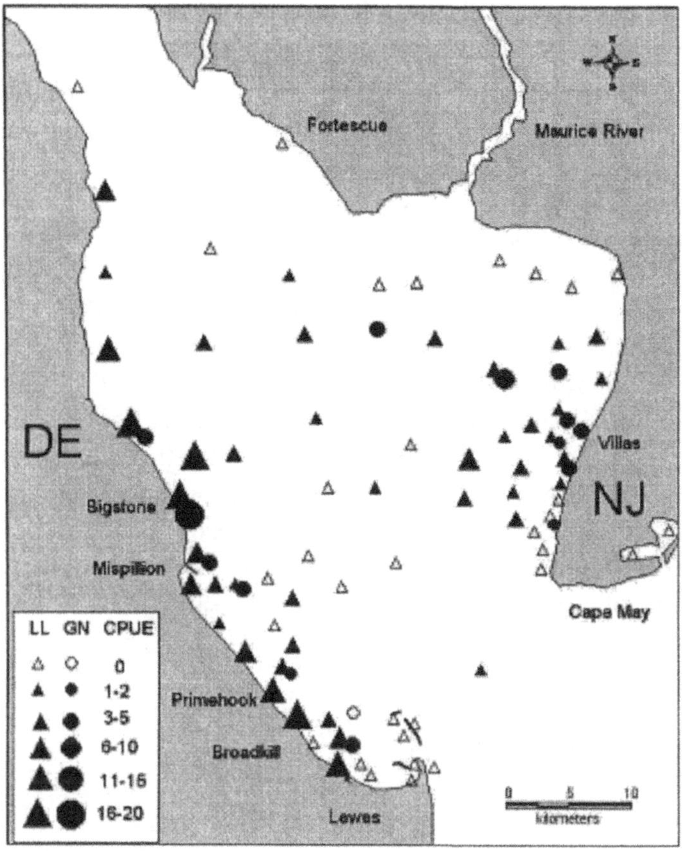

Figure 5. Catch per unit effort (CPUE) by longline (LL) and gillnet (GN) fishing in Delaware Bay. Units of CPUE for longline (triangles) and gillnet (circles) in scale are sharks per hour.

Gillnets are nonselective and often result in a high mortality for sandbar sharks when deployed in core areas, whereas a very high proportion of sharks caught on rod and reel can be released alive (Pratt, unpublished data). Therefore, rod and reel fishing that targets other species of fish may be acceptable in core areas if local fishing regulations require release of sharks in a specific area or during a specific time of the year.

An additional consideration for enhancement of shark populations in Delaware Bay is a limit on degradation of essential shark habitat. Nursery areas may be negatively impacted by human activities such as coastal development (and associated sedimentation and freshwater runoff), pollution, dredging, and dumping of dredge tailings. The relatively limited movements of sandbar sharks in Delaware Bay increase the feasibility of protection from degradation of small, discrete areas that serve as important habitat in the shark nursery.

5. SUMMARY

This study demonstrates the usefulness of telemetry in providing data that can be integrated into management policies. Our tracking data have revealed patterns of fine-scale movements and habitat use of sandbar sharks in the Delaware Bay nursery. This in turn has helped to identify specific areas that are heavily utilized by sharks, which were confirmed by CPUE data. These areas of concentrated shark activity are relatively small, and area closures in even these limited locations would likely reduce fishing pressure on large numbers of sharks and enhance population growth with a minimum of disturbance to angling activity in the bay. It appears that the effectiveness of area closures as a management technique would differ between the Delaware and New Jersey sides of the bay since movement patterns of sharks appear to differ between the two sides. In addition to providing information vital to the establishment of effective 'no fishing zones,' our study has identified areas where human disturbance should be restricted to achieve maximum conservation of essential habitat for juvenile sandbar sharks in Delaware Bay.

ACKNOWLEDGEMENTS

Funding for this project was provided by the NOAA/NMFS/ NEFSC, the Highly Migratory Species Management Division of NOAA/NMFS, and the National Research Council. Logistic support was provided by Art Sundberg and the University of Delaware Lewes Marine Operations facility, the Coast Guard Station at Cape May, NJ, Betsy Archer and the Delaware National Estuarine Research Reserve, and the Mallard Lodge of the Delaware Division of Fish and Wildlife. We thank Nancy Kohler, Carolyn Mostello, Krista Sheilds, Joe Mello, Brandon Ducharme, Matt Thyng, Abby Spargo, Kara Dwyer, Kathy St. Laurent, Ryan Campbell, Mike Smith, Lisa Webster, Greg Bodnar and the many volunteers that assisted in tracking sharks. Very special thanks to Frank Fuhr for extensive help with this project. We also thank John Roberts and Ron Guilimette from the NMFS-Narragansett lab for their logistical support and help in preparing the research vessels.

REFERENCES

Cailliet, G.M., Holts, D.B,. and Bedford, D. (1993) A review of the commercial fisheries for sharks on the west coast of the United States. In Pepperell, J., West, J. and Woon, P. eds. Shark Conservation: Proceedings of an International Workshop on the Conservation of Elasmobranchs Held at Taronga Zoo, February 24, 1991, Sydney, Australia.

Camhi, M. (1998) *Sharks on the Line: a State-by-State Analysis of Sharks and their Fisheries.* Living Oceans Program, National Audubon Society, Islip, NY, 157 pp.

Carlson, J.K. (1999) Occurrence of neonate and juvenile sandbar sharks, *Carcharhinus plumbeus*, in the northeastern Gulf of Mexico. *Fish. Bull.* **97**, 387-391.

Casey, J.G., and Natanson, L.J. (1992) Revised estimates of age and growth of the sandbar shark (*Carcharhinus plumbeus*) from the western North Atlantic. *Can. J. Fish. Aquat. Sci.* **49**, 1474-1477.

Casey, J.G., Pratt, H.L. Jr., and Stillwell, C.E. (1985) Age and growth of the sandbar shark *Carcharhinus plumbeus*, from the western North Atlantic. *Can. J. Fish. Aquat. Sci.* **42**, 963-975.

Castro, J.I. (1993) The shark nursery of Bulls Bay, South Carolina, with a review of the shark nurseries of the southeastern coast of the United States. *Env. Biol. Fish.* **38**, 37-48.

Castro, J.I., Woodley, C.M., and Brudek, R.L. (1999) A preliminary evaluation of the status of shark species. *FAO Fisheries Technical Paper*, No. 380. Rome, FAO 72 pp.

Compagno, L.J.V. (1984) FAO species catalogue. Vol. 4. *Sharks of the World. An Annotated and Illustrated Catalogue of Shark Species Known to Date.* Parts 1 and 2. FAO Fish. Syn. (125) 4 (1&2), 655 pp.

Cortes, E. (1999) A stochastic stage-based population model of the sandbar shark in the western North Atlantic. *Am. Fish. Soc. Symp.* **23**, 115-136.

Goldman, K.J., and Anderson, S.D. (1999) Space utilization and swimming depth of white sharks, *Carcharodon carcharias*, at the South Farallon Islands, central California. *Env. Biol. Fish.* **56**, 351-364.

Gruber, S.H., Nelson, D.R., and Morrissey, J.F. (1988) Patterns of activity and space utilization of lemon sharks, *Negaprion brevirostris*, in a shallow Bahamian lagoon. *Bull. Mar. Sci.* **43**, 61-76.

Holland, K.N., Peterson, J.D., Lowe, C.G., and Wetherbee, B.M. (1993a) Movements, distribution and growth rates of the white goatfish *Mulloides flavolineatus* in a fisheries conservation zone. *Bull. Mar. Sci.* **52**, 982-992.

Holland, K.N., Wetherbee, B.M., Peterson, J.D., and Lowe, C.G. (1993b) Movements and distribution of hammerhead shark pups on their natal grounds. *Copeia* 1993, 495-502.

Holland, K.N., Lowe, C.G., and Wetherbee, B.M. (1996) Movements and dispersal patterns of blue trevally (*Caranx melampygus*) in a fisheries conservation zone. *Fish. Res.* **25**, 279-292.

Kohler, N. E., Casey, J. G., and Turner, P.A. (1998) NMFS Cooperative Shark Tagging Program, 1962-93: An Atlas of Shark Tag and Recapture Data. *Mar. Fish. Rev.* **60** (2), 1-87.

Kraft, J.C. (1991) Geology. In Bryant, T.L., and Pennock, J.R., eds. *The Delaware Estuary: Rediscovering a Forgotten Resource.* Univ. Del. Sea Grant College Prog., Newark, DE, pp. 31-41.

Lent, R. (1999) *Final fishery management plan for Atlantic tunas, swordfish, and sharks.* Vol. 2. National Oceanic and Atmospheric Administration/ National Marine Fisheries Service Highly Migratory Species Division, US Department of Commerce.

MacDonald, D.W., Ball, F.G., and Hough, N.G. (1980) The evaluation of home range size and configuration using radio tracking data. In Amlaner, C.J. and MacDonald, D.W., eds. *Handbook of Biotelemetry and Radio Tracking.* Oxford: Permagon Press, pp. 405-426.

McKibben, J.N., and Nelson, D.R. (1986) Patterns of movement and grouping of gray reef sharks, *Carcharhinus amblyrhynchos*, at Enewetak, Marshall Islands. *Bull. Mar. Sci.* **38**, 89-110.

Medved, R.J., and Marshall, J.A. (1981) Feeding behavior and biology of young sandbar sharks, *Carcharhinus plumbeus* (Pisces, Carcharhinidae), in Chincoteague Bay, Virginia. *Fish Bull.* **79**, 441-447.

Medved, R.J., and Marshall, J.A. (1983) Short-term movements of young sandbar sharks, *Carcharhinus plumbeus* (Pisces, Carcharhinidae). *Bull. Mar. Sci.* **33**, 87-93.

Medved, R.J., Stillwell, C.E., and Casey, J.G. (1985) Stomach contents of young sandbar sharks, *Carcharhinus plumbeus*, in Chincoteague Bay, Virginia. *Fish. Bull.* **83**, 395-402.

Merson, R., and Pratt, H.L. Jr. (2001 Distribution, movements and growth of young sandbar sharks (*Carcharhinus plumbeus*) in the nursery grounds of Delaware Bay. *Fish. Bull.* (in press).

Musick, J.A., Branstetter, S., and Colvocoresses, J.A. (1993) Trends in shark abundance from 1974 to 1991 for the Chesapeake Bight region of the U.S. Mid-Atlantic Coast. In Branstetter, S., ed. *Conservation Biology of Elasmobranchs*. NOAA Tech. Rep. NMFS 115, 1-18.

NMFS. (1993) *Fishery management plan for sharks of the Atlantic Ocean*. National Oceanic and Atmospheric Administration/ National Marine Fisheries Service, US Department of Commerce, 167 pp.

NMFS. (1998) *Report of the shark evaluation workshop*. National Oceanic and Atmospheric Administration/ National Marine Fisheries Service Southeast Fisheries Science Center, US Department of Commerce, 109 pp.

Nelson, D.R. (1990) Telemetry studies of sharks: a review, with applications in resource management. In Pratt, H.L. Jr., Gruber, S.H. and Taniuchi, T., eds. Elasmobranchs as Living Resources: Advances in the Biology, Ecology, Systematics, and the Status of the Fisheries. NOAA Tech. Rep. NMFS 90, 239-256.

Price, K.S., Beck, R.A., Tweed, S.M., and Epifanio, C.E. (1991) Fisheries. In Bryant, T.L. and Pennock, J.R., eds. *The Delaware Estuary: Rediscovering a Forgotten Resource*. Univ. Del. Sea Grant College Prog., Newark, DE, pp. 71-93.

Sharp, J.H. (1991) Dynamics. *In*: Bryant, T.L. and Pennock, J.R., eds. *The Delaware Estuary: Rediscovering a Forgotten Resource*. Univ. Del. Sea Grant College Prog., Newark, DE, pp.43-53.

Sminkey, T.R., and Musick, J.A. (1996) Demographic analysis of the sandbar shark, *Carcharhinus plumbeus*, in the western North Atlantic. *Fish. Bull.* **94**, 341-347.

Springer, S. (1960) Natural history of the sandbar shark, *Eulamia milberti*. *Fish. Bull.* 61, 1-38.

Stillwell, C.E. and Kohler, N.E. (1993) Food habits of the sandbar shark *Carcharhinus plumbeus* off the U.S. northeast coast, with estimates of daily ration. *Fish. Bull.* **91**, 138-150.

Weil, C.B. (1976) A model for the distribution, dynamics and evolution of Holocene sediments and morphologic features of Delaware Bay. Ph.D. Thesis, Univ. Delaware, Newark, 311 pp.

Weil, C.B. (1977) Sediments, structural framework, and evolution of Delaware Bay, a transgressive estuarine delta. Univ. Delaware Sea Grant Tech. Rep., DEL-SG-5-75, 5pp.

Winter, J.D., and Ross, M.J. (1982) Methods of analyzing fish habitat utilization from telemetry data. In Armantrout, N., ed. *Proceedings of the Symposium on Acquisition and Utilization of Aquatic Habitat Inventory Information*. Am. Fish. Soc. West. Div., pp. 273-279.

Orientation and Swimming Speed of Plaice Migrating by Selective Tidal Stream Transport

Preliminary Results from Acoustic Tracking and ADCP Measurements

Ainsley Buckley and Geoff Arnold
The Centre for Environment, Fisheries, and Aquaculture Science, Lowestoft Laboratory, Suffolk, UK

Key words: plaice, selective tidal stream transport, midwater orientation, swimming speed and direction, acoustic tags, ADCP

Abstract: We observed selective tidal stream transport in 10 large female plaice tracked in the southern North Sea and eastern English Channel during the winters of 1994 to 1999. We tagged the fish with 300 kHz transponding acoustic tags and followed them using sector-scanning sonar. We determined the position of the research vessel by differential GPS and measured the speed and direction of the tidal stream using an Acoustic Doppler Current Profiler (ADCP). We are in the process of estimating the swimming speed and direction of the fish by vector subtraction, using data from 53 midwater excursions. We present here the data for one representative track, which agree with theoretical predictions that migrating plaice should head downtide in midwater and swim at a speed close to their optimum. Mean headings for the selected fish were mostly within ± 35° of the mean direction of the tidal stream, all were within ± 60°, and the average down-tide swimming speed was approximately 0.6 fish lengths s^{-1}. Summary data for the remaining nine fish do not differ significantly from these values. We discuss our results in relation to the reconstruction of the geographical tracks of fish from pressure records obtained from data storage (archival) tags and the sensory mechanisms involved in the migration of demersal fish in tidal seas. We also compare them with the results of an earlier analysis based on whole-tide ground tracks and average tidal stream vectors derived from navigation charts.

J.R. Sibert and J.L. Nielsen (eds.), Electronic Tagging and Tracking in Marine Fisheries, 263–277.
© 2001 *British Crown copyright. Printed in the Netherlands.*

1. INTRODUCTION

Tracking studies and midwater trawling experiments (Greer Walker et al., 1978; Harden Jones et al., 1979; Arnold and Metcalfe, 1996) have shown that plaice (*Pleuronectes platessa* L.) use selective tidal stream transport to migrate between their summer feeding grounds and winter spawning grounds. Theoretically, this behaviour saves energy (Weihs, 1978), which can then be used for growth or reproduction (Harden Jones, 1980; Metcalfe et al., 1990). Calculations based on fluid dynamic principles show that selective tidal stream transport could reduce the metabolic cost of migration by up to 40% for adult plaice (Weihs, 1978). To achieve this saving, however, fish must swim at their optimal speed and be precisely orientated down-tide. Experimental results and metabolic calculations (Metcalfe, et al., 1990) indicate that plaice do save energy by using tidal stream transport but reduce the cost of migration by only about 20%. This conclusion was, however, based on whole-tide ground tracks and average tidal stream vectors derived from navigation charts rather than on direct measurements. This paper presents the preliminary results of experiments in which we measured the speed and direction of the tidal current, whilst simultaneously tracking migrating fish. Our aim was to obtain statistically validated estimates of orientation and swimming speed (a) to test theoretical predictions about selective tidal stream transport directly and (b) to reconstruct ground tracks of fish from pressure measurements made by data storage tags.

2. MATERIALS AND METHOD

2.1 Fish Capture and Tagging Procedure

We carried out an extensive plaice-tracking programme in the southern North Sea and English Channel between 1994 and 1999. We tagged 25 female plaice ranging in length between 35 and 65 cm with a 'longlife' version of the 300 kHz transponding acoustic tag developed by Mitson and Storeton West (1971). Ten of these fish exhibited tidal stream transport, the subject of this analysis; the behaviour of the other fish will be described elsewhere. The fish were followed by RV *Corystes* using sector scanning sonar (Voglis and Cook, 1966; Mitson and Cook, 1971; Holley et al., 1975) and tracking was carried out between November and February. These months include most of the period of the annual pre-spawning migrations of plaice that spawn in the Southern Bight and eastern English Channel and the early part of their post-spawning migrations.

The fish were caught by beam trawl from commercial fishing boats and research vessels. Each fish was tagged with a 'saddle' (Figure 1) made by extending the stainless steel wire used to attach standard Petersen disc tags (McFarlane et al., 1990). Most fish were tagged at the laboratory prior to the cruise, although during later cruises fish were caught at the start of the trip, tagged and held in deck tanks for a few days to recover before release. Fish are often thought to exhibit atypical behaviour for some time after release and this may be evident in the early vertical movements of some fish in this study. In the absence of an independent, objective measure of the length of this refractory period, however, our analysis includes all midwater periods that meet the criteria of selective tidal stream transport that we define in the next section.

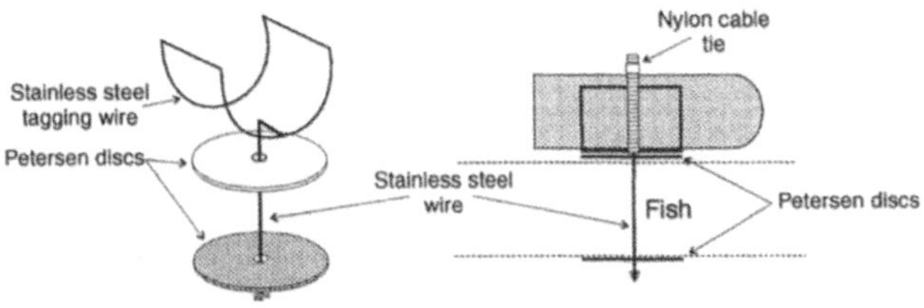

Figure 1. Diagram of modified Petersen tag with a wire 'saddle' above the upper disc. The acoustic tag rests in the cup formed by the extended wire and is held in place by a nylon cable tie. A nylon cord (not shown) is used as a secondary restraint.

2.2 Tracking Procedure

Immediately prior to the release of a fish, an acoustic tag (15 x 60 mm) was attached to the saddle using a cable tie and fine nylon cord (Figure 1). With a weight of 7.5 g in seawater the tag was almost 50% heavier in water than the original version of the transponding tag developed by Mitson and Storeton West (1971). It was also about 50% larger in diameter and exerted a proportionately greater drag than the original tag, whose dimensions were 10 x 51 mm. The precise factor is difficult to quantify because the saddle and cable tie also added to the overall drag. However, the original tag was shown not to impede the fish in any significant way (Arnold and Holford, 1978) and the behaviour of fish tagged with the larger tag was very similar to that observed previously (Greer Walker et al., 1978).

The tag returned a 3-ms pulse when insonified by the sonar (Figure 2) and produced an unambiguous signal on the sonar screen. This signal was kept under constant surveillance during the track, using the tracking procedures described by Greer Walker et al. (1978), Arnold et al. (1994) and McCleave and Arnold (1999). The ship manoeuvred to keep the tag at a mean range of 175 m ± 30 m (S.D.) (Arnold et al., 1994) and adopted a position up- or down-tide, or to one side of the fish, depending on the speed and direction of the wind relative to the tidal stream. Horizontal and vertical positions of the fish relative to the ship were recorded routinely every 10 minutes. The horizontal track of the fish was reconstructed from the geographical track of the ship, whose position was determined by differential GPS and recorded automatically at the same time as the range and bearing of the acoustic tag. The relative depth of the fish in the water column was estimated by simple trigonometry from the measured range and the corresponding tilt angles of tag and seabed.

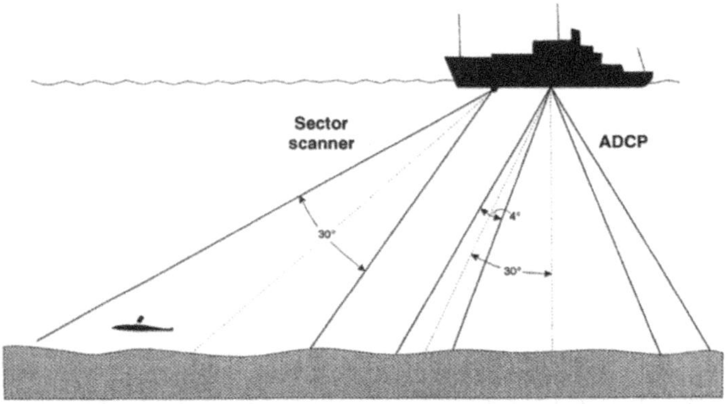

Figure 2. Diagram to show simultaneous operation of the high-resolution 300 kHz sector scanning sonar and the broad band 150 kHz ADCP. (During later tracks the transmissions of the two instruments were synchronised to avoid interference between the two frequencies.)

The errors associated with estimating the depth of a target with the sonar are moderate near the ship but increase substantially at greater ranges (Arnold and Greer Walker, 1992). To minimise these errors we estimated the depth of the fish by multiplying the proportional depth in the water column by the actual depth of water below the ship, which we measured with the ADCP. We assumed that there was no difference in depth between the position of the ship and the position of the fish. We also assumed that there was no difference in the speed and direction of the tidal stream between these two positions.

We assumed that a fish showed selective tidal stream transport if it moved into midwater for more than 1 hour during each of two successive tides flowing in the same geographical direction and returned to the bottom for the duration of the intervening tide flowing in the opposite direction. We defined a fish as being in midwater when the tag signal was clearly separated from the bottom by a tilt angle of 1° or more. At the mean tracking range of 175 m this corresponds to a height of 3 m, at a range of 250 m to a height of 4.4 m.

RV *Corystes* is an unusually quiet ship (Mitson, 1995) and is unlikely to have affected the orientation of the tracked fish significantly. Trials with RV *Clione*, a much noisier ship, showed no change in heading of compass-tagged plaice in midwater when the ship circled at typical tracking ranges (Harden Jones and Arnold, unpublished observations). Similarly, we observed no effect on the direction of any of our fish tracks as a result of the ship changing from an up-tide to a down-tide position, or otherwise manoeuvring around the fish.

2.3 Acoustic Doppler Current Profiler (ADCP)

During tracking (Figure 2) we recorded the speed and direction of the tidal stream using a vessel-mounted broadband ADCP (RD Instruments, San Diego, California) transmitting at a frequency of 150 kHz. The speed of the ship through the water varied with weather conditions, but was highest when the fish was immobile on the seabed and the ship was stemming the tide (3-4 knots maximum). The speed was usually much lower when the fish was swimming down-tide in midwater, although it increased periodically as the ship compensated for the wind and manoeuvred to remain within tracking range.

As we were using the ADCP in relatively shallow water (maximum depth 80 m), we set a bin depth of 4 m and a maximum depth of 100 m; we recorded water vectors at 10-min intervals. The ADCP software uses a number of components to estimate the proportion of good data in each bin (Lee Gordon, 1996); we only used data that scored over 80%. Residual currents are slow in this area of the North Sea and the tidal stream is the predominant component of the measured total current vector.

2.4 Data Combination Procedure and Vector Calculations

We calculated the speed and direction of the fish through the water for each 10-min period by subtracting the tidal stream vector from the equivalent fish vector calculated from the track of the fish over the ground (Metcalfe et al., 1990; their Figure 2). We wrote a computer program to extract the tidal stream vector from the ADCP depth bin corresponding to the estimated depth of the fish for the matching 10-min interval. We calculated a mean through-water swimming speed for each midwater period and expressed it non-dimensionally in terms of fish lengths s^{-1}.

We plotted the through-water directions of the fish on a circular stick plot for each midwater excursion and did the same for the tidal stream data. We calculated the mean heading of fish and tide for each midwater period, together with the corresponding mean angular deviations and 95% confidence limits (Batschelet, 1981, p. 9). We used the Rayleigh test (Batschelet, 1981, p. 54) to calculate r and P and estimate if there was any significant departure from a random heading. We calculated the angular difference between the heading of the fish and the direction of the tide for each midwater period and also the heading of the fish with respect to the direction of the tide. We plotted the normalised headings (treating the direction of the tidal stream as 0° True (0° T)) for all midwater periods on a single circular frequency plot for the whole track.

3. RESULTS

The ten plaice that showed selective tidal stream transport undertook 53 midwater excursions that met the criteria we defined above. We have not yet finished analysing these data and for the purposes of this preliminary account have chosen one fish, whose track appears to be representative of the full data set. This fish (number 4) undertook 15 excursions into midwater on successive north-going tides.

3.1 The Fish Track

Fish 4 was a 40-cm long maturing female. It was caught in the Southern Bight of the North Sea near Smiths Knoll (52° 43.5'N, 02° 20.0'E) in November 1996 and tagged at the laboratory prior to the cruise. It was released about 14 nautical miles due south of Smiths Knoll (Figure 3) at 1840 h on 6 December 1996. The fish moved a net distance of 153 km in a northerly direction over the next 12 days. During this period it made 15 midwater excursions (Figure 4), although we were able to analyse only 12 of these. We rejected the other three because the fish did not come off the bottom for more than 1 hour during the first excursion, not enough data were recorded during the second, and the percentage of good ADCP data was less than 80% for most of the third. During midwater periods 2-8 (Figure 4) vertical movements were closely synchronised with each north-going tidal stream and had a periodicity of 12 hours. Following the midwater period 1 the fish remained in midwater for about half the ensuing south-going tide. During the last 6 days of the track (midwater periods 9-12, Figure 4) the fish exhibited a 24-hour periodicity in its vertical movement. It remained on the bottom during the daytime and only used the north-going transporting tide at night.

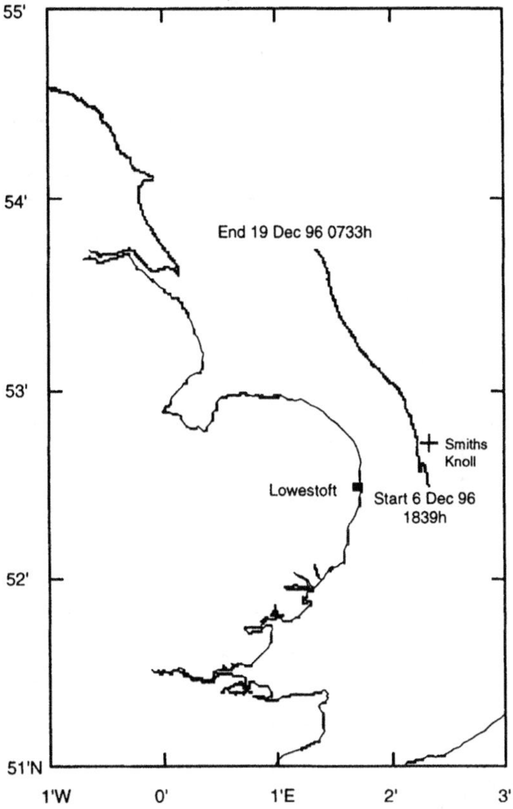

Figure 3. Track of a 40-cm female plaice fitted with a 300 kHz long-life transponding acoustic tag and released at 1840 h, 6 December 1996. The fish was released at 52° 29.7'N, 02° 19.8'E and tracked until 0730 h, 19 December 1996.

3.2 Tidal Stream Direction

For most of the 12 periods when the fish was in midwater during a north-going tide the measured 10-min tidal stream vectors showed a low angular deviation (Table 1). During midwater periods 1-3 (Figure 4) the fish remained in midwater whilst the tide was turning and this is reflected in an increased angular deviation for midwater periods 2 and 3. The effect was less pronounced for period 3 when the fish returned to the seabed rather sooner at the end of the north-going tide than it did at the end of midwater period 2. Because the fish remained in midwater for several hours after the tidal stream reversed, we split the data for the midwater period 1 into separate components for the north-going and south-going tidal streams, respectively (Table 1).

Table 1. Orientation and swimming speed of plaice 4 in midwater during an extended period (12 days) of migration by selective tidal stream transport.

Mid-water period	No. of 10-min. periods	Mean direction of tidal stream		Mean heading of fish						Average swimming speed	
	(n)	θ_t (°T)	Angular deviation (°T)	(°T)	Angular deviation (°T)	95% confidence limits	*r*	*P*	θ_t - θ_f (°T)	Fish lengths s^{-1}	±SD
1[a]	29	008	16.6	320	29.5	12	0.87	<0.001	-48	0.74	0.16
1[b]	(19)	(173)	(17.0)	(315)	33.3	17	-	-	-	0.66	0.31
2	42	014	81.0	341	28.6	8	0.88	<0.001	-33	0.65	0.33
3	30	350	23.4	332	29.5	12	0.87	"	-18	0.49	0.16
4	11	340	3.8	321	35.7	25	0.81	"	-19	0.42	0.16
5	14	322	5.4	019	47.5	31	0.66	"	57	0.43	0.23
6	21	319	1.4	331	43.5	23	0.71	"	12	0.47	0.14
7	25	324	1.8	299	64.1	47	0.37	<0.031	-25	0.7	0.72
8	14	327	6.6	322	35.3	22	0.81	<0.001	-4	0.69	0.32
9	27	322	2.5	354	42.0	19	0.73	"	32	0.51	0.24
10	20	338	2.7	352	27.9	13	0.88	"	14	0.63	0.15
11	22	350	22.5	024	26.7	12	0.89	"	34	0.52	0.22
12	19	324	6.0	356	33.5	17	0.83	"	32	0.49	0.14
Overall	274	340	28.5	341	41.7	7	0.74	<0.001	1.0	0.58	0.32

[a]North-going tide
[b]South-going tide

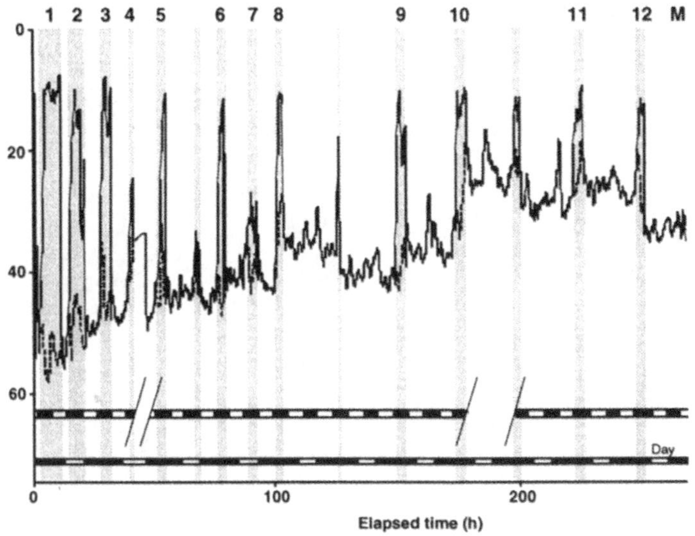

Figure 4. Vertical track of a 40-cm female plaice released at 1840 h, 6 December 1996 (Figure 3). The upper horizontal scale shows periods of consecutive north- and south-going tidal streams; the lower horizontal scale shows times of day and night. Breaks in the tide bar indicate where the ADCP was stopped to avoid acoustic interference with the sonar display. (The two instruments were not synchronised during this track.)

3.3 Fish Heading

Fish 4 was never randomly orientated during its 12 excursions into midwater on north-going tides (Figure 5, Table 1) and in all but one case P was less than 0.001. For most (60%) of these midwater periods the angular deviation was less than ±35° from the mean heading. In the case of midwater period 7, there was a wider scatter of headings and the angular deviation was ±64°. The corresponding 95% confidence limits for this period were ±47° and for the rest of the data between ±8° and ±31°. The heading of the fish was more variable than the direction of the tidal stream (see previous section).

Fish 4 generally headed in a downtide direction when in midwater. Its mean heading ranged from 299° to 024° T with an overall average of 341° T (Table 1). The corresponding directions of the tidal stream were 319° to 014° T with an overall average of 340° T. During 5 midwater periods (4, 6, 7, 8 and 10) the mean heading of the fish was very close to the mean direction of the tidal stream and statistically the two did not differ significantly. During the other 7 midwater periods there was a wider divergence and statistically there was a significant difference between the two means. Overall the difference in the mean heading of the fish and the direction of the tidal stream ranged from -048° to + 057° (Figure 6, Table 1), although for most (83%) of the 12 midwater periods it did not exceed ±34°. The divergence was equally distributed left and right of the tidal stream direction and the average difference was less than 1° (Table 1).

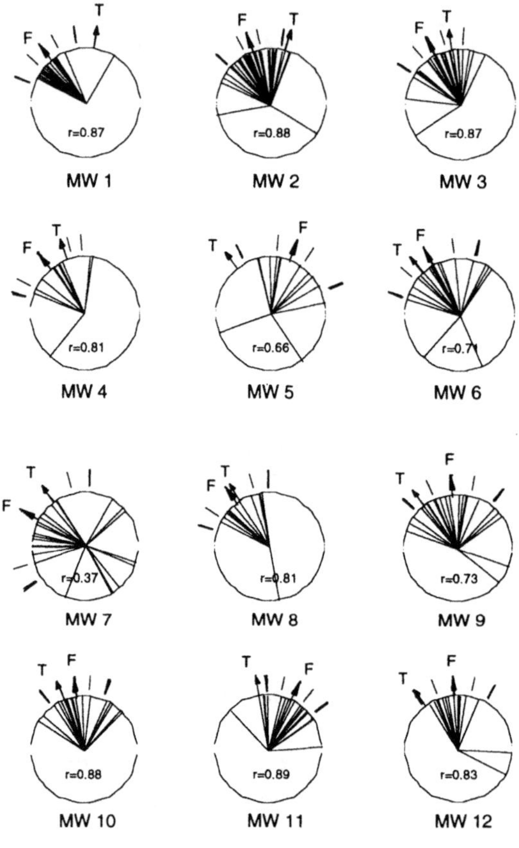

Figure 5. Orientation of the fish during midwater periods 1-12 (MW1 etc.). The letters F and T indicate the mean heading of the fish and the mean direction of the tidal stream (both in degrees true—°T), respectively. The angular deviations (bold dashes), 95% confidence limits (light dashes), and *r*-values (Raleigh test) for the fish headings are also shown (see text and Table 1 for further details).

3.4 Swimming Speed of Fish

The average speed of the fish through the water for each midwater period ranged from 0.42 to 0.74 lengths s^{-1}, with an overall mean of 0.58 ±0.32 lengths s^{-1} (Table 1). Swimming speed was generally higher during the first three midwater periods after release (0.6-0.7 lengths s^{-1}) than later in the track, although the speed during periods 7 and 8 was also high (0.7 lengths s^{-1}).

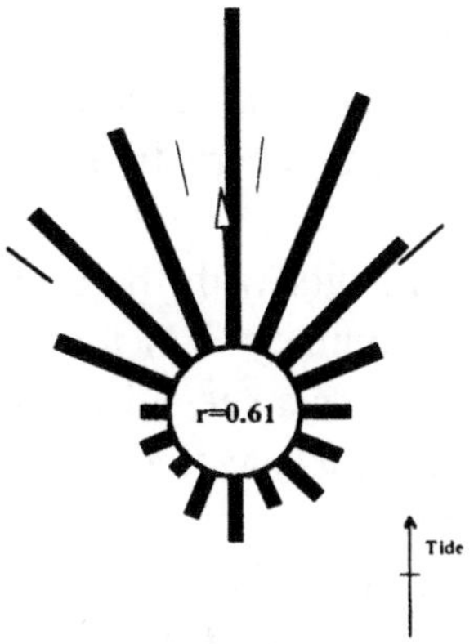

Figure 6. The combined headings of the fish for midwater periods 1-12 normalised to the direction of the tide (shown as 360°). The white arrowhead indicates the mean heading of the fish. Angular deviations (bold dashes) and 95% confidence limits (light dashes) are also shown. The data have been grouped in 22.5° sectors.

3.5 Summary Data for the Other Tracks

We have not yet analysed separately the data for each of the 41 midwater excursions shown by the remaining nine fish. We have, however, analysed the combined midwater data for each fish and these summary results lead us to believe that fish 4 is representative of the full data set. None of the nine fish was randomly orientated ($P < 0.001$). The angular deviation of the mean headings ranged between ±35° and ±60°, 95% confidence limits between ±6° and ±22°, and *r*-values between 0.45 and 0.82. Overall, the difference between the mean heading of the fish and the direction of the tidal stream ranged from -56° to +48°, although for most (73%) of the fish it did not exceed ±18°. Statistically, the average speed for fish 4 (0.58 ±32 length s^{-1}) is not significantly different from the overall mean for all ten fish (0.65 ±0.16 lengths s^{-1}).

4. DISCUSSION

We carried out an extensive series of tracking experiments in the southern North Sea and eastern English Channel between 1994 and 1999. Our first aim was to establish whether plaice migrating by selective tidal stream transport orientate and swim downtide in midwater and thereby save energy, as predicted on fluid dynamic principles alone (Weihs, 1978). We used a broadband ADCP to measure the speed and direction of the tidal stream at the same depth as the fish. Our approach was thus similar to that of Brill et al. (1993), who used an ADCP to measure ocean currents whilst tracking striped marlin (*Tetrapturus audax*) near the Hawaiian Islands and showed that the current had only a limited influence on swimming speed and direction. Our conclusions for a highly energetic tidal environment are, however, rather different.

We have illustrated our results with the track of a single fish (fish 4), which—on the basis of the summary data for the remaining nine fish—we believe to be typical of the full data set. Fish 4 maintained a consistent orientation when in midwater during selective tidal stream transport and generally swam in a downtide direction. In most cases it maintained a heading within ±35° of the mean direction of the north-going tidal stream and swam downtide at a mean swimming speed of 0.58 ±0.32 fish lengths s^{-1}. Clearly, therefore, large female plaice do migrate in an energy efficient way. However, because the observed swimming speed is only about half the optimal swimming speed of 1.3 lengths s^{-1} (Priede and Holliday, 1980) and downtide orientation is far from precise, it is equally clear that the energy savings must be significantly lower than those predicted on fluid dynamic principles (Weihs, 1978).

Our second aim was to obtain estimates of the orientation and mean swimming speed of migrating plaice for the practical purpose of reconstructing the ground tracks of fish from pressure measurements made by data storage tags (Metcalfe and Arnold, 1997). At present we assume a downtide swimming speed of 0.6 ±0.29 fish lengths s^{-1}, a value that was derived from an analysis of whole-tide ground tracks and average tidal stream speeds taken from navigation charts (Metcalfe et al., 1990). Because our new estimate based on real-time measurements is not significantly different from the earlier estimate, it appears unlikely that we will have to adopt a different value. On completion of our analysis, however, we will have a validated estimate of the mean and variance of both orientation and downtide swimming speed and will be able to put confidence limits to the reconstructed tracks. We will also be able to estimate the geographical dispersion expected for a population of fish migrating by selective tidal stream transport, and take account of stochastic variation as we develop the rules of a general model of behaviour for plaice migration and distribution on the European continental shelf.

Although we observed a rather lower maximum deviation (57°) between the heading of the fish and the mean direction of the north-going tidal stream than the value (72°) reported by Metcalfe et al. (1990), it appears that downtide orientation in tidally transporting plaice is nevertheless far from perfect. It may, however, be biologically significant because metabolic calculations indicate that 90% of the maximum potential saving can be achieved with relatively poor orientation (Metcalfe et al., 1990). On a typical round-trip migration (560 km), it is calculated that this degree of orientation will save approximately 30% (~186 kJ) of the energy content of the annual egg production of a 35 cm (0.5 kg) female plaice (Metcalfe et al., 1993).

It is instructive to compare our observations of the midwater orientation of migrating plaice with those of Metcalfe et al. (1993), who analysed the heading of non-migrating plaice fitted with transponding acoustic compass tags (Mitson et al., 1982). For these fish, which were tracked during summer months and did not exhibit tidal stream transport, orientation was non-random at night but random during the daytime. The heading of fish 4 was non-random by day as well as during the night, however, which suggests that there could be a systematic difference in behaviour between migrating and non-migrating fish. Interestingly, though, the heading of fish 4 was rather more variable during the midwater period 7 (which occurred in the daytime) than during the other periods (most of which occurred by night) and several 10-min vectors were opposed to the tidal stream (Figure 5).

We do not yet have the appropriate data to identify the physical clue, or clues, that plaice use first to adopt and then maintain a non-random heading in midwater. Because consistent headings occur at night in the apparent absence of both visual and tactile information it is likely, however, that plaice maintain their heading by reference to some geophysical clue (Metcalfe et al.,1993). Our observation that many downtide headings are not precisely aligned with the tidal stream, but have no overall bias to left or right of the tidal stream axis, may suggest that the compass sense is rather imprecise.

ACKNOWLEDGEMENTS

Our thanks are due to the captain and crew of RV 'Corystes' for assistance at sea; Juan Brown and Liam Fernand for help with setting up the ADCP and analysing the data it produced; and various other scientific colleagues who helped at sea, in particular Julian Metcalfe, who was Scientist-in-Charge for several cruises. The work was funded by the Ministry of Agriculture, Fisheries and Food under its Marine Fisheries R&D Commission (Contract MF0130).

REFERENCES

Arnold, G.P. and Holford, B.H. (1978) The physical effects of an acoustic tag on the swimming performance of plaice and cod. *J. Cons. Int. Explor. Mer* **38,** 189-200.

Arnold, G.P. and Greer Walker, M. (1992) Vertical movements of cod (*Gadus morhua* L.) in the open sea and the hydrostatic function of the swimbladder. *ICES. J. Mar. Sci.* **49,** 357-372.

Arnold, G.P. and Metcalfe, J.D. (1996) Seasonal migrations of plaice (*Pleuronectes platessa*) through the Dover Strait. *Mar. Biol.* **127**, 151-160.

Arnold, G.P., Greer Walker, M., Emerson L.S. and Holford, B.H. (1994) Movements of cod (*Gadus morhua* L.) in relation to the tidal streams in the southern North Sea. *ICES J. Mar. Sci.* **51,** 207-232.

Batschelet, E. (1981) *Circular Statistics in Biology*. Academic Press, London.

Brill, R. W., Holts, D.B., Chang, R.K.C., Sullivan, S., Dewar, H. and Carey, F.G. (1993) Vertical and horizontal movements of striped marlin (*Tetrapterus audax*) near the Hawaiian Islands, determined by ultrasonic telemetry, with simultaneous measurement of oceanic currents. *Mar. Biol.* **117,** 567-576.

Greer Walker, M., Harden Jones, F.R. and Arnold, G.P. (1978) The movements of plaice (*Pleuronectes platessa* L.) tracked in the open sea. *J. Cons. Int. Explor. Mer* **38,** 58-86.

Harden Jones, F.R. (1980) The Nekton: production and migration patterns. *In* Barnes, R.K. and Mann, K.H. (eds.) *Fundamentals of Aquatic Ecosystems,* Blackwell Science, Oxford, pp. 119-142.

Harden Jones, F.R., Arnold, G.P., Greer Walker, M. and Scholes, P. (1979) Selective tidal stream transport and the migration of plaice (*Pleuronectes platessa* L.) in the southern North Sea. *J. Cons. Int. Explor. Mer* **38,** 331-337.

Holley, M.L., Mitson, R.B. and Pratt, A.R. (1975) Developments in sector scanning sonar. *IERE Conf. Proc.* **32,** 139-153.

Lee Gordon, R. (1996) Acoustic Doppler current profiler: principles of operation, a practical primer. Second edition for Broadband ADCPs. RD Instruments, San Diego, California.

McCleave, J.D. and Arnold, G.P. (1999) Movements of yellow- and silver-phase European eels (*Anguilla anguilla* L.) tracked in the western North Sea. *ICES. J. Mar. Sci.* **56,** 510-536.

McFarlane, G.A., Wydoski, R.S. and Prince, E.D. (1990) Historical review of the development of external tags and marks. *In* Parker, N.C., Giorgi, A.E., Heidinger, R.C., Jester, D.B., Prince, E.D. and Winans, G.A. (eds.) Fish-Marking Techniques, American Fisheries Society Symposium (7), Bethesda, Maryland.

Metcalfe, J.D. and Arnold, G.P. (1997) Tracking fish with electronic tags. *Nature* **387**, 665-666.

Metcalfe, J.D., Arnold, G.P., and Webb, P.W. (1990) The energetics of migration by selective tidal stream transport: an analysis for plaice tracked in the southern North Sea. *J. Mar. Biol. Assoc. U.K.* **70,** 149-162.

Metcalfe J.D., Holford, B.H. and Arnold, G.P. (1993) Orientation of plaice (*Pleuronectes platessa*) in the open sea: evidence for the use of external directional clues. *Mar. Biol.* **117,** 559-566.

Mitson, R.B. (1995) Underwater noise of research vessels. Review and recommendations. ICES Cooperative Research Report (209), 1-61.

Mitson, R. B. and Cook, J. (1971) Shipboard installation and trials of an electronic sector-scanning sonar. Radio electron. Engng. **41,** 339-350.

Mitson, R.B. and Storeton West, T.J. (1971) A transponding acoustic fish tag. *Radio Electron. Eng.* **41,** 483-489.

Mitson, R.B., Storeton-West, T.J. and Pearson, N.D. (1982) Trials of an acoustic transponding fish tag compass. *Biotelem. Pat. Mon.* **9**, 69-79.

Priede, I.G. and Holliday, F.G.T. (1980) The use of a new tilting respirometer to investigate some aspects of metabolism and swimming activity of the plaice (*Pleuronectes platessa* L.). *J. Exp. Biol.* **85**, 295-309.

Voglis, G.M. and Cook, J.C. (1966) Underwater applications of an advanced acoustic scanning equipment. *Ultrasonics* **4,** 1-9.

Weihs, D. (1978). Tidal stream transport as an efficient method for migration. *J. Cons. Int. Explor. Mer* **38,** 92-99.

Notes About the Ecology of Ocellate Puffer, *Takifugu rubripes*, Using Archival Tags

Hiroshi Nakajima[1] and Akira Nitta[2]
[1] *Fisheries Research Institute of Mie, 3564-3 Hamajima shima, Mie 517-0404, Japan*
[2] *Japan NUS Co., Ltd. 3-9-15 Kaigan Minato-ku, Tokyo 108-0022, Japan*

Keywords: tagging experiment, archival tag, ocellate puffer, swimming behavior, burrowing behavior

Abstract: We examined movement behavior of the ocellate puffer using electronic archival data storage tags. Four out of eight tagged ocellate puffers were recaptured after an average of 202 days at liberty within a coastal area off Mie, Aichi and Shizuoka prefectures. Only two of the tags could be successfully downloaded to retrieve the archived data. Using these archived data, swimming and burrowing behavioral patterns were analyzed. Diel swimming patterns apparently changed seasonally. In winter, the fish swam near the surface, shallower than 10 m, but the maximum diving depth was in the range 40-90 m. During the spawning season, male specific spawning-related behavior patterns were possibly indicated. Burrowing behavior was frequently observed and it is suggested that it serves an important ecological function.

1. INTRODUCTION

Ocellate puffer *Takifugu rubripes* is found in the Pacific Ocean from southern Hokkaido to Taiwan. It is one of the most expensive fish species and it is especially popular in the winter season in Japan. The main fishing method is by longline and the preferred fishing grounds are located in the Tokai district (Mie, Aichi, Shizuoka), Seto Inland Sea, off Northern Kyushu, in the Yellow Sea and the East China Sea.

Recently, the dwindling catch of this fish has highlighted the need for fishery management and conservation of this species. For a rational stock management plan to be viable for ocellate puffers, critical life history information such as migration, growth and behavior is needed. In the past, a large

J.R. Sibert and J.L. Nielsen (eds.), Electronic Tagging and Tracking in Marine Fisheries, 279–287.
© 2001 *Kluwer Academic Publishers. Printed in the Netherlands.*

amount of research on ocellate puffers was conducted using conventional tags to obtain information such as migration and homing behavior (Tagawa *et al.*, 1996, Sato *et al.*, 1999, Ito *et al.* 1999). In this study, we circumvent the limitations of conventional tags by using state-of-the-art electronic archival data storage tags that were useful in the investigation of tunas and chum salmon (Itoh *et al.*, 1997, Block *et al.*, 1998, Wada and Ueno, 1999) to gain information into the vertical and horizontal movement patterns of these species.

Puffer fish hide in bottom sediments, this behavior is well documented (Takai *et al.*, 1959 Katayama *et al.*, 1965), and the retrieved archived data confirm this behavior. For the first time, we chronicle and document burrowing behavior exhibited by this species.

2. MATERIALS AND METHODS

The Northwest Marine Technology (NMT) archival tags used in this study were 16 mm diameter, 110 mm long with a sensor stalk 160 mm long, and weighed 52 g in air. Nominal battery life is at least 7 y. The rate of data acquisition was set to record and store data every 128 s for a total of 675 readings per day. The measured parameters included light intensity, internal temperature (°C), external temperature (°C), and pressure, converted to depth (m). At this rate of data acquisition, the available memory capacity was 80 days. The memory was divided into two sections, A and B. The former is never overwritten but the latter is overwritten with the latest data after the memory limit is exceeded. Thus, each memory section was allocated 40 days to store the first and last forty days of archival data for the four parameters over the course of the experiment.

The archival tags were inserted into the peritoneal cavity by surgical operation and the incision was closed by suture. Implantation of the tag took about 5 minutes including the time of anesthesia using 2-phenoxyethanol with a concentration of 400 ppm. The fish was also double tagged with a disk tag directly at the base of the dorsal fin. Eight ocellate puffers caught by longline were tagged and released in the coastal waters off Mie prefecture in October 1997 (Fig. 1). They ranged in size from 39.0 to 52.5 cm TL (Table 1). Prior to burrowing analysis, burrowing behavior was defined using a combination of invariant depth and light condition that showed zero light intensity (of the archival tag) in spite of daytime (Fig. 2). However, owing to the difficulty of discrimination, data that were recorded only at night were excluded from the analyses.

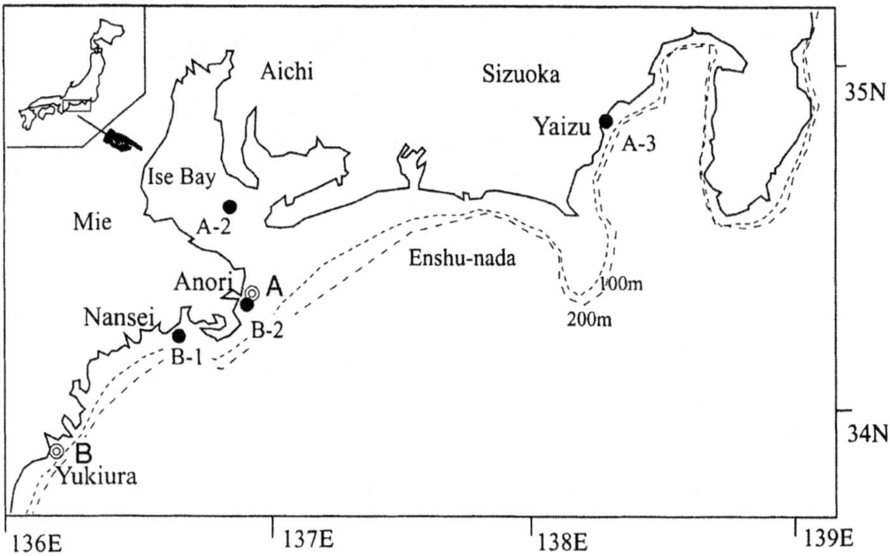

Figure 1. Locations and areas where ocellate puffer implanted with archival tags were recovered. Double circles and dotted circles indicate release and recapture locations, respectively.

Table 1. Released and recaptured ocellate puffer with archival tags implanted

Release					
Tag No.	Date	TL(cm)*	BW(g)*	Location	Fishing Method
A-1	10/01/97	46.0	2,040	Off Anori	Longline
A-2	10/06/97	51.0	2,760	Off Anori	Longline
A-3	10/10/97	50.0	2,160	Off Anori	Longline
A-4	10/10/97	45.0	1,550	Off Anori	Longline
A-5	10/15/97	52.5	2,560	Off Anori	Longline
B-1	10/09/97	39.0	1,260	Off Yukiura	Longline
B-2	10/13/97	39.5	1,200	Off Yukiura	Longline
B-3	10/14/97	46.0	1,800	Off Yukiura	Longline
Recapture					
Tag No.	Date	TL(cm)	Days-at-liberty	Location	Fishing Method
A-1					
A-2	08/19/98	•	317	Mouth of Ise Bay	Small trawl net
A-3	04/01/98	48.0	173	Off Yaizu	Set net
A-4					
A-5					
B-1	03/01/98	41.0	143	Off Nansei	Longline
B-2	04/06/98	40.4	175	Off Anori	Purse seine net
B-3					

* TL and BW means total length and body weight, respectively.

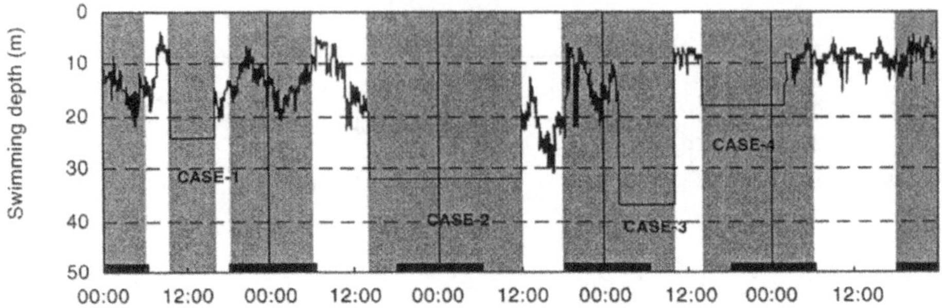

Figure 2. Illustration of Cases 1-4 for the judgments of burrowing behavior of ocellate puffer defined by using a combination of invariant depth with zero light intensity. Bottom black bars indicate nighttime. The gray shaded zones show the times when tag sensor could not sense light intensity.

Case-1: Burrowing behavior during daytime.
Case-2: Burrowing behavior from the daytime continued to the following daytime.
Case-3: Burrowing behavior from the nighttime continued to the following daytime.
Case-4: Burrowing behavior from the daytime continued to the following nighttime.

3. RESULTS

Of the eight tagged puffers, four tagged individuals were recaptured within the coastal area off Mie, Aichi and Shizuoka prefectures by longline, purse seine net, small trawl net and set net (Table 1, Fig. 1). This migration pattern was similar to the results of previous conventional tagging experiments (Ito *et al.*, 1999). Times at liberty ranged from 143 to 317 d (average 202 d). The fish carrying tag B-1 was caught off the coastal area at Nansei on March 1 by longline and was found at a puffer fish restaurant 7 days later. Dissection revealed it to be a female. Fish B-2 was caught at the spawning ground off Anori on April 6 by purse seine net and was identified as a male. In the case of the other two returned tags (A-2, A-3), one tag's data could not be downloaded and in the other, only the light sensor stalk was returned. Reasons for failure to download data from the returned tag are not known. Upon recapture, there was no obvious physical manifestation to discriminate among implanted and non-implanted puffers. Owing to the tag implantation, we could see no demonstrable evidence that the implant adversely affected the biology or behavior of these fishes. Archived data from only two tagged individuals (B-1 and B-2) could be successfully downloaded and analyzed.

3.1 Swimming Behavior

Typical diel change of varying swimming depth was thought to be seasonal (Fig. 3). In autumn, the main occupied swimming layer ranged from the surface to about 30 m in depth and fluctuated slightly with vertical movements or occasional dives to about 40 m. In winter, the fish swam near the surface, shallower than 10 m, and exhibited similar diving patterns as those observed in autumn but the maximum diving depth became deeper, usually in the range of 40-90 m.

During the spawning season from the end of March to the beginning of April when mature fish were caught in early April in 1998, vertical preferences changed drastically. The average swimming depth of B-2 was about 35 m and the diving behavior consisted of repeated rising to the subsurface at around 10 m. The bottom depth of the spawning ground off Anori is almost 30-40 m. The fish (B-2) was assumed to remain near the same bottom from March 27 to April 6 because vertical movement did not exceed 35 m during this period.

3.2 Burrowing Behavior

An example of burrowing behavior is shown in Fig. 4. Table 2 shows the results of confirmed burrowing behavior. Burrowing behavior occurred on about 40-65% of the days analyzed. Average hours of burrowing per day varied from 6 h 36 m to 13 h 39 m. The duration of burrowing fluctuated from 2 min. to about 33 h with average of 4 h 58 min in October and November. In January, February, March, and April, burrowing behavior lasted from 2 min. to about 75 h (more than 3 days) with an average of 9 h 17 min. The average depth at burrowing varied seasonally from 11.8 to 33.2 m and tended to be deeper in January–February than in October–November. It is noted that burrowing also occurred during spawning.

4. DISCUSSION

The vertical swimming patterns in winter were different from those in autumn. The former were characterized by the shallower swimming layer and deeper diving. There are two kinds of fishing methods known to exploit puffer fish, bottom longline fishing and surface longline fishing. The gear of the latter is usually suspended from near the surface to about 20 m in winter. As reported by the archival tag data, the current fishing methods cover the habitat of this species. It is well documented that some puffers such as the ocellate puffer and grass puffer (*Fugu niphobles)* exhibit sex-specific behaviors in that females leave the spawning ground soon after spawning is over. On the other hand, males remain at the spawning ground to participate in multiple spawning activities (Kusakabe *et al.*, 1962, Uno, 1955). One of the tagged fish (B-2) was a male and approximately 2 years old. The age of

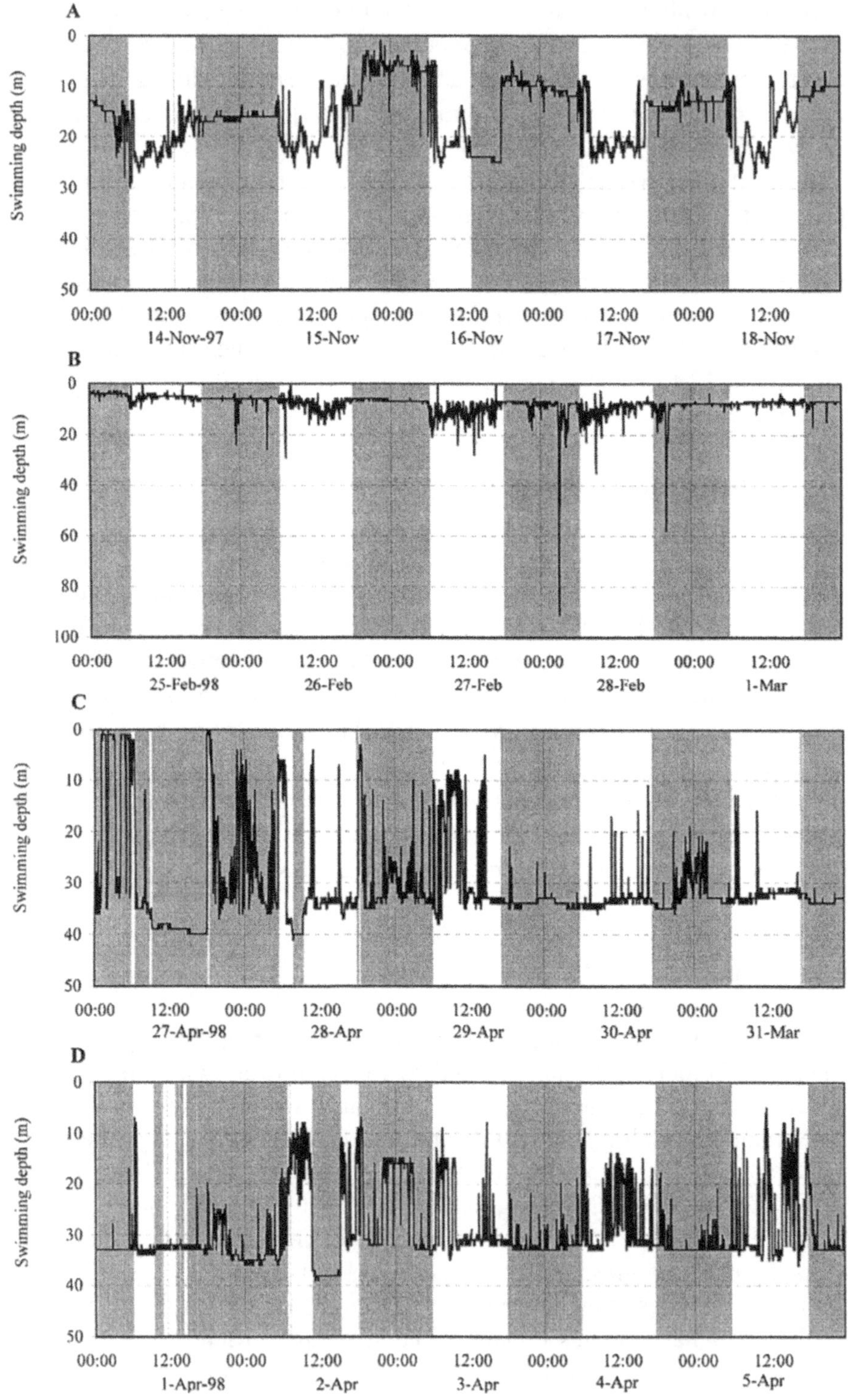
A
Swimming depth (m)
0
10
20
30
40
50
00:00 12:00 00:00 12:00 00:00 12:00 00:00 12:00 00:00 12:00
14-Nov-97 15-Nov 16-Nov 17-Nov 18-Nov
B
Swimming depth (m)
0
20
40
60
80
100
00:00 12:00 00:00 12:00 00:00 12:00 00:00 12:00 00:00 12:00
25-Feb-98 26-Feb 27-Feb 28-Feb 1-Mar
C
Swimming depth (m)
0
10
20
30
40
50
00:00 12:00 00:00 12:00 00:00 12:00 00:00 12:00 00:00 12:00
27-Apr-98 28-Apr 29-Apr 30-Apr 31-Mar
D
Swimming depth (m)
0
10
20
30
40
50
00:00 12:00 00:00 12:00 00:00 12:00 00:00 12:00 00:00 12:00
1-Apr-98 2-Apr 3-Apr 4-Apr 5-Apr

Figure 3. (left page) The illustration of typical diel changes in fish depth of ocellate puffer by season. A: autumn (Tag No. B-1) B: winter (Tag No. B-2) C-D: spring (Tag No. B-2). The gray shaded zones shown at equal spaces from the time of about 17 h to 6 h indicate night-time. The others show burrowing behavior.

←

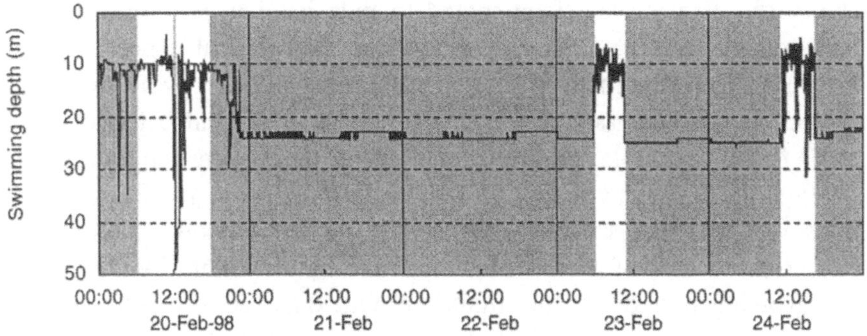

Figure 4 Typical burrowing behavior of ocellate puffer (Tag No. B-1). The gray shading zones are the time when tag sensor could not sense light intensity. Invariant depth with no light intensity shows burrowing behavior.

Table 2. Results of observed burrowing behavior

Tag No.	B-1		B-2	
	ALog	BLog	ALog	BLog
Analysis days	40 d	33 d	40 d	40 d
Analysis duration	Oct.10-Nov.18	Jan.27-Feb.28	Oct.14-Nov.22	Feb.25-Apr.5
Burrowing days	26 d	17 d	16 d	16 d
Burrowing ratio	65%	52%	40%	40%
Average burrowing period per day	4 h 18 min	7 h 06 min	3 h 54 min	3 h 30 min
Burrowing duration				
Average	4 h	19 h 27min	6 h 44 min	5 h 13 min
Min.-Max.	2 min-15 h 42 min	6 min-74 h 52 min	2 min-32 h 36 min	2 min-41 h 04 min
Burrowing depth				
Average	25.5 m	32.0 m	11.8 m	33.2 m
Min.-Max.	15-32 m	24-71 m	6-21 m	10-53 m

maturity for males of this species is 2 years (Tokai *et al.*, 1993). The results obtained suggest that implanting archival tags into puffers does not adversely alter the behavior or ecology. Moreover, the individual with tag B-2 remained at the same location for at least ten days just before being caught in April, which could indicate the first documented case of male spawning-related behavior under natural sea conditions. Regularity in diel diving patterns was not apparently observed in ocellate puffers.

The puffer fish is well documented to hide itself in bottom sediments. Why puffers burrow is not clearly understood, and several hypotheses have been proposed: concealment from predators, resting phase, avoidance of unsuitable environmental factors, effect of disease or diurnal behavior (night burrowing). For example, grass puffers kept in tanks were reported to burrow frequently in the daytime but not at night (Katayama and Fujita, 1965). Our observations also suggest that ocellate puffers burrow quite often in spite of no direct evidence of this behavior and the impossibility of detecting short burrowing periods during nighttime. Our data seem to prove this behavior indirectly by variation in light intensity recorded by the archival tag light sensor during periods of invariant depth during daytime. Moreover, burrowing seemed to occur even at nighttime from the definition of burrowing behavior. Burrowing habit seems to be different among puffer groups. Burrowing behavior was also observed under unsuitable temperature conditions, higher than 28°C or lower than 10°C, (Takai *et al.*, 1959). In this study, there was no strong correlation between burrowing behavior and ambient temperature, which indicates burrowing behavior is a usual habit. We conclude that rest is the most reasonable explanation of burrowing among the reasons mentioned because it was observed most often. The burrowing depth tended to be deeper in winter than in autumn, which reflected the seasonal differences of the habitat. The above observations were the first records of the long-term burrowing behavior of ocellate puffer and suggest that this is an important ecological adaptation.

The recapture rate for archival tags was as high as 50% in this study, which is similar to the results for conventional tags. Therefore, we suspect that individuals tagged with archive tags did not exhibit significantly higher mortality than conventionally tagged fishes. It is suggested that fishing intensity for this species is fairly high in this area. Though only two individuals were analyzed in this study, useful ecological information was obtained. It is expected that further tagging experiments using archival tags and ecological information derived from them will be used to formulate the optimum fisheries management and conservation of this species.

ACKNOWLEDGMENTS

The authors wish to thank Mr. Kume of Japan NUS Co., Ltd. for helpful discussion. We also thank Dr. Michael K Musyl of University of Hawaii at Manoa and Dr. Pierre Freon of IRD Research Associate at MCM for their critical reviews of the manuscript.

REFERENCES

Block, B.A., Dewar, H., Williams, T., Prince, E.D., Farwell, C., Fudge, D. (1998) Archival tagging of Atlantic bluefin tuna (*Thunnus thynnus thynnus*). *Mar. Tech. Soc. J.* **32**, 37-46.

Ito, M.,Yasui, M.,Tsukui, F., and Tabeta O. (1999) Migration of Adult Ocellate Puffer *Takifugu rubripes* in Enshu Nada, central Japan, estimated from tagging experiments. *Nippon Suisan Gakkaishi.* **65**, 175-181.

Itoh,T., Tsuji, S., Nitta, A. (1997) Experiment of archival tag on the Pacific bluefin tuna juvenile. In: Proceedings of the 47th annual tuna conference. 37 pp. Lake Arrowhead, CA: Inter-American Tropical Tuna Commission, National Marine Fisheries Service.

Katayama, M., and Fujita, S. (1965) Ecological studies on the puffer, *Fugu niphobles* (JORDAN *et* SNYDER) . On the habit of burrowing into the bottom sand, *Bull. Fac. Ed., Yamaguchi Univ.,* **15**, 77-84.

Kusakabe, D., Murakami, Y., and Onbe, T. (1962) Fecundity and spawning of a puffer, *Fugu rubripes* (TEMMINCK *et* SCHLEGEL) in the central waters of the Inland Sea of Japan. *J. Fac. Fish. Hurb. Hiroshima Univ.*, **4**, 47-79.

Sato, R., Suzuki, N., Shibata, R., and Yamamoto, M. (1999) Evidence of homing behavior to the natal spawning ground, Mekari-Seto in the Seto Inland Sea from adult ocellate puffer *Takifugu rubripes* recaptured. *Nippon Suisan Gakkaishi.* **65**, 689-694.

Tagawa, M., Ito, M. (1996) Migration of ocellate puffer, *Takifugu rubripes*, based on tagging experiments in the East China Sea and Yellow Sea. *Bull. Seikai Natl. Fish. Res. Inst.* **74**, 73-83.

Takai, T., Mizokami, A., and Matsui, I. (1959) Study on the fisheries biology and culture of the puffer, *Fugu rubripes* (TEMMINCK et SCHLEGEL) 1. On the pond culture of the puffer. *Bull. Fac. Fish.* **8**, 91-99.

Tokai, T., Sato, R., Ito, H., and Kitahara, T. (1993) Year-class strength of the ocellate puffer around a spawning area in the Inland Sea of Japan. *Nippon Suisan Gakkaishi*, **59**, 245-252.

Uno, Y. (1955) Spawning habit and early development of a puffer *Fugu* (Torafugu) *niphobles* (JORDAN et SNYDER). *J. Tokyo Univ. Fish.*, **41**, 169-183.

Wada, K., and Ueno, Y. (1999) Homing behaviour of chum salmon determined by an archival tag. NPAFC Doc. 425.

A Kayak Method for Tracking Fish in Very Shallow Habitats

Carl G. Meyer[1,2] and Kim N. Holland[2]
[1]*Department of Zoology, University of Hawaii at Manoa*
[2]*Hawaii Institute of Marine Biology, University of Hawaii at Manoa*

Key words: acoustic telemetry, kayak, marine reserve, surgeonfish

Abstract: Acoustic telemetry techniques were adapted for use with a kayak to facilitate tracking of species that range into shallow habitats that are inaccessible to even small outboard motor powered vessels. Low cost and portability make the tracking kayak a potentially useful tool for tracking aquatic animals in a wide variety of marine and freshwater environments. In the present study, blue-spined unicornfish (*Naso unicornis, Acanthuridae*) were successfully tracked for periods of up to 22 days in shallow high-energy fringing reef habitat in a Hawaii marine reserve (Waikiki Marine Life Conservation District-MLCD). To date all unicornfish tracked were highly site-attached to well-defined home ranges and showed similar diel patterns of behaviour; refuging within holes in the reef by night and moving to a separate daytime area. Unicornfish home ranges were completely encompassed by the boundaries of the 0.32 km^2 MLCD.

1. INTRODUCTION

Acoustic telemetry can provide detailed fish movement data essential to the formulation of sound management and conservation policies. The current interest in using no-fishing marine reserves (MR) as possible tools for mitigating the negative effects of human activity on coral reefs provides a case in point. Many of the anticipated benefits of MR are dependent upon movement patterns and dispersal rates of target species that are currently largely unquantified. Consequently, several recent studies of marine reserves have identified empirical fish movement data as an important research goal (DeMartini, 1993; Appeldoorn et al., 1997; Sladeck Nowlis and Roberts, 1997).

J.R. Sibert and J.L. Nielsen (eds.), Electronic Tagging and Tracking in Marine Fisheries, 289–296.
© 2001 *Kluwer Academic Publishers. Printed in the Netherlands.*

Development of small boat tracking techniques for coastal fish by Holland et al. (1992) facilitated initial studies of coral reef fish movements in sheltered patch reef habitat (Holland et al., 1993, 1996; Meyer and Holland, unpublished). Results of these studies revealed that the targeted coral reef fish were highly site-attached to well-defined home ranges that were completely encompassed within the boundaries of a small (<0.5 km^2) marine reserve. These results suggest that even small marine reserves may be effective in protecting important target species such as jacks and goatfish. The results also raise a question of whether the insular layout of the patch reef habitat may significantly influence fish movements by favouring retention of fish on individual reefs. This is an important concern in Hawaii where contiguous, high-energy fringing reefs are the dominant coastal habitat type and patch reef habitat is rare. In order to address the question of whether reef fish movement patterns differ between patch reef and fringing reef habitats, a marine reserve situated in high-energy fringing reef habitat at Waikiki, Oahu, Hawaii, was chosen as a site for further tracking and tagging studies.

Tracking fish in the fringing reef habitat at Waikiki presented a variety of challenges that were not encountered in the sheltered patch reef environment. For example, because of very shallow water, jagged reef and breaking waves, more than half of the protected area at Waikiki is completely unnavigable by small outboard motor powered boats of the type used in previous reef fish tracking studies. Fixed hydrophone arrays are also unsuitable for these shallow areas because of signal blocking by the highly rugose habitat and the high probability of wave energy damaging the equipment. Other Waikiki shoreline areas are no boating zones that are used daily by thousands of beachgoers. These unnavigable areas may include important fish habitat, and so a new method of tracking fish into these areas was required. This paper describes the tracking equipment and techniques developed for use with a one-man kayak and presents examples of the type of data that can be obtained from this system.

2. METHODS AND MATERIALS

2.1 Study Site

The Waikiki Marine Life Conservation District (MLCD) is situated at the eastern end of Waikiki on the south coast of Oahu. Removal of marine life has been prohibited since 1988 in the MLCD. Signs posted at intervals along the shoreline of the reserve show the zone boundaries, regulations, and a telephone number for reporting violations. The MLCD covers a 700-m stretch of shoreline and the seaward boundary of the zone runs parallel to the shore at a distance of 500 yds (455 m) offshore from the high-water mark, enclosing an area of approximately 0.32 km^2.

Marine habitat in the MLCD consists of high-energy fringing reef which is subdivided into four well-defined habitat zones: 1) A sheltered back reef area consisting of shallow (<1 m) reef interspersed with deeper sandy areas and backed by two sandy beaches, 2) an extensive shallow flat reef bench seaward of the back reef area, 3) a narrow band of high rugosity reef crest forming the seaward margin of the reef bench, and 4) a deeper (>4 m) area of large sand patches and flat reef located seaward of the reef crest.

2.2 Tracking System

A kayak was chosen as the tracking platform for the fringing reefs at Waikiki. The three main advantages of the kayak were a very shallow draft (approx. 20 cm when fully loaded), high maneuverability and near silent passage enabling the kayak to unobtrusively negotiate the entire protected area. A 4-m Cobra kayak was donated to the study by a kayak retailer (Go Bananas, Kapahulu, Oahu, Hawaii) (Figure 1). This model proved to be highly stable during initial field tests and also had several large hatches giving access to the kayak's interior for equipment storage.

Telemetry equipment configured to fit the kayak consisted of a Communication Associates Inc., CR-40 acoustic receiver powered by two Yuasa 12 volt gel cell batteries and connected to a Vemco V10 50-80 kHz directional hydrophone. A waterproof housing was built from Plexiglas to protect the receiver, and batteries were housed in a plastic container. Both receiver and batteries fitted inside the stern section of the kayak through the rear hatch, where they were held firmly in place on a wooden board by loops of surgical tubing. Waterproof connectors were installed in the rear hatch so that equipment could be operated with the hatch closed. The hydrophone was mounted on a PVC staff and bracket, which slotted into holes molded into the front end of the kayak (Figure 1). As a precaution against hydrophone damage, the hydrophone mounting bracket incorporated a link which rotated upward if the hydrophone struck the reef. The kayak operator wore headphones to monitor the signal and turned the kayak to locate the direction of maximum signal strength. The hydrophone and headphone cables plugged into the connectors mounted in the rear hatch.

A chart of the study site incorporating a metric grid was created from high-resolution aerial photos of the Waikiki coastline. A waterproof version of the chart was used to determine position from landmarks every fifteen minutes during tracks. Positional data were also collected from a handheld Global Positioning System unit (Garmin 12XL). The kayak was maneuvered to within 5 m of the fish whenever positional fixes were taken. The veracity of this distance estimate was periodically corroborated by comparing signal strength with the position of a snorkeler in visual contact with a transmitter-equipped fish. Fish were tracked continuously for up to 72 h, during which time kayak operators were switched every two hours either from the shore or by using a second kayak as a shuttle.

Figure 1. Tracking kayak. The mounting staff and bracket for the directional hydrophone can be seen at the front end of the kayak. The kayak operator turns the kayak to localise the signal which is monitored through headphones. The cable from the hydrophone connects to the receiver inside the kayak via connecters in the rear hatch cover.

The kayak tracking system was first tested on the large surgeonfish *Naso unicornis* (blue-spined unicornfish) which is abundant on the reefs at Waikiki and is an important food fish in Hawaii and elsewhere. Anatomical examination of several unicornfish revealed that the abdominal cavity was tightly packed with viscera and lacked the space observed in goatfish in which transmitters have been placed intraperitoneally (Holland et al., 1993, 1996; Zeller, 1997). For this reason external attachment of the transmitters was selected as the optimum method.

Unicornfish were captured using surround nets and anaesthetized with MS-222 (0.75 g/l). Sonic transmitters (V8-2L, 0.8 cm diameter, 3.0 cm long, Vemco, Halifax, Nova Scotia) with individualized frequencies (ranging between 65.4 kHz and 76.8 kHz), and a 21 day nominal battery life were attached to the dorsal surface of the animal (Figure 2). Dorsal attachment of the transmitter was similar to that previously used successfully with several species of jack and tuna (Holland et al., 1990, 1996). The cylindrical transmitters were attached to the unicornfish with stainless steel wrapping wire. Two mounting wires were installed by passing a 0.3-cm O.D. hypodermic needle completely through the dorsal musculature 1 cm below the dorsal surface of the fish, inserting the tip of the wire into the tip of the needle and then retracting the needle. In this way, the wire could be pulled completely through the fish. One wire was passed through a ring on the front end of the transmitter and then the ends of the wire were twisted together, forming a complete loop. A second piece of wire was inserted in similar fashion at the other end of the transmitter to stop the device from wobbling as the fish swam. When in place, the transmitter lay snugly against the side of the dorsal fin. Each fish was also tagged with a serially numbered, 8.0-cm plastic dart identification tag (Hallprint, South Australia). Unicornfish were released at point of capture immediately after resuscitation. No tracking data were collected during the first five days after release as initial observations indicated that newly released unicornfish remained in holes for several days.

Figure 1. Naso unicornis with dorsally attached Vemco V8-2L acoustic transmitter.

3. RESULTS

The kayak tracking technique proved to be a viable method for acquiring detailed, continuous positional data for periods of more than 48 h and spanning 22 days. Shallow draft and high maneuverability of the kayak permitted the fish to be tracked in all areas of the fringing reef at Waikiki. The quick release PVC hydrophone mount allowed the hydrophone to be rapidly raised in order to negotiate very shallow (<30 cm deep) areas. A single 12V gel cell battery powered the receiver for over 48 h. To date, data have been analyzed from four unicornfish (338-446 mm TL) captured inside the Waikiki MLCD, and additional tracks are currently being acquired

External transmitter attachment worked well with unicornfish. Only one transmitter was shed by a fish during tracking, and this was after two weeks at liberty. Snorkel observations of fish equipped with transmitter revealed normal orientation and swimming and no signs of ulceration around the mounting wires or transmitter.

Each unicornfish was tracked on three or four occasions over periods spanning between 9 and 22 days. Total track duration ranged between 24 and 118 hours. The size and shape of the home ranges of three tracked fish were determined using convex polygons incorporating all positional fixes (Figure 3). The home ranges of all four fish were encompassed entirely by the boundaries of the Waikiki MLCD. Home range habitat included areas of high rugosity reef in the reef crest and back reef area but did not include the deeper flat reef and sand habitat seaward of the reef crest. Unicornfish were very strongly associated with high relief habitat at all times. Home range size calculated from convex polygons varied from 325 to 7650 m^2 (mean = 4667 m^2). Most of the difference in home range size was due to the distance that each fish traveled between daytime habitat and nighttime refuge holes. Home range size was apparently unrelated to fish size and all fish had daytime activity spaces of similar size.

A clear diel pattern of movements was apparent for all unicornfish that were tracked. During daytime, the fish moved slowly around in a limited (<600 m^2) area of reef, and snorkel observations revealed that transmitter-equipped fish were often associated with large schools of conspecifics. Fish typically showed high fidelity to their daytime habitats, returning consistently to the same area of reef each day. Within 15-30 minutes after sunset, each unicornfish left its daytime area and moved up to several hundred meters to a nighttime refuge hole where it remained quiescent until dawn when it returned directly to daytime habitat. Movements between daytime and nighttime habitat were very direct. Fish typically showed very high fidelity to nighttime habitat, repeatedly using the same nighttime hole throughout the tracking period.

4. DISCUSSION

External transmitter attachment proved satisfactory even though unicornfish were presumably at risk of dislodging transmitters when refuging in holes at night. Only one of the four tracked fish shed the transmitter before the battery died and this was after nearly two weeks at liberty. None of the unicornfish tracked showed any signs of ulceration caused by the external transmitter attachment. This contrasts with Zeller's (1997) results for coral trout (*Plectropomus leopardus*, Serranidae) which developed severe skin ulcerations when transmitters were attached externally. One unicornfish recaptured five months after being tracked still had one of the stainless steel mounting wires in situ and no sign of skin ulceration. Surgeonfish have extremely tough skin, which may explain why they were resistant to ulceration.

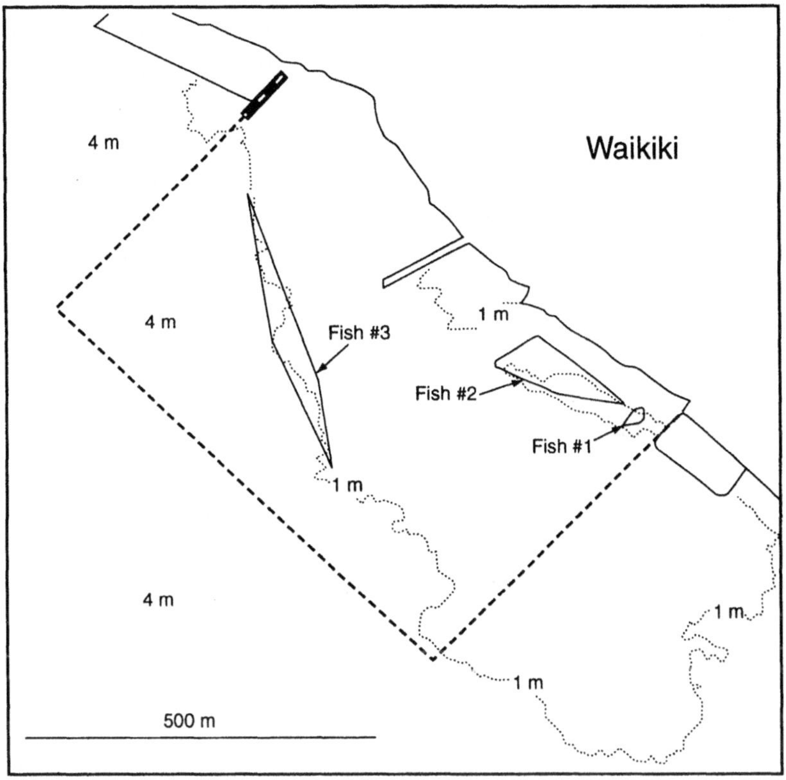

Figure 3. Home ranges of three unicornfish tracked in Waikiki MLCD illustrated as minimum convex polygons. Perimeter of shallow (<1 m) reef (dotted line). MLCD boundary (dashed line).

The movement patterns of unicornfish tracked at Waikiki had a number of broad similarities to the behaviour of goatfish and jack species tracked on patch reef habitats at Coconut Island, Kaneohe Bay, Oahu, Hawaii (Holland et al. 1993, 1996, Meyer and Holland, unpublished). For example, all species tracked had predictable, repeated diel patterns of behaviour and utilized separate daytime and nighttime areas within well defined home ranges. Unicornfish also showed extremely high fidelity to refuge holes as did the goatfish *Parupeneus porphyreus* (Mullidae) at Coconut Island (Meyer and Holland, unpublished). Unicornfish home ranges were completely encompassed by the boundaries of the Waikiki MLCD as were goatfish and jack home ranges at Coconut Island. The strong site attachment of unicornfish to small home ranges in the Waikiki MLCD was surprising given the contiguous high relief habitat in the area. Unicornfish daytime activity space was also surprisingly small. Small home ranges and limited diurnal movements may help to minimize metabolic costs and reduce food requirements.

The kayak tracking method can produce high-quality, low-cost fish movement data from areas inaccessible to even small, outboard motor powered vessels. The tracking kayak could be used to obtain movement data from aquatic animals in a wide variety of shallow marine and freshwater habitats, such as lakes, mud flats, estuaries or beaches. Silent operation makes this method ideal for tracking timid species that are easily startled by engine noise. The portable, modular design of the kayak tracking system makes it ideal for transportation to remote study sites where it can be rapidly assembled and deployed from shore or from a small boat. Crew changes can be carried out from a second kayak or larger vessel, or from shore in the tracking kayak itself depending upon local topography and fish activity.

ACKNOWLEDGEMENTS

Thanks to Go Bananas kayak shop and Cobra kayaks for donating the kayak used in this research. This work was also funded by Sea Grant and the Hawaii Department of Land and Natural Resources.

REFERENCES

Appeldoorn R.S., Recksiek, C.W., Hill, R.L., Pagan, F.E., and Dennis, G.D. (1997) Marine protected areas and reef fish movements: the role of habitat in controlling ontogenetic migration. Proc. 8th Int.Coral Reef Symp.

DeMartini, E.E. (1993) Modeling the potential of fishery reserves for managing Pacific coral reef fishes. *Fish. Bull.* **91**, 414-427.

Holland, K.N., Brill, R.W., and Chang, R.K.C (1990) Horizontal and Vertical Movements of Yellowfin and Bigeye Tuna Associated with Fish Aggregating Devices. *Fish. Bull.* **88**, 493-507

Holland, K.N., Lowe, C.G., Peterson, J.D., and Gill, A. (1992) Tracking Coastal Sharks with Small Boats: Hammerhead Shark Pups as a Case Study. *Aus. J. Mar. Fresh. Res.* **43**, 61-66.

Holland K.N., Peterson, J.D., Lowe, C.G., and Wetherbee, B.M. (1993) Movements, distribution and growth rates of the white goatfish *Mulloides flavolineatus* in a fisheries conservation zone. *Bull. Mar. Sci.* **52** (3), 982-992.

Holland, K.N., Lowe, C.G., and Wetherbee, B.M. (1996) Movement and dispersal patterns of blue trevally (*Caranx melampygus*) in a fisheries conservation zone. *Fish. Res.* **25**, 279-292.

Sladeck Nowlis, J., and Roberts, C.M. (1997) You can have your fish and eat it too: Theoretical approaches to marine reserve design. Proc. 8th Int. Coral Reef Symp. 1-4.

Zeller, D.C. (1997) Home range and activity patterns of the coral trout *Plectropomus leopardus* (Serranidae). *Mar. Ecol. Prog. Ser.* **154**, 65-77.

Evaluating Differential Pressure in the European Sea Bass (*Dicentrarchus labrax*) as a Telemetered Index of Swimming Speed

D. M. Webber,[1,2] G. P. McKinnon,[1] and G. Claireaux[3]
[1] *Vemco Ltd., Shad Bay, N.S., Canada*
[2] *Department of Biology, Dalhousie University, Halifax, N.S., Canada*
[3] *CREMA-L'Houmeau, CNRS-IFREMER, B.P.5, L'Houmeau, 17137 France*

Key words: telemetry, swimming speed, tail beat pressure, fish, sea bass

Abstract: We investigated the feasibility of measuring differential pressure of tail undulation of 6 European sea bass, *Dicentrarchus labrax*, averaging 0.25 kg in the laboratory and of two 1.5-kg sea bass in a salt marsh pond. Tail beat pressure was determined for sea bass swimming in a recirculating "Brett style" respirometer. Average or integrated and maximum-minimum (max-min) pressure increased as a power function with swimming speed (r^2>0.95). Pressure ranged from 0 to 100 pascals (Pa) and 0 to 1000 Pa for average and for max-min pressure respectively for speeds up to 0.45 m s^{-1} or 2 body length s^{-1}. Small changes in swimming speed of less than 0.05 m s^{-1} were readily detected as differences in tail beat pressure. We designed an ultrasonic transmitter that sampled the tail pressure and put out 10 millisecond acoustic pulses. The time between pulses was correlated to average and max-min pressure generated by the caudal peduncle.

1. INTRODUCTION

For many years biologists have been interested in measuring activity and energetics of marine and freshwater organisms. There are a few techniques available but all have shortcomings. Heart rate is the best energetics parameter available but because it is influenced by energetic, physical (i.e., temperature, oxygen concentration, salinity, etc.), neural and hormonal inputs it may have several relationships with oxygen consumption within one species (Thorarensen *et al.*, 1996). Also, many fish species vary stroke volume

J.R. Sibert and J.L. Nielsen (eds.), Electronic Tagging and Tracking in Marine Fisheries, 297–313.
© 2001 *Kluwer Academic Publishers. Printed in the Netherlands.*

more than heart rate in response to exercise and energetic demand (Stevens and Randall, 1967; Randall 1968; Kiceniuk and Jones, 1977; Webber *et al.*, 1998); however, there are a few species, e.g. tuna, that have a reasonably large range of heart rate (Korsmeyer *et al.*, 1997). Webber *et al.* (1998) described cardiac output as a superior correlate compared to heart rate, however, a useable ultrasonic cardiac output transmitter has not yet been marketed. Cardiac output or heart rate used alone is not sufficient to provide detailed measurements of the activity and behaviour that accompany the various metabolic states.

In the past a number of techniques have been used to measure swimming speed of fish in nature. They include straight-line measurements between poorly defined positions (O'Dor *et al.*, 1993; Nakamura, 1993), 3-D time positioning using sophisticated radio-buoy systems (O'Dor *et al.*, 1998; Sauer *et al.*, 1997), rotating paddle wheels (Block *et al.*, 1992) and the measurement of axial muscle electromyograms (EMG) (Briggs and Post, 1997; Økland *et al.*, 1997; Kaseloo *et al.*, 1992; Weatherley *et al.*, 1982). Two- and 3-D high resolution positioning systems are limited to slow moving animals and positioning accuracy is highly dependent on sea swell, wave action, tidal movements, currents and wind (O'Dor *et al.*, 1998; Webber, pers. obs.). Our experience using the paddle wheel technique indicates that it is limited because the paddle wheel stalls at velocities of 0.5 m s^{-1} or less and adds to the drag of the fish. The moving paddle wheel can also stall as a result of accumulation of particulate matter. Finally, the EMG technique appears to be the best technique currently available. This approach incorporates radio transmission of EMG signals of contracting red, white and mosaic muscle fibers of a fish's axial muscles. It is assumed that oxygen consumption of the axial musculature is highly correlated to electrolytic potentials produced by muscle contraction (Weatherley *et al.*, 1982). This technique involves implanting electrodes into the lateral musculature and transmitting radio pulses whose period is a measure of the time it takes for the EMG voltage to sum to a preset threshold (Økland *et al.*, 1997). The intensity of exercise and corresponding EMGs are correlated to swimming speed (Briggs and Post, 1997; Økland *et al.*, 1997; Kaseloo *et al.*, 1992; Weatherley *et al.*, 1982). The correlation coefficients (r^2) of the EMG versus speed relationship for Atlantic salmon ranged between 0.35 and 0.76 for individual fish and 0.75 for pooled data after correcting for individual differences in resting EMGs (Økland *et al.* 1997).

There are, however, limitations to the EMG technique. It is difficult to measure the EMG from all muscle types at once and the relationship between the timing of EMGs and force generation of muscle is not entirely clear (Videler, 1993). Also, the signal is an integration of an unknown number and type of muscle fibers surrounding the electrodes. It must be assumed that this relationship does not change throughout the weeks of measurements (electrode shifting, scar tissue impeding signals) and that the electro-chemical nature of the muscle fibers does not change (e.g., training). Also there is very little known of the effect of water temperature on the relationship between EMG signals and swimming speed. The power of contracting muscle fibers decreases with temperature, therefore the fish recruits more fibers to swim the same speed (Rome, 1990; Rome *et al.*, 1990).

Another approach to measuring swimming speed is to measure tail beat frequency by attaching a magnetized pivoting vane that opens and closes a reed switch with each sweep of the tail (Lowe *et al.*, 1998). This technique has been used successfully on sharks but the device is relatively large and it incurs extra drag limiting it to large fish. This technique has the advantage that it is less invasive than the EMG technique.

In this study we present an alternative or complementary technique to the EMG method. We investigated tail beat pressure as a new approach to predict swimming speed in free-ranging fish. A number of studies have shown that a strong relationship exists between tail beat frequency and swimming speed for a number of different species (Bainbridge, 1958, 1960; Hunter and Zweifel, 1971; Wardle *et al.*, 1989; Scharold *et al.*, 1989). However, amplitude of the tail beat varies considerably below 2 to 5 tail beats s^{-1} at low swimming speeds (Webb, 1971). Thus, amplitude should be considered because it is generally known that in nature animals most often swim at low speeds (Priede and Young, 1977). Integrating tail beat pressure during a tail beat should incorporate both frequency and amplitude of the tail. Also, if a fish is accelerating or turning, tail beat pressure should be a better correlate of power than measuring tail beat amplitude or frequency alone. Regardless of which muscle fiber types are being used (aerobic versus anaerobic) and regardless of the water temperature, tail beat pressure should be related to the power the whole tail exerts during a tail beat stroke. In the last ten years there have been rapid advances in the design and miniaturization of pressure sensors (gauge, absolute and differential) (Aitken *et al.*, in press). Their low mass in water of less than 1 g and high sensitivity to low pressure changes make these sensors suitable for attachment to the caudal peduncle of small and large fish.

The purpose of this study was to determine the feasibility of using tail beat pressure as a predictor of swimming speed in small and large European sea bass and to design a miniature ultrasonic transmitter that could transmit pressure on free-swimming fish.

2. MATERIALS AND METHODS

Forced swimming trials were conducted in November and December 1999 using 6 European sea bass (0.19 to 0.54 kg; avg. 0.25 kg; 0.23 to 0.30 m fork length; avg. = 0.258 m) originating in the Bay of Biscaye near La Rochelle, France. The sea bass were maintained at 14°C in holding tanks and were swum in a "Brett style" swimming respirometer (Figure 1). In the respirometer, temperature was controlled to 14 ± 0.25°C. Fish were swum in 0.1 m s^{-1} speed increments for periods of 30 min to 1 h starting from 0.05 m s^{-1} until exhaustion. The fish were given time to adjust to each new water speed by slowly increasing speed in 0.01 m s^{-1} increments over a 5 min period. Fish were occasionally prodded but generally they swam immediately and would swim constantly for hours. Voltage, representing the revolutions min^{-1} and water speed of the water pump, was digitized and stored on hard disk.

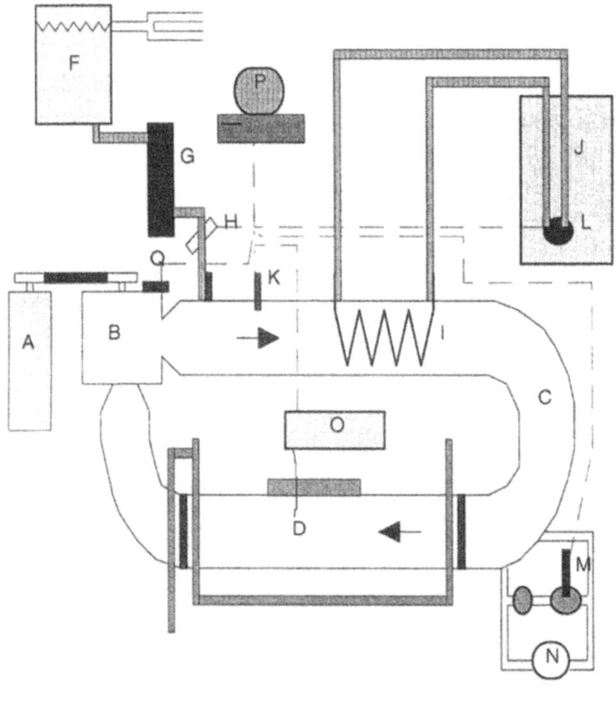

Figure 1. Diagrammatic representation of a "Brett style" respirometer used in this study to exercise European sea bass. The arrows indicate the direction of water flow. (A) 5 HP 440 volt AC electric motor. (B) Centrifugal pump that pumps water through the respirometer pipe. (C) Acrylic pipe, 10.0 cm i.d. (D) Swimming test section, 1.0 m length. (E) Fresh sea water inlets, 14 °C. (F) Header supply tank, 1.0 m^3. (G) De-gassing column, 8 l. (H) Solenoid valve, computer controlled. (I) Chilling coil. (J) Chilling water bath. (K) Temperature sensor (L) Water pump. (M) Oxygen polarographic electrode. (N) Oxygen pump. (O) Data acquisition hardware. (P) Computer.

Six fish were swum with pressure sensors mounted on their tails. The pressure sensor was mounted on a fish 12 to 24 hours prior to exercise. A miniature differential pressure sensor was modified by severing each pressure port at the base of the sensor body and reducing the size of the main sensor body. An 18 gauge needle was glued in the positive port of the sensor.

The fish were anaesthetized in seawater containing ethyl-m-aminobenzoate (MS-222). The transducer was mounted on one side of the caudal peduncle at the position of the second last vertebral centrum approximately 1 cm anterior of the base of the caudal fin rays and approximately 0.5 cm dorsal to the vertebra (Figure 2). The 18 gauge needle, attached to the positive port of the sensor, was inserted through the body of the caudal peduncle between adjacent neural spines. The needle exited the other side such that the sensor would measure the pressure difference across the caudal peduncle. The sensor was sutured firmly to the side of the caudal peduncle to minimize pressure fluctuations as a result of differences in momentum of the sensor and tail. The 4 wires leading from the sensor were inserted through three or four 0.5-cm diameter suture loops along the length of the caudal peduncle. The wires exited the swimming chamber from the base of the dorsal fin of the sea bass. The wires were then connected to the data acquisition system. The surgery was completed in less than 5 min and there was minimal blood loss.

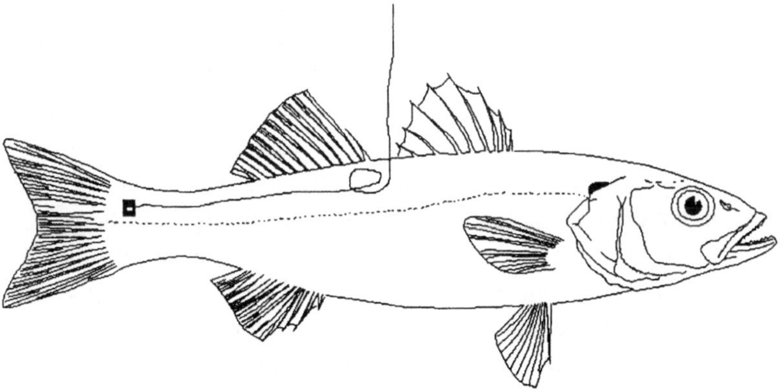

Figure 2. Schematic of a European sea bass, illustrating the placement of a differential pressure sensor on the caudal peduncle

The pressure sensors used in this study are manufactured in miniature plastic housings with a significant reduction in mass and size but with increases in sensitivity. The sensor diaphragm is a silicone wafer of a few microns in thickness embedded with a traditional "Wheatstone Bridge" that when deformed changes the electrical resistance of the bridge. These diaphragms are coated with a very thin layer of silicone that allows the sensor bridge to be exposed to seawater. Full scale pressure ranged from -6 to 6 kilo pascals (kPa). Electrical power consumption is less than 1 milliamp at 3-6 volts DC making these sensors ideally suited to telemetry applications where low electrical power consumption is necessary. The sensor was powered by a ±12 volt regulated circuit. The sensor output (0 to 10 millivolts) was amplified 500 times with a model AD624C (Analog Devices, Inc.) high precision instrumentation operational amplifier. The output from the AD624C was connected to a successive approximation A/D converter. The signal was sampled at 6350 to 15000 Hz and every 100 samples were averaged to give a digital value stored on disk at 63.5 to 150 Hz. We used a sampling rate that would sample appropriately the pressure pulse generated from the tail beat amplitude and frequency of the fish (1 to 4 tail beats s^{-1}). The pressure data were displayed on a CRT computer monitor in real time (Figure 3). The resolution of the system was 1.85 Pa or 0.2 mm water pressure at 4°C. Two pressure sensors were used throughout the study. They were calibrated daily against a column of water and the calibration never changed. Through each day the 0 pressure voltage value from the output of the AD624C drifted up or down by a maximum of 0.05 volts or 37 Pa. The voltage drift was not of significance to the results because differential pressure rather than absolute pressure was the variable of interest.

Beat-to-beat tail beat pressure was analysed using a FFT (fast-Fourier transform) analytical software program (ADCFOUR.EXE) written by DMW. The computer program analysed consecutive data windows of 1024 data. Tail beat pressure was calculated two ways. The lowest and highest values for pressure were recorded every 190 to 420 measurements (3 s; depending on A/D sampling rate). A three-second period was used because it exceeded the time interval between pressure peaks for the lowest tail beat frequency observed. The average of all minimum and maximum pressure values for a swimming trial was calculated and the minimum pressure value was subtracted from the maximum pressure value to give a value termed 'pressure difference' (*PD*). A digital filter algorithm was used to estimate the baseline of the positive and negative going electrical signal from the pressure sensor. The total area of raw pressure data above and below the baseline was calculated and averaged and is termed 'integrated pressure' (*IP*).

An ultrasonic transmitter (Figure 4) was designed to output acoustic time periods proportional to tail beat pressure every 3 s. Three seconds was chosen to ensure that the transmitter would sample a complete tail beat pressure cycle at the lowest tail beat frequency of a 1.5-kg sea bass. The pressure sensor and transmitter were both powered by a single 3.6 V lithium cell. The transmitter functioned as follows:

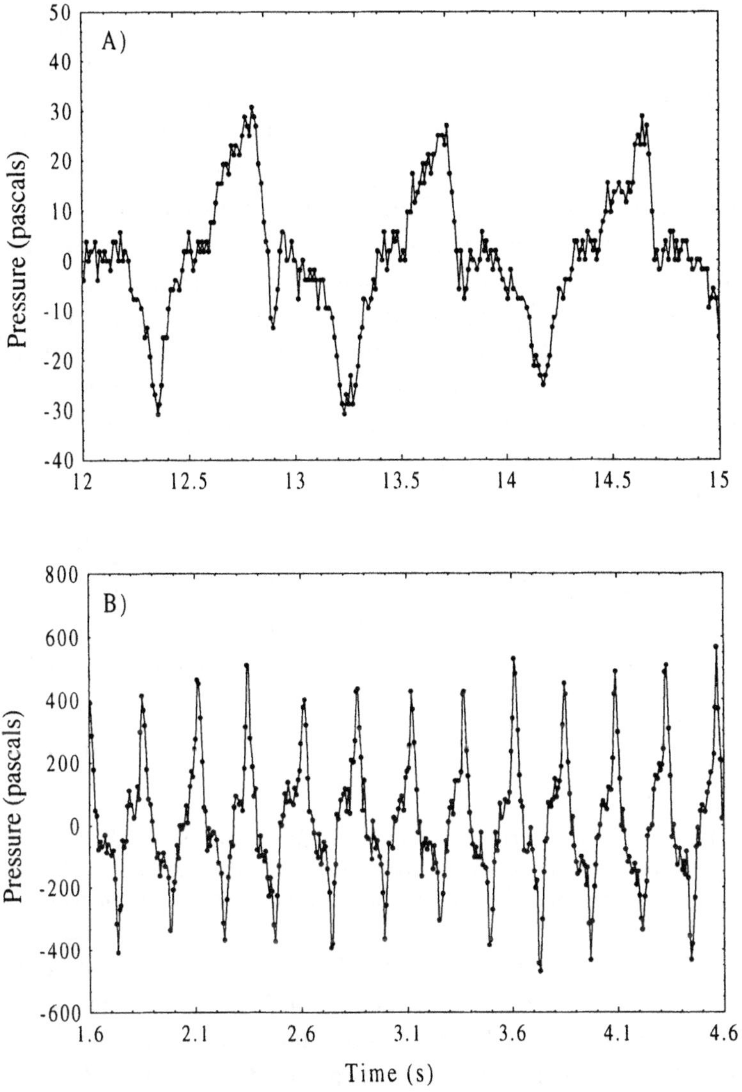

Figure 3. The pressure signal from an undulating European sea bass, swimming (A) 0.1 and (B) 0.5 m s^{-1}. Each tail beat cycle is represented by the peak to peak signal. Sampling rate of the data acquisition system was 130 Hz. Note the difference in pressure and tail beat frequency.

All timing operations, A/D data acquisitions and data manipulations were performed by a microprocessor. A voltage signal proportional to differential pressure was amplified, rectified and then digitised by a 12-bit successive approximation A/D converter. The converter sampled at approximately 40Hz depending on the frequency of the clock oscillator which controlled the acoustic frequency. Each new A/D conversion was input into a digital filter routine that computed a new value for the zero-crossover baseline. The A/D conversion and new baseline value were then used in an integration algorithm to compute the area under the tail beat pressure curve. Also the maximum and minimum A/D values were stored during the 3 s data acquisition period. At the end of each 3 s period of data acquisition, the microprocessor output three 10 ms ultrasonic pulses, the first one corresponding to a constant interval of 1120 ms and the next two corresponding to acoustic periods that represent *IP* and *PD* respectively. The sensitivity to pressure was determined by the amplification characteristics of the amplifier, the range of the A/D converter and by the ability of the ultrasonic receiver to detect the acoustic signal. This detection sensitivity varied depending on many factors but primarily on the level of background acoustic noise in the surrounding environment. Generally, the sensitivity of the transmitter in detecting variations in pressure ranged between 0.18 and 0.34 Pa.

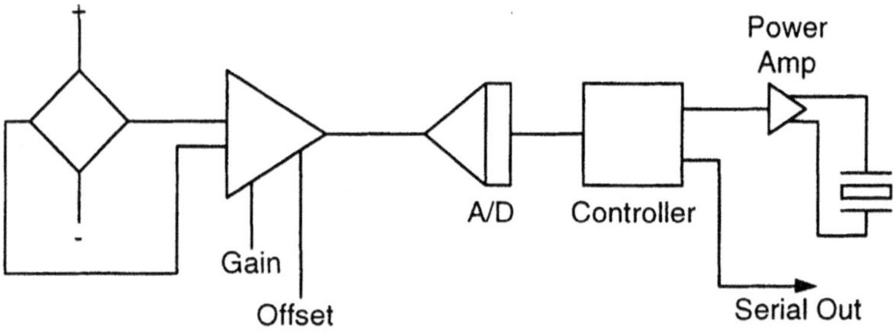

Figure 4. Diagram of the components of an ultrasonic transmitter. From the left, the components are pressure strain bridge; operational amplifier; 12-bit analog to digital voltage converter; microprocessor; power amplifier and acoustic output and serial data output.

An RS232C serial transmit capability was also incorporated into the transmitter to permit direct capture of each A/D conversion as well as the values for zero baseline, *IP* and *PD* computed by the microprocessor. The sampling rate of the A/D converter was reduced from 40 to 16 Hz when data were serially transmitted.

The transmitter was designed to operate in one of four modes. The operating mode was selected by applying a small magnet close to a magnetic reed switch under the surface of the transmitter. Modes can be switched in air or under water. Mode 1 output a synchronization pulse (1120 ms), followed by acoustic pulses representing *IP* and *PD* (Figure 5). Mode 2 is identical to mode 1 with the addition of serial data transmission. Mode 3 outputs a continuous stream of acoustic pulses proportional to "raw" differential pressure. The A/D converter takes an instantaneous measurement (microseconds) and the transmitter outputs an acoustic pulse after a time delay proportional to pressure. The advantage of this mode is that the "raw" pressure signal is recorded. Mode 4 is called "sleep mode" and is a power saving mode. In this mode the microprocessor significantly reduces power consumption from approximately 1 mA (averaged over 3 s) down to a few :A by shutting down power to the analog circuitry and A/D converter. The experimenter can choose between mode 1, 2 or 3 depending on the nature of the

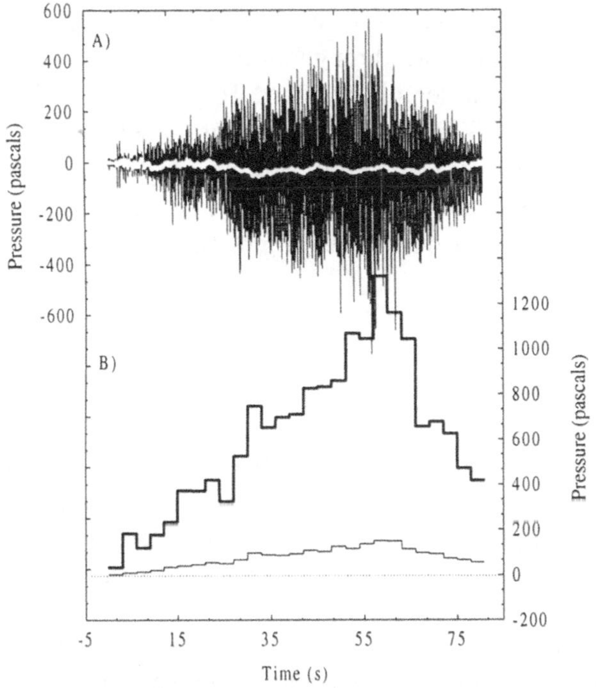

Figure 5. A) Sample of differential pressure data measured by an ultrasonic transmitter, from the tail of a European sea bass accelerating and decelerating from 0 to 0.6 to 0.2 m s^{-1}. The data were sampled from the serial port of an ultrasonic transmitter. The thick line through the center of the "raw" signal is the computed baseline. B) The thick line is the computed pressure difference (PD) (maximum minus minimum pressure) values for each 3-second period. The thin line is the computed integration pressure (IP) above and below the baseline. The dashed line indicates 0 pascals.

pressure signal being measured. Modes 1 and 2 are best suited to applications when the pressure pulse is quickly oscillating and can only be adequately measured using continuous A/D sampling. Ideally mode 3 should be used for slowly changing signals (<=1 Hz) or when the pressure pulse can be accurately characterised by the instantaneous acoustic periods.

Calibration of the transmitter with known pressure was accomplished by two methods. The transmitter was calibrated in mode 1 by recording acoustic data from an ultrasonic receiver into one serial communications port (RS232) of a PC and recording the A/D conversion data by connecting the transmitter RS232 to another RS232 port of the PC. Positive and negative pressure pulses were applied to one or both of the two sensor pressure ports. In mode 3 calibration was achieved simply by applying columns of water of known density and height (converted to Pa) to one of the pressure sensor ports while recording the acoustic time intervals.

To test the robustness of the prototype transmitters we instrumented two sea bass (1.5 kg), released them into a marsh pond (15 m x 20 m x 1 m deep) and measured tail beat pressure throughout a period of 16 days. The transmitter (1.6 cm dia., 8.0 cm length) weighed 16 g in water. The transmitter was placed on one side of the dorsal fin by suturing between skeletal fin ray supports. Foam rubber (1.5 cm wide, 8.0 cm length, 0.2 cm depth) was placed between the skin and transmitter. The fish were in excellent condition after 16 days. The transmitters output pressure data as time periods representing *PD* and *IP*.

3. RESULTS

Tail beat pressure of laboratory swum fish was very consistent for each swimming speed. Amplitude of the pressure signal varied only when the fish did not maintain position (i.e., if it lost or gained position in the swimming chamber). This was especially evident when the sea bass reached maximum sustainable velocities. In this experiment water velocity was increased or decreased in steps of 0.05 m s^{-1}, however pressure amplitude changes were clearly visible when velocity was changed by as little as 0.01 to 0.02 m s^{-1}. Burst swimming, observed as maximum accelerations from the downstream to upstream grids of the swimming chamber, resulted in tail beat pressures considerably larger (>1000 Pa) than pressures measured at maximum sustained constant swimming velocities.

For speeds between 0.1 and 0.9 m s^{-1} a linear regression of tail beat frequency *versus* speed produced the equation

$$TBF = 34.0 + 120.1(U),\ r^2 = 0.91,\ P = 0.0,\ n = 34 \qquad (1)$$

where *TBF* = tail beat frequency in beats min^{-1} and U = swimming speed in body length s^{-1} (L s^{-1}) (Figure 6).

'Pressure difference' increased by almost 30-fold over the range of speeds and *IP* increased by approximately 20 fold (Figure 7). Both pressure variables were highly correlated to swimming velocity. A Bernoulli power relationship provided the best predictive model. The relationships between *PD*, *IP* and swimming speed were

$$PD = (8.166\ U^2)/(2\ (1/1023)),\ r^2 = 0.95,\ p = 0.0,\ n = 31 \qquad (2)$$

$$IP = (1.377\ U^2)/(2\ (1/1023)),\ r^2 = 0.97,\ p = 0.0,\ n = 31 \qquad (3)$$

where *PD* and *IP* = pressure in Pa, U is swimming speed in m s^{-1}, and 1023 is seawater density in N s^2 m^{-4}.

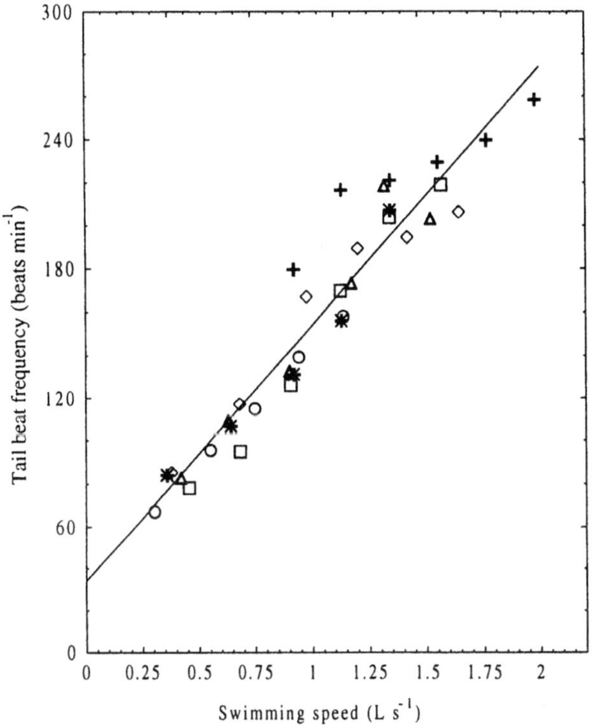

Figure 6. Tail beat frequency versus swimming speed for 6 European sea bass, instrumented with differential pressure sensors. Different symbols indicate different fish.

The pressure data calculated from acoustic transmission (mode 1) of *IP* and *PD* from a 1.5 kg sea bass over a 16 d period (approx. 15000 data) are illustrated in Figure 8. Both methods gave similar results except near 240 hours when either *IP* was too high or *PD* was too low. *PD* was approximately five times higher than IP. There were many periods of activity ranging from minutes to hours in duration. Activity was highest during the first few days of monitoring and was lowest during the coldest period at approximately 5°C. Also, a 24-h rhythm of activity was observed (Figure 8C). Activity was lowest during the morning and rose throughout the day and

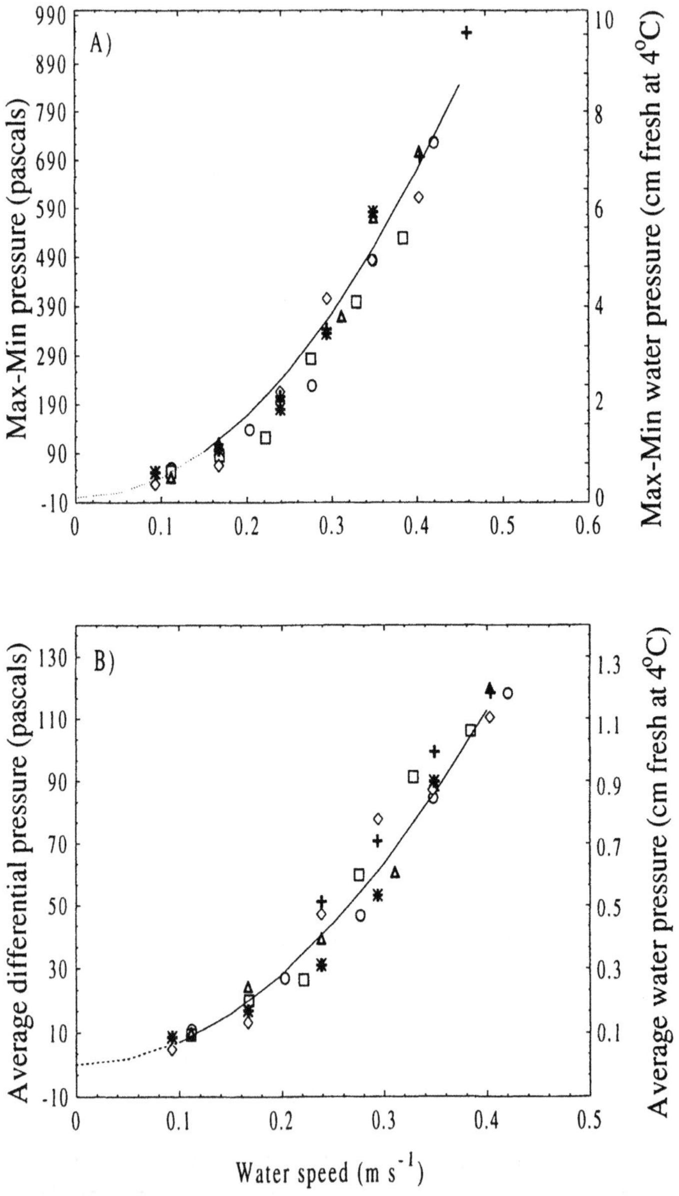

peaked during the night between 2000 h and 0600 h.

Figure 7. The relationship between pressure difference (PD), integrated pressure (IP) and swimming speed (U) of 6 European sea bass instrumented with differential pressure sensors. Different symbols indicate individual fish.

4. DISCUSSION

The results of this study indicate that tail beat pressure is an excellent predictor of swimming speed in sea bass. This technique can be applied to any creature under water that undulates and produces pressure as their major form of locomotion. We have shown that it can be used on small fish and it has been proven for medium sized fish (Webber, 1999). Additional studies need to be carried out on other fish species; however, we expect that the technique should accurately predict swimming speed for large animals (tunas, sharks, whales, dolphins, divers) as well. European sea bass of 0.25 kg produce differential pressures very similar to swimming cod of an order of magnitude greater mass (Webber, 1999).

The EMG technique is currently restricted to low salinity field applications (freshwater and brackish water) because the signal is transmitted on an FM radio carrier. The tail beat pressure technique should complement the existing EMG technique because ultrasonic telemetry can be used in freshwater and marine applications. An advantage of measuring tail beat pressure is that pressure is produced by all tail muscle fibers firing through a tail beat cycle regardless if they are red, white or mosaic fibers. If the surface area of the caudal peduncle does not significantly change from slow to fast swimming speeds then pressure will be highly correlated to power produced by the tail to overcome frictional and induced or pressure drag. The electrical signal of the EMG technique is a product of an unknown number of muscle fibers firing within a localized region of the total tail musculature. It is too early to state that the tail beat pressure technique will correlate to swimming speeds in all modes of locomotion, however, we expect that tail beat pressure will accurately reflect all tail flexes from constant low speed swimming to burst swimming. Additional studies are warranted.

A decrease in ambient temperature reduces the power output of muscle fibers and the animal recruits more fibers to swim the same speed (Rome, 1990; Rome *et al.* 1990). The tail beat pressure technique should be immune to this effect and the measured pressure should be the same at a specific speed regardless of the temperature. At the moment it is not known how temperature affects EMG measurements but we expect that different relationships would exist for swimming speed and EMG voltage at different temperatures. Temperature should not affect *IP* and *PD* measurements because the signal from the pressure sensor is treated as an alternating waveform (positive to negative to positive pressure) (Figure 3) and the computed baseline moves up or down as the DC offset of the sensor changes with temperature.

As is the case for EMG, tail beat pressure should be useful to predict energy expenditure of axial muscles. If temperature and swimming speed are

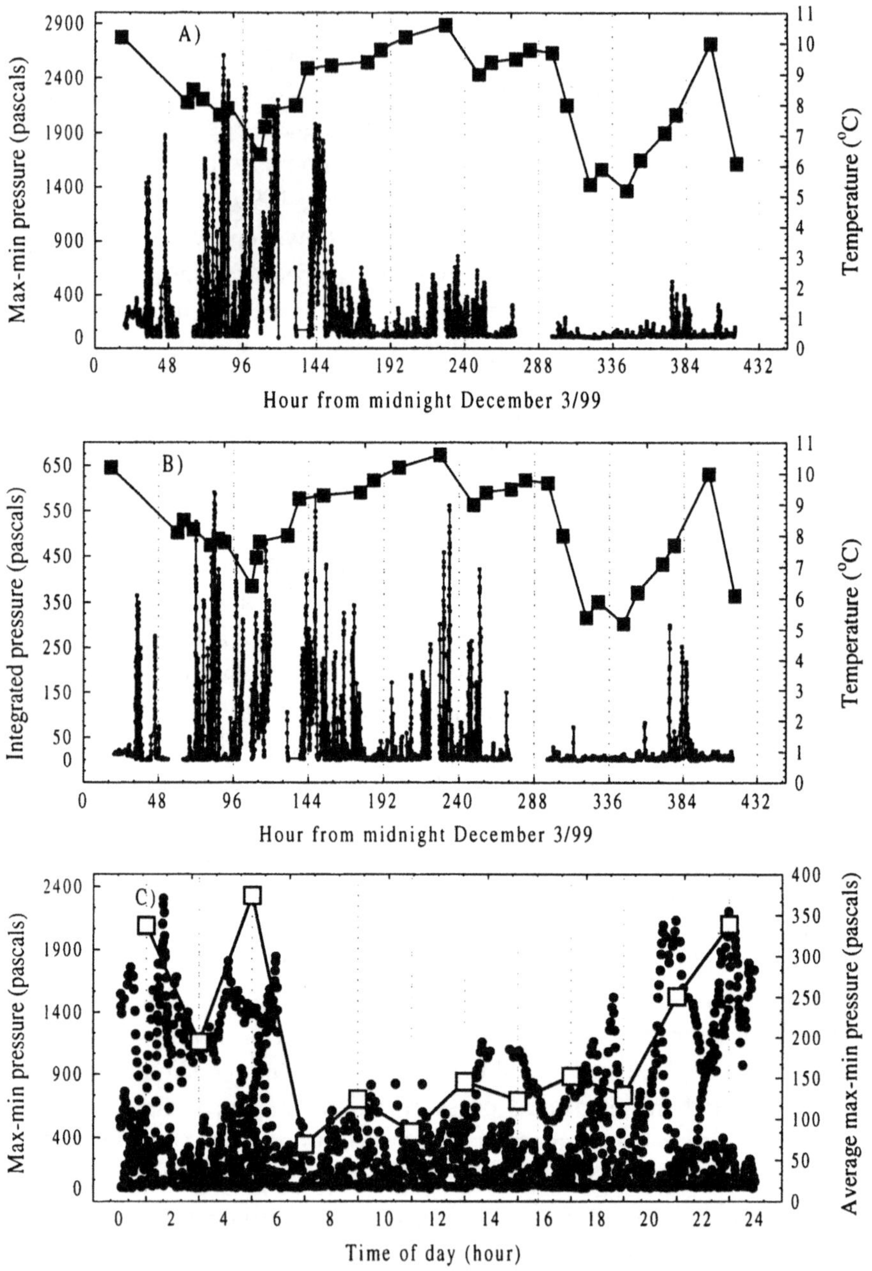

Figure 8. Tail-beat pressure vs swimming speed of a 1.5-kg European sea bass in a marsh pond. Temperature (filled squares) ranged between 5 and 11°C. (A) peak-to-peak pressure difference (*PD*). (B) Average or integrated pressure (*IP*). (C) Difference pressure (filled circles) vs hour of day over a 12-day period from December 7 to 19, 1999, illustrating a daily activity rhythm. The open squares represent 2-hr averages of pressure data.

known, standard metabolism can be estimated from laboratory measure measurements and it is possibly to estimate the activity portion of the bio-energetic model. We emphasize that activity is only one component of an animal's energy budget (Kerr, 1982); however, good long-term estimates of activity in nature would give considerable insight into behaviour and activity patterns of fish in nature.

The ultrasonic transmitters, attached to the 1.5-kg sea bass in the salt marsh pond, continuously transmitted pressure data throughout the 16 day period. Transmitter battery life was 30, 30 and 14 days in modes 1, 2 and 3 respectively. We removed the sea bass from the marsh after 16 days because of low water temperatures. Activity, averaged for 2 hour periods each day for 12 days, ranged from 25 to 69 Pa and from 69 to 374 Pa for *IP* and *PD*, respectively (Figure 8C). Equations 2 and 3 from Figure 7 predict these pressures will give swimming speeds from 0.19 to 0.31 m s^{-1} (0.27 to 0.44 L s^{-1}) and 0.13 to 0.30 m s^{-1} (0.18 to 0.43 L s^{-1}) for *IP* and *PD*, respectively. The sea bass with the transmitter was almost ten times greater in mass than the sea bass used in the laboratory study. Since we do not yet know how mass affects tail beat pressure with swimming speed we cannot with high confidence use the relationships from Figure 7 to predict swimming speed for the 1.5 kg sea bass. However, given that *PD* is approximately five times greater than *IP* from laboratory swum fish (Figure 7) and that PD is 3 to 5 times greater than *IP* for the marsh fish the values for *IP* and *PD* for the 1.5 kg sea bass are reasonable. The water temperature was near the low end of temperatures experienced by this species and sea bass metabolic scope is quite low at this temperature (Claireaux and Lagardère, 1999). Our predicted swimming speeds were low in accordance with low water temperatures. We also observed many activity cycles and a relationship between activity and time of day. Generally fish exhibit nocturnal and diurnal activity patterns and cycles suggesting that the data we observed from the transmitter were real.

We have presented a limited data set from two fish equipped with transmitters. The impact of the transmitter and sensor on the tissue was minimal and the animals did not exhibit any outward signs of physiological stress. Cod carrying tail beat transmitters in a 1-month study exhibited similar signs of minimal tissue degradation (Webber *pers. obs.*).

Many physiological events occur over short time periods, i.e., fractions of seconds, and the raw data cannot be reported in real time on an ultrasonic frequency. This is because ultrasonic transmission through water takes considerable power. This new transmitter design using a microprocessor permits measurement of fast events by computing onboard values of *PD* and *IP*. This avoids the problem of short transmitter life from energy demanding acoustic transmission of raw data. In the marsh pond we were able to collect data for over two weeks compared to a few days.

In conclusion, this study demonstrates that differential pressure measurements of tail beat undulation will be very useful in measuring swimming speeds and activity patterns of fish in nature and should complement predictions of swimming speed from EMG measurements. If such pressure measurements can be coupled with a cardiac output transmitter it should be possible to define the energetic expenditures of free-ranging fish in nature with greater precision than attained previously.

ACKNOWLEDGEMENTS

Financial support to GC by IFREMER and the Conseil Régional Poitou-Charentes is gratefully acknowledged.

REFERENCES

Aitken, J., Webber, D.M., O'Dor, R.K., and Voegeli, F.A. (2000) Acoustic telemetry of integrated differential pressures: calibration and applications. Proc. 3rd Conf. Fish Telemetry Europe, 7 pp. (in press).

Bainbridge, R. (1958). The speed of swimming of fish as related to the size and to the frequency and amplitude of the tail beat. *J. Exp. Biol.* **35**, 109-133.

Bainbridge, R. (1960). Speed and stamina in three fish. *J. Exp. Biol.* **37**, 129-153.

Block, B.A., Booth, D., and Carey, F.G. (1992). Direct measurement of swimming speeds and depth of blue marlin. *J. Exp. Biol.* **166**, 267-284.

Briggs, C.T., and Post, J.R. (1997). Field metabolic rates of rainbow trout estimated using electromyogram telemetry. *J. Fish Biol.* **51**, 807-823.

Claireaux, G., and Lagardère, J.P., (1999). Influence of temperature, oxygen and salinity on the metabolism of the european sea bass. *J. Sea Res.* **42**, 157-168.

Hunter, J.R., and Zweifel, J.R. (1971). Swimming speed, tail beat frequency, tail beat amplitude, and size in Jack Mackerel, *Trachurus symmetricus*, and other fishes. *Fish. Bull.* **69**(2), 253-266.

Kaseloo, P.A., Weatherley, A.H., Lotimer, J., and Farina, M.D. (1992). A biotelemetry system recording fish activity. *J. Fish Biol.* **40**, 165-179.

Kerr, S.R. (1982). Estimating the energy budgets of actively predatory fishes. *Can. J. Fish. Aquat. Sci.* **39**, 371-379.

Kiceniuk, J.W., and Jones, D.R. (1977). The oxygen transport system in trout (*Salmo gairdneri*) during sustained exercise. *J. Exp. Biol.* **69**, 247-260.

Korsmeyer, K.E., Lai, N.C., Shadwick, R.E., and Graham, J.B. (1997). Heart rate and stroke volume contributions to cardiac output in swimming yellowfin tuna: response to exercise and temperature. *J. Exp. Biol.* **200**, 1975-1986.

Lowe, C.G., Holland, K.N., and Wolcott, T.G. (1998). A new acoustic tailbeat transmitter for fishes. *Fish. Res.* **36**, 275-283.

Nakamura, Y.(1993). Vertical and horizontal movements of mature females of *Ommastrephes bartramii* observed by ultrasonic telemetry. In Okutani, T., O'Dor, R. K., and Kubodera, T., eds. *Recent Advances in Fisheries Biology*. Tokai University Press, Tokyo.
O'Dor, R.K., Andrade, Y., Webber, D.M., Sauer, W.H.H., Roberts, M.J., Smale, M.J., and Voegeli, F. M. (1998). Applications and performance of radio-acoustic positioning and telemetry (RAPT) systems. *Hydrobiologia* **371/372**, 1-8.
O'Dor, R.K., J. Forsythe, D.M. Webber, Wells, J., and Wells, M.J. (1993) Activity levels of *Nautilus* in the wild. *Nature, Lond.* **362**, 626-628.
Økland, F., Finstad, B., McKinley, R.S., Thorstad, E.B., and Booth, R.K. (1997). Radio-transmitted electromyogram signals as indicators of physical activity in Atlantic salmon. *J. Fish Biol.* **51**, 476-488.
Priede, I.G., and Young, A.H. (1977) The ultrasonic telemetry of cardiac rhythms of wild free-living brown trout (*Salmo trutta* L.) as an indicator of bioenergetics and behaviour. *J. Fish Biol.* **10**, 299-318.
Randall, D.J. (1968) Functional morphology of the heart in fishes. *Am. Zool.* **8**, 179-189.
Rome, L.C. (1990) Influence of temperature on muscle recruitment and muscle function in vivo. *Am. J. Physiol.* **259**. (*Regulatory Integrative Comp. Physiol.* 28), R210-R222.
Rome, L.C. Funke, R.P., and Alexander, R.M. (1990) The influence of temperature on muscle velocity and sustained performance in swimming carp. *J. Exp. Biol.* **154**, 163-178.
Sauer, W.H.H., Roberts, M.J., Lipinski, M.R., Smale, M.J., Hanlon, R.T., Webber, D.M., and O'Dor, R.K. (1997) Choreography of the squid's 'nuptial dance'. *Biol. Bull.* **192**, 203-207.
Scharold, J., Chin Lai, N., Lowell, W.R., and Graham, J.B. (1989) Metabolic rate, heart rate, and tailbeat frequency during sustained swimming in the leopard shark *Triakis semifasciata*. *Exp. Biol.* **48**, 223-230.
Stevens , E.D., and Randall, D.J. (1967) Changes in gas concentrations in blood and water during moderate swimming activity in rainbow trout. *J. Exp. Biol.* **46**, 329-337.
Thorarensen, H., Gallaugher, P.E., and Farrell, A.P. (1996) The limitations of heart rate as a predictor of metabolic rate in fish. *J. Fish Biol.* **49**, 226-236.
Videler, J. J. (1993) *Fish Swimming*, Chapman and Hall, London, U.K, 260 pp.
Wardle, C.S., Videler, J.J., Arimoto, T., Franco, J.M., and He, P. (1989) The muscle twitch and the maximum swimming speed of the giant bluefin tuna, *Thunnus thynnus* L. *J. Fish Biol.* **35**, 129-137.
Weatherley, A.H., Rogers, S.C., Pincock, D.G., and Patch, J.R. (1982) Oxygen consumption of active rainbow trout, *Salmo gairdneri* Richardson, derived from electromyograms obtained by radiotelemetry. *J. Fish Biol.* **20**, 479-489.
Webb, P.W. (1971) The swimming energetics of trout I. Thrust and power output at cruising speeds. *J. Exp. Biol.* **55**, 489-520.
Webber, D.M., Boutilier, R.G. and S.R. Kerr. (1998) Cardiac output as a predictor of metabolic rate in cod *Gadus morhua*. *J. Exp. Biol.* **201**, 2779-2789.
Webber, D.M. (1999) Respiratory and cardiovascular correlates of activity metabolism in the Atlantic cod (*Gadus morhua*). PhD thesis. Cambridge University. U.K.

Geolocation by Light Level

The Next Step: Latitude

Roger D. Hill and Melinda J. Braun

Wildlife Computer, 16150 NE 85th Street, Suite 226, Redmond, WA 98052

Key words: geolocation, atmospheric refraction, latitude, zenith angle, location by light level

Abstract: Methods to calculate an animal's position from light-level data collected by archival tags (geolocation by light levels) have been employed by wildlife researchers and the engineers who design the tags for over a decade. The problem of estimating longitude proved easy to solve, but accurate latitude estimates remain elusive. This paper addresses the absolute accuracy in estimating latitude (as defined by physical constraints) that is achievable using the astronomical equations and offers a new approach to minimize the variability of latitude estimations.

1. INTRODUCTION

Although sea-surface temperatures and bathymetry can be used to determine an animal's location, a more robust method involves calculating location from collected light-level data. A simplified description of the method is as follows: Times of "sunrise" and "sunset" for a given day, based on the tag's collected light-level data, are identified. "Standard" astronomical equations are used to determine longitude from the midpoint of the rise-set events (local noon or midnight, which varies only with longitude), and latitude is determined from the time between the rise-set events (e.g., "day length," which varies only with latitude). However, determining latitude near the spring and autumn equinox cannot be done, as there is very little variation in day length.

The times of sunrise and sunset do not have to correspond with the times when the sun appears/disappears over the horizon. Any position of the sun

J.R. Sibert and J.L. Nielsen (eds.), Electronic Tagging and Tracking in Marine Fisheries, 315–330.
© 2001 *Kluwer Academic Publishers. Printed in the Netherlands.*

that can be readily identified from the light-level data can be used. The sun's position is usually defined using an angle, referred to as the zenith angle (z), which measures the angle between the vertical and the observed center of the sun. The time when the center of the sun crosses a given z is used to determine the start (and end) of the rise and set events for that z. Thus, one can say there is no single definition of sunrise, sunset or day length, but that they depend upon which z is used.

Published methods (Delong et al., 1992, Wilson et al., 1992, Welch and Eveson, 1999) to calculate position from light-level data describe finding the best way to identify the time of rise and set events by processing the light-level curves, then optimizing a single zenith angle to be used in the astronomical algorithms to match up with these events. The methods of determining the time of the rise and set events and optimum zenith angle are often empirically verified using buoy experiments.

There are many factors that influence accuracy of geolocation estimates. There may be variability in the sensitivity and range of light-level sensors, not only between tags from different manufacturers, but even between individual tags from the same manufacturer. The recorded light levels may be externally attenuated, for example by the diving behavior of the animal, changing water turbidity and weather conditions, and biofouling of the tag. The location of the tag may be different during the rise and set events (e.g., the animal moved a significant distance). Finally, the standard astronomical equations are complex, often poorly understood, and subject to significant error if misapplied. This paper only addresses the absolute accuracy (as defined by physical constraints) that is achievable using the astronomical equations. The actual accuracy may be worse if it is degraded by these other factors.

Included in this paper as appendices are the annotated routines written in Microsoft Visual Basic™ by Wildlife Computers. These files are available as AppendixA_Eqn_of_Time.txt and AppendixB_Night_Length.txt at http://www.soest.hawaii.edu/PFRP/elec.tagdata/tagdata.html. These routines are used by Wildlife Computers for geolocating their instruments. The astronomical equations are implemented so that they can be used as straightforward called functions in computer programs. Appendix A contains the routines for applying the equation of time to improve the estimate of longitude from the time of local noon/midnight. Appendix B contains the subroutine to calculate night length (for latitude determination). Although these routines are not presented with any guarantee of accuracy, they have been empirically verified by Wildlife Computers. They are offered in this paper as a starting point for other researchers to work upon and improve. Corrections and comments to Wildlife Computers regarding these routines are encouraged.

2. BACKGROUND

Some definitions:

- ***Zenith angle***, or *z*, of the sun is defined as the angle between a vertical line from the observer and a line between the observer to the center of the sun. When $z = 0°$ the sun is directly overhead, and when $z = 90°$ the center of the sun is at the horizon (ignoring atmospheric refraction).
- ***Day length*** is the time between the sun's ascent (rise) and descent (set) through a given zenith angle.
- ***Local noon*** and ***local midnight*** are the times midway between rise and set. Measuring these times in Universal Time (UT, which is very close to GMT) simplifies subsequent calculations of longitude.
- ***Equinox*** is the date when the day length based upon a zenith angle of 90° is the same throughout all latitudes.

The most straightforward estimate of longitude is based on the time of local noon (or local midnight) measured in UT hours. Longitude is derived by multiplying this time by 15. This assumes that the earth has a perfectly circular orbit around the sun. A correction, called the equation of time, can be applied to make the longitude estimate much more accurate (see Appendix A).

Latitude estimates are based on day length. For centuries, mariners have used a sextant and an accurate timepiece to determine their location. Using the sextant and timepiece, the mariner notes the sun's zenith angle at some specified time. Because the astronomical equations are complex, navigational tables, e.g. the Nautical Almanac, have been produced to facilitate the calculation of latitude and longitude.

Since it is difficult to coerce an animal to make sextant observations, archival tags that measure and record light levels have been developed to locate the animal. Depending upon the archival tag's sensitivity, it is possible to determine the rise/set times for zenith angles from 87° (the sun is 3° above the horizon) to 102° (end of nautical twilight). The actual methods used by researchers to determine these points vary widely (Delong et al., 1992, Wilson et al., 1992, Welch and Eveson, 1999), and are beyond the scope of this paper.

The refraction of light by the earth's atmosphere allows the light of the sun to be seen even after it passes below the horizon. The standard definition of sunrise and sunset is when the sun is at a zenith angle of 90.83°. This is when the limb of the sun first appears/disappears over the horizon. Of this angle, 0.25° is the observed radius of the sun, and 0.58° is attributed to atmospheric refraction at the horizon (Meeus, 1991).

Twilights occur when the sun's zenith angle is between 90.83° and 108°. This is the period when light scattered by the upper atmosphere illuminates the earth. There are several periods of twilight that have been commonly defined. Civil twilight ends at $z = 96°$, nautical twilight ends at $z = 102°$, and astronomical twilight ends at $z = 108°$ (Duffett-Smith, 1988). During civil twilight, objects can be seen by the human eye. During nautical twilight, the horizon can be discerned. At the end of astronomical twilight, there is no appreciable light from the sun in the sky.

This effect of atmospheric refraction is not fixed. This variability is the reason it is generally stated that the exact time of a rise/set event cannot be predicted with an accuracy of better than ±2 minutes (Doggett et al., 1990). Schaefer and Liller (1990) described this phenomenon by demonstrating that the observed zenith angle of the sun at sunrise and sunset differs from the expected angle of 90.83° by ±0.32° with a 95% confidence interval. This variability between the observed and actual z will result in a corresponding variability in the observed day lengths. The variability depends upon the latitude of the point of observation, and the day of year.

For a given day of year, the day length (DL) is a function of only zenith angle (z) and latitude (L):

$$DL = f(z, L)$$

Variability in atmospheric refraction will generate a range of day lengths with a minimum and maximum day length (DL_{min} and DL_{max}):

$$DL_{min} = f(z-0.32, L)$$
$$DL_{max} = f(z+0.32, L)$$

Since DL also varies with latitude, the same DL_{min} and DL_{max} can be obtained:

$$DL_{min} = f(z-0.32, L) = f(z, L_{min})$$
$$DL_{max} = f(z+0.32, L) = f(z, L_{max})$$

Therefore, L_{min} and L_{max} are the limits of the range of latitudes that are predicted if a constant z is used. Even though L_{min} and L_{max} may not vary symmetrically about the true latitude, we will simply use $|L_{min} - L_{max}| / 2$ to define the Expected Variability in Latitude. An equivalent Expected Variability in Longitude can also be defined. This is simply 0.32°, year-round, regardless of location.

We set out to quantify the expected variability in latitude for varying values of z, for different latitudes, and for different days of the year.

3. MATERIALS AND METHODS

There is no one standard astronomical equation to generate the rise/set times of the sun. The authors culled through various equations (Duffett-Smith, 1988, Meeus, 1991, Muneer, 1998) to determine the most straightforward approach that could be implemented using PC-based computer programming. The selected equations and routines, plus their references, are detailed in the code (written in Microsoft Visual Basic™) presented in Appendices A and B.

These routines were used to calculate the expected variability in latitude for the family of graphs in Figure 1. Fourteen graphs are presented, each representing a different day of year. The graphs span a two-month period around the March equinox. Each graph shows the expected variability in latitude as a function of latitude for six different zenith angles. The angles used are 87°, 90° (dawn/dusk), 93°, 96° (end of civil twilight), 99° and 102° (end of nautical twilight).

The routines were then used in a similar manner to calculate the expected variability in latitude for a series of northern latitudes over an entire year, using a fixed zenith angle of 96°. These results are plotted in Figure 2.

Expected variability in latitude were calculated in a manner similar to that used for Figure 2, except the variability in latitude was calculated using all the zenith angles in the 87° to 102° range. Figure 3 shows the smallest variability for each latitude/day (the generating zenith is not depicted).

4. RESULTS

The graphs in Figure 1 illustrate the following:

- The choice of zenith angle can make a significant difference in the expected variability in latitude for a given latitude/day of year, especially around the equinoxes. For example, on 4 April at latitude -40°, a zenith angle of 96° has an uncertainty of 6°, while a zenith angle of 87° has an uncertainty of less than 2°.
- If summer is defined as the dates from the spring to autumn equinox, and winter from the autumn to spring equinox, the following general observations can be made:
    - It is best to use the largest z possible during the summer, and the smallest z possible during the winter. This is true in both the northern and southern hemispheres, as the seasons are reversed.
    - The dates of greatest variability in latitude are on the winter side of the equinoxes when using a large z.

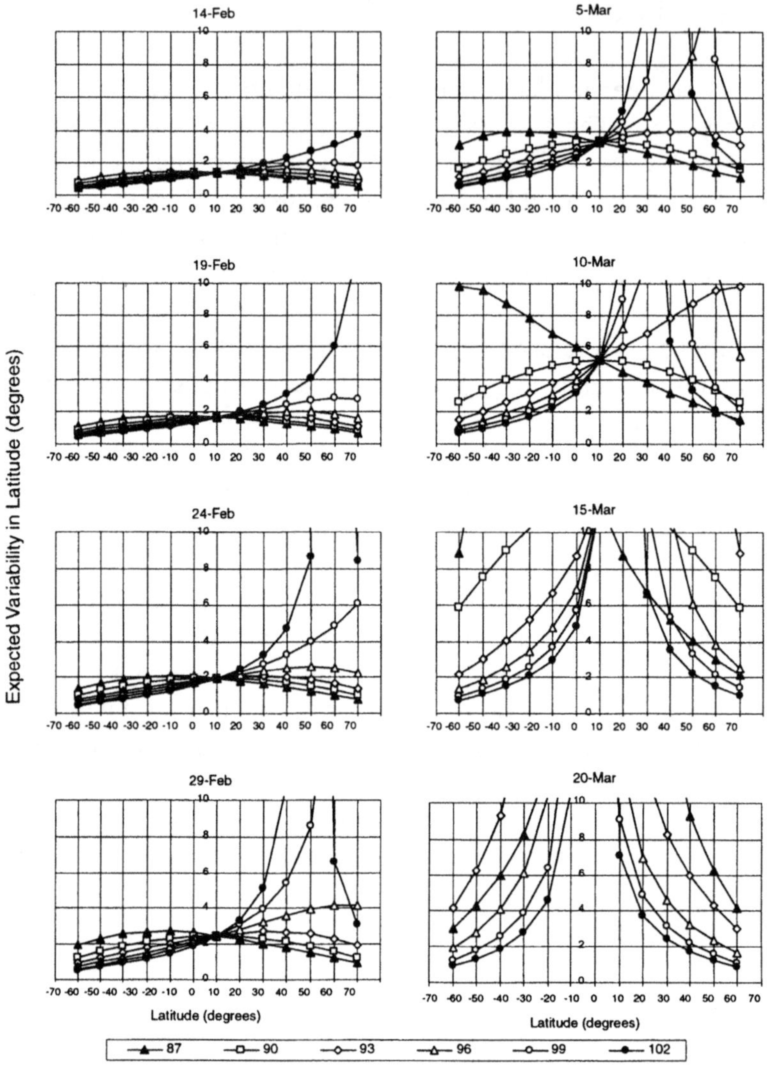

Figure 1. Expected variability in latitude for zenith angles 87° to 102° for fourteen days surrounding the March equinox.

Within dates where there is the greatest variability in latitude, the *z* that generates the least variability in latitude is not obvious. For example, on 15 Mar at +30°, the best *z* to use is 102°. However, using that *z* may not be possible if the light sensor is not sensitive enough to record such low light levels. Contrary to intuition, the next best choice for *z* is 87°, followed by 99°.

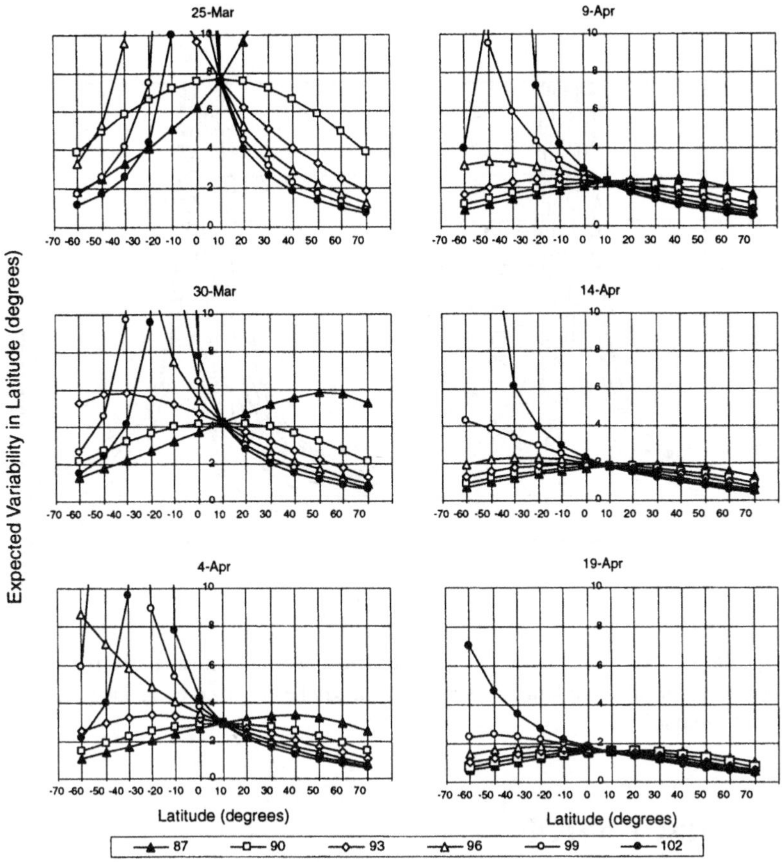

Figure 2. Expected variability in latitude for a fixed zenith angle of 96°. These are the errors in latitude estimates that can be expected using a single zenith angle technique.

Figure 2 shows the effects of using a single *z* of 96° in the northern latitudes:

- Note the asymmetry about the equinoxes for all latitudes except 0° (the equator).
- The expected variability in latitude is greater at the winter solstice than the summer solstice.
- The periods of greatest variability in latitude precede the spring and follow the autumn equinox.

Figure 3 shows these findings:

- It is possible to significantly reduce the variability in latitude by carefully choosing the best *z* for a given latitude/day of year, except at the equator.
- The improvement is greater at higher latitudes. At +60°, the expected variability remains below 3° through the equinox periods.
- The expected variability in latitude is never less than 0.7°.

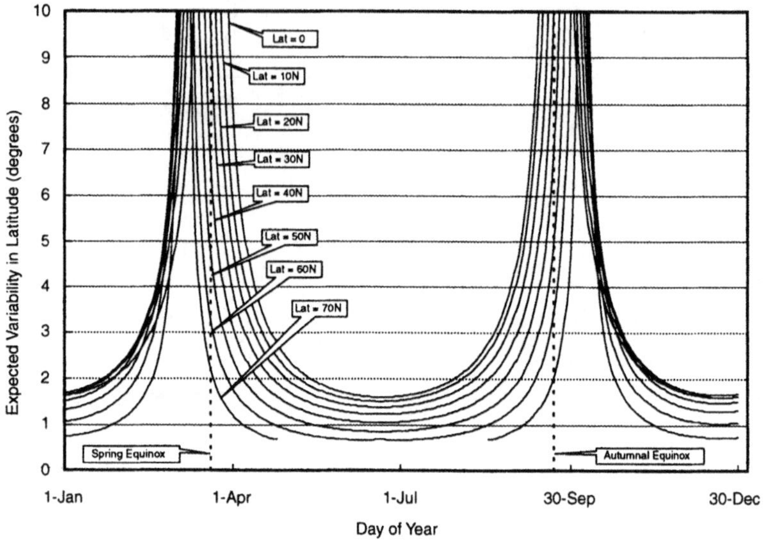

Figure 3. Minimum expected variability in latitude using zenith angles in the range of 87° to 102°. These are the errors in latitude estimates that can be expected by using a multiple zenith angle technique.

5. DISCUSSION

A single optimum zenith angle derived from data taken at a fixed position (buoy) cannot be applied across all latitudes. For example, on 4 April, the expected variability in latitude for a zenith angle of 96° is less than 1° at a latitude of +60°, but is greater than 8° at -60°.

When using a fixed zenith angle, there are ranges of days for which there will be a high variability in estimated latitude. Selecting a different fixed zenith merely moves where the range occurs. These ranges should be identified when using a fixed zenith angle for latitude analysis.

In order to achieve the best possible results, careful choice of zenith angle should be made (on a day-by-day basis) to avoid ranges of days for which there is a high estimated variability in latitude. However, if one does not know where an instrument is on a given day (i.e. the tag is on an animal, not a buoy), it is not possible to choose the best single zenith angle. However, this problem can be avoided if more than one point on each light-level curve is associated to a corresponding fixed zenith angle. Then a family of day lengths can be determined, one for each zenith angle, and the latitude that best predicts all these day lengths is the best estimate. Wildlife Computers generalizes this technique by fitting all the day lengths corresponding to zenith angles from 87 to 102°. Finally, no matter how ingenious the geolocation analysis, the expected variability in latitude will never be less than 0.7°, while the expected variability in longitude will be a constant 0.32°.

REFERENCES

Delong, R.L., Stewart, B.S. and Hill, R.D. (1992) Documenting migrations of northern elephant seals using day length. *Mar. Mamm. Sci.* **8**, 155-159.

Doggett, L.E., Tangren, J.T., and Panossian, S.P. (1990) *Almanac for Computers 1990.* U.S. Government Printing Office, Washington DC.

Duffett-Smith, P. (1988) *Practical Astronomy with your Calculator,* 3rd edition. Cambridge University Press.

Hill, R. (1994) Theory of geolocation by light levels. *In* LeBouef, B.J. and Laws, R.M. (eds.) *Elephant Seals: Population Ecology, Behaviour, and Physiology.* University of California Press, Berkeley, CA, 227-236 pp.

Meeus, J. (1991) *Astronomical Algorithms.* Willmann-Bell, VA, 429 pp.

Schaefer, B.E., and Liller, W. (1990) Refraction near the horizon. *Publ. Astron. Soc. Pac.* **102**, 796-805.

Welch, D.W., and Eveson, J.P. (1999) An assessment of light-based geoposition estimates from archival tags. *Can. J. Fish. Aquat. Sci.* **56**, 1317-1327.

Wilson, R.P., Ducamp, J.-J., Rees, W.G., Culik, B.M., and Niekamp, K. (1992) Estimation of location: global coverage using light intensity. *In* Priede, I.G. and Swift, S.M. (eds.) *Wildlife Telemetry.* Ellis Horwood, NY. pp. 131-134.

APPENDIX

Appendix A: BASIC subroutine to determine the Equations of Time (EqT) in minutes of time

This is used in the calculation of longitude to correct for the elliptical nature of the earth's orbit, and the tilt of the earth's axis, in the following manner:

East Longitude = 360 – (time of midnight (hours)) × **15 – (eqT** (minutes)) ÷ **4**

```
Dim MidnnightGMT as Date, correctedyear as Integer, T as Double
Dim correctedmonth as integer, G as Double, C as Double
Dim I as Double, alpha as Double

If Month(MidnightGMT) > 2 Then
  correctedYear = Year(MidnightGMT)
  correctedMonth = Month(MidnightGMT) - 3
Else
  correctedYear = Year(MidnightGMT) - 1
  correctedMonth = Month(MidnightGMT) + 9
End If
```

```
  T = (MidnightGMT - Int(MidnightGMT) + Day(MidnightGMT)
  T = T + Int(30.6 * correctedMonth + 0.5)
  T = T + Int(365.25 * (correctedYear - 1976)) - 8707.5) / 36525

  G = (357.528 + 35999.05 * T)
  C = 1.915 * Sin(DTR * (G)) + 0.02 * Sin(DTR * (G * 2))
  l = (280.46 + C + 36000.77 * T)
  alpha = 2.466 * Sin(DTR * (2 * l)) - 0.053 * Sin(DTR * (4 * l))
  EqT = (alpha - C) * 4
```

Code is derived from equations of Muneer (1998)

Appendix B. Set of BASIC subroutines to calculate night length

Subroutine NightLevelInit is first called to initialize the public variables. Function NightLength is then called and returns the number of hours for the sun to travel from zenith to zenith on date d at location Longitude and Latitude. Functions GetUTRise and GetUTSet can then be called to get the times of sunrise and sunset in UT (formerly GMT) hours.

Source of equations are marked, for example, as [PDS 25] for Duffet-Smith (1988) page 25, or [JM 10] for Meeus (1991) page 10.

```
Public DTR As Double                    'constant degrees-to-radians
Public RTD As Double                    'constant radians-to-degrees
Public Pi As Double                     'constant 3.14159...
Public TwoPi As Double                  'constant 6.28318...
Public DaysPerYear As Double            'constant 365.25...
Public Epoch As Date                    'epoch used for sun motion equations
Public EclipticLongitudeAtEpoch As Double
Public EclipticLongitudeOfPerigreeAtEpoch As Double
Public EccentricityAtEpoch As Double
Public ObliquityOfTheElliptic As Double
Public UTrise As Single                 'Time of sunrise in UT
Public UTset As Single                  'time of sunset in UT

Sub NightLevelInit()
  DTR = 3.14159265358979 / 180
  RTD = 180 / 3.14159265358979
  Pi = 3.14159265358979
  TwoPi = 2 * Pi
  DaysPerYear = 365.242191
  Epoch = #12/31/1989#
  EclipticLongitudeAtEpoch = 279.403303
  EclipticLongitudeOfPerigreeAtEpoch = 282.768422
  EccentricityAtEpoch = 0.016713
  ObliquityOfTheElliptic = 23.441884
End Sub
```

```
Function NightLength(Latitude As Single, Longitude As Single, d As Date, _
            Zenith As Single) As Single
```

```
Dim RA1 As Double              'Right ascension of sun on day d at 00UT
Dim HA1 As Date                'Hour angle of sun on day d at 00UT
Dim Declination1 As Double     'Declination of sun on day d at 00UT
Dim d2 As Date                 'Date of day afer d
Dim RA2 As Double              'Right ascension of sun on day d+1 at 00UT
Dim HA2 As Date                'Hour angle of sun on day d+1 at 00UT
Dim Declination2 As Double     'Declination of sun on day d+1 at 00UT
Dim N As Double                'Julian Date of day d at 00UT
Dim R As Double                'Atmospheric refraction in degrees

Dim LST1r As Double   'Local sidereal time of sunrise calculated from sun's' position on midnight before
Dim LST1s As Double   'Local sidereal time of sunset calculated from sun's position on midnight before
Dim GST1r As Double   'Grwich sidereal time of sunrise calculated from sun's position on midnight before
Dim GST1s As Double   'Grwich sidereal time of sunset calculated from sun's position on midnight before

Dim LST2r As Double   'Local sidereal time of sunrise calculated from sun's position on midnight after
Dim LST2s As Double   'Local sidereal time of sunset calculated from sun's position on midnight after
Dim GST2r As Double   'Grwich sidereal time of sunrise calculated from sun's position on midnight after
Dim GST2s As Double   'Grwich sidereal time of sunset calculated from sun's position on midnight after

Dim GSTr As Double    'Greenwich sidereal time of sunrise calculated from the sun's interpolated position
Dim GSTs As Double    'Greenwich sidereal time of sunset calculated from the sun's interpolated position
Dim T00 As Double     'Greenwich sidereal time of 00UT on day d at Greenwich
Dim T00p As Double    'Greenwich sidereal time of 00UT on day d at observer's longitude
Dim v As Double

N = Int(d) + 2415018.5            'Calculate Julian Date
R = GetRefraction(Zenith)         'When sun is actually at Zenith, it appears to be R higher
Call getRAD(d, RA1, Declination1) 'For day d, get Right Ascension and Declination of sun at 00UT
LST1r = GetLSTr(Latitude, Declination1, RA1, Zenith + R)
                        'Calculate local sidereal time of sunrise at angle Zenith
If LST1r < 0 Then
  NightLength = LST1r             'special case
  Exit Function
End If
LST1s = GetLSTs(Latitude, Declination1, RA1, Zenith + R)
                        'Calculate local sidereal time of sunset at angle Zenith
If LST1s < 0 Then
```

```
                              'Calculate local sidereal time of sunset at angle Zenith
 If LST1s < 0 Then
   NightLength = LST1s   'special case
   Exit Function
 End If
 GST1r = Range0_24(LST1r + Longitude / 15)    'Convert to Greenwich Sidereal Time
 GST1s = Range0_24(LST1s + Longitude / 15)    'Convert to Greenwich Sidereal Time
 d2 = d + 1                                   'For the next day, calculate the same parameters
 Call getRAD(d2, RA2, Declination2)
 LST2r = GetLSTr(Latitude, Declination2, RA2, Zenith + R)
 If LST2r < 0 Then
   NightLength = LST2r
   Exit Function
 End If
 LST2s = GetLSTs(Latitude, Declination2, RA2, Zenith + R)
 If LST2s < 0 Then
   NightLength = LST2s
   Exit Function
 End If
 GST2r = Range0_24(LST2r + Longitude / 15)
 GST2s = Range0_24(LST2s + Longitude / 15)

 If GST1r > GST2r Then GST2r = GST2r + 24
 If GST1s > GST2s Then GST2s = GST2s + 24

 T00 = UT_to_GST(d)           'Calculate the Greenwich sidereal time at Greenwich at 00UT
 T00p = T00 + 1.002737909 * Longitude / 15    'Greenwich sidereal time at observer at 00UT
 If T00p < 0 Then T00p = T00p + 24
 If GST1r < T00p Then          'Use this time to update Greenwich sidereal times of rising, if
necessary
   GST1r = GST1r + 24
   GST2r = GST2r + 24
 End If
 If GST1s < T00p Then          'Use this time to update Greenwich sidereal times of setting, if
necessary
   GST1s = GST1s + 24
   GST2s = GST2s + 24
 End If
 GSTr = (24.065709816*GST1r - T00*(GST2r - GST1r)) / (24.065709816 + GST1r - GST2r)
 GSTs = (24.065709816*GST1s - T00*(GST2s - GST1s)) / (24.065709816 + GST1s - GST2s)
                         'interpolate times

 UTrise = GST_to_UT(d, GSTr)
 UTset = GST_to_UT(d, GSTs)

 NightLength = Range0_24((GSTr - GSTs) * 0.9972695663)
                         'night length is time difference converted from sidereal time
End Function
```

```
Function GetUTRise() As Single
  GetUTRise = Utrise              'Return already calculated UT of rise (hours)
End Function
```

```
Function GetUTSet() As Single
  GetUTSet = UTset                'Return already calculated UT of set (hours)
End Function
```

Supporting Astronomical Subroutines

```
Function GetLSTr(Latitude As Single, Declination As Double, RAscension As Double, _
          Zenith As Single) As Double
'
' Get local sidereal time of sun rising at angle Zenith with Declination and RightAscension
'
'  [PDS 52, modified for arbitrary Zenith from JM 98]
'
' Returns special cases of -1 for sun never gets high enough in the sky
'                        and -2 for sun never gets low enough in the sky

  Dim H As Double
  H = (Cos(Zenith * DTR) - Sin(Latitude * DTR) * Sin(Declination * DTR)) / _
    (Cos(Latitude * DTR) * Cos(Declination * DTR))
  If H >= -1 And H <= 1 Then
    GetLSTr = 24 - RTD * ACos(H) / 15 + 24 * RAscension / 6.28318530717959
    While GetLSTr > 24
      GetLSTr = GetLSTr - 24
    Wend
  ElseIf H > 0 Then
    GetLSTr = -1
  Else
    GetLSTr = -2
  End If
End Function
```

```
Function GetLSTs(Latitude As Single, Declination As Double, RAscension As Double, _
          Zenith As Single) As Double
'
' Get local sidereal time of sun setting at angle Zenith with Declination and RightAscension
'  [PDS 52, modified for arbitrary Zenith from JM 98]
'
' Returns special cases of -1 for sun never gets high enough in the sky
'                        and -2 for sun never gets low enough in the sky

  Dim H As Double
  H = (Cos(Zenith * DTR) - Sin(Latitude * DTR) * Sin(Declination * DTR)) / _
    (Cos(Latitude * DTR) * Cos(Declination * DTR))
  If H >= -1 And H <= 1 Then
    GetLSTs = RTD * ACos(H) / 15 + 24 * RAscension / 6.28318530717959
```

```
   While GetLSTs >= 24
    GetLSTs = GetLSTs - 24
   Wend
  ElseIf H > 0 Then
   GetLSTs = -1
  Else
   GetLSTs = -2
  End If
End Function
```

```
Sub getRAD(d1 As Date, RightAscension As Double, Declination As Double)
'
' Calculation of the position of the sun at 00UT on date d1 - returns RightAscension and Declination
'
' First we determine the Longitude of the sun (lambda) using standard values for date Epoch from [PDS 86]
' This value is calculated in the ecliptic plane's coordinate set [PDS 30] (the plane of the orbit of the
' earth around the sun), thus the Latitude of the sun (in these coordinates) is zero.
' These are converted to equitorial units [PDS 27] to facilitate determination of position of sun in earth-based
' units.

  Dim d As Long, N As Double, Mdot As Double
  Dim Ec As Double, lambda As Double
  Dim x As Double, y As Double

  d = Int(d1 - Epoch)
  N = d * 360 / DaysPerYear
  While N < 0
   N = N + 360
  Wend
  While N > 360
   N = N - 360
  Wend
  Mdot = N + EclipticLongitudeAtEpoch - EclipticLongitudeOfPerigreeAtEpoch
  If Mdot < 0 Then Mdot = Mdot + 360
  Ec = Sin(Mdot * DTR) * EccentricityAtEpoch * 360 / Pi
  lambda = Mdot + Ec + EclipticLongitudeOfPerigreeAtEpoch
  If lambda > 360 Then lambda = lambda - 360
  lambda = lambda * DTR
  Declination = ASin(Sin(DTR * ObliquityOfTheElliptic) * Sin(lambda)) * RTD
  y = Sin(lambda) * Cos(DTR * ObliquityOfTheElliptic)
  x = Cos(lambda)
  RightAscension = ATan2(x, y)
End Sub
```

```
Function GetRefraction(Zenith As Single) As Double
'
' Atmospheric refraction [JM 102] to be added to Zenith
' when calculating LSTr's and LSTs's.  Refraction varies with Zenith angle up to a max of .553
degrees at horizon
'
 Dim R As Double
 If Zenith <= 90 Then
  R = 1 / (60 * Tan(DTR * (90 - Zenith + 7.31 / (94.4 - Zenith))))
  GetRefraction = R - 0.06 * Sin(DTR * (14.7 * R + 13))
 Else
  GetRefraction = 0.552687141853696                    'use value for zenith = 90
 End If
End Function
```

```
Function GST_to_UT(d As Date, GST As Double) As Double
'
' Convert Greewich Sidereal Time GST on date d to Universal Time [PDS 18]
'
 Dim S As Double, T As Double, T0 As Double, N As Double, H As Single
 N = Int(d) + 2415018.5
 S = N - 2451545
 T = S / 36525
 T0 = 6.697374558 + 2400.051336 * T + 0.000025862 * T * T
 T0 = Range0_24(GST - T0)
 GST_to_UT = T0 * 0.9972695663

End Function
```

```
Function UT_to_GST(d As Date) As Double
'
'  Convert Universal Time d to Greewich Sidereal Time [PDS 17]
'
 Dim S As Double, T As Double, T0 As Double, N As Double, H As Single
 H = 24 * (d - Int(d))
 N = Int(d) + 2415018.5
 S = N - 2451545
 T = S / 36525
 T0 = 6.697374558 + 2400.051336 * T + 0.000025862 * T * T
 T0 = T0 + H * 1.002737909
 T0 = Range0_24(T0)
 UT_to_GST = T0
End Function
```

Trigonometric Subroutines

```
Function ACos(x As Double) As Double
'
' returns arccosine in radians of x
```

```
'
 ACos = Atn(-x / Sqr(1 - x * x)) + 2 * Atn(1)
End Function
```

```
Function ASin(x As Double) As Double
'
' returns arcsine in radians of x
'
 ASin = Atn(x / Sqr(1 - x * x))
End Function
```

```
Function ATan2(x As Double, y As Double) As Double
'
' Returns arctangent (y/x) with results from 0 to 360; also tolerates x = 0
'
 Dim Pi As Double
 Pi = 3.14159265359

 If x = 0 Then
  If y > 0 Then
   ATan2 = Pi / 2
  Else
   ATan2 = -Pi / 2
  End If
 Else
  ATan2 = Atn(y / x)
  If x < 0 Then
   If y >= 0 Then
    ATan2 = ATan2 + Pi
   Else
    ATan2 = ATan2 - Pi
   End If
  End If
 End If
 If ATan2 < 0 Then ATan2 = ATan2 + 2 * Pi
End Function
```

```
Function Range0_24(a As Double) As Double
 Range0_24 = a
 While Range0_24 > 24
  Range0_24 = Range0_24 - 24
 Wend
 While Range0_24 < 0
  Range0_24 = Range0_24 + 24
 Wend

End Function
```

Summary Report of a Workshop on Daylight Measurements for Geolocation in Animal Telemetry

J. D. Metcalfe
The Centre for Environment, Fisheries and Aquaculture Science, Lowestoft Laboratory, Pakefield Road, Lowestoft, Suffolk, NR33 0HT, UK

1. INTRODUCTION

The ability to monitor the detailed geographical movements of fish in the open sea is problematic. Active tracking (e.g., from a research vessel using acoustic devices) is expensive and time-consuming, and radio-based satellite tracking is not normally an option because of the inability to transmit radio signals through sea water. In recent years, various groups of scientists have been using ambient daylight recorded by electronic archival (data storage) tags to estimate the geographical position of fish (Gunn et al., 1994), birds (Wilson et al., 1992), and mammals (Delong et al., 1992; Hill, 1994).

The method has the advantage that it not only obviates the need for radio access to airborne satellite systems, but the instrumentation can be small. In principle, the method is also simple, using the difference between the times of local noon and noon at Greenwich to estimate longitude and day length to estimate latitude. However, in practice there are various complicating factors that compromise accuracy and utility.

Because of the rapid increase in use of archival tags that use ambient daylight measurements to calculate geoposition, there is a need to exchange experiences, evaluate errors and, where appropriate, standardise methods and procedures. To facilitate such an exchange, a workshop was convened and the first meeting held at the Lowestoft Laboratory of the Centre for Environment, Fisheries and Aquaculture Science (CEFAS) June 17-18, 1999. A subsequent meeting was held at the East-West Centre, University of Hawai'i, Honolulu February 3-4, 2000.

J.R. Sibert and J.L. Nielsen (eds.), Electronic Tagging and Tracking in Marine Fisheries, 331–342.
© 2001 *British Crown copyright. Printed in the Netherlands.*

This report describes the aims and objectives of the workshop, summarises the discussions, and presents the conclusions and recommendations. It is stressed that this summary is not meant to be a definitive description of the methodology. Rather, it is an account of the current state of technology and methodology, and indicates how the participants thought that techniques are likely to develop in the future.

2. WORKSHOP OBJECTIVES

The aim of the workshop was to bring together a small number of biologists and physicists with relevant experience with these objectives:

- discussing basic physical principles and existing methods;
- identifying any developments that could improve accuracy and precision;
- proposing research programmes to realise such developments and identify sources of funding;
- producing a workshop report.

3. WORKSHOP MEETINGS

There were 13 participants at the first meeting held in Lowestoft and 11 at the subsequent meeting held in Honolulu. In the following text, conclusions and actions agreed by the participants are presented indented and in italics to differentiate them from the general text.

3.1 The Basic Astronomical Principles

3.1.1 The astronomical algorithm

To date, most geolocation methods that use ambient daylight involve identifying the times when the sun is at a precisely known zenith angle (e.g., sunrise and sunset). Although this report is not intended to provide a detailed description of the methodology, a description of the basic astronomical algorithm used to determine geolocation is presented here for completeness. However, different implementations are used by different practitioners, and the absolute accuracy of various methods has yet to be critically compared. The reader is therefore referred to more detailed accounts presented in this volume (e.g., Hill and Braun; Musyl et al.) and elsewhere (Hill, 1994; Nautical Almanac Office, 1991)

3.1.2 Sunrise, sunset, and twilight

The times of sunrise and sunset are defined as those when the Sun's upper edge (technically referred to as the "limb"), as affected by refraction, is on the true horizon of an observer at sea-level. Assuming the mean refraction to be 34', and the Sun's semi-diameter to be 16', sunrise and sunset occur when the true zenith distance of the Sun's centre is at 90° 50' (i.e. 90° + 34' + 16'). At this time, the Sun's centre is 50' below the true horizon, and the upper limb is 34' below the horizon, but is brought to the horizon by refraction of the sunlight through the Earth's atmosphere. (Fig. 1).

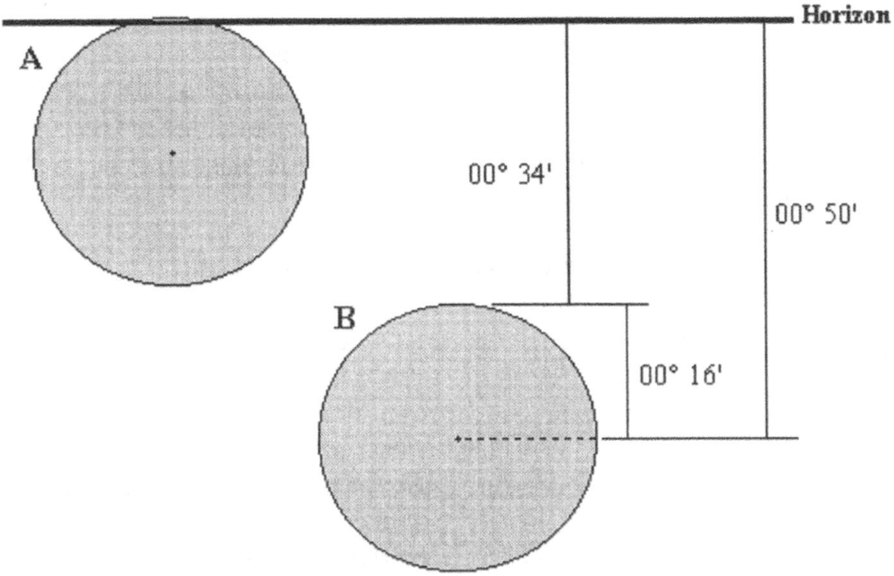

Figure 1. The apparent (A) and true (B) positions of the Sun relative to the horizon of an observer at sunrise or sunset. When the Sun's centre is 50' below the horizon, its upper limb is 34' below the horizon (B); refraction by the earth's atmosphere brings the upper limb to the horizon (A)

Twilight is the period before sunrise and after sunset when the sky is light. Three zenith angles (96°, 102° and 108°) are denoted as civil, nautical and astronomical twilight respectively.

Given an accurate clock, whose time is set to a known reference location (e.g., Greenwich), and a cloudless horizon, a human observer at sea can estimate the time sunrise to within a few seconds by observing when the Sun's limb appears at the horizon. Similarly our observer could estimate the time of sunset by observing when the Sun's upper limb disappears from the horizon.

From accurate estimates of the time of sunrise and sunset, and by understanding of the true position of the Sun at these times, an observer can calculate his geoposition using the basic astronomical variables and algorithms given below.

3.1.3 The variables

The Julian date (N) where 1 January = 1 and 31 December = 365 (or 366 in a leap year)

Time of sunrise (T_r), in minutes Universal Time (UT*);

Time of sunset (T_s), in minutes UT;

The sun's declination (d), is the angular distance of the sun north or south of the Earth's Equator for the date in question. Because the earth's equator is tilted 23.45° with respect to the plane of the earth's orbit around the sun, the declination varies between 23.45° N and 23.45° S through the year. Declination is calculated as

$$d = 23.45 * \sin((360/365) * (284 + N))$$

The Sun's altitude (or elevation) angle at dawn/dusk (A) is 50' or 0.8333°.

The equation of time (E), compensates for the Earth's elliptical orbit around the sun and its axial tilt for the date in question. Although more sophisticated algorithms are available, E can be approximated, in minutes, by

$$E = (9.87 * \sin(2B)) - (7.53 * \cos(B)) - (1.5 * \sin(B))$$
$$\text{Where } B = 360*((N-81)/365)$$

3.1.4 Calculating geoposition

The longitude (L) in degrees east of Greenwich is calculated by

$$L = 180 - (T_r+T_s)/8 - E/4$$

The latitude (B), in degrees north of the equator is calculated by iteration from the formula that relates the altitude angle (A, above) to the Sun's hour angle (H).

The hour angle at sunrise and sunset is given by

* The term Universal Time (which is the same as current Greenwich Mean Time) has been used by astronomers since 1928 to avoid the confusion arising from 0000h GMT being taken from noon prior to 1 January 1925, but from midnight after that date.

$$H = \pi\,(T_s\text{-}T_r)\ 1440 \text{ (in radians) or}$$
$$H = (T_s\text{-}T_r)/8 \text{ (in degrees)}$$

and latitude is then calculated by iteration from

$$\sin(A) = \sin(d)\sin(B) + \cos(d)\cos(B)\cos(H)$$

The accuracy of this method is neither constant throughout the year, nor the same at all latitudes (see Hill and Braun, this volume, for a fuller description). In particular latitude is most difficult to determine around the time of the equinoxes, when day-length is almost independent of latitude. However, experiments indicate that reasonable latitudinal locations can be obtained for periods five or more days from the equinoxes.

The workshop concluded that, although the astronomical algorithms appear to be fully developed, it is possible that some of the terms used may be simplified approximations because of the lack of a need for more accurate estimations in other applications. It was noted, for instance, that there are several different algorithms for calculating the Equation of Time.

Action: The workshop agreed to seek advice on the precision of the terms used.

3.2 Existing Geolocation Methods

A critical aspect in the geolocation calculations is the ability accurately to associate a time with a particular zenith angle. In the case above for a human observer, sunrise and sunset are useful because the zenith angle (⊥ 90° 50') can be identified precisely as the upper limb of the sun crosses the horizon. The observer could equally calculate geoposition at any other time (e.g., at civil or nautical twilight) if the zenith angle is known with equal accuracy.

However, for an electronic device that records ambient light levels the situation is less straightforward. The sky is light well before sunrise, gets brighter as the sun crosses the horizon, and continues to get brighter still after sunrise. Because they occur during periods of rising and falling light levels, sunrise and sunset are therefore not simple to identify or define. The problem then becomes "what levels of ambient light correspond with particular sun angles (e.g., sunrise and sunset)?" The workshop adopted the term "reference method" to describe this approach of relating a "reference" light level to a particular zenith angle. Other methods are, however, being developed (see 3.2.3 below).

3.2.1 The "fixed reference" method

The simplest "reference method" involves "calibrating" each device by gathering light data for 5-10 days prior to deployment at a precisely known location. The times of sunrise and sunset are calculated for the particular location, and the reference light levels that equate to sunrise and sunset are determined. These reference light levels are then used to identify the times of sunrise and sunset, and hence location, during the subsequent deployment (see Wilson et al., 1992).

3.2.2 The "variable reference" method

A variation of the reference method is to define the reference light level as some relative value (e.g., 70%) of the maximum (k_{max}) light intensity measured around noon each day (see Welch and Eveson, 1999).

3.2.3 The "reflection" method

This method involves reflecting the "dawn" light curve around noon onto the "dusk" light curve (see Musyl et al., this volume). Geoposition is then derived by comparing the dawn and dusk light curves with a series of curves generated for different latitudes.

> *The workshop agreed that estimates of longitude can be achieved simply and reliably by determining the geometric mid-point between arbitrary sunrise and sunset. However, a variety of methods are used for estimating the times of sunrise and sunset, and hence calculating latitude from day length. The workshop agreed that no particular technique had yet been shown unequivocally to be consistently more accurate than any other, and that methodologies were still developing. It was therefore not possible to identify a common method.*

> *The workshop agreed that, although the reflection method has not yet been critically evaluated, it had the advantages of providing an objective technique for rejecting light data that was too "noisy" to give a reliable estimate of geolocation. Also, the method should be inherently more reliable because it uses more of the available light data than either of the reference methods.*

Action: The workshop agreed to assemble a compendium of methods for estimating geolocation and a library of appropriate algorithms.

3.3 Improving Accuracy

While the geolocation methods above are, in principle, quite simple, there are various technical problems that confound the measurement of ambient light level, and hence the calculation of geolocation, which the workshop addressed.

3.3.1 Light sensors in air

3.3.1.1 Signal amplification

Most archival tags currently use a silicon semi-conductor photoreceptor that converts the impinging light energy into electrical energy. Because ambient light levels increase through approximately 9 orders of magnitude from midnight to midday, linear amplification would not give sufficient dynamic range to cover the full span of ambient light levels. Consequently logarithmic, rather than linear, amplification is most appropriate. However, it was noted that the circuits required to generate a logarithmic output are temperature-sensitive and that, as yet, most light circuits in archival tags were not fully temperature-compensated.

3.3.1.2 Wavelength response

As the sun rises the spectral composition of day light changes with shorter wavelengths (blue) being predominant before sunrise, and longer wavelengths (red) being predominant after sunrise. Because blue light is more energetic than red light, the voltage output from a broad band photoreceptor does not necessarily have a simple relationship to sun angle as dawn or dusk progresses. To reduce this effect, the spectral response of the photoreceptor should be restricted with a narrow band light filter. Because blue light predominates at the sun angles most appropriate for geolocation (just before sunrise and just after sunset–see later), and shorter wavelengths are attenuated less by water than longer wavelengths, most practitioners use a light-blue filter (e.g., Schott BG23, $\sim 450 \pm 50$ nm).

3.3.2 Problems associated with geolocating in water

Water attenuates light. Even in clear, open ocean, water there is usually insufficient light available for geolocating with currently available devices below depths of about 300 m. Above this depth, the light level measured will be lower than that measured at the same time at the sea surface. Therefore, to identify the times of the reference sun angle, it is necessary to correct the measured light for depth. This requires an on-board depth (pressure) sensor and an estimate of the attenuation coefficient for the particular water mass. In theory, attenuation of light with depth should differ at different times of the day because different sun angles result in different light paths through the water. However, in practice, attenuation curves at dawn and around noon have very similar slopes. Therefore, for a given water mass, it is possible to use a single attenuation coefficient to correct the light level at a given depth back to what it would be at the surface.

If the general location of the animal is known it is possible to make an assumption about the optical classification of the water mass (see Jerlov, 1976) and so to estimate a suitable attenuation coefficient. Alternatively, a more precise on-board estimate can be made. For an hour of so either side of noon (clouds and weather notwithstanding), the level of ambient daylight can be assumed to be relatively constant. If the animal makes suitable vertical movements through the water column during this time, light attenuation coefficients can be derived directly from the light and depth data.

3.3.3 Sources of stochastic error

Although some sources in variation of light level are systematic and can, to some extent, be corrected (e.g., water depth, above), there are others that are random and unpredictable and can lead to substantial problems in calculating geoposition.

3.3.3.1 Clouds

In air the most ubiquitous source of random error is weather and cloud cover. Experiments show that light intensity can vary by a factor of up to four solely as a result of clouds. However, water vapour attenuates red light more than it does blue light, therefore the accuracy of geolocation can be improved by placing a blue filter over the photoreceptor.

Experiments (Wilson, unpublished) show that variability in light intensity due to clouds is reduced when the sun is below the horizon, Thus, instead of using the times of sunrise and sunset (i.e. sun angles of 90° 50'), calculating geoposition is improved if the sun angles used are about 4° to 5° below the horizon (i.e. ~95°, or similar to the times of civil twilight).

Estimates of geolocation using photoreceptors with blue filters show no apparent correlation between the accuracy of fixes and cloud cover. However, positional fixes became considerably less reliable when sunlight coming from the sun below the horizon is reflected from clouds. In such circumstances, the rate of change in light intensity around dawn and dusk does not follow a smooth curve but, typically, is characterised by irregularities and by an inflection in the curve at a point where the light suddenly stops being reflected off the clouds. Despite the use of a blue filter, thick and extensive cloud cover influences the form and timing of light intensity around dawn and dusk. However, experiments (Wilson, unpublished) show that if a subjective 'quality assessment step' is made, in which the regularity of the rate of change of light intensity at each dawn and dusk is ranked between 1 (very regular) and 5 (very irregular), fixes derived from lower ranking dawns and dusks are generally more accurate than those derived from higher ranking ones.

Problems associated with clouds will also affect geolocation in water.

3.3.3.2 Waves

Wave action was considered very likely to affect the amount of light incident at the sea surface that reaches a tag on a fish that is below the sea surface. However, the workshop did not consider in detail how wave effects might be taken into account in calculating geoposition. Jerlov (1976) states that at low solar elevations, reflectance of direct sunlight is reduced by wave action. In contrast, the reflectance of skylight (i.e., the light being detected by a device when the sun is still below the horizon) is less affected by wave action than direct sunlight.

3.3.3.3 Animal behaviour

A number of behavioural factors can affect the light levels detected by the photoreceptor, these include photoreceptor attitude (i.e., the angle at which the sensor views the sky); animal orientation; parts of the animal covering the sensor (not usually a factor with fish, but sometimes significant with birds); effects of burrow-nesting (again not usually a factor with fish, but sometimes significant with birds); and diving. The latter can be very significant for fish, particularly open-ocean species that make significant vertical movements around the time of dawn and dusk.

3.3.3.4 Fouling

Although every care is taken during the manufacture of archival tags to ensure the best transmission of light to the photoreceptor, once the tag is deployed on a free-ranging animal there is potential for the casing to become fouled and this will effect its light gathering properties. Fouling does not appear to be a significant problem with seabirds, but significant fouling has been observed on tags that have been deployed on fish in shelf seas.

3.3.3.5 Geographic movement

Although in many instances it is unlikely that an animal will move sufficiently far in a day to affect the length of the day it experiences, geographic movement between dawn and dusk may be a significant factor for some highly mobile animals. One method of reducing any effect of movement is to calculate location from either day-length or night-length, whichever is the shorter.

The workshop agreed that, inevitably, there would be fundamental limits to the accuracy of the geolocation methodology, but that "limits to accuracy" are not the same as "error."

The workshop agreed that there was a need to improve understanding of the importance of sensor type, and/or the most appropriate wavelength(s) to measure. It was also agreed that improvements may accrue from making simultaneous measurements at two or more wavelengths.

The workshop agreed that the variability in light intensity due to cloud cover is reduced when the sun is below the horizon. Therefore, associated errors could be reduced by calculating geolocation from diffuse skylight at a time when the sun's zenith angle is somewhere between 5° and 4° below the horizon.

The workshop agreed that water vapour attenuates red light more than blue, and therefore that in air, errors associated with varying cloud cover can be reduced by incorporating a blue filter with the photoreceptor.

It was noted that light intensity at astronomical sunset is about three-fold greater than at astronomical sunrise, and that the degree of polarisation changes from about 75% when the sun is at the horizon to 0% when the sun is at its zenith. The workshop did not identify how these factors might affect the geolocation calculations.

The workshop concluded that, unlike the situation with depth, it is likely to be very difficult to develop methods for correcting stochastic errors. The workshop agreed that there was therefore merit in developing techniques that could discriminate between "good days" and "bad days."

This provides a method of estimating geolocation only when stochastic errors are minimised.

The workshop agreed that there would always be finite limits to the accuracy of daylight-based methods and that the use of other environmental data (e.g., sea surface temperature (Smith and Goodman, 1986; Metcalfe et al., *1994; Metcalfe and Arnold, 1997), depth, tidal data (Metcalfe and Arnold, 1997), angle of magnetic dip, etc. as appropriate) in "expert systems" would significantly enhance non-radio based geolocation methods. In this respect the workshop agreed that having a standard data format could greatly enhance progress in this area. It was suggested that there might be a need to discuss such issues with the engineers involved in tag design.*

3.3.4 Comparison of archival tags and geolocation methods

The workshop agreed that there was still a need to

- carry out definitive mooring tests to identify the effects of water type and depth on light at different wavelengths and that there was ample scope for devising experimental programs directed at improving methods;
- evaluate critically different aspects of geolocation methods, e.g., best zenith angle and geolocation algorithms.
- develop decision rules for rejecting suspect geoposition estimates.

While no specific joint research programme was identified at this stage, the workshop agreed that there was a need to continue to exchange data and ideas. It was agreed to form an open study group comprising the present participants and others as appropriate.

3.3.5 Summary of agreed actions

The workshop agreed that it would

- draft a report of the workshop;
- assemble a library of geolocation methods and appropriate algorithms and to seek advice on the precision of the terms used;
- seek to develop decision rules for rejecting unrealistic position estimates (outliers);
- continue to exchange data and ideas;

- hold a third meeting critically to evaluate different geolocation methods.

4. PARTICIPANTS

Julian Metcalfe (Chairman, both meetings)

Gareth Rees[1]
Geoff Arnold[1,2]
David Welch[1,2]
Grant West[1,2]
Jason Hartog[2]
Christofer Boggs[2]
Rory Wilson[1]
Håkan Westerberg[1,2]
Paige Eveson[1]
Richard Brill[2]
Melinda Braun[2]
Simon Wotherspoon[1]
John Gunn[1,2]
Molly Lutcavage[1]
Mike Musyl[2]
Roger Hill[2]

[1]Attended the Lowestoft meeting.
[2]Attended the Honolulu meeting.

Ainsley Buckley was raporteur for the meeting at Lowestoft, and Ewan Hunter was raporteur for the meeting in Hawai'i.

REFERENCES

Gunn, J., Polachek, T., Davis, T., Sherlock, M., and Betlehem, A. (1994). The development and use of archival tags for studying the migration, behaviour and physiology of southern bluefin tuna, with an assessment of the potential for transfer of the technology to groundfish research. ICES C.M. Mini 2.1, 22 pp

Delong, R. L., Stewart, B. S. and Hill, R. D. (1992). Documenting migrations of northern elephant seals using day length. *Mar. Mamm. Sci.* **8,** 155-159

Hill, R. D. (1994). Theory of geolocation by light levels. *In* Le Boeuf, B.J., and Laws, R.M. (eds.) *Elephant Seals: Population Ecology, Behavior, and Physiology*, University of California Press, Berkeley, California, pp. 227-236.

Jerlov, N.G. (1976) *Marine Optics*, 2nd ed. Elsevier, Amsterdam, Oxford, New York, p. 131.

Metcalfe, J.D., Arnold, G.P., and Holford, B.H., (1994) The migratory behaviour of plaice in the North Sea as revealed by data storage tags. ICES CM 1994/Mini, 11, 8 pp.

Metcalfe, J.D., and Arnold, G.P. (1997) Tracking fish with electronic tags. *Nature* **387**, 665-666.

Nautical Almanac Office (1991) *Almanac for Computers.* Washington, D.C., United States Naval Observatory.

Smith, P., and Goodman, D. (1986) Determining fish movements from an "archival" tag: precision of geographical positions made from a time series of swimming temperature and depth. NOAA Tech. Mem. NOAA-TM-NMFS-SWFC-60, 14 pp.

Welch, D.W., and Eveson, J.P. (1999) An assessment of light-based geoposition estimates from archival tags. *Can. J. Fish. Aquatic Sci.* **56**, 1317-1327.

Wilson, R. P., Ducamp, J.-J., Rees, W. G., Culik, B. M., and Niekamp, K. (1992) Estimation of location: Global coverage using light intensity. *In* Priede, I. G., and Swift, S. (eds.) *Wildlife Telemetry*, Ellis Horwood, New York, pp. 131-134.

Ability of Archival Tags to Provide Estimates of Geographical Position Based on Light Intensity

Michael K. Musyl[1*], Richard W. Brill[2], Daniel S. Curran[1], John S. Gunn[3], Jason R. Hartog[3], Roger D. Hill[4], David W. Welch[5], J. Paige Eveson[5,3], Christofer H. Boggs[2], and Russell E. Brainard[2]

[1]*Joint Institute for Marine and Atmospheric Research, University of Hawaii at Manoa, 2570 Dole St., Honolulu, HI 96822, U.S.A.*

[2]*National Marine Fisheries Service, Honolulu Lab/Southwest Fisheries Science Center, 2570 Dole St., Honolulu, HI 96822, U.S.A.*

[3] *CSIRO Marine Research, PO Box 1538, Hobart, Tasmania, 7000, AUSTRALIA*

[4]*Wildlife Computers, 16150 NE 85th St., Suite #226, Redmond, WA 98052, U.S.A.*

[5]*Fisheries and Oceans Canada, Pacific Biological Station, Nanaimo, British Columbia, V9R 5K6, CANADA*

**Corresponding author: Mike Musyl, Joint Institute for Marine and Atmospheric Research, University of Hawaii at Manoa, 2570 Dole St., Honolulu, HI 96822, U.S.A., Email: mmusyl@honlab.nmfs.hawaii.edu Phone: 808/5928305 FAX: 808/5928300*

Key words: archival tags, ambient light level, geoposition algorithm, light attenuation, positional errors, crepuscular, Equinox, pelagic fishes

Abstract: We tested the ability of archival tags and their associated algorithms to estimate geographical position based on ambient light intensity by attaching six tags (three tags each from Northwest Marine Technologies [NMT] and Wildlife Computers [WC]) at different depths to a stationary mooring line in the Pacific Ocean (approx. 166°42'W, 24°00'N), for approximately one year (29-Aug-98 to 16-Aug-99). Upon retrieval, one tag each from the two vendors had malfunctioned: from these no data (NMT) or only partial data (WC) could be downloaded. An algorithm onboard the NMT tag automatically calculated geographical positions. For the WC tags, three different algorithms were used to estimate geographical positions from the recorded light intensity data. Estimates of longitude from all tags were significantly less variable than those for latitude. The mean absolute error for longitude estimates from the NMT tags ranged from 0.29 to 0.35°, and for the WC tags from 0.13 to 0.25°. The mean absolute error in latitude estimates from the NMT tags ranged from 1.5 to 5.5°, and for the WC tags from 0.78 to 3.50°. Ambient weather conditions and water clarity will obviously introduce errors into any geoposition algo-

J.R. Sibert and J.L. Nielsen (eds.), Electronic Tagging and Tracking in Marine Fisheries, 343–367.
© 2001 *Kluwer Academic Publishers. Printed in the Netherlands.*

rithm based on light intensity. We show that by applying objective criteria to light level data, outliers can be removed and the variability of geographical position estimates reduced. We conclude that, although archival tags are suitable for questions of ocean basin-scale movements, they are not well suited for studies of daily fine scale movement patterns because of the likely magnitude of position estimate errors. For studies of fine scale movements in relation to specific oceanographic conditions, forage densities and distance scales of 100 km or less, other methods (e.g. acoustic tracking) remain the tool of choice.

1. INTRODUCTION

Archival tags are electronic data storage devices that are either surgically implanted or attached to the outside of a fish with an anchoring device (Klimley et al., 1994). Current generations of archival tags can store data from up to four sensors (e.g., internal temperature, external temperature, pressure (i.e., depth), and light intensity) taken at one minute intervals for almost one year. At a slower rate of data acquisition, data can be recorded for up to eight to ten years, depending on battery life and amount of memory. Archival tags have been used to chronicle horizontal and vertical movement patterns, residence time in particular areas, and location of possible spawning areas for a variety of marine fishes (Gunn et al., 1994; Metcalfe and Arnold 1997; Block et al., 1998a; 1998b; Lutcavage et al., 1999).

Although archival tags are valuable and welcome tools for fisheries research, an important consideration in evaluating their utility is the accuracy of the estimates of geographical positions. The use of light data (i.e., times of dawn and dusk) to estimate longitude (from local noon) and latitude (from day length) is not a new concept (Smith and Goodman, 1986; Hill, 1994; Bowditch 1995; Sobel, 1995). However, the application of this idea to estimating geographical positions of fishes carrying archival tags can be problematic because of systematic and random errors resulting from equinoxes, light attenuation with depth, water clarity, weather patterns, accuracy of astronomical algorithms, clock errors, resolution of light sensor, and behavior of animal. Welch and Eveson (1999) recently estimated the accuracy of geographical positions based on light data for archival tags placed on a moving vessel and on submerged mooring lines in the northwest Pacific. Based on their findings, they calculated that average positional errors for tagged animals would be about 140 km (i.e., standard deviations of 0.9° in longitude and 1.2° in latitude). Gunn et al. (1994) compared position estimates based on light readings from archival tags attached to southern bluefin tuna being towed in a sea cage in the open ocean to the true position of the cage at midday. As the maximum depth of the cages was less than 20 m and the experiment was conducted over a short period in January, the effect of deep diving and variations around the equinoxes were not examined in their experiments. However, their calculated positional error was 1.52° (standard deviation of 0.22°) for latitude and 0.54° (standard deviation of 0.09°) for longitude.

To define more accurately the ability of available hardware and software to estimate geographic position from light data, we submerged archival tags from two manufacturers at varying depths on a stationary mooring line in the subtropical North Pacific. We then tested the efficacy of four different algorithms to provide estimates of geographical location from the light level data. We thus define some of the errors inherent in the procedures, and provide a measure of positional error estimates for tags submerged at varying depths. Our intention is to aid researchers in interpreting data recorded from swimming animals.

2. METHODS AND MATERIALS

Three archival tags from Northwest Marine Technologies (NMT, Shaw Island, WA, USA, serial Nos. 587, 621, 935) and three from Wildlife Computers (WC, Redmond, WA, model MK 7, serial nos. 97-218, 97-219, 97-224) were attached at approximately 20-m intervals to a subsurface oceanographic mooring deployed on a 76-m bank in the subtropical North Pacific Ocean (approx. 166°42'W, 24°00'N). Tags were housed in PVC plastic pipe. The light sensor stalk protruded toward the surface through a hole drilled in the pipe. The tags were deployed August 29, 1998, and retrieved August 16, 1999.

For the NMT tags, data from each sensor (two temperatures, pressure, and light intensity) were recorded every 8.5 min and stored as eight-bit binary integers (i.e., with possible values between 0 and 255). The light sensor was most sensitive to wavelengths near 450 nm. Onboard correction algorithms adjusted recorded light data for temperature variation and light attenuation with depth. The corrected light intensity values were scaled logarithmically, with 15 points per decade intensity (e.g., high noon at sea level on a clear day is about 150 units on this scale). From the light curves, we estimated the effective range of 100 nW cm^{-2} to at least 10 mW cm^{-2} (bright sunshine) with a sensitivity of 15 readings/decade. Geolocations for the NMT tags were automatically estimated onboard the tags by a proprietary algorithm. This algorithm employed times of sunrise and sunset to find local noon and thus estimate longitude. The light level data used were taken every 128 s but recorded in the tag's memory only every 8.5 min. Latitude estimates were based on known temperatures at standard depths (Smith and Goodman, 1986). Poor time resolution of the stored light level data, especially during crepuscular transitions (i.e., dawn and dusk), prevented any post-processing of the recorded light level data.

The WC archival tags were set to record data from each of the four channels (depth, light, and two temperatures) every 60 s. Data could be collected at this rate for 353 days before filling the memory, the approximate duration of the experiment. Temperature and depth were measured with 12-bit accuracy, but compressed to 8 bits before storing. Temperature had a resolution of 0.05°C, depth had a resolution of 0.5 m, and the light sensor recorded values in the range from 0 to 255, scaled logarithmically with 18 points per decade. The effective range of the light sensor was from bright sunshine to a level 10 decades less bright (150 readings lower). This covered the range from 1 pW cm^{-2} to 20 mW cm^{-2} (bright sunshine is approximately 10 mW cm^{-2} at sea level). Interference filters were used to make the light sensor sensitive to light only in the 450 ± 50 nm wavelength range, and light level was internally corrected for temperature, but not depth. Light level data were downloaded from the WC tags, and geographical positions estimated using three different methods.

2.1 Method 1: Fixed Reference Light Level Method

This method was developed in 1993 by CSIRO Marine Research (Hobart, Tasmania, Australia) based on techniques described in Nielsen (1963) and Wilson et al. (1992). It uses a "reference light level" to define times of dawn or dusk. In other words, when light levels equaled the reference light level, this defined a specific "zenith angle" (i.e., the angle between the horizon and the center of the sun's disk). Dawn and dusk were initially determined by finding the periods when light levels were changing most rapidly. Next, an ordinary least squares routine was used to fit a second order polynomial equation to the light data between 100 and 155 light units. The time when light level was equal to the reference light level was determined from the fitted curve. The equation changes daily but remains a second order polynomial with the same independent (i.e., time) and dependent (i.e., light) variables. Standard astronomical formulae were then used to determine position from a specified zenith angle (Wilson et al., 1992, Hill, 1994).

The estimation software was tuned by determining the reference light level and corresponding zenith angles that resulted in estimates closest to the known location of the tags. This empirical optimization was conducted using data collected over ten days of the experiment in November 1998. This time period was chosen because it preceded any biofouling of the light sensor (Fig. 1) and was temporally distant from the autumnal equinox (Sept. 21).

Once established, the optimal reference light level was kept constant. A light level of 135 units corresponded to a zenith angle of 97°.

Method 1 requires light levels recorded by the individual tags be corrected back to the light level at the surface. This was not possible in this situation because light-depth attenuation coefficients have to be specific for each tag. The data required to generate these coefficients were not available because the tags were kept stationary at specific depths. As a substitute, light data were normalized, so that each tag read similar light values at the same time using the following relationships:

$$\hat{L}_{20} = L_{20} + 15 \qquad (1)$$

$$\hat{L}_{40} = L_{40} + 21 \qquad (2)$$

$$\hat{L}_{60} = L_{60} + 28 \qquad (3)$$

where

$\hat{L}_{20}$ = estimated surface light intensity by the tag at 20 m,

$\hat{L}_{40}$ = estimated surface light intensity by the tag at 40 m,

$\hat{L}_{60}$ = estimated surface light intensity by the tag at 60 m,

L_{20} = light level recorded by the tag at 20 m,

L_{40} = light level recorded by the tag at 40 m,

L_{60} = light level recorded by the tag at 60 m.

Because accuracies of geographical position estimates depend on times of sunrise and sunset, any drift in the tags' real-time clocks must be corrected. The real-time clocks in the tags at 20 and 40 m drifted 2 min 38 s and 2 min 18 s over the 12 month period (respectively), which are within the expected range. A linear drift of time error was assumed. However, the tag at 60 m drifted 2 h, 22 min, and 24 s. The tag's real-time clock clearly malfunctioned 146 days after deployment. We presumed that the major proportion of the clock error occurred after the 146th day and, for simplicity, we did not correct for drift in the real-time clock of this tag throughout the portion analyzed.

2.2 Method 2: Dawn and Dusk Symmetry Method

This method was developed by Roger Hill of Wildlife Computers and is explained in greater detail elsewhere in this volume. Longitude was determined by finding the time of midnight, such that the dawn and dusk curves were maximally symmetrical, allowing for small amounts of relative light level change from dawn to dusk caused by changing cloud cover. By matching the entire curves, this technique avoids errors caused by spurious readings at some standard reference light level. The time of midnight is translated to longitude by multiplying the time by 15, and factoring in the Equation of Time (Meeus, 1991). Latitude was calculated by using data from a day close to the start of the deployment (September 15, 1998) for which the light curves were relatively clean, and creating a translation of observed light level with known angle of sun as measured from the vertical (zenith angle). The zenith angle of the sun can be calculated at specific times of day for a given location from standard astronomical equations (Duffet-Smith, 1988; Meeus, 1991). This translation from light level to zenith angle can then be applied to the rest of the data set, so that for each day there is a plot of zenith angle against time. Latitude was determined by finding the latitude that best matches those zenith angles at the measured times. The advantage of this method is that it can provide estimates of latitude throughout the year (including the equinoxes), although in equatorial waters (as in this experiment), the error in the latitude estimate at the equinoxes becomes large. At higher latitudes, good estimates can be expected all year. Corrections of position estimates for drift in the real-time clocks were performed as in Method 1.

2.3 Method 3: Variable Reference Light Level Method

This method was developed by Welch and Eveson (1999). The method is similar to the fixed reference light level method described above, but reference light levels were independently estimated for each day by taking 70% of that day's average mid-day light level. For each day, the light versus time data taken from a two hour period around sunrise and sunset were fitted with a 4th order polynomial using least squares. Initial estimates of the times of dawn and dusk were taken to be the times at which the fitted curve equalled the reference light level. The estimated times of dawn and dusk were were then refined by repeating the above procedure but this time fitting a 4th order polynomial to data that extended just 15 min before and after the times of dawn and dusk determined in the first step. A zenith angle of 94° was assumed to correspond to the derived reference light level. This value was determined without making any reference to the known mooring location, but was taken from a previously published study in which it was found to give reasonably good estimates of geographical position (Welch and Eveson 1999). As in Method 1, based on inspection of light curves for 10 days in November, we later optimised the zenith angle to 94.7°. In this third method, no corrections were applied for light attenuation with depth, as it did not significantly improve location estimates. Corrections of position estimates for drift in the real-time clocks were performed as in Method 1.

2.4 Comparison of Geographical Estimates

In order to compare the two types of archival tags and the three different methods used to estimate positions from the WC tags, we grouped the data by depth and used only data that were common to tags at comparable depths. In other words, the NMT tag at 58 m was grouped with the WC at 60 m, and the NMT tag at 24 m was grouped with the WC tag at 20 m. As a result, fewer data points were used from the NMT tag at 58 m because comparable data from the WC tag at 60 m were unavailable owing to tag failure 225 days after deployment. In Table 1, the only data points rejected were those approximately 10 days before and after the autumnal and spring equinoxes. In Table 2, data points around the autumnal and spring equinoxes were rejected, as well as outliers, based on days with significant asymmetry of light level readings (e.g., days with highly variable cloud cover). The latter criteria for rejection and its application are described in detail elsewhere in this volume. To determine if this technique actually improved estimates, we tested for significant differences between mean absolute errors (i.e. |true position-estimate|) for latitude for all data in Table 1 to equivalent data in Table 2 using Student's *t*-test, where $P<0.05$ was considered significant. We also calculated the raw errors (i.e., true position-estimate) to gain insight into the mean positional bias of the overall estimates. Mean raw errors were also tested against a hypothetical mean of zero using a two-tailed test at the $P<0.05$ level (Sokal and Rohlf 1981). Next, to test whether the distribution of raw errors conformed to a normal distribution, we used the Kolmogorov-Smirnov D test for goodness of Fit (Sokal and Rohlf 1981) at the $P<0.05$ level. Lastly, a nearest neighbor clustering algorithm with Euclidean distances of raw positional errors was used to examine the tag/method combinations for patterns (Pielou 1984).

3. RESULTS

Upon retrieval, we found that one NMT tag had failed completely and no data could be downloaded. The reason for failure is unknown. Also, upon retrieval, ship personnel reported that all tags were covered with an algal film and some fouling organisms. The diminution of light level over time is evident in Figure 1 and suggests that biofouling was indeed accumulating on the light stalks. At the end of 12 months submergence, the maximum daily light levels in units defined and recorded by the NMT tag at 24 m had decreased from 139 to 120. The maximum daily light levels in units defined and recorded by the NMT tag at 58 m had decreased from 125 to 102. Equivalent reductions in maximum daily light levels in units defined and recorded by WC tags were 244 to 233, 239to 206, and 236 to 213 for the tags at 20, 40, and 60 m, respectively.

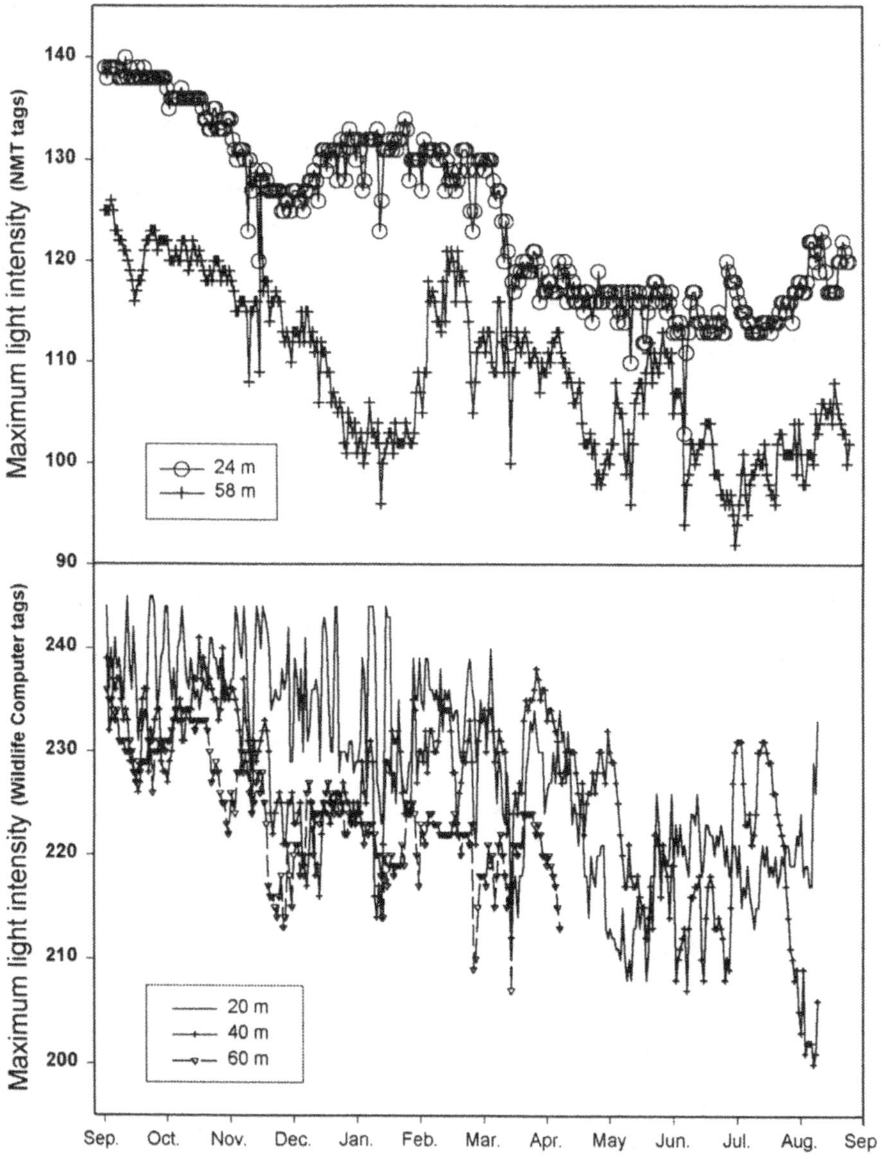

Figure 1. Diminution of maximum light intensity over time for the Northwest Marine Technologies and Wildlife Computers tags.

3.1 NMT Tags

Geographical position estimates from the NMT tags over the entire experimental period are shown in Figure 2. The tags' proprietary algorithm was not able to derive latitude estimates for approximately 18 days around the spring and autumnal equinoxes (March 21 and September 21, respectively). There was more variability in latitude estimates from the tag at 58 m than the tag at 24 m. The former never fully "recovered" after the spring equinox of 1999. In the temperature records after the spring equinox (not shown), a large discrepancy occurred between readings taken by the thermistor located within the tag body itself and the external thermistor located at the very tip of the fiber optic stalk. The external thermistor was found to be missing when the tag was recovered. It is highly likely that the fiber optic light stalk was also damaged and that the light sensor data were compromised (P. Ekstrom, pers. com.). Damage to the fiber optic light stalk may therefore explain the higher variability in latitude estimates after March on the deeper tag. Another tenable explanation for the variability of geolocations provided by the NMT tag at 58 m is the sensitivity of the light sensor, which covers about a 5-decade range, as opposed to the WC tag which covers about a 10-decade range.

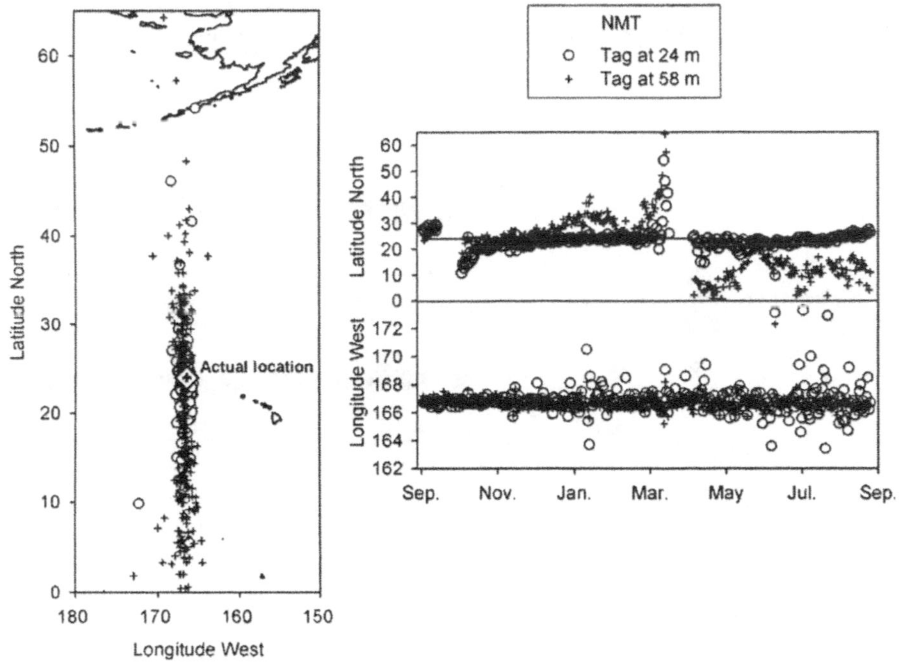

Figure 2. Estimated positions from the Northwest Marine Technologies tags.

The mean absolute error for latitude was approximately five times greater than that for longitude for the tag at 24 m (Table 1). For the tag at 58 m this ratio was approximately ten to one. The tag at 24 m had a significantly smaller (Student's *t*-test, $P < 0.05$) mean absolute error for latitude than the tag at 58 m. Mean absolute errors for longitude for the two NMT tags were similar.

3.2 WC Tags

Figure 3 shows longitude and latitude estimates for the three WC tags calculated using the fixed reference light level method (Method 1) over the entire experimental period. Figure 4 shows equivalent data derived from the dawn and dusk symmetry method (Method 2), and Figure 5 shows equivalent data from the variable reference light level method (Method 3). As with positional algorithms within NMT tags, all three algorithms applied to the data from the WC tags produced larger positional errors for latitude around the equinoxes than during the remainder of the year. Method 2 exhibited smaller equinox-induced latitude errors than the other two algorithms.

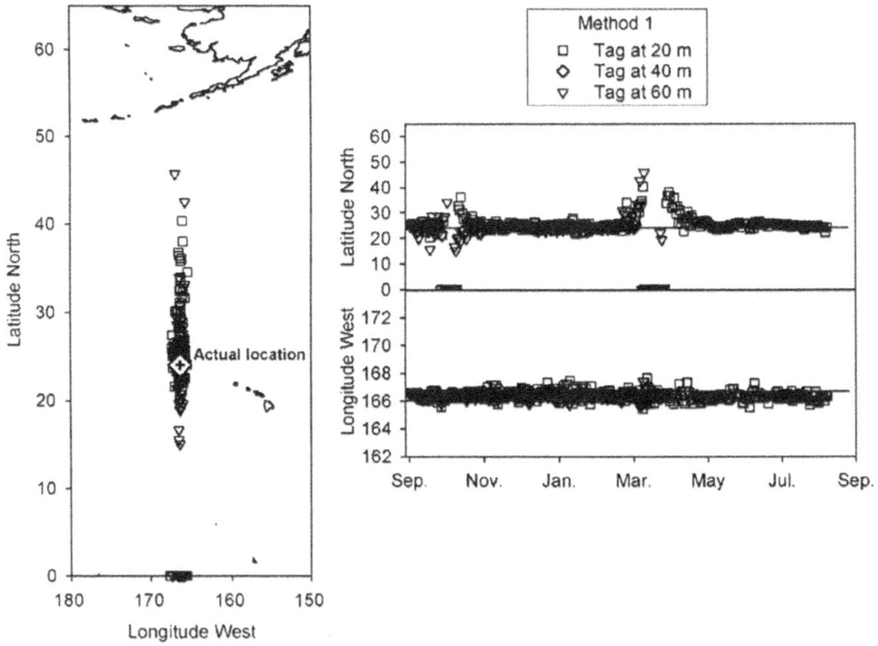

Figure 3. Estimated positions based on data from the Wildlife Computers tags calculated by Method 1.

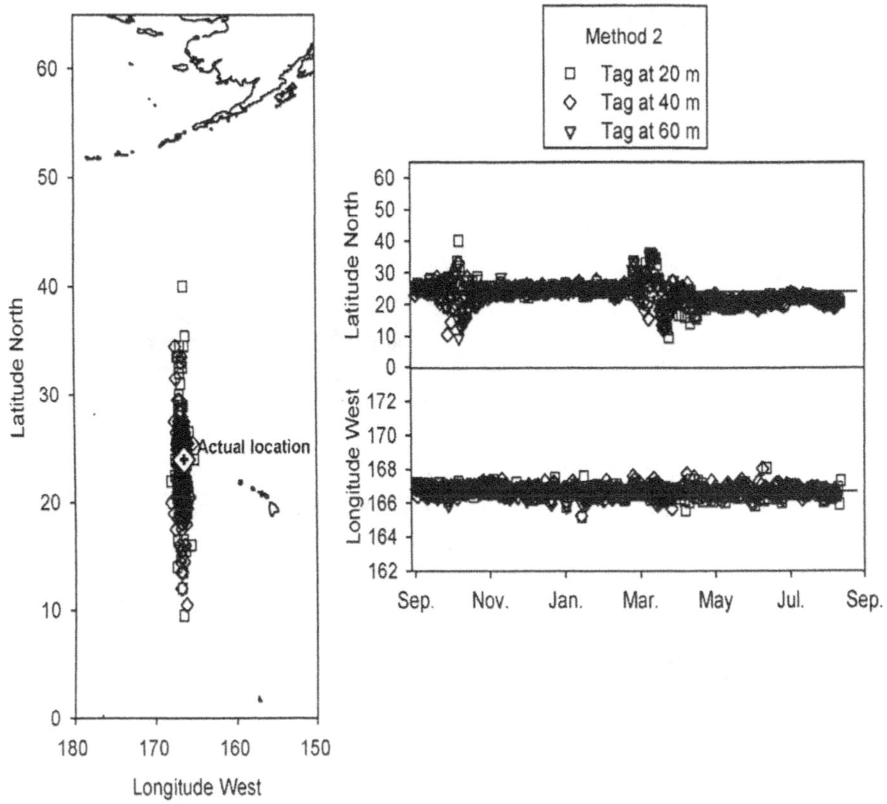

Figure 4. Estimated positions based on data from the Wildlife Computers tags calculated by Method 2.

As with the NMT tags, mean absolute errors for latitude were approximately an order of magnitude larger than those for longitude (Table 1). When data were averaged across the three methods for the WC tags, there were significant differences in errors for latitude by depth (one-way ANOVA, $P << 0.001$). However, there was no trend by depth. For example, the tag at the intermediate depth had the largest mean absolute error (40 m; 2.66±3.20 [mean±SD]), whereas the deepest (60 m; 1.68±3.58) and the shallowest tag (20 m; 1.68±1.93) were virtually identical (*t*-test; $P>0.05$).

Table 2 provides a summary of the mean absolute errors with outliers rejected. By objectively rejecting data based on asymmetry between the dawn and dusk light level curves, the mean absolute errors for latitude were significantly reduced in six of the eleven possible comparisons of equivalent data in Tables 1 and 2. Significant reductions were not observed for the NMT tags at 24 and 58 m; WC tags at 20 and 40 m were calculated using Method 1, and the WC tag at 20 m was calculated using Method 3. Mean absolute errors for latitude (e.g., Methods 1 and 3) can be reduced to less than 1°, although in most instances mean absolute errors for latitude are reduced only to approximately 1.5 to 3°.

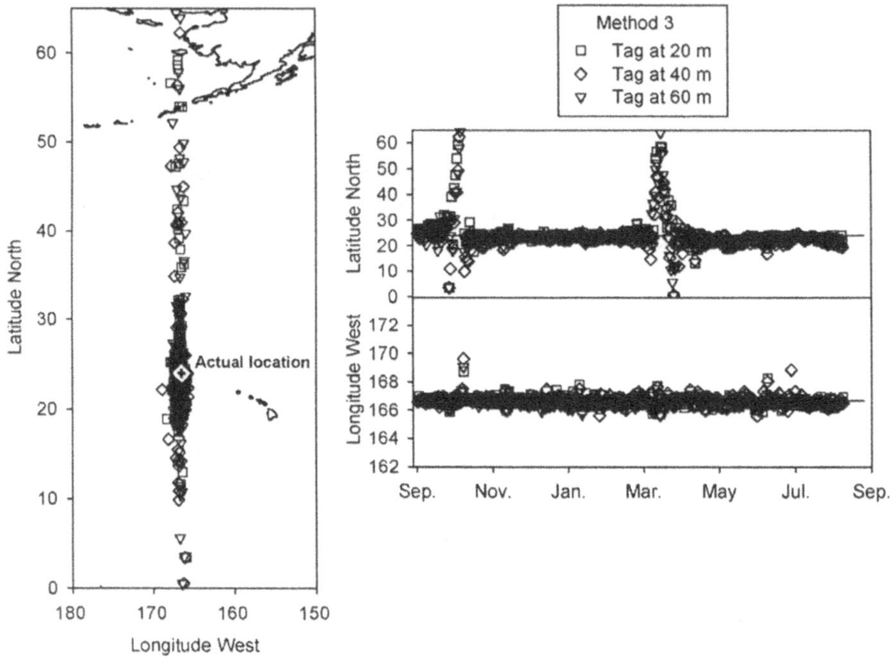

Figure 5. Estimated positions based on data from the Wildlife Computers tags calculated by Method 3.

The mean positional biases of the raw errors for each tag/method combination are presented in Tables 1 and 2. In Appendix I, we provide a detailed statistical summary of this data. In all tag/method combinations (Tables 1 and 2), mean positional raw errors were not significantly different from a hypothesized mean of zero (two-tailed test, $t<t_{0.05\ (2),\ df=n-1}$). However, in only 6 of 22 cases did the raw longitude errors conform to a normal distribution (Kolmogorov-Smirnov D test for goodness of fit, $P>0.05$). None of the raw latitude errors were normally distributed. Except in one instance (i.e., Method 3, tag at 60 m), the outlier rejection technique used herein did not alter the mean positional bias of the tag/method combinations (Table 2). It did, however, quite dramatically impact the distribution of raw errors in 6 out of 17 cases including three latitude estimates (NMT tag at 58 m; Method 1, tag at 60 m; Method 3, tag at 20 m) and three longitude estimates (Method 1, tag at 20 m and two of the tags using Method 3). Longitude errors using Method 2 were normally distributed before and after outlier rejection, as was the case for two of the tags using Method 1 and one tag for Method 3 (Table 2).

Table 1. Summary of the positional errors for the tags/methods using all comparable days. Absolute errors (|true position-estimate|) and raw errors (|true position-estimate|) are given. Distribution of raw errors were tested for conformity to the normal distribution using the Kolmogorov-Smirnov D test for Goodness of Fit (Sokal and Rohlf, 1981) at the $P<0.05$ level. Unless indicated as being non-significant (***bold-italicised***), the remaining tests were significant at $P<0.05$ (16 of 22 cases).

			NMT		Method 1		Method 2		Method 3	
Depth of tag (m)	Common lat. days (no.)	Common lon. days (no.)	Mean ± SD lat. errors (°)	Mean ± SD lon. errors (°)	Mean ± SD lat. errors (°)	Mean ± SD lon. errors (°)	Mean ± SD lat. errors (°)	Mean ± SD lon. errors (°)	Mean ± SD lat. errors (°)	Mean ± SD lon. errors (°)
					Absolute					
60	163	202			1.49 ± 2.72	0.15 ± 0.12	1.83 ± 2.35	0.16 ± 0.16	1.70 ± 5.06	0.25 ± 0.25
58	163	202	5.55 ± 6.40	0.35 ± 0.48						
40	317	333			3.50 ± 4.17	0.19 ± 0.18	2.15 ± 1.94	0.23 ± 0.21	2.33 ± 2.93	0.25 ± 0.28
24	286	337	1.74 ± 2.00	0.32 ± 0.38						
20	286	337			1.53 ± 2.16	0.20 ± 0.16	2.10 ± 1.99	0.23 ± 0.21	1.41 ± 1.51	0.25 ± 0.23
					Raw					
60	163	202			-0.40 ± 3.08	***-0.03 ± 0.24***	-1.20 ± 2.73	***-0.01 ± 0.23***	-0.76 ± 5.29	0.14 ± 0.32
58	163	202	-4.69 ± 7.06	-0.14 ± 0.58						
40	317	333			-0.23 ± 5.44	***-0.04 ± 0.34***	1.19 ± 2.64	***-0.03 ± 0.31***	1.42 ± 3.46	0.01 ± 0.37
24	286	337	0.95 ± 2.47	0.04 ± 0.50						
20	286	337			-1.30 ± 2.31	0.07 ± 0.30	0.89 ± 2.75	***0.0 ± 0.31***	0.81 ± 1.90	***0.0 ± 0.33***

Table 2. Summary of positional errors for the tags/methods using outlier rejection criteria. Absolute errors (|true position-estimate|) and raw errors (true position-estimate) are given. Distribution of raw errors were tested for conformity to the normal distribution using the Kolmogorov-Smirnov D test for Goodness of Fit (Sokal and Rohlf, 1981) at the $P<0.05$ level. Unless indicated as being non-significant (***bold-italicized***), the remaining tests were significant at $P<0.05$ (10 of 22 cases).

			NMT		Method 1		Method 2		Method 3	
Depth of Tag (m)	Common lat. days (no.)	Common lon. days (no.)	Mean ± SD lat. errors (°)	Mean ± SD lon. errors (°)	Mean ± SD lat. errors (°)	Mean ± SD lon. errors (°)	Mean ± SD lat. errors (°)	Mean ± SD lon. errors (°)	Mean ± SD lat. errors (°)	Mean ± SD lon. errors (°)
Absolute										
60	139	150			0.91 ± 0.91	0.13 ± 0.09	1.27 ± 1.25	0.15 ± 0.13	0.78 ± 0.74	0.23 ± 0.17
58	139	150	4.42 ± 3.57	0.29 ± 0.39						
40	263	263			3.08 ± 3.74	0.19 ± 0.18	1.72 ± 1.28	0.21 ± 0.18	1.84 ± 2.25	0.21 ± 0.19
24	216	241	1.50 ± 1.30	0.29 ± 0.21						
20	216	241			1.52 ± 2.21	0.19 ± 0.15	1.74 ± 1.82	0.20 ± 0.18	1.16 ± 1.26	0.23 ± 0.19
Raw										
60	139	150			***-0.10 ± 1.29***	***-0.06 ± 0.21***	-1.09 ± 1.41	***-0.01 ± 0.20***	***0.04 ± 1.07***	***0.17 ± 0.23***
58	139	150	***-3.91 ± 4.13***	-0.08 ± 0.48						
40	263	263			-0.63 ± 4.81	***-0.06 ± 0.34***	0.87 ± 1.96	***-0.03 ± 0.27***	1.40 ± 2.54	***0.05 ± 0.28***
24	216	241	0.64 ± 1.88	0.08 ± 0.34						
20	216	241			-1.32 ± 2.33	***0.05 ± 0.29***	0.42 ± 2.48	***0.04 ± 0.27***	0.53 ± 1.63	***0.11 ± 0.27***

The outlier rejection technique also significantly altered the correlations between raw latitude and longitude errors in 5 of 11 cases for each method (Table 3) except for Method 2. Before and after outlier rejection, there were no obvious patterns in the correlations between latitude and longitude errors for tags by depth or method. We examined the distribution of position estimates with box-and-whiskers plots using the culled dataset (Figure 6). For both latitude and longitude estimates, is apparent that tab/method combinations exhibited mixtures of potential bias and varying degrees of skewness, particularly for the latitude estimates (see also Appendix I). The positional biases of the longitude estimates were similar (Fig. 6).

Table 3. Correlation coefficients between raw latitude and longitude errors for each of the tags/methods where *=P<0.05, **=P<0.01, and ***P<<0.001. "All days" indicates all available days were used as in Table 1. Outliers were removed as in Table 2.

NMT	Tag depth	24 m		58 m
	All days	-0.1756**		-0.1180*
	outliers removed	0.0133		-0.1762**
Method 1	Tag depth	20 m	40 m	60 m
	All days	-0.1149	0.0699	-0.0549
	outliers removed	-0.1678*	0.0200	0.1958*
Method 2	Tag depth	20 m	40 m	60 m
	All days	0.1019	-0.0003	0.0384
	outliers removed	0.0834	-0.0487	0.0706
Method 3	Tag depth	20 m	40 m	60 m
	All days	0.0172	0.0598	-0.3260***
	outliers removed	0.0203	0.0981	0.0726

4. DISCUSSION

We assume biofouling accounted for the reduction of maximum light levels recorded over the entire length of the experiment (Fig. 1), although the changes did not appear correlated with depth or season. Compared to the errors induced by the spring and autumnal equinoxes (Fig. 2-5), biofouling did not significantly impede the ability to derive accurate geographical positions. For example, Method 3 uses 70% of the ambient light to estimate positions and thus would not be affected by biofouling. Moreover, we have not found biofouling to be a significant problem in archival tags attached to bigeye tuna (*Thunnus obesus*), even when fish were at liberty for 272 days (M. Musyl and R. Brill, unpublished observations).

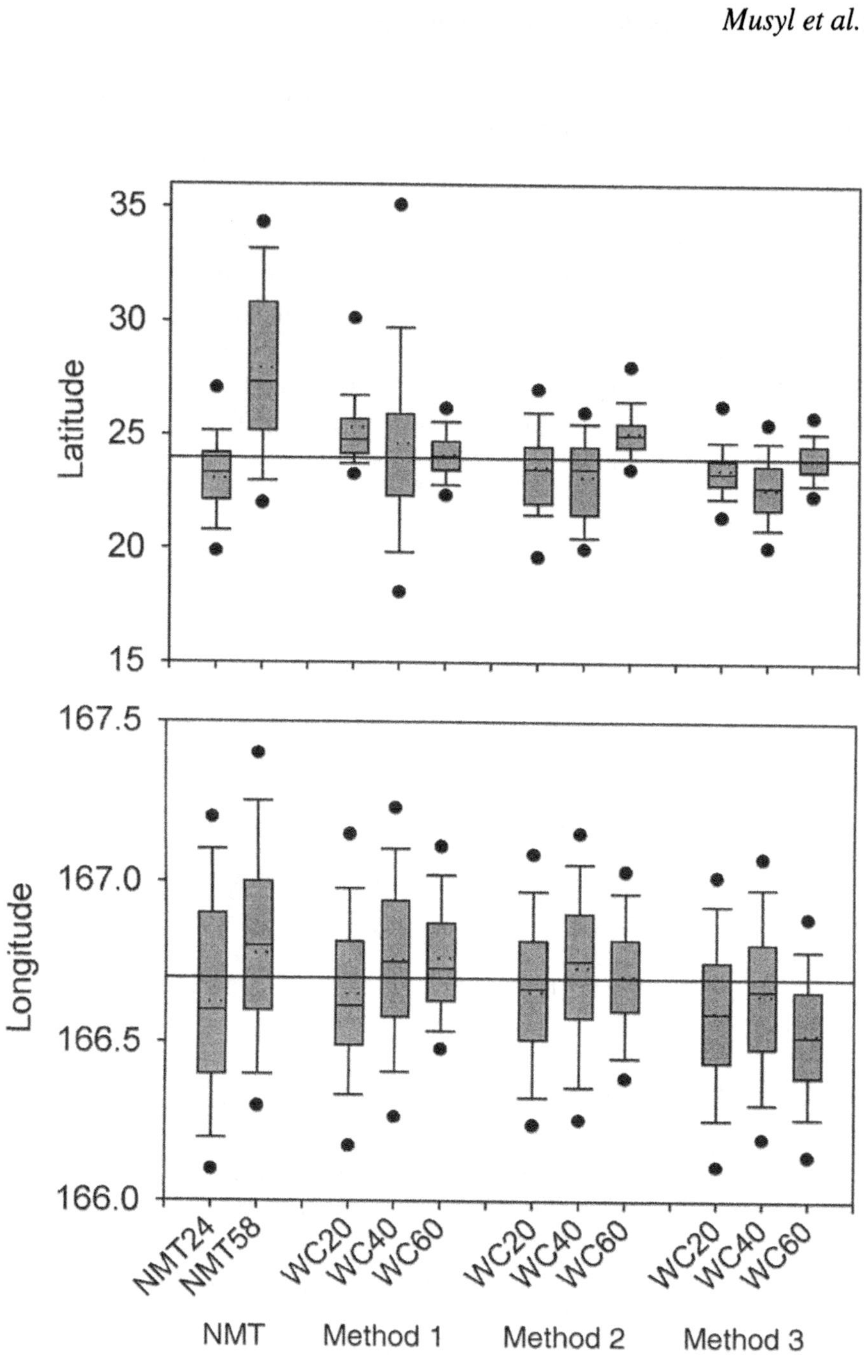

Figure 6. Box and Whiskers plots for the estimated latitude and longitude positions for each tag/method using outlier rejection (as in Table 2). The location of the mooring line is indicated by the horizontal bar across the plots. In the boxes, the horizontal line is the median and the dashed line is the mean. The extremities of the box are the 25th and 75th percentiles and the capped "whiskers" either side represent the 10th and 90th percentiles. The dots represent the 5th and 95th percentiles.

For all tags and all methods, longitude is always more accurately estimated than latitude. This occurs for several reasons. The estimates of longitude require only an accurate measure of the time of local noon (i.e., the midpoint between dawn and dusk), so any random errors in estimating times of dawn and dusk (e.g., from variable zeniths) will tend to cancel each other. That is, if the rate of data acquisition is set for 60 s (as for the WC tags), taking the average of dawn and dusk will improve the resolution to 30s assuming the errors for the dawn and dusk estimates are uncorrelated. However, the more likely scenario is that because of varying cloud cover, the light sensors may respond in a similar manner such that the dawn and dusk errors are serially correlated and N is not 2 independent degrees of freedom, but some smaller number. For example: SE=SD/sqrt (N), therefore SE=0.71*SD; so the SE will lie somewhere between 0.71 and 1, but we are not sure where. Also, an error of 1 minute in estimating the time of local noon results in a longitude error of only 0.25°, regardless of latitude or day of year. In contrast, determination of latitude requires a very precise measurement of day length. For example, at 24° N on February 1, an error of 1 minute in day length results in a latitude estimate error of approximately 1°. The situation becomes worse towards the equator and around spring and autumnal equinoxes (Hill, 1994; Welch and Eveson 1999).

The equinox problem occurs because day length is almost the same at all latitudes. As a result, even very small errors in estimating times of dawn and dusk translate into large errors in estimated latitude. This situation is clearly reflected in our data (Figs. 2-5). Longitude estimates remain stable over the 12 month period, whereas latitude estimates vary dramatically about 12 days approximately before and after the equinoxes. Indeed, NMT tags' onboard algorithm could not estimate latitude during these times (Fig. 2).

4.1 NMT Tags

In the generation of NMT tags used for this study, no post-processing of the light data is possible. This represents a clear tradeoff. The data returned by the tags already contain geographical position estimates and the storage methodology makes better use of the tag's limited memory. Our results suggest a possible effect of depth on the stability (Fig. 2) and magnitude of mean absolute errors of latitude (Table 1). The NMT tag at 24 m clearly performed better than the tag at 58 m. However, because of the failure of the third NMT tag, we cannot be definitive about systematic errors occurring with depth and our results may simply reflect inter-tag variability.

With moving fish, data will have to be rejected based on some *a priori* criteria such as a reasonable distance the animal could have moved over a given time period. For example, median speed over ground estimated for acoustically tagged tunas (~55 to 165 cm FL, Brill et al., 1999) ranged from about 80 to 120 cm s^{-1} (i.e. 1.55 to 2.33 nmi h^{-1}). At these speeds, fish traveling in a straight line for 24 h could cover between 37 and 55 nmi. In the best of circumstances, several days' of travel in a straight line would be required to cover distances measurable by the current generation of archival tags. In other words, the degree of resolution for current generation archival tags and resultant geopositions cannot distinguish whether a tuna is associated or away from a FAD. But for many fisheries management related questions weekly geopositions are likely to be adequate (e.g., Hilborn and Walters, 1992; Bakun, 1996). Given reasonable fish speeds, weekly positions can easily differ by several degrees of latitude and longitude. Based on these assumptions, we argue that a resolution of 0.5° longitude and 2° latitude is acceptable in many instances. For example, using these criteria, mean absolute longitude estimates for all tag/method combinations are within this range or below (Table 2). These spatial scales also have about the same resolution of frontal weather systems and oceanographic eddies (Steele, 1995). Assuming tags are to be used to resolve questions of ocean basin scale movements, latitude estimates are acceptable for the NMT tag at 24 m and for WC tag/method combinations at 20 and 60 m, in that mean absolute errors were below 2° (120 nmi or 222 km) (Table 2).

4.2 WC Tags

The WC tags record light levels and thus allow the user to choose an algorithm for estimating geographical positions after retrieval. Post-recovery analysis allows further refinement of position estimates as better algorithms and software are developed. However, this philosophy means that algorithms need to be validated through studies such as ours. To this end, our light level data are available from the Pelagic Fisheries Research Program web site (http://www.soest.hawaii.edu/PFRP/elec.tagdata/tagdata.html) for investigators wishing to test new geographical positioning algorithms.

Although there were significant effects of depth on mean absolute errors of latitude provided by the WC tags (Table 1), there was no trend. From our data it is not possible to determine whether this result is due to depth, to intrinsic properties of the tag (e.g., light sensor function, voltage stability, etc.), or to some external factor such as different orientation of the light sensor pods, differential shading, etc. Clearly, the possible effects of depth on the accuracy of geographical position estimates need further investigation.

Like the algorithm within the NMT tags, all three algorithms used to process the light level data from the WC tags were significantly poorer at estimating latitude around the equinoxes (Figs. 3-5). If fish movements are known to be constrained to a defined geographic coastal area, then it may be possible to anchor tags at fixed locations in order to refine zenith angles for that specific study site. This would, in turn, reduce mean absolute errors for latitude for all of the methods presented herein. For example, when we tuned the zenith angle in Method 3 from 94° to 94.7°, mean absolute errors for latitude were more than halved (Appendix II). However, the reader is cautioned that the accuracy of the results presented in this report may be more dependent on the specific zenith angle chosen than the true merits of the algorithms. For the purpose of tracking migration of oceanic pelagic fishes any of the three methods used with the WC tags would provide a satisfactory track of movements, particularly if used in combination with water temperature at-depth measurements to improve latitude estimates.

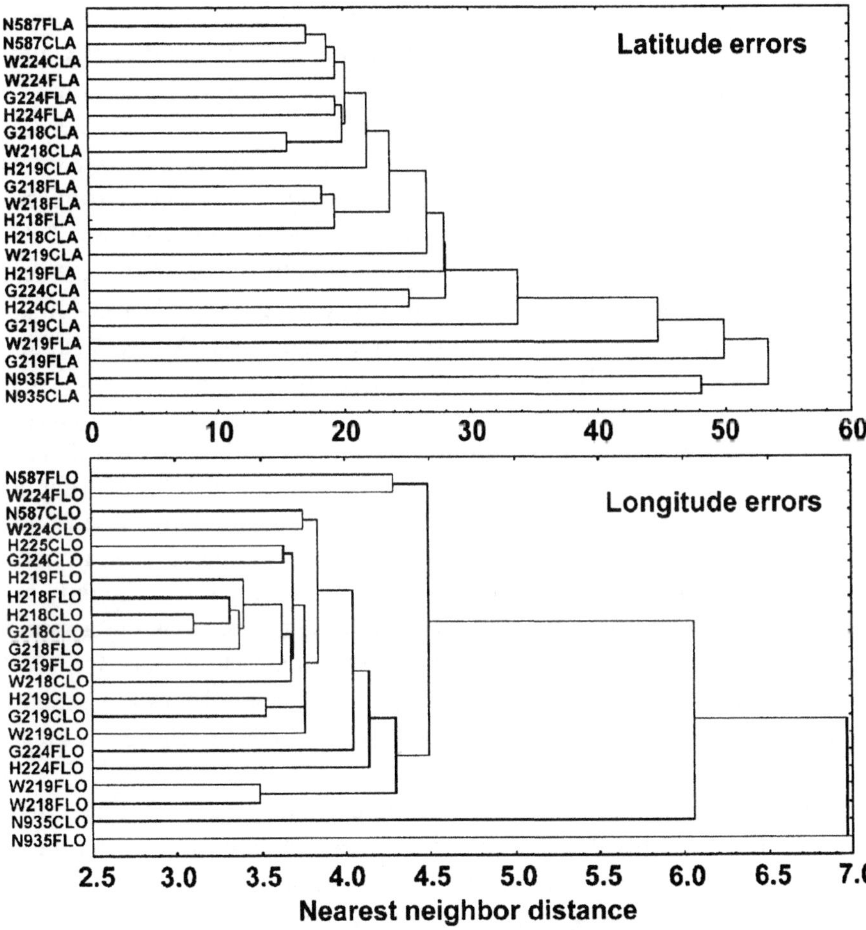

Figure 7. Nearest neighbor clustering using euclidean distance for the raw positional errors. Abbreviations for the tag/method combinations follow those described in Appendix I.

In this report, we examined 44 possible tag/method combinations and the positional data derived from them (Tables 1 and 2; Appendix I). To explore for patterns in the raw positional errors, we subjected the data to nearest neighbor clustering using Euclidean distance (Fig. 7). In general, methods using the WC tags tended to form the tightest clusters (i.e., branches with smaller distances) for both raw latitude and longitude errors (Fig.7). Yet, it was apparent that there were some striking inter-tag differences among the different tag/method combinations deployed at different depths as indicated in Tables 1 and 2. Still, there was no obvious systematic pattern in the errors.

Whilst it is beyond the scope of this paper to investigate in greater detail the nature of the systematic errors reported herein for each tag/method combination, a few general conclusions can be drawn. Firstly, the outlier rejection techniques used herein significantly reduced mean absolute errors in approximately half the cases. Outlier rejection did not significantly alter the positional bias. However, outlier rejection techniques can dramatically change the distribution of the errors (6 out of 17 cases, Table 2). Outlier rejection may be advantageous in certain instances in that it normalizes the distribution of some of the errors. That advantage may be compromised owing to significant correlation between latitude and longitude errors (Table 3). However, if weather and/or ambient oceanographic conditions were responsible for correlation, both latitude and longitude positional errors might be expected to degrade in unison. Correlation among errors might suggest an external cause for which adjustments could be made. We suggest investigators examine different outlier rejection techniques to find which ones complement their algorithms and minimize the amount of bias in their data.

In this experiment we were unable to examine the effect of depth and diving patterns of fish on the accuracy of position estimates based on light. We know from archival tag data collected by a number of researchers that diving, particularly around the times of dusk and dawn, significantly affects the quality of light data collected by archival tags, and thus the accuracy of position estimates based on light (J. Gunn, unpublished data for southern bluefin tuna, *Thunnus macoyii*). Our data represent a best-case scenario in estimating position; we stress that much work remains to those working with "real fish data."

We conclude that for questions of fine scale daily movement patterns, currently available archival tags, light data processing algorithms, and the resultant geographical position estimates are not suitable. Resolutions of latitude to only 2° are likely to be greater than the distance fish will travel over a 24 h period. As a result, horizontal movement patterns will be undetectable. Acoustic telemetry remains a better technology for resolving questions of fine-scale movements in relation to specific oceanographic conditions and forage densities (Brill et al., 1999; Dagorn et al., 2000). Archival tags, however, are clearly suitable for answering questions of ocean basin scale migration patterns, residency in specific fishing areas, spawning areas, fidelity to natal spawning areas, and possible stock structure. Archival tags are also excellent for recording vertical movement patterns in relation to oceanographic conditions. From these data, conclusions and models of physiological abilities and tolerances can be derived (Brill and Lutcavage, 2001). As both hardware and software improve, we suspect that the use of archival tag technology will become an increasingly important tool in fisheries research.

ACKNOWLEDGEMENTS

This project was funded by Cooperative Agreements NA37RJ0199 and NA67RJ0154 between the National Oceanic and Atmospheric Administration and the Joint Institute for Marine and Atmospheric Research, University of Hawaii (UH). We thank Drs. Phil Ekstrom of Northwest Marine Technologies and Suzanne Hill of Wildlife Computers for providing comments on an earlier draft. Two anonymous referees provided comments on an earlier draft and Dr. J.R. Sibert, UH, provided comments and suggestions that significantly improved the manuscript. The mooring was deployed from the NOAA ship *Ka'imi moana* courtesy of Pacific Marine Environmental Laboratory (PMEL) and recovered using the NOAA ship *Townsend Cromwell* courtesy of NMFS Honolulu Laboratory. We thank officers and crews of both vessels for their outstanding support. We thank Dr. Pierre Flament of UH and Bill Parker, Hugh Milburn, and Rick Miller of NOAA-PMEL for kindly providing mooring equipment, instrumentation, and design assistance. The authors or their agencies do not necessarily approve, recommend, or endorse any proprietary hardware or software mentioned in this publication. The views expressed herein are those of the authors and do not necessarily reflect the views of their agencies.

REFERENCES

Bakun, A. (1996) *Patterns in the Ocean: Ocean Processes and Marine Population Dynamics.* California Sea Grant College System, NOAA, 323 pp.

Block, B.A., Dewar, H., Farwell, C., and Prince, E.D. (1998a) A new satellite technology for tracking the movements of the Atlantic bluefin tuna. *Proc. Natl. Acad. Sci. USA* **95**, 9384-9389.

Block, B.A., Dewar, H., Williams, T., Prince, E.D., Farwell, C., and Fudge, D. (1998b) Archival tagging of Atlantic bluefin tuna (*Thunnus thynnus*). *Mar. Tech. Soc. J.* **32**, 37-46.

Bowditch, N. (1995) *The American Practical Navigator: An Epitome of Navigation* Originally by Nathaniel Bowdich (1802), Pub. No. 9. Defense Mapping Agency Hydrographic/Topographic Center, Bethesda, MD, 873 pp. [also available online http://pollux.nss.nima.mil/pubs/].

Brill, R.W., Block, B.A., Boggs, C.H., Bigelow, K.A., Freund, E.V., and Marcinek, D.J. (1999) Horizontal movements and depth distribution of large adult yellowfin tuna (Thunnus albacares) near the Hawaiian Islands, recorded using ultrasonic telemetry: Implications for the physiological ecology of pelagic fishes. *Mar. Biol.* **133**, 395-408.

Brill, R., and Lutcavage, M. (2001) Understanding environmental influences on movements and depth distributions of tunas and billfishes can significantly improve population assessments. *In* Sedberry, G. ed. *Islands in the Stream: Oceanography and Fisheries of the Charleston Bump*. American Fisheries Society Symposium 25, Bethesda, MD (in press).

Dagorn, L., Bach, P., and Josse, E. (2000) Movement patterns of large bigeye tuna (Thunnus obesus) in the open ocean determined using ultrasonic telemetry. *Mar. Biol.* **136**, 361-371.

Duffet-Smith, P. (1988) *Practical Astronomy with your Calculator*, 3rd edition. Cambridge University Press. 129 pp.

Gunn, J., Polachek, T., Davis, T., Sherlock, M., and Betlehem, A. (1994) The development and use of archival tags for studying the migration, behaviour and physiology of bluefin tuna, with an assessment of the potential for transfer of the technology for groundfish research. *ICES C.M. Mini*, 21.

Hilborn, R. and Walters, C.J. (1992) *Quantitative Fisheries Stock Assessment: Choice, Dynamics & Uncertainty.* Chapman and Hall, NY, 570 pp.

Hill, R. (1994) Theory of geolocation by light levels. In: LeBoeuf, B.J. and Laws, R.M. eds. *Elephant Seals: Population Ecology, Behaviour, and Physiology.* University of California Press, Berkeley, CA, pp. 227-236.

Klimley, A.P., Prince, E.D., Brill, R.W., and Holland, K. (1994). Archival Tags 1994: Present and Future. NOAA Technical Memorandum NMFS-SEFSC-357. US Department of Commerce, 42 pp.

Lutcavage, M.E., Brill, R.W., Skomal, G.B., Chase, B.C., and Howey, P.W. (1999) Results of pop-up satellite tagging of spawning size class fish in the Gulf of Maine: do North Atlantic bluefin tuna spawn in the mid-Atlantic? *Can. J. Fish. Aquat. Sci.* **56**, 173-177.

Meeus, J. (1991) *Astronomical Algorithms.* Willman-Bell, VA, 429 pp.

Metcalfe, J.D., and Arnold, G.P. (1997) Tracking fish with electronic tags. *Nature (Lond.)* 387, 665-666.

Nielsen, E.T. (1963) Illumination at twilight. *Oikos* **14**, 9-21.

Pielou, E.C. (1984) *The Interpretation of Ecological Data.* John Wiley, NY, 263 pp.

Sobel, D. (1995) *Longitude: The True Story of a Lone Genius Who Solved the Greatest Scientific Problem of His Time.* Penguin Books, NY, 184 pp.

Sokal, R.R., and Rohlf, F.J. (1981) *Biometry.* W.H. Freeman and Company, NY, 859 pp.

Smith, P., and Goodman, D. (1986) Determining fish movements from an "archival" tag: precision of geographical positions made from a time series of swimming temperature and depth. NOAA Technical Memorandum NMFS, NOAA-TM-NMFS-SWFC-60. Southwest Fisheries Science Center, La Jolla, CA, 92038.

Steele, J.H. (1995) Can ecological concepts span the land and ocean domains? *In* Powell, T.E., and Steele, J.H., eds. *Ecological Time Series.* Chapman & Hall, NY, pp. 5-19.

Welch, D.W., and Eveson, J.P. (1999) An assessment of light-based geoposition estimates from archival tags. *Can. J. Fish Aquat. Sci.* **56**, 1317-1327.

Wilson, R.P., Ducamp, J.-J., Rees, W.G., Culik, B.M., and Neikamp, K. (1992) Estimation of location: global coverage using light intensity. *In* Priede, I.G., and Swift, S.M., eds. *Wildlife Telemetry: Remote Monitoring and Tracking of Animals.* Ellis Horwood, NY, pp. 131-134.

Appendix I: Summary statistics for the raw position errors. Abbreviations are as follows: FLA=full latitude days and FLO= full longitude days as in Table 1; CLA=culled latitude days and CLO=culled longitude days as in Table 2; N587=NMT tag at 24 m; (*continued*⇒)

ID	N	Mean	C. I. ± 95%	C. I. ± 95%	Median	Sum	Min	Max
N587FLA	286	0.95	0.66	1.24	0.70	272.10	-6.50	18.50
N587CLA	216	0.64	0.39	0.90	0.70	139.00	-6.50	4.90
N935FLA	163	-4.70	-5.79	-3.60	-3.30	-765.40	-53.60	9.70
N935CLA	139	-3.91	-4.60	-3.22	-3.30	-543.50	-17.70	3.20
G224FLA	286	-1.30	-1.57	-1.03	-0.76	-372.22	-16.33	2.40
G224CLA	216	-1.32	-1.63	-1.01	-0.80	-284.82	-16.33	2.40
G219FLA	317	-0.23	-0.83	0.37	0.18	-73.52	-22.79	24.00
G219CLA	263	-0.63	-1.21	-0.05	0.00	-165.46	-22.79	9.65
G218FLA	163	-0.40	-0.87	0.08	-0.05	-64.67	-21.73	8.94
G218CLA	139	-0.10	-0.32	0.11	-0.03	-14.37	-5.23	4.33
H224FLA	286	0.89	0.57	1.21	1.00	255.00	-11.50	10.00
H224CLA	216	0.42	0.08	0.75	0.50	90.00	-11.50	10.00
H219FLA	317	1.19	0.89	1.48	1.00	376.00	-7.50	13.50
H219CLA	263	0.87	0.63	1.11	0.50	229.50	-3.50	6.00
H218FLA	163	-1.20	-1.63	-0.78	-1.00	-196.00	-12.50	11.50
H218CLA	163	-1.20	-1.63	-0.78	-1.00	-196.00	-12.50	11.50
W224FLA	286	0.81	0.59	1.03	0.81	232.67	-12.16	11.00
W224CLA	216	0.53	0.31	0.75	0.67	114.36	-12.16	6.53
W219FLA	317	1.42	1.04	1.80	1.41	450.49	-18.06	23.44
W219CLA	317	1.40	1.09	1.71	1.29	368.95	-6.66	23.44
W218FLA	163	-0.76	-1.58	0.06	-0.01	-123.89	-53.91	6.39
W218CLA	139	0.42	-0.14	0.22	0.05	5.80	-3.07	4.10
N587FLO	337	0.04	-0.01	0.10	0.10	14.90	-5.60	1.50
N587CLO	241	0.08	0.03	0.12	0.10	18.20	-1.10	1.10
N935FLO	202	-0.14	-0.22	-0.06	-0.10	-27.80	-3.80	3.00
N935CLO	150	-0.08	-0.15	0.00	-0.10	-11.30	-1.80	3.00
G224FLO	337	0.07	0.03	0.10	0.10	21.95	-1.22	0.99
G224CLO	241	0.05	0.01	0.09	0.09	11.85	-1.03	0.99
G219FLO	333	-0.04	-0.07	-0.00	-0.02	-12.47	-2.23	1.12
G219CLO	263	-0.06	-0.10	-0.02	-0.05	-14.93	-2.23	1.12
G218FLO	202	-0.03	-0.07	-0.00	-0.03	-6.82	-0.99	0.69
G218CLO	150	-0.06	-0.10	-0.03	-0.03	-9.34	-0.86	0.57
H224FLO	337	0.05	0.02	0.09	0.04	17.53	-1.36	1.55
H224CLO	241	0.04	0.01	0.08	0.03	10.33	-0.87	1.13
H219FLO	333	-0.03	-0.06	0.00	-0.04	-9.70	-1.31	1.45
H219CLO	263	-0.03	-0.07	-0.00	-0.05	-8.86	-1.01	1.03
H218FLO	202	-0.01	-0.04	0.02	-0.01	-1.86	-0.59	0.89
H218CLO	150	-0.01	-0.04	0.03	-0.00	-0.84	-0.58	0.76
W224FLO	337	0.08	0.04	0.11	0.08	25.61	-2.03	0.99
W224CLO	241	0.11	0.07	0..14	0.11	25.59	-0.79	0.84
W219FLO	333	0.01	-0.03	0.05	0.02	3.95	-2.92	1.09
W219CLO	263	0.05	0.02	0.09	0.04	13.58	-0.80	1.09
W218FLO	202	0.14	0.10	0.19	0.16	29.01	-2.35	1.03
W218CLO	150	0.17	0.14	0.21	0.18	26.05	-0.58	0.98

(*Appendix* continued⇒) N935=NMT tag at 58 m; for the Wildlife Computers= (WC) tags; G=Method 1, H=Method 2, W=Method 3; 224=WC tag at 20 m, 219=WC tag at 40 m, 218=WC tag at 60 m.

Lower Quartile	Upper Quartile	Range	Quartile Range	Vari-ance	S. D.	S. E.	Skew.	S. E. Skew.	Kurt.	S. E. Kurt.
-0.20	1.90	25.00	2.10	6.11	2.47	0.15	1.79	0.14	12.23	0.29
-0.30	1.60	11.40	1.90	3.53	1.88	0.13	-0.87	0.17	2.01	0.33
-7.40	-0.90	63.30	6.50	49.88	7.06	0.55	-3.14	0.19	17.58	0.38
-6.80	-1.20	20.90	5.60	17.02	4.13	0.35	-0.74	0.21	0.74	0.41
-1.65	-0.12	18.73	1.53	5.33	2.31	0.14	-2.89	0.14	11.09	0.29
-1.67	-0.17	18.73	1.50	5.44	2.33	0.16	-3.19	0.17	13.21	0.33
-1.96	2.01	46.79	3.97	29.55	5.44	0.31	-0.40	0.14	4.54	0.27
-1.94	1.69	32.44	3.63	23.13	4.81	0.30	-1.54	0.15	4.40	0.30
-0.95	0.60	30.67	1.55	9.47	3.08	0.24	-3.49	0.19	22.00	0.38
-0.72	0.56	9.56	1.28	1.65	1.29	0.11	-0.18	0.21	3.24	0.41
-0.50	2.50	21.50	3.00	7.57	2.75	0.16	-0.68	0.14	3.49	0.29
-0.50	2.00	21.50	2.50	6.15	2.48	0.17	-0.98	0.17	5.71	0.33
-0.50	3.00	21.00	3.50	7.00	2.64	0.15	0.82	0.14	2.54	0.27
-0.50	2.50	9.50	3.00	3.84	1.96	0.12	0.17	0.15	-0.89	0.30
-2.00	-0.50	24.00	1.50	7.47	2.73	0.21	-0.36	0.19	7.86	0.38
-2.00	-0.50	24.00	1.50	7.47	2.73	0.21	-0.36	0.19	7.86	0.38
0.15	1.47	23.16	1.33	3.60	1.90	0.11	-0.60	0.14	10.97	0.29
0.10	1.19	18.68	1.09	2.65	1.63	0.11	-2.33	0.17	18.29	0.33
0.30	2.50	41.50	2.20	11.98	3.46	0.19	0.05	0.14	15.70	0.27
0.34	2.25	30.10	1.91	6.48	2.55	0.16	4.19	0.15	33.70	0.30
-0.72	0.56	60.30	1.28	28.03	5.29	0.42	-7.36	0.19	66.49	0.38
-0.57	0.56	7.17	1.12	1.15	1.07	0.09	0.42	0.21	2.28	0.41
-0.20	0.30	7.10	0.50	0.25	0.50	0.03	-4.49	0.13	49.04	0.26
-0.20	0.30	2.20	0.50	0.12	0.34	0.02	-0.14	0.16	0.33	0.31
-0.30	0.10	6.80	0.40	0.33	0.58	0.04	-1.09	0.17	13.02	0.34
-0.30	0.10	4.80	0.40	0.23	0.48	0.04	1.27	0.20	12.13	0.39
-0.10	0.24	2.21	0.34	0.09	0.30	0.02	-0.60	0.13	2.07	0.26
-0.11	0.21	2.02	0.32	0.08	0.29	0.02	-0.44	0.16	1.78	0.31
-0.24	0.16	3.35	0.40	0.12	0.34	0.02	-0.89	0.13	5.49	0.27
-0.24	0.12	3.35	0.36	0.12	0.34	0.02	-1.01	0.15	6.96	0.30
-0.17	0.09	1.68	0.25	0.06	0.24	0.02	-0.29	0.17	2.35	0.34
-0.18	0.07	1.43	0.24	0.04	0.21	0.02	-0.77	0.20	2.26	0.39
-0.12	0.21	2.92	0.34	0.09	0.31	0.02	0.17	0.13	2.94	0.26
-0.12	0.19	2.00	0.30	0.07	0.27	0.02	0.31	0.16	1.83	0.31
-0.20	0.13	2.76	0.34	0.09	0.31	0.02	0.23	0.13	2.68	0.27
-0.20	0.12	2.05	0.32	0.07	0.27	0.02	0.28	0.15	1.08	0.30
-0.13	0.10	1.48	0.23	0.05	0.23	0.02	0.57	0.17	2.20	0.34
-0.12	0.10	1.34	0.22	0.04	0.20	0.02	0.26	0.20	1.44	0.39
-0.08	0.26	3.02	0.34	0.11	0.33	0.02	-1.30	0.13	6.75	0.26
-0.06	0.26	1.63	0.32	0.08	0.27	0.02	-0.14	0.16	0.82	0.31
-0.15	0.21	4.01	0.37	0.14	0.37	0.02	-2.06	0.13	14.68	0.27
-0.11	0.22	1.89	0.33	0.08	0.28	0.02	0.28	0.15	1.42	0.30
0.03	0.30	3.38	0.27	0.10	0.32	0.02	-2.41	0.17	18.03	0.34
0.03	0.30	1.56	0.27	0.05	0.23	0.02	0.04	0.20	1.44	0.39

Appendix II. Summary of positional errors for the WC tags using Method 3 with a zenith angle of 94°. All days (Table 1) and outlier rejection criteria (Table 2) are as before. Absolute errors (|true position-estimate|) and raw errors (|true position-estimate|) are given. Distribution of raw errors were tested for conformity to the normal distribution using the Kolmogorov-Smirnov D test for Goodness of Fit (Sokal and Rohlf, 1981) at the $P<0.05$ level. Unless indicated as being non-significant ***(bold-italicised)***, the remaining tests were significant at $P<0.05$ (9 of 12 cases).

Absolute Errors				
	Method 3, all days		Outliers removed	
Depth of tag (m)	Mean± SD lat. errors (°)	Mean± SD lon. errors (°)	Mean± SD lat. errors (°)	Mean± SD lon. errors (°)
60	3.32±3.89	0.28±0.48	2.73±1.44	0.23±0.17
40	4.36±5.78	0.24±0.24	3.34±3.15	0.21±0.19
20	2.69±2.11	0.29±0.83	2.93±2.09	0.23±0.19

Raw Errors				
	Method 3, all days		Outliers removed	
Depth of tag (m)	Mean± SD lat. errors (°)	Mean± SD lon. errors (°)	Mean± SD lat. errors (°)	Mean± SD lon. errors (°)
60	1.89±4.76	0.18±0.52	2.15±2.22	***0.17±0.23***
40	0.16±7.24	0.01±0.34	0.65±4.55	***0.05±0.28***
20	1.23±3.19	0.12±0.87	1.27±3.37	***0.10±0.27***

Recent Progress in Estimating Geoposition Using Daylight

David W. Welch[1] and J. Paige Eveson[1,2]
[1]*Fisheries and Oceans Canada, Pacific Biological Station, Nanaimo, BC*
[2]*CSIRO, Hobart, Tasmania, Australia*

Key words: migration, geoposition, archival tags

Abstract: In this paper we report on the results from a new experiment where we placed tags at three distinct depths on a fixed subsurface mooring. The accuracy of the daily geoposition estimates is substantially improved over our earlier work, with average positional errors of 30 km in longitude and 44 km in latitude. We attribute the result to improvements in both the tags and our software. From the resulting data, we were able to estimate a relationship between light and depth. We generated a simulated variable-depth light record, corrected the light values to a constant depth using the estimated relationship, and then applied our estimation algorithm to the depth-corrected values. This yielded an average absolute positional error of 42 km in longitude and 68 km in latitude, a significantly better result than previously reported. Finally, we provide a theoretical assessment of the expected differences that can be obtained by shielding the light sensor with a blue filter. Although this helps to linearize the relationship between light intensity and depth, the combination of the broad-band spectral sensitivity of the light sensor and the surface light spectrum means that a simple linear response should not be expected. Attempts to correct the light record for variations caused by changes in depth of a tagged animal are probably not substantially improved by filtering the light sensor. More attention needs to be given to the combined effect of the spectral sensitivity of the light sensor and the light spectrum present at twilight.

1. INTRODUCTION

Light-depth-temperature recording archival tags continue to be improved by manufacturers and to gain the interest of marine scientists, particularly those involved in research on animals that remain continuously submerged

J.R. Sibert and J.L. Nielsen (eds.), Electronic Tagging and Tracking in Marine Fisheries, 369–383.
© 2001 *Kluwer Academic Publishers. Printed in the Netherlands.*

in seawater where conventional radio-telemetry or GPS methods will not work. Changes in the intensity of light through the day can be used to determine geoposition and, in combination with simultaneously collected depth data, provide an opportunity to track the movements of marine animals in three dimensions across ocean basins. There are now a number of studies in which tags were attached to free-ranging fish and successfully recovered after basin-scale movements (DeLong et al., 1992; Itoh et al., 1999; Lutcavage et al., 1999; Wada and Ueno, 1999); however, difficulties with analysing the light data still exist and new methods continue to be explored (Hill, this volume).

For several years we have been investigating the geolocation capability of archival tags for use on Pacific salmon (Welch and Eveson, 1999; Eveson and Welch, 2001). In our initial experiments, we deployed tags at either a special-purpose subsurface ocean mooring or off a surface meteorological buoy in order to develop estimation techniques and evaluate the accuracy of the position estimates obtained (Welch and Eveson, 1999). The results of these studies were encouraging, but they represented an idealised situation since the tags remained not only at a constant latitude and longitude, but also at a constant depth. When tags are attached to salmon, they will undergo continual changes in depth and, consequently, in the amount of light that reaches the tag's sensor.

Light attenuation through seawater increases with depth, but the exact relationship depends on properties of the light sensor as well as on the clarity of the water. If the relationship was known, we could use it to correct the entire light record to a constant depth. We report here the results of a new experiment in which we placed tags at three depths on a fixed subsurface mooring. We estimated the light-depth relationship and evaluated the improvement gained in our position estimates by correcting for depth from this new data set. We also report on a theoretical evaluation of the benefits of using a blue or blue-green filter with the light sensor in order to improve the response with depth, as this has been suggested to significantly improve the geolocation capabilities of light-based tags.

2. MATERIALS AND METHODS

On May 27, 1998, we deployed six Wildlife Computers (16150 NE 85th St. #226, Redmond, WA, USA 98052) MK7 tags at the fixed subsurface mooring ENP8-B in the eastern North Pacific Ocean (48°48.9′N, 128°38.6′W). This mooring is off the continental shelf west of Vancouver Island and has an unobstructed horizon to both the east and west. Two tags were secured to each of three circular PVC plates so that their light sensors faced up toward the surface, and the plates were attached to the mooring cable at depths of 80 m, 45 m, and 15 m. Data from a previous mooring of the same design showed that the top of the mooring experienced vertical movements of less than 5 m during the deployment. The mooring was recovered four months later on September 27 and the tags retrieved. The amount of algal growth on the surface of the tags was quite heavy on the those at 15 m, somewhat less on those at 45 m, and negligible on those at 80 m.

The temperature, depth, light, and time data were downloaded from each tag, and the amount of clock drift determined. The average clock error was 80 seconds (range: 17 to 153 seconds), and we corrected the individual time records assuming a consistent linear drift over the duration of the experiment. Each of the data records was plotted and examined visually for obvious errors and characteristics of interest.

In order to investigate the relationship between light and depth, we calculated the average daytime light level and the average depth for each tag, using data from all days between the times 07:15 and 16:45 (Pacific Standard Time). Using least-squares estimation, we fit a curve of the form $L = \alpha + \beta \log(Z)$ through the data, where L is light in logarithmic units, as measured by the tags, and Z is depth in meters (Figure 1). Parameter estimates are given on the figure. Although theory suggests the relationship between log light and depth is linear, we found that using log depth gave a better fit. Possible reasons for this are given in the discussion. We then derived an equation describing the difference in light between any two depths, $\Delta L(z_1, z_2) = \beta(z_1/z_2)$, so that the light values could be corrected to a constant depth. For example, we could correct all light measurements to a depth of 15 m by increasing the measurements taken at 45 m by a value of $\Delta L(15,45)$ and increasing the measurements taken at 80 m by $\Delta L(15,80)$.

Even though the tags were placed at different depths, each individual light record was created at an unchanging depth. Thus, the results obtained from applying our geoposition algorithms to the unmodified data are still significantly unrepresentative of those we would obtain from a tagged fish. To make the results more comparable, we simulated a depth record which varied between the three experimental depths according to a random process, and then used the existing light data to create a light record corresponding to the simulated depth record (Figure 2a and b, grey line). The random process used to generate the depth record had three states representing depths 15 m, 45 m, and 80 m. The probabilities of moving between states (depths) were defined by the following transition matrix:

$$P = \begin{matrix} 97 & 3 & 0 \\ 5 & 93 & 2 \\ 3 & 7 & 90 \end{matrix}$$

The transition matrix was designed such that once at a given depth, there is a tendency to remain at that depth, with this tendency being strongest at 15 m and lessening as the depth increased. Our intent was to simulate depth variations somewhat representative of a salmon which spent most of its time in the upper surface layer (at 15 m) (Ogura and Ishida, 1995) and made fairly frequent excursions to greater depths (45 m) as well as occasional deep dives (to 80 m).

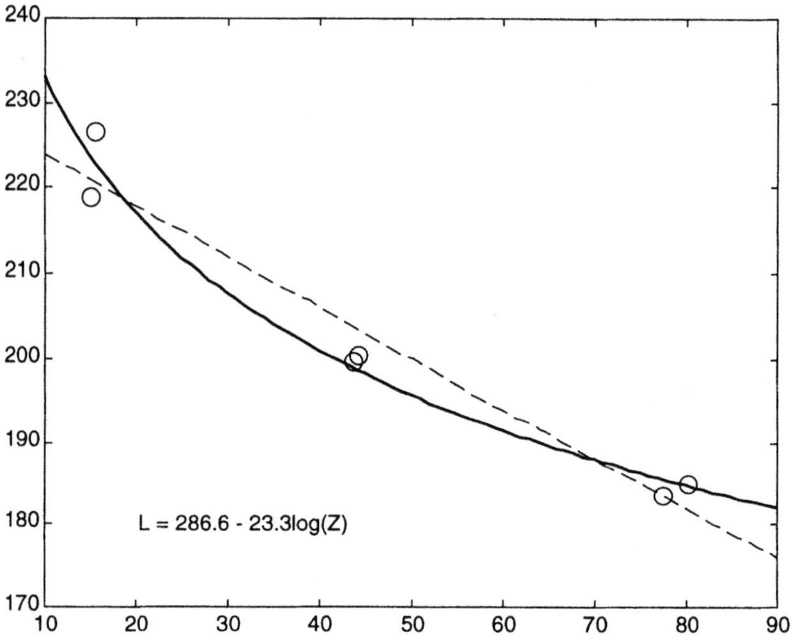

Figure 1. Average logarithmic light level (L, as recorded by the tags) vs. average depth, Z, for the six tags. The solid line shows the fitted curve between L and log(Z); the equation of the curve is given. The dashed line shows the estimated linear fit between L and Z.

After creating the variable-depth light record, we corrected the light values to a uniform depth of 15 m using the procedure discussed above (Figure 2b, black line). The k_{max} light level method (Welch and Eveson, 1999) was then applied to the simulated light record, both before and after correcting for depth. In addition, we applied the k_{max} method to the light data from the two tags which had been deployed at 15 m in order to compare the accuracy of the position estimates derived from data corrected to 15 m and data which were actually generated at 15 m.

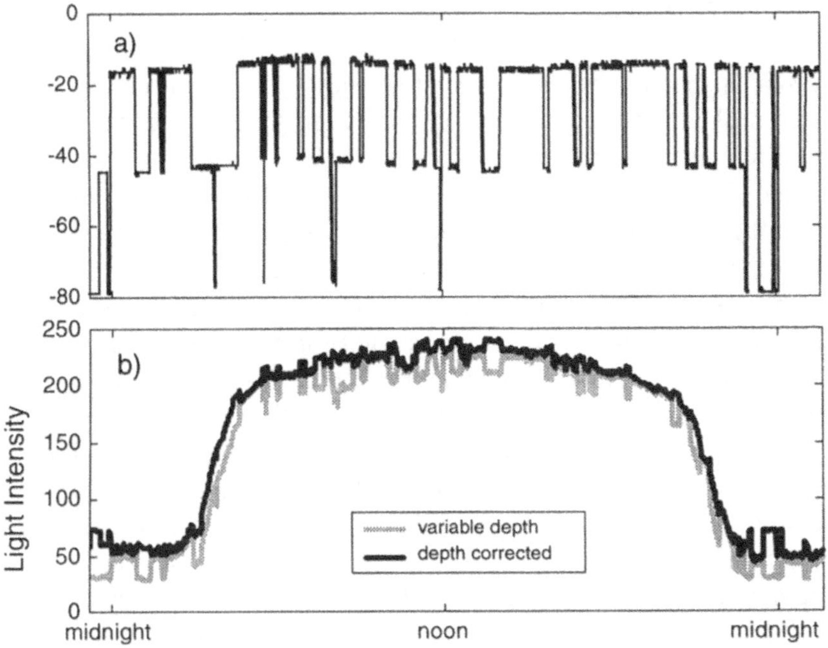

Figure 2. a) One day of the simulated depth record, and b) the corresponding light record before (grey line) and after (black line) correcting for depth.

The details of the k_{max} method can be found in Welch and Eveson (1999). The basic principle is to calculate a reference light level as some proportion of the average midday (maximum) light level, then estimate the times of dawn and dusk as the times at which this reference level is achieved. Navigational routines can then be used to convert the estimated times of dawn and dusk into daily estimates of latitude and longitude. Because the experiment was conducted at a known location, we were able to estimate the value of k_{max} that minimized the average mean squared error between the true location and our estimates of geoposition.

The astronomical theory to determine latitude based on daylength cannot be applied successfully near the equinoxes when daylength is roughly equivalent at all latitudes. The September equinox occurred one week prior to the termination of our experiment, so we could not generate reliable geoposition estimates for a time period at the end of the study. To avoid problems due to the equinox, we used only the data up to September 1 in deriving the results reported here.

3. RESULTS

Visual inspection of the temperature, depth, and light data revealed no obvious problems for any of the tags. The temperature records for the tags at 80 m remained approximately constant at 9°C for the duration of the experiment. This suggests that the tags were below the seasonal thermocline. The average temperature of the tags at 15 m and 45 m increased roughly linearly through summer, from about 12 to 17°C for the tags at 15 m and from about 10 to 15°C for the tags at 45 m. The recorded temperatures were highly variable for the tags at 45 m, suggesting that they were situated near the seasonal thermocline and that the rapid temperature changes were due to internal waves changing the depth of the thermocline.

Summary statistics on depth (Table 1) confirm that the mooring experienced only minor vertical excursions, and all depth records showed similar variability. The two tags at each of the two shallower depths had very similar depth records whereas one of the two deepest tags (Tag 045) tended to record a depth about 3 m greater than the other (Tag 044). The difference is not significant for our purposes, and could likely have been corrected by recalibrating the depth sensors after the experiment. Despite small discrepancies between the design depths and the depths recorded by the tags, we will continue to refer to the tags as being deployed at 15 m, 45 m, and 80 m for simplicity.

Table 1. Depth statistics (in meters).

Tag ID	Min.	Average	Max.
040	11	15.1	19
041	12	15.4	19
042	40	43.6	48
043	41	44.2	49
044	74	77.5	81
045	77	80.3	83

The light sensors have a logarithmic response that records values in the range 0 to 2^8 (=256), such that they are most sensitive to low light values. Time series plots of the light records show that the night-time values rise and fall in response to the moon's cycle, with the effect being greatest near the surface. The light records from the two tags at a design depth of 80 m are almost identical; the same holds true for the tags at 45 m except for a small discrepancy during the nights. The two tags at 15 m show a small but consistent difference in light measurements which persists for about the first 75 days of the study, then diminishes. The reason for this is unknown.

A reviewer noted that the tags record uncalibrated light data, unlike the depth and temperature sensors. For a given ambient light level L, the recorded level is linearly related to the ambient light level, $L_{obs} = a + bL$, with the values of L_{obs} ranging from 0-255. As specific values for *a* and *b* for individual tags were not measured by the manufacturer, light levels experienced by different tags are not strictly comparable. However, the differences between paired tags are generally small, as indicated by the similar average light values recorded by different tags (see Figure 1). As a result, we use the data from the different tags as if it represented a common light level. This results in the small variation between tags in recorded light levels contributing slightly to the variance in the Monte Carlo results on the effect of depth on geoposition accuracy that are reported below. This effect is relatively small, however, because the depth effects on measured light level are larger than the between-tag variation at a given depth.

A closeup of the light readings for all tags for one day gives an indication of how the values are affected by depth (Figure 3). During the daytime, the absolute difference in light level between depths remains fairly constant and the pattern of light variation is highly correlated between tags. The night-time response is more complicated, varying across tags as well as days. Generally, the difference between the light measured by the tags closest to the surface and the light measured by the deeper tags becomes greater than it is during the daytime; however, the difference is very irregular. A possible explanation is that the deep scattering layer rises during the night to sit somewhere between 15 m and 45 m, causing the four deeper tags to "see" less light than those near the surface. As the timing and distance of the vertical migration of the deep scattering layer varies from night to night, this explanation would be consistent with the irregular patterns we obtained.

In applying the k_{max} light level method to the light records of the tags at 15 m (tags 040 and 041), we found the optimal value of k_{max} to be 65%, and the optimal zenith angle for converting the times of dawn and dusk into latitudes and longitudes to be 95.2°. The accuracy of the geoposition estimates was very good (Table 2), with a mean absolute error in latitude of 44 km and in longitude of 30 km—significantly better than attained in our previous studies (Welch and Eveson, 1999).

Using the optimal parameters derived for the constant 15 m depth case (k_{max} = 65% and zenith angle = 95.2°), we applied the k_{max} estimation method to the simulated variable-depth light record and the depth-corrected light record. This is representative of how we would run a real tagging study, by placing tags at one or more fixed locations within the study area in order to tune the estimation parameters to appropriate values. Summary geoposition results are given in Table 2. As expected, the positional errors are higher for the variable-depth case, with mean errors and standard deviations in latitude and longitude almost three times greater than for the constant depth case. The results for the depth-corrected light record are much better (Figure 4, Table 2). Although the errors are still larger than in the ideal constant depth situation, they are a significant improvement over the variable-depth results, and well within acceptable range for our objectives. Furthermore, there is no apparent temporal pattern in the daily estimates from the mooring (Figure 5), suggesting that neither large-scale weather systems nor fouling had a significant effect on the near-surface tags in degrading our position estimates over time.

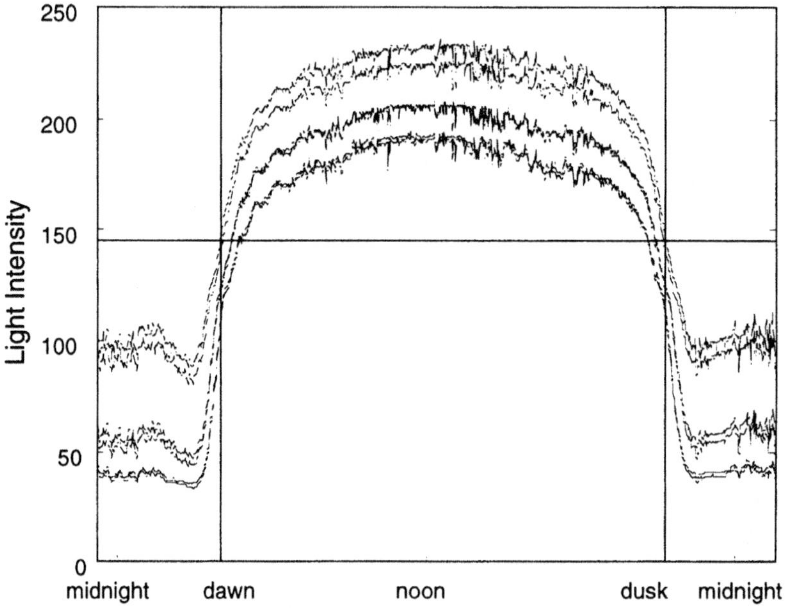

Figure 3. A close-up of the light records from all tags for one day. The horizontal line shows the 65% reference light level for the tags at 15 m and the vertical lines show the corresponding estimated times of dawn and dusk.

Table 2. Geoposition results using k_{max}=65%. Mean errors were calculated by taking the absolute value of the calculated daily positions minus the true position of the mooring. Bracketed values show the standard deviations on the means. As the mean error is less than one standard deviation away from zero any systematic bias is small relative to the variability over time.

Light Record	Mean Error Latitude (km)	Mean Error Longitude (km)
Constant depth (15 m)	44 (± 54)	30 (± 38)
Simulated, variable-depth	119 (± 144)	85 (± 110)
Simulated, depth-corrected	68 (± 59)	42 (± 53)

3.1 Effect of Filters

The amount of light in the ocean decreases exponentially with depth. However, the degree of attenuation is dependent on wavelength, with blue-green light being attenuated least (Jerlov, 1976). As a result, total irradiance at depth is not a simple exponential function of the amount of light at the surface, as often assumed. Rather it is a summation of exponential functions. It is not possible to linearize the change in light level with depth by taking logarithms because different spectral components of the light do not decrease uniformly (Ivanoff, 1977). As both the response of the light sensor and the amount of light reaching the surface of the ocean also show strong frequency-dependence, assessing the effect of depth variation on geoposition becomes a complicated proposition.

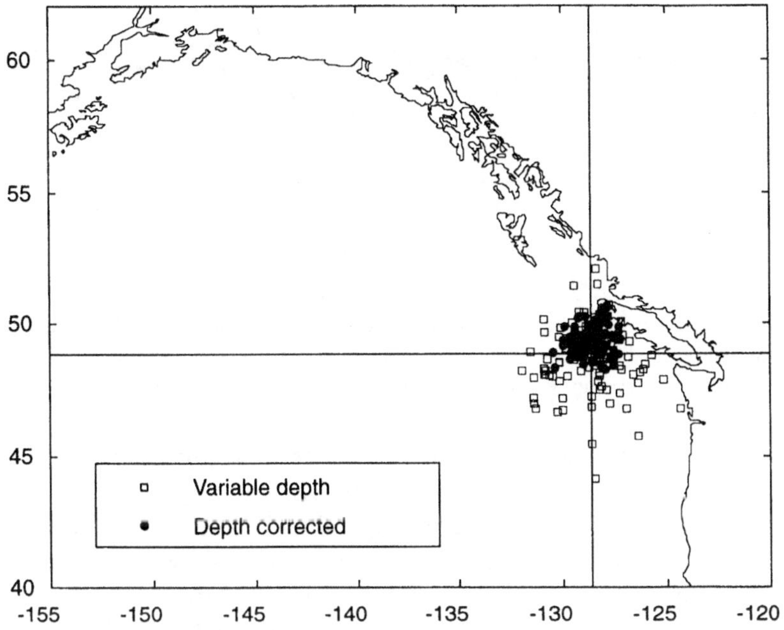

Figure 4. A map of the eastern North Pacific showing the position estimates derived from the simulated light record before correcting for depth (open squares) and after correcting for depth (filled circles). The horizontal and vertical lines mark the true latitude and longitude.

There have been several suggestions that the use of a blue-green filter on the light sensors for archival tags should improve the light response and geoposition estimates obtained from archival tags. Most commercially available archival tags now use some form of filter, but it is uncertain how much improvement in the light response is really possible. A critical point is that water is itself a blue-green filter that sharply attenuates other wavelengths. It is unclear how much more efficient a very thin (< 1mm) blue-green filter capable of use in small archival tags is once it is submerged under a few metres of seawater.

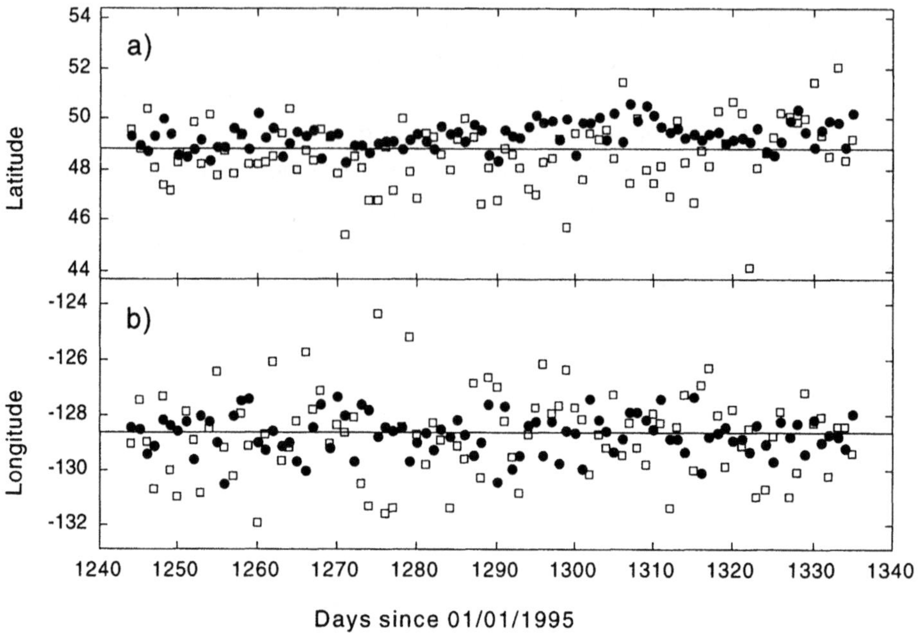

Figure 5. Time series plots of estimated a) latitudes and b) longitudes obtained from the simulated light record before correcting for depth (open squares) and after correcting for depth (filled circles). The horizontal lines indicate the true latitude and longitude.

To examine these questions, we compared the published transmission characteristics of a commercially available BG39 blue-green filter as a function of depth and surface light level (Figs. 6 and 7). A typical spectrum for irradiance at the sea surface (in W/m^2; Jerlov, 1976) shows a relatively broad spectrum, with the peak of the spectrum at slightly longer wavelengths than that of blue-green light (0.480 μm; i.e., towards the red end of the spectrum). In contrast, the sensor response (mA/W) is most sensitive near the red end of the spectrum. The result is that their product, the effective sensor response in (mA/m^2) is sensitive to a wide range of wavelengths. The BG39 filter shows two transmission windows, one centred in the blue-green, and the other in the far infrared. A filtered sensor results in a response to light over a narrower spectral range, but not one strictly limited to the blue-green.

Ivanoff (1977; Fig. 5.10) examined the dependence of light extinction on wavelength for three ocean regions ranging from very clear open ocean water to turbid coastal regions. The amount of light reaching any depth is complicated because the rate of light attenuation depends on both the clarity of the water and the wavelength of the light. The response of sensors used in most archival tags to light is logarithmic—what difference does filtering make to the expected response to light with depth? As Figure 7 indicates, except in exceptionally clear seawater very near the surface, filtering has little effect on the predicted logarithmic response of irradiance to depth. Thus, although it has been informally suggested that blue-green filtering should improve the light response, it seems that except for the top 1-2 metres in the open ocean, the transmissivity of seawater achieves the same effect.

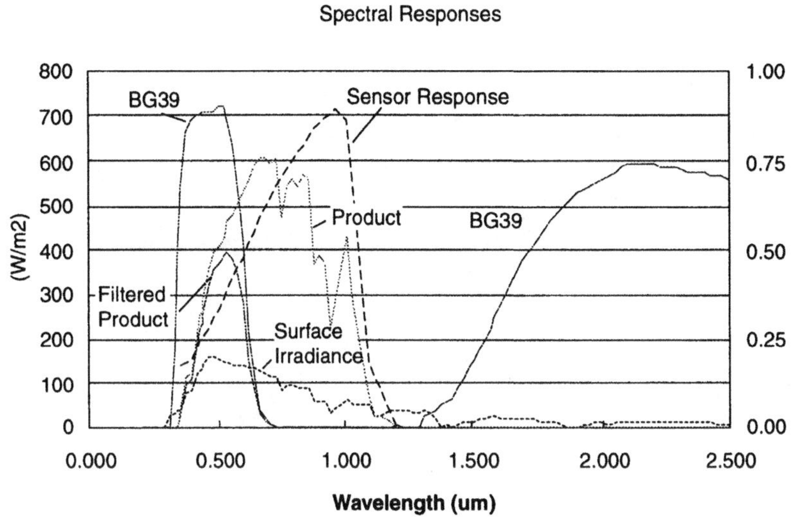

Figure 6. Comparison of a typical spectrum for irradiance at the sea surface (in W/m^2; Jerlov, 1976) with the spectral response of a photodiode light sensor (in mA/W) and their product. The spectral response of the blue-green filter (BG39; right axis) shows two peaks. The predicted response of the filtered sensor to surface illumination (Filtered Product) shows a narrower but still relatively broad-band response.

4. DISCUSSION

The general relationship between light and depth is of the form $l = \alpha \exp^{-\beta z}$, where l is light, z is depth, and β is the extinction coefficient. Because the tag's light response is logarithmic, we expected a linear relationship between the recorded light and depth. We found, however, that a linear relationship between the measured light profile and the logarithm of depth provided a better fit. The reason for this is probably that the water clarity varies with depth. The water below the seasonal thermocline is clearer, so we should expect to observe a lower rate of light extinction. Because the tags at 80 m were below the seasonal thermocline, they should "see" more light than if the water clarity remained the same as above the thermocline. This explanation is consistent with the data, and would mean that the true relationship should be a piece-wise linear fit with a changepoint at the thermocline. Without data at different depths and without knowing the exact location of the thermocline, we were unable to fit such a relationship. In an actual deployment on fish, knowing the correct theoretical relationship is not essential since thousands of data points over a large range of depths will be available for determining the best-fitting curve.

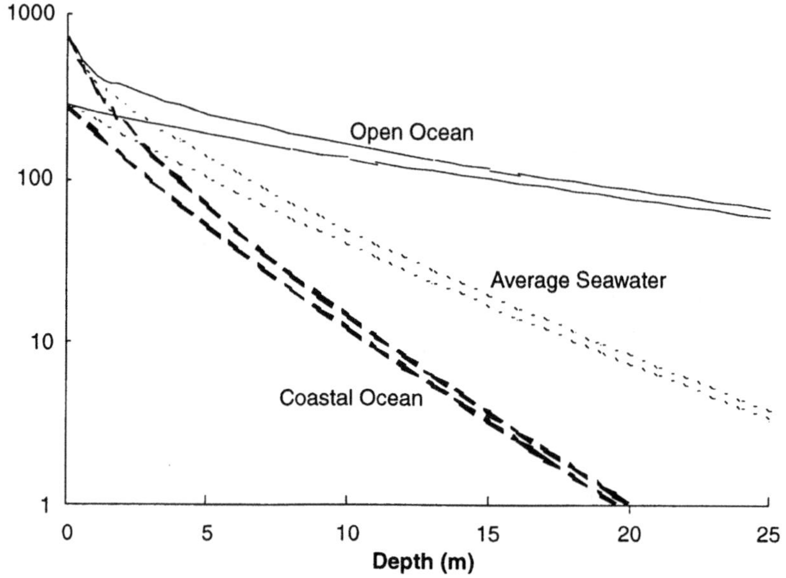

Figure 7. Effect of depth on total irradiance (log scale). For simplicity, values have not been multiplied by the sensor response, but exclude light response in the UV. Three ocean regions are considered, using the spectral variation in light extinction reported by Ivanoff (1977, Fig. 5.10). The lower of each pair of curves represents the predicted total irradiance after filtering. Except in exceptionally clear seawater very near the surface, filtering has little effect on the predicted logarithmic response of irradiance to depth.

The results we obtained in this study are substantially better than the results from our previous fixed depth study at the North Nomad buoy (Welch and Eveson, 1999). Both studies used Wildlife MK7 tags. One concrete reason for our improved results was a change made to the navigational routines used in the earlier study to correct for an error in which an iterative equation was being solved using only one iteration rather than iterating to convergence. Still, this does not account for the entire difference. When we reran the previous analysis of the Nomad North data using the new navigational algorithms, the average error in latitude improved from 114 km to 95 km and the error in longitude improved from 67 km to 46 km, but these results are still not as good as those achieved in our latest study. The remaining difference could be due to different environmental conditions between experiments such as water clarity, cloud cover, and sea state and to different tag deployment methods. The way in which the tags were deployed in the previous study probably contributed significantly to degrading the position estimates since the tags were subject to large movement from surface waves acting on the buoy. Latitudinal position of the tag is also a factor to consider since a small mis-specification in daylength results in a larger error in latitude closer to the equator; however this is not a valid explanation in this case because the North Nomad buoy is at a higher latitude than the ENP8-B mooring.

The issue of water clarity is an important consideration. Light is attenuated less with depth when the water is clearer, and the actual form of the light-depth relationship may be affected. This brings about the practical issue of correcting a light record for a fish which is moving through changing water conditions, since the appropriate light-depth relationship will vary. The correction can still be carried out either using a single functional relationship in the manner that we have described or as a series of piece-wise log-linear relationships applicable to different depth strata. In either case it will be necessary to re-estimate the relationships as water conditions change. For example, as a tagged animal moves from clear offshore waters to more turbid continental shelf waters, both the thickness of the mixed layer and the light extinction coefficient will vary, requiring changes in the light correction algorithm.

Recent studies have shown that the behaviour of tagged salmon may take several days to recover after tagging, apparently reflecting an initial period of handling stress associated with the tagging operation (Walker et al., 2000). Walker et al. (2000) found that Pacific salmon tagged with a temperature recording tag experienced wide thermal variations during their homing migrations, and concluded that these observations indicated that the animals appeared to be undertaking substantial vertical migrations during daylight hours (but not at night). Although this information is of considerable interest, the ability to track the movement of fish in three dimensions made possible by light recording tags is an important consideration.

Examination of oceanographic data shows that the thermal structure of the ocean along the straight line path between release and recovery sites in the Walker et al. study changed radically. For example, -1°C water may be encountered within 10 m of the surface in some parts of the Okhotsk Sea, which lies along the shortest migration path. Thus attempts to infer depth distribution from temperature-only tags can be potentially misleading. Even tags that record both temperature and depth can be equally limiting in the conjectures that the investigator can draw, because they only trace out the temperature-depth profile for the region of the water column that the tagged animal visits. In the case of salmon, which appear to be strongly surface-oriented, resolution of questions as to whether salmon ever significantly penetrate the seasonal thermocline (and why) depends on knowing what the temperature-depth profile is below the region that is visited.

Without geoposition data to fix the general area of the ocean where the animal is located, it is not possible to use the extensive archives of oceanographic data to examine the physical structure of the ocean below the region that they do exploit, or to develop an understanding of how they migrate either horizontally or vertically. It is clear from our current results, however, that it is possible to determine the position of a tagged animal to an accuracy that is either equivalent or more precise than the resolution of the best eddy-resolving ocean models of surface currents. Given this accuracy, it should be possible to begin addressing whether salmon migrations compensate for the large inter-annual variations in surface currents that characterize the North Pacific (Bourque et al., 1999; Dat et al., 1995; Peterman et al., 1994; Thompson et al., 1994). In addition, recent suggestions that greenhouse-induced warming of the surface mixed layer may exclude entire species of Pacific salmon from the North Pacific within a few decades make a direct evaluation of the avoidance behaviour of salmon to measured thermal limits of pressing importance (Welch et al., 1995, 1998a,b). Answers to both questions require an improved understanding of how animals such as Pacific salmon exploit the structure of the ocean.

ACKNOWLEDGEMENTS

We are indebted to Wildlife Computer for contributing the tags used in this study, and Tom Juhasz of the Institute of Ocean Sciences, Fisheries and Oceans Canada for deploying and retrieving them for us. We also wish to acknowledge the pioneering efforts of many of our colleagues in promoting the basic technology and for graciously sharing their advice, particularly John Gunn, Geoff Arnold, and Julian Metcalfe.

REFERENCES

Bourque, M-C., LeBlond, P.H. and Cummins, P.F. (1999) Effects of tidal currents on Pacific salmon migration: Results from a fine-resolution coastal model. *Can. J. Fish. Aquat. Sci.* **56**, 839–846.

Dat, C., LeBlond, P.H., Thomson, K.A., and Ingraham, W.J. (1995) Computer simulations of homeward-migrating Fraser River sockeye salmon: Is compass orientation a sufficient direction-finding mechanism in the north-east Pacific Ocean? *Fish. Oceanogr.* **4**, 209–216.

DeLong, R.L., Stewart, B.S., and Hill, R.D. (1992) Documenting migrations of northern elephant seals using day length. *Mar. Mammal Sci.* **8**, 155-159.

Eveson, J.P., and Welch, D.W. (2001) Evaluation of techniques for attaching archival tags to salmon: Influence on growth and survival. *In* Moore, A. and Russell, I. (eds.) Fish Telemetry: Proceedings of the 3rd Conference on Fish Telemetry in Europe. MAFF Technical Report (in press).

Itoh, T., Tsuji, S. and Nitta, A. (1999) Trans-Pacific migration of bluefin tuna observed with archival tags. *In* Proceedings of the 50th Annual Tuna Conference, Lake Arrowhead, CA, May 24-27, 1999. Inter-American Tropical Tuna Commission.

Ivanoff, A. (1977) Oceanic absorption of solar energy. *In* Kraus, E.B. (ed.) *Modelling and Prediction of the Upper Layers of the Ocean*, Pergamon Press, Oxford, pp. 47-71.

Jerlov, N.G. (1976) *Marine Optics*. Elsevier Oceanography Series, Vol. 14. Elsevier Scientific Publishing Company, Amsterdam. 231 pp.

Lutcavage, M.E., Brill, R.W., Skomal, G.B., Chase, B.C., and Howey, P.W. (1999) Results of pop-up satellite tagging of spawning size class fish in the Gulf of Maine: Do North Atlantic bluefin tuna spawn in the mid-Atlantic? *Can. J. Fish. Aquat. Sci.* **56**, 173-177.

Ogura, M., and Ishida, Y. (1995) Homing behaviour and vertical movements of four species of Pacific salmon (*Oncorhynchus spp.*) in the central Bering Sea. *Can. J. Fish. Aquat. Sci.* **52,** 532-540.

Peterman, R.M., Marinone, S.G., Thomson, K.A., Jardine, I.D., Crittenden, R.N., LeBlond, P.H., and Walters, C.J. (1994) Simulation of juvenile sockeye salmon (*Oncorhynchus nerka*) migrations in the Strait of Georgia, British Columbia. *Fish. Oceanogr.* **3**, 221–235.

Thomson, K.A., Ingraham, W.J., Healey, M.C., Leblond, P.H., Groot, C., and Healey, C.G. (1994) Computer simulations of the influence of ocean currents on Fraser River sockeye salmon (*Oncorhynchus nerka*) return times. *Can. J. Fish. Aquat. Sci.* **51**, 441–449.

Wada, K., and Ueno, Y. (1999) Homing behaviour of chum salmon determined by an archival tag. North Pacific Anadromous Fish Commission Document 425, 29 pp.

Walker, V.R., Myers, K.W., Davis, N.D., Aydin, K.Y., Frieland, K.D., Carlson, H.R., Boehlert, G.W., Urawa, S., Ueno, Y. and Anma, G. (2000) Diurnal variation in thermal environment experienced by salmonids in the North Pacific as indicated by storage tags. *Fish. Oceanogr.* **9**, 171-186.

Welch, D.W., Chigirinsky, A.I., and Ishida, Y. (1995) Upper Thermal Limits on the Oceanic Distribution of Pacific Salmon (Oncorhynchus spp.) in the Spring. *Can. J. Fish. Aquat. Sci.* **52**, 489-503

Welch, D.W., Ishida, Y., and Nagasawa, K. (1998a) Thermal limits and ocean migrations of sockeye salmon (Oncorhynchus nerka): long-term consequences of global warming. *Can. J. Fish. Aquat. Sci.* **55**, 937-948.

Welch, D.W., Ishida, Y., Nagasawa, K., and Eveson, J.P. (1998b) Thermal limits on the ocean distribution of steelhead trout (*Oncorhynchus mykiss*). *In* Welch, D.W., Eggers, D.M., Wakabayashi, K. and Karpenko, V.I. (eds.) Assessment and status of Pacific Rim salmonid stocks. N. Pac. Anadr. Fish Comm. Bull. No. 1, 396-404.

Welch, D.W., and Eveson, J.P. (1999) An assessment of light-based geoposition estimates from archival tags. *Can. J. Fish. Aquat. Sci.* **56**, 1317-1327.

Improving our Understanding of Tropical Tuna Movements from Small to Large Scales

Laurent Dagorn, Arnaud Bertrand, Pascal Bach, Michel Petit,[1] Erwan Josse[2]
Pôle halieutique méditerranéen et tropical, BP 171, 34203 Sète Cedex, France
[1]Maison de La Télédétection, France
[2]IRD, BP 70, 29280 Plouzané, France

Key words: tuna, behaviour, tags, movements, scaling, models

Abstract: One tactical objective of behavioural studies in fisheries sciences is to understand the 3-D movements of fish, with the ambitious strategic objective of being able to explain some large-scale movements and distributions from knowledge of small-scale behaviour. The present paper reviews different possibilities to improve our understanding of tropical tuna movements, from small (days) to large scales (weeks and months). We propose some ideas for better observations of fine-scale movements of fish (sonic tags) and of the surrounding environment of the tagged fish. After determining behaviour rules, appropriate modelling should be developed in order to extrapolate results to larger-scale movements. This process can be achieved only if we can simultaneously observe the relevant factors of the environment at appropriate scales and the large-scale movements of fish (using pop-up archival tags), in order to force the models to reproduce the observed movements. This paper shows the importance of (i) small-scale studies, (ii) appropriate observations of large-scale movements and environment, and (iii) models of behaviour in order to extract relevant processes necessary to predict tuna dynamics.

1. INTRODUCTION

Levin (1992) clearly stated that understanding patterns in terms of the processes that produce them is the "essence of science" and is the key to the development of principles for resource management. He went on to say that without an understanding of mechanisms, one must evaluate each new stress on each system *de novo*, without any scientific basis for extrapolation. Behavioural ecology studies focus on the interactions between animals and

J.R. Sibert and J.L. Nielsen (eds.), Electronic Tagging and Tracking in Marine Fisheries, 385–405.
© 2001 *Kluwer Academic Publishers. Printed in the Netherlands.*

their environment. One important tactical objective of behavioural studies in fisheries sciences is to be able to explain some large-scale movements and distribution from knowledge of small-scale behaviour. For such purpose, it is necessary to extract and abstract the fine-scale features that are relevant for the phenomena studied on other scales.

The purpose of this paper is not to draw a general framework for any study aimed at understanding fish movements from small to large scales, to present the state-of-the-art of electronic tags, or to review the different techniques that exist to model fish movements. The objective is to propose ideas in order to improve our understanding of tropical tunas movements. To do this, we studied the specific biology and behaviour of skipjack (*Katsuwonus pelamis*), yellowfin (*Thunnus albacares*), albacore (*T. alalunga*), and bigeye tuna (*T. obesus*). Tunas are extensively distributed throughout world's oceans and support one of the world's largest commercial fisheries. Tropical tunas can be characterized as apex predators living in the pelagic environment. Reviewing the literature on energetics and physiological abilities of tunas, Brill (1996) showed that it is likely that the high performance physiology of tunas depends on (i) rapid rates of somatic and gonadal growth, (ii) rapid rates of digestion, and (iii) rapid rates of recovery from exhaustive exercise rather than on their exceptional swimming abilities, as was previously surmised. Tropical tunas are known to travel long distances, to swim between the surface and depths of 500 m or greater, and to school. These behavioural abilities are important considerations for stock assessment and management of related fisheries. The movements of tunas and the resulting distributions are one of the key biological problems in the management of tuna fisheries (Hunter *et al.*, 1986). Collecting data on temporal and spatial movement patterns has therefore been a priority for tuna fishery biologists for decades. Hunter *et al.* (1986) have clearly identified that our lack of understanding of distribution and movement of tropical tunas has constrained the development of rational management policy for many fisheries.

Observing movements of fishes like tunas implies the use of tags. Conventional tags have been widely used to estimate tuna movements, as well as some population parameters (fishing mortality, natural mortality, growth) of primary importance for stock assessment (Kleiber *et al.*, 1987; Hilborn, 1989; Deriso *et al.*, 1991; Kleiber and Hampton, 1994; Bertignac *et al.*, 1999; Holland *et al.*, 1999; Sibert *et al.*, 1999; Hampton, 2000). Ultrasonic transmitters attached to the animals make it possible to collect data on the fine-scale movements of tunas (Yuen, 1970; Carey and Olson, 1981; Cayré and Chabanne, 1986; Holland *et al.*, 1990; Cayré, 1991; Holland *et al.*, 1992; Block *et al.*, 1997; Josse *et al.*, 1998; Marsac and Cayré, 1998; Brill *et al.*, 1999; Dagorn *et al.*, 2000a), usually with information on the fish's depth (with pressure sensors) and sometimes on the fish's body temperature (Holland *et al.*, 1992). The large sizes of bluefin tunas have allowed scientists to equip them with large electronic tags that make it possible to collect large-scale data on their movements and biology. Data storage tags (archival tags) have been developed for monitoring the geoposition (based on ambient light levels), thermal physiology, and diving behaviour of bluefin tunas (Gunn *et al.*, 1994; Block *et al.*, 1998a; Block *et al.*, this volume). Finally, pop-up satellite tags have been employed to study the large-scale movements and physiology of bluefin tunas (Block *et al.*, 1998b, Lutcavage *et al.*, 1999, Block *et al.*, this volume).

Large-scale movements of tunas only concern horizontal movements, as their vertical movements are limited to a few hundreds of meters. Therefore, this study will focus on horizontal movements, although vertical movements should not be ignored when interpreting small-scale behaviour. We assume that small-scale processes are responsible for some part of large-scale movements of tropical tunas; i.e., if we extract behaviour rules that occur at fine scales in relation to features of the environment at the same scales, fisheries biologists will have the appropriate knowledge to transfer the results to larger scales. All the physiology and behaviour characteristics of tropical tunas need to be considered, as well as recent technological and modelling improvements, to propose frameworks for studying tuna movements. Using the approach of Hunter *et al.* (1986), we discuss the use of presently available techniques to observe tuna behaviour at scales from days to weeks and months. First, we will explore the different designs of tracking studies and the importance of these designs to the results. We will also review the different factors that are likely to affect tuna behaviour and identify the gaps that should be filled in order to improve our understanding of the spatial dynamics of tunas, from small-scale studies and the determination of behaviour rules to modelling of behaviour and large-scale movements. Finally, an attempt to include behaviour in stock assessment studies is proposed, following the work of Fréon and Misund (1999).

2. DESCRIPTIVE, CORRELATIONAL AND MANIPULATIVE STUDIES

As in any scientific study, the use of tags should be clearly defined and should correspond to the questions addressed by the study. White and Garrott (1990) divided wildlife tracking studies into three basic conceptual designs: descriptive, correlational, and manipulative. Descriptive studies aim to observe natural behavioural processes, usually with no attempt at hypothesis formulation or testing before the data are collected. These studies, however, are limited to learning what an animal does but not why the animal does it.

This approach is appropriate for exploring a wholly unknown subject. Correlational studies represent a more informative approach and consist of formulating one or more hypotheses and observing the behaviour of animals to determine if it supports the hypotheses. For example, one study might measure a factor in the habitat while collecting information on the movements of the animal. Correlations can be obtained between this factor and features of the movement of the animal. The investigator, however, must be aware that in a correlational study a relationship between two or more variables does not always imply cause and effect. Manipulative studies require that the system be perturbed in order to demonstrate cause-and-effect relationships, i.e., behavioural mechanisms. These three approaches are usually developed in a phased approach. The descriptive phase corresponds to the first observational studies to determine what an animal does. The correlational phase should provide information on the hypotheses to understand how and why an animal moves in its habitat. Manipulative studies can then be developed in order to understand why the animal does what it does and to design an effective way to predict movements of animals. Then those results can be applied to different scales. Manipulative studies, however, are not always easy to develop, especially for species living in pelagic habitats. One solution is to conduct correlational studies using a comparative approach, which can provide results close to those obtained from manipulative studies. These correlational experiments use identical protocols but in different physical or biological environments. Such a design provides different values of factors and can be used to define empirical cause-and-effect relationships.

Ultrasonic tags have been widely deployed on tropical tunas (Yuen, 1970; Carey and Olson, 1981; Cayré and Chabanne, 1986; Holland *et al.*, 1990; Cayré, 1991; Holland *et al.*, 1992; Block *et al.*, 1997; Josse *et al.*, 1998; Marsac and Cayré, 1998; Brill *et al.*, 1999; Dagorn *et al.*, 2000a) in descriptive and correlational approaches to fine-scale movements of individuals. These studies correspond to the first phase: observing what a tuna does, what movements it can achieve, and whether these movements can be explained by environmental factors. Their logical conclusion is that hydrological and biological environment play significant roles in tuna movements. Water temperature is easily measured and tunas are known to thermoregulate (review in Cayré *et al.*, 1988; Holland *et al.*, 1992). Therefore, temperature is the most common variable measured during tracking experiments. Other hydrological factors such as dissolved oxygen or salinity are also recorded (occasionally before or after the tracking operation). Recently, the prey environment has been observed during tracking experiments in French Polynesia (Josse *et al.*, 1998, Dagorn *et al.*, 2000a). These factors, however, are commonly used to interpret the vertical behaviour of these fish and are rarely applied to their fine-scale horizontal movements. The horizontal movements of tunas are known to be affected by well-defined physical structures such as floating objects, outer reef drop-offs, or seamounts, because these features are easily observed during telemetry experiments. Cause-and-effect relationships, however, have seldom been obtained (Dagorn *et al.*, 2000b). The lack of such results, especially on the horizontal movements of tunas, prevent biologists from extrapolating results on behaviour to data usable for stock assessment.

3. DETERMINATION OF SMALL-SCALE PROCESSES: IDENTIFYING FACTORS THAT AFFECT TUNA MOVEMENTS

The small-scale movements exhibited by a tuna are motivated by both internal and external factors. Internal factors like age, stomach fullness, condition factor (fat content), and reproduction state are likely to affect the behavioural motivation of fish and thus their movements. In addition to the diel cycle, the local environment of fish represents the main external stimuli that influence fine-scale tuna movements: (i) oceanography (sea temperature, dissolved oxygen, currents), (ii) biological environment (prey, predators, competitors, and conspecifics, with the special case of individuals being members of a school), and (iii) aggregative structures or geolocational environment (floating objects, seamounts, islands). In behavioural ecology, it is therefore necessary to collect information on individuals (movements, internal stimuli) and on their local environment (external stimuli) to understand the role of the different factors on the behaviour of animals. Our purpose here is to examine the different techniques available to simultaneously observe the paths of tunas and their environment.

3.1 Individuals

Telemetry experiments on tunas have provided more significant results on vertical movements of tunas than on their horizontal movements. We can identify two main reasons for that. First, horizontal movements are not really limited, whereas vertical movements are limited to a few hundred meters. The observations of the movements and of the surrounding environment are therefore easier in the vertical dimension. Second, currently available ultrasonic telemetry receiving equipment provides accurate depths and accurate directions but only rough estimates of the distance between the transmitter (fish) and the receiver. During these experiments, the tracks are those of the tracking vessel, not those of the fish. An alternative would be to use dead reckoning to determine animal positions. Dead reckoning necessitates sending (or storing, with loggers) information on animal speed, animal vertical displacement and animal heading at regular intervals. This system is fairly recent (Wilson, 1998) and has not yet been tested on fishes. While accurate depth sensors exist and speed sensors have recently been improved (Wolcott, 1995; Wilson, 1998), designing a small and reliable compass to determine animal headings is still problematic. Precise data on horizontal movements are necessary to improve our analyses of the fine-scale behaviour of tunas, as it is done on terrestrial species. Such data are necessary to extract exact patterns in speed and sinuosity and their relation to environmental features, which can be exploited in population models.

Improvements in sonic tags in studying tuna behaviour require increasing the distance of detection between the tag and the receiver, which will help to increase the duration of the tracking operations and decrease the possible influence of the tracking platform on the behaviour of the tagged fish (Dagorn *et al.*, 2001). Increasing the distance of detection is essential and might be possible using specific receivers (R. Brill, pers. comm.).

One other area of possible improvement might be to use research vessels only for complementary observations (environment) rather than for tracking operations. The idea is to design autonomous tracking systems that eliminate the use of humans and research vessels in tracking studies, which would reduce the costs of operations. For instance, Goudey *et al.* (1998) have developed a robotic kayak for autonomous fish tracking. The performance of this surface platform still needs improvement to ensure successful tracking operations in the oceans. The research vessel could then deploy a fleet of autonomous tracking platforms and achieve complementary observations of the environment (Fig. 1). A more expensive alternative would be to use autonomous underwater vehicles capable of 3-D motion and able to track and record fish movements for several days/weeks. Such an underwater tracking system would be useful for species occupying deep waters during daytime, such as bigeye tuna. A submarine vehicle could make underwater observations that are not always possible from surface platforms, especially on the deep scattering layer (see Fernandes *et al.*, 2000). One of the advantages of such autonomous systems, whether surface or underwater vehicles, would be that the vehicle can carry sensors to observe the fishes environment (e.g., temperature sensors, sonars and echosounders, ADCP, videos), thus eliminating the need for sensors mounted on the fish itself.

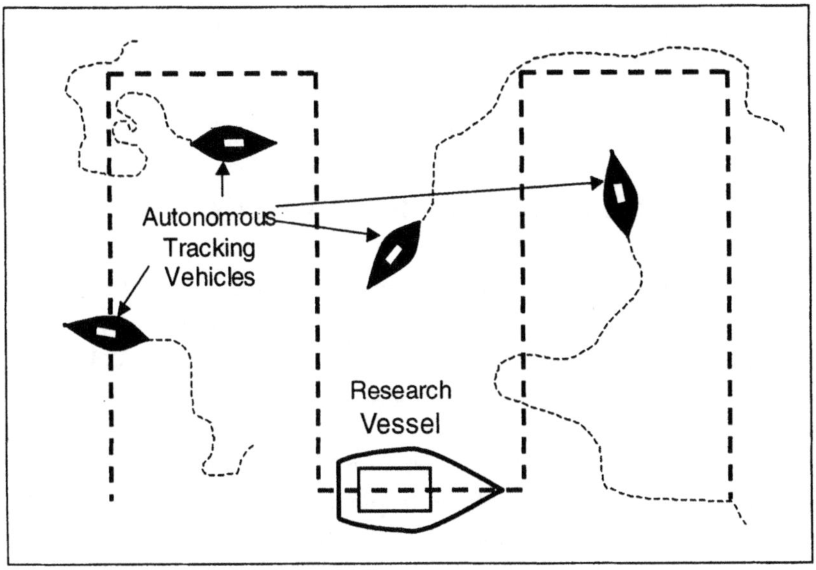

Figure 1. A schematic of complementary observations of the local environment and tracking of fish by using one research vessel and a fleet of autonomous surface vehicles. The bold dashed line represents an acoustic traverse made by the research vessel observing the local environment. The light dashed lines represent tracks by four autonomous surface vehicles, each tracking a fish.

3.2 Schooling

Tunas are known to school, and the effects of a school on the movements of individuals are not known. Tracking a school is possible using a sonar, and this has been done on small pelagic fishes for a maximum of a few hours (Misund, 1993; Pitcher *et al.*, 1996). A school is not a stable closed structure, and tracking a school corresponds to tracking all the individuals that are members of a school during a given period. The dynamics of tunas within a school, the mixing and splitting events, are not known. Some studies (Bayliff, 1988; Hilborn, 1991; Klimley and Holloway, 1999) have estimated tuna school cohesion using long-term tagging studies, while others examined the characteristics of tuna school sizes (Dagorn *et al.*, 1997a; Bonabeau *et al.*, 1999). However, none of these studies provided clear conclusions on tuna schooling behaviour; further experiments, especially at a fine scale, are needed. A very few experiments tracked tunas in one school (Block *et al.*, 1997). Tracking fish from the same school, all equipped with acoustic tags, with the possibility of temporary (or permanent) divergent movements, is not realistic from large research vessels, as it would practically necessitate one research vessel per tagged fish. One solution would be to use the large vessel as the tagging platform and the fleet of small autonomous vehicles to track the tagged fishes (Fig. 1).

Even though studying schooling behaviour using only a few fish from a school might not be sufficient to understand schooling dynamics, combining individual tagging and acoustics (sonars and echosounders) is currently the best available method to observe small-scale schooling behaviour of tunas.

3.3 Internal Environment of Individuals

A key point for interpreting movements of tunas is to determine their behavioural motivation, e.g., to determine when fish are hungry, by observing-the feeding events. Available sensors have been used on different species to measure tailbeat frequency, heart frequency, temperature of the viscera, pH, etc. (see Wolcott, 1995). For instance, Gunn *et al.* (this volume) have estimated rations of southern bluefin tuna from visceral temperature using archival tags. No *in situ* experiments on tropical tunas, however, have yet been reported with telemetered information on feeding activity; precise laboratory experiments are needed to determine the appropriate sensor to observe feeding events in tropical tunas.

3.4 Hydrological Environment

Different techniques can be utilized to collect hydrological data during small-scale *in situ* studies. Sensors on the tag provide information on the environmental sphere around the fish. Sensors for temperature already exist but not for dissolved oxygen, which is also a key factor for tuna movements (especially vertical movements).

Oceanographic stations (CTD or XBT) onboard research vessels or even on autonomous vehicles (Goudey *et al.*, 1998) can provide data on the hydrological environment of the area where the fish is swimming. It would also be necessary to observe the local area of tracking through remote sensing data. Remote sensing is a major tool to observe the environment of the oceans. Satellite infrared imaging, narrow spectral bands, radar backscatter sensors, and radar altimeters operationally record daily global data of the ocean surface. Analysing and modelling of these data can provide daily to weekly sea surface temperature, integrated column chlorophyll content, surface winds plus associated parameters (vorticity), and sea level anomalies associated with geostrophic near surface currents. All of these products are not available everywhere in the world but in an optimum context (receiving station, agreement with space agencies), it is realistic to expect a description of "oceanicscape" with a space/time basis resolution of one-mile/three-days. In the framework of small-scale studies, it could seem to be a poor resolution. The repetitive and global nature of these products, however, insure coherence in the description of geophysical processes. Thanks to this coherence, it is possible to know the oceanicscape that corresponds to the location and the time of the experiment. Even if the use of airborne sensors instead of satellites could dramatically improve this space/time resolution, remote sensors are now limited to time and surface area; depth remains a future but essential prospect. These combined data would yield a precise view of the oceanographic conditions during tracking operations.

3.5 Biological Environment

At present, hydroacoustics represents the major direct technique to characterize the biological environment. It allows direct observations of tunas (Bertrand *et al.*, 1999a; Josse *et al.*, 1999; Bertrand and Josse, 2000; Josse and Bertrand, 2000) and their biological environment (Marchal and Lebourges, 1996; Josse *et al.*, 1998, Bertrand, 1999; Bertrand *et al.*, 1999b), and it can be used simultaneously during acoustic tracking (Josse *et al.* 1998; Dagorn *et al.*, 2000a). The use of multi-frequency and multi-beam echosounders will improve the quality of data. Further studies on target strength are needed, especially to distinguish conspecifics, competitors, and predators. Moreover, it is necessary to extend the observations of the environment (biological as well as physical) around a tagged tuna. As stated above, it would be appropriate to use a fleet of autonomous vehicles for tracking the fish and have the research vessel for simultaneous survey in the area of tracking (Fig. 1).

Video cameras fitted to animals themselves is another complementary tool that could be used in the future . This technology has been used on seals (Parrish *et al.*, 2000), sea turtles, sharks and whales, and can record feeding events. The existing technology, however, cannot be used on fishes like tunas but can represent one direction for the future.

Simultaneous tagging of conspecifics or competitors associated with the tagged fish represents an appropriate method for studying the associative behaviour of tunas. Tagging of species that are associated with tunas, like dolphins or sea birds, can allow the investigation of these associations. Dolphins associated with tunas in the Eastern Pacific Ocean, or sea birds in general, can be equipped with satellite tags (Argos) so that the animals are tracked simultaneously while they are associated. The outcomes of such experiments can provide insights into ecological relationships. If other associated and tagged animals can carry sensors that tunas cannot, this approach can open a new field wherein associated animals become observers of the tunas' environment.

Such animals could also represent appropriate mobile platforms to carry listening stations (see Census of Marine Life) to places where tunas go, such as common feeding areas.

3.6 Geolocational Environment

The positions of islands, seamounts, and anchored FADs are constant and easily known. Drifting FADs, however, are difficult to locate and can significantly affect the movements of tunas. These objects can be easily observed using sea surveys, but further developments are needed to observe them by remote techniques (see next section).

4. OBSERVING LARGE-SCALE MOVEMENTS OF TUNAS WITH FINE-SCALE INFORMATION

Pop-up archival tags are certainly the best tool to observe large-scale movements of tunas. The geographic positions provided by such tags are generally obtained at a relatively broad scale, i.e. one position per day (Block *et al.*, 1998b; Lutcavage *et al.*, 1999; Block *et al.*, this volume) and a relatively low geographic precision (Welch and Eveson, 1999). These geolocational data indicate the areas visited by the fish, but improvements of geolocational data are clearly needed.

The development of a network of automated "listening" stations that can record the presence of fish equipped with data storage tags in many different places should be encouraged. The Census of Marine Life[1] has made this one of its major goals. Of course, one must keep in mind that these buoys can modify the movements of tunas, as can any floating object. These stations can be fixed or drift with the current. Such stations could also be put onboard fishing vessels if a fleet is operating in the zone, the fishing vessels becoming the listening stations.

Real-time daily positions of fish equipped with pop-up archival tags are not available and therefore preclude the use of research vessels for simultaneous observations of the environment. We would like to emphasize our focus on the observation of the biological environment and the possible use of tunas as observers of their environment. Information about the structure of the environment with respect to the total biomass of prey is very important in interpreting fish movements (Bertrand, 1999; Swartzman *et al.*, 1999). If internal sensors that reveal tuna feeding behaviour were available, small-scale studies could allow us to determine if fish adopt a typical feeding behaviour according to the structure of the biological environment. For instance, if prey are dispersed and distributed through classical scattering layers, it is likely that tunas feed over a large period, picking prey opportunisti-

[1] (See Census of Marine Life, http://core.cast.msstate.edu/censhome.html)

cally (Fig. 2-a). If large prey aggregations are present in the habitat, tuna may forage more efficiently when they find these patches and be rapidly satisfied (Fig. 2-b). Therefore, if fish exhibit a particular feeding behaviour for each structure of the environment, it is possible to link these two components (feeding behaviour/structure of the environment). Complementary information on prey abundance is also needed. Wilson (1998) stated that it is possible to derive measures of prey abundance by appropriate and careful design of tagging operations. The idea is to apply the catch per unit effort (CPUE) concept, well-known by fisheries scientists. The first step is to be able to record catch, i.e. feeding events, using internal sensors with tags. Effort will be defined from movements of tagged tunas. This original protocol has been used with penguins (Wilson, 1998) and, even if not applicable now on tunas, it merits further attention because it can provide an estimate of the abundance of prey. Using appropriate sensors with pop-up archival tags, the tagged tunas would become perfect "autonomous submarine vehicles" to observe the structure and the abundance of prey on a large-scale basis.

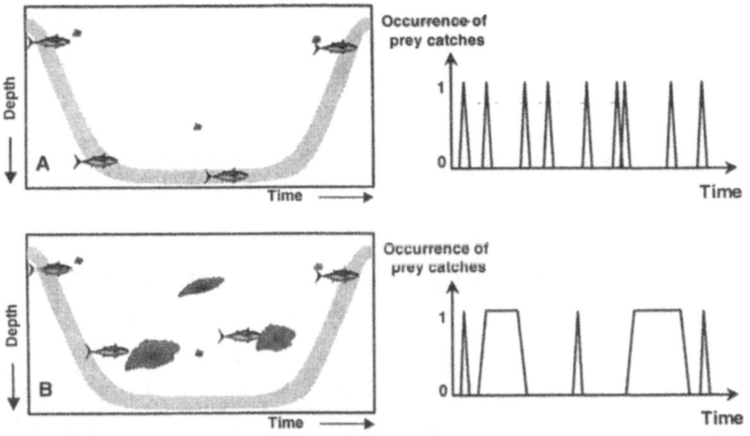

Figure 2. Using feeding sensors to estimate the structure of the habitat. Feeding activity of a predator can be related to a particular structure of the habitat. For instance, a predator feeding behaviour in environment A with only the presence of the DSL is likely to be different than the same predator feeding in environment B with patches of prey located at intermediate depths.

Pop-up archival tags also seem to be a promising tool to study schooling behaviour. The main difficulty (as with any multiple tagging experiment) is trying to tag a large number of fish from the same school. Obtaining quasi-continuous fine-scale data on associations of individuals on a long-term basis will be of primary importance to determine the frequency of splitting and mixing events. Such scientific operations should greatly be encouraged.

Satellites remain the main technique that should be used to observe the environment on a long-term basis (Petit *et al.*, 1994). The vertical structure of the ocean, however, is not yet well observed using satellites, and this technique requires further development. The new generation of remote sensors/satellites, however, will improve capabilities through the development of hyperspectral and high-resolution technology. Such satellites could also collect original information on the spatial distribution of floating objects, which is an important challenge for tuna fisheries management. Moored arrays provide vertical and horizontal data on the hydrological structure of the ocean. The main problem is that these moored stations are fixed and can only be used if the fish swim in the vicinity of these structures. Such a system does not seem to be really exploitable, except for the global understanding of the dynamics of the ocean, which is of course necessary to improve our *in situ* observations. The new technology of autonomous profiling floats enables description of most of the inner ocean features to 2000 meters, which complements surface remote observations. These floats, developed over ten years in the WOCE/ALACE framework and for the future GODAE/ARGO program, could be adapted for observing areas during movements of tagged individuals.

5. MODELLING BEHAVIOUR, CHANGING SCALES

Interpreting tag data in terms of why and how the tagged fish did what they did is not always easy. Modelling can play a powerful role in exploring the possible consequences of individual factors that cannot be easily separated experimentally and in relating fine-scale data to broad-scale patterns (Levin, 1992; Kirby, this volume). Models are also appropriate when combining several factors in continuous space and time (Giske *et al.*, 1998). Different modelling methods have allowed for great advances in ecological theory. Fishery biologists have been developing mathematical models for years in order to reproduce the general patterns of movements of tunas observed from conventional tagging or CPUE analyses (Kleiber *et al.*, 1987; Hilborn, 1989; Deriso *et al.*, 1991; Hampton, 1991; Kleiber and Hampton, 1994; Lehodey *et al.*, 1997; Bertignac *et al.*, 1998; Holland *et al.*, 1999; Sibert *et al.*, 1999; Maury *et al.*, 2001) and more recently from sonic tagging data (Sibert, this volume). An other branch of modelling implements rules for individuals: the individual-based models or IBM. This family of models has been applied to terrestrial ecology (see Huston *et al.*, 1988; Hogeweg and Hesper, 1990) and fish ecology (Tyler and Rose, 1994; Shin and Cury, 1999, 2001) and are founded on the role of individual differences in ecology and population dynamics. In particular, this technique has recently been applied on tunas (Dagorn and Fréon, 1999; Dagorn *et al.*, 2000b,c; Newlands and Lutcavage, this volume).

Our objective is not to review the different modelling techniques that exist to study the spatial dynamics of fish (see Tyler and Rose, 1994; Giske *et al.*, 1998 for reviews; Kirby, this volume, specifically for tunas) but rather to focus on some modelling techniques from the artificial life field that can be used to pass from behavioural observations to large-scale patterns, by exploiting present knowledge and exploring the effects of new behaviour rules. Artificial life models are especially appropriate for exploratory simulations where modellers and biologists look for plausible behavioural mechanisms to reproduce fish movements and/or examine the effects of changes in the environments. The general principle is that by generating "simple" agents capable of adaptive processes, behaviour can emerge that allows scientists to generate and evaluate hypotheses for behavioural mechanisms for tunas. Artificial neural networks (ANNs) (Rumelhart and McClelland, 1986) and genetic algorithms (GAs) (Holland, 1992) are among the major techniques used for producing adaptive behaviour. These techniques are particularly appropriate for representing the evolution and learning processes that shape a behaviour. ANNs apply neurobiological principles of synaptic brain activity to model behavioural output by differential weighting of input variables. The genetic algorithm or, more generally, genetic programming, is a technique that applies the principle of evolution by natural selection to search for adapted (sometimes optimal) solutions to a problem. These two methods can be combined, i.e. genetic programming can be used to determine the appropriate characteristics of ANNs that make them efficient for the task fixed by the modeller, allowing non-supervised learning. These techniques have only been recently applied for modelling behaviour of pelagic fish (Dagorn *et al.*, 1995; Dagorn *et al.*, 1997b; Huse and Giske, 1998, Giske *et al.*, 1998; Dagorn *et al.*, 2000c; Huse *et al.*, 1999).

We think it is important to mention the difference between adaptation and optimisation (Menczer and Belew, 1996). An individual adapted to its environment is not necessarily the best "living solution" to live in its habitat. All the modelling techniques that search for optimality can be discussed; these techniques are very useful for theoretical biology to understand what evolution could find favourable, but might not be appropriate to reproduce real behaviours. The objective is not to find the "best tunas" of the world, but to reproduce movements exhibited by real tunas, which are adapted to their habitat, not necessarily optimal.

5.1 Modelling the Environment

In the adaptation approach, the representation of the environment is a key component of the modelling process. Artificial animals will be equipped with adaptive tools so that they will be able to adapt to the task fixed by the modeller. The environment will "shape" the emergent behaviour of the artificial tunas. An appropriate model of tuna environment should include abiotic and biotic data. The challenge is to use essential variables and to represent the spatial and temporal dynamics of these variables for the duration of the studied period. The best approach is to use real observations of the studied area, which is easier with abiotic data such as daily SST obtained from remote sensing (Dagorn *et al.*, 1997b; Humston *et al.*, 2000). Remote sensing is likely to play a major role in this part, but particular attention should be paid to the fine-scale resolution of data, both in space and time. The use of acoustic data for prey distribution and dynamics is not yet a common practice and should be encouraged (Brandt and Mason, 1994). When such data are not available, particular attention should be paid to the choice of the model of the environment. For instance, Lehodey *et al.* (1997, 1998) have coupled a bio-geochemical model with a general circulation model for modelling tuna forage in the equatorial Pacific Ocean.

5.2 Modelling the Movements

The approach adopted here corresponds to those presented by Dagorn *et al.* (1997b), but is extended to the use of archival tags rather than conventional tags or fisheries data for representing large-scale movements. The artificial tuna is able to sense data in its local environment, make a decision and move. The mapping from sensory states to motor actions is represented using an ANN, and the unsupervised learning can be achieved through genetic programming. From an appropriate dynamic model of the pelagic environment, the objective is to find the possible behaviour solutions through the adaptive process of the ANN that allows reproduction of large-scale patterns from interactions with the environment and under the constraints of the knowledge on the physiology/behaviour of the species (small-scale studies). Equipped with behavioural abilities (swimming performances, reactions to some local conditions, etc.) obtained from small-scale studies, artificial tunas will have to determine behaviour rules to reproduce paths obtained from pop-up archival tags. An artificial tuna will be considered as adapted to its dynamic environment when the "distance" between its path and those of a tagged fish is very short or null (Dagorn *et al.*, 1997b). At the beginning of a simulation, the artificial tuna only uses its behavioural abilities and some random rules to move. The adaptive process of the genetic programming leads to changes in the behaviour rules, so that after some simulations, the artificial tuna has found the appropriate rules to reproduce the paths of the real tagged fish. The modeller therefore has a model of small-scale behaviour that can be used to reproduce large-scale movements from interactions with the environment.

Schooling behaviour should be taken into account while modelling the movements of tropical tunas. The schooling behaviour of fish has been successfully modelled by several authors (Aoki, 1982, 1986; Huth and Wissel, 1992, Reuter and Breckling, 1994; Romey, 1996; Vabo and Nottestad, 1997; Stocker, 1999), but the exact effects of schools on individual movements on a small- and a large-scale basis has not been clearly established. Newlands and Lutcavage (this volume) present a methodological framework wherein modelled movements of bluefin tuna integrate the schooling behaviour. Further theoretical and experimental studies are needed to understand the possible effects of school size and school composition on movements of the members of the school. Schooling should be one of the major questions addressed by modellers and biologists in order to better assess the role of schooling behaviour on large-scale movements of fishes like tropical tunas. IBMs appear to be an appropriate modelling technique to study this particular behaviour.

5.3 Predicting Tuna Dynamics

After conducting small-scale studies to determine behaviour rules, models that exploit this knowledge and are able to reproduce large-scale movements are established. The next step is prediction, which is the general objective of those combined studies. In the field of pelagic fish dynamics, prediction means forecasting where individuals go in their environment. Following the behavioural ecology approach, this implies the ability to predict the habitat of tropical tunas. This very ambitious project can only be realized through modelling and the use of various data collected from satellites, sea surveys, moored arrays, and merchant ships. As fish forage appears to be relevant for tuna dynamics (Lehodey *et al.*, 1997, 1998), we consider that, in the future, more efforts should be devoted to "routine" observations of this variable. One idea is to put echosounders on moored buoys and merchant ships, as these are already equipped with tools to observe the physical environment. A great effort is made to collect physical data of the ocean, which contributes to the development of more accurate models of the oceans. From the standpoint of modelling tuna movements, one weakness of ocean models is the lack of attempts to predict forage abundance. Collecting data on the biological components of the oceans is highly necessary and is possible on a routine basis, as is now done for physical ocean parameters. Without such effort, it will not be possible to develop the best synthetic model of tuna movements.

6. CONCLUSION

Table 1 summarises the technical improvements that we consider as necessary to improve our understanding of tuna movements. Of course, tags can be used for other purposes, in particular pop-up archival tags provide valuable information on population models. In this paper, we restricted our discussion to the field of behavioural ecology of tropical tunas.

Table 1. Perspectives for future technical developments to study the movements of tropical tunas

	Necessary improvements
Acoustic telemetry	Greater distance of detection
	Accurate position relative to the receiver (dead reckoning?)
	Intelligent autonomous vehicles to track the fish, possible observations of the surrounding environment
	Internal tags for observing tuna activities (feeding)
Pop-up archival Tags	Accuracy of geographic positions
	Transfer of real-time daily positions
	Internal tags for observing tuna activities (feeding) (same as for sonic tags)
Environment	Satellites (hydrology, biology, geolocational environment)
	Tagged tunas or other animals living in the same niche as observers of their environment
	Routine observations of the biological environment (echosounders on moored buoys, merchant ships, ...)

Including behaviour in a study of population dynamics is similar to scaling problems. Levin (1992) stated that the key to scaling and interrelating phenomena at different scales is knowing what fine detail is relevant at the higher levels and what is noise. We can connect phenomena on different scales if we can extract and abstract the appropriate details of fine-scale processes responsible for larger scales phenomena. Fréon and Misund (1999) have identified different weak points related to stock assessment. One of these points concerns small-scale temporal and spatial data collection. These authors stated that "nowadays, our limits in stock assessment are more on the side of basic knowledge than on methods of representation and modelling". Behavioural studies must seek the major processes that drive fish movement and evaluate the individual differences. Magurran (1993) states that individual differences can be viewed as the product of three mechanisms: (1) a variable environment, (2) phenotypic differences, and (3) behaviour of other individuals. The major issue for behavioural ecologists is therefore to extract the rules that govern the behaviour of individuals and to estimate how phenotypic differences can generate different movement patterns. One practical problem is to observe a sufficient number of situations in order to extract the major rules used by most of the individuals of the population to move in their environment. But even if all individuals were equal, variable environments should generate different patterns. One of the main objectives in the future will be to observe movements of fish independently of human activities, i.e. of fishing and research vessels. By deploying autonomous vehicles able to track fish to collect fine-scale data on their movements and by developing the use of pop-up archival tags as well as data storage data with listening stations, it will be possible to determine the spatial dynamics of tunas in and out of fishing areas. Appropriate modelling (e.g., using artificial life techniques and maybe combined with mathematical models) should allow us to pass from observations of individuals to scenarios of spatial dynamics of stocks. Such information could then be used to determine appropriate conservation and fishery management. One major issue in transferring information is our ability to observe the environment at appropriate scales. It seems to us that particular attention should be focused on the simultaneous use of data from complementary techniques such as satellites, aerial and sea surveys, moored arrays, autonomous vehicles or floats, and of course, electronic tags.

ACKNOWLEDGEMENT

We would like to sincerely thank John Sibert for his valuable suggestions.

REFERENCES

Aoki, I. (1982) A simulation study on the schooling mechanisms in fish. *Bull. Jpn. Soc. Sci. Fish.* **48**, 1081-1088.

Aoki, I. (1986) A simulation experiment on individual differences in schooling behaviour of fish. *Bull. Jpn. Soc. Sci. Fish.* **52,** 1115-1119.

Bayliff, W.H. (1988) Integrity of schools of skipjack tuna, *Katsuwonus pelamis*, in the Eastern Pacific Ocean, as determined from tagging data. *Fish. Bull.* **86**(4), 631-643.

Bertignac, M., Lehodey, P., and Hampton, J. (1998) A spatial population dynamics simulation model of tropical tunas using a habitat index based on environmental parameters. *Fish. Oceanogr.* **7**(3-4), 326-334.

Bertignac , M., Hampton, J., and Coan, A.L. (1999) Estimates of exploitation rates for north Pacific albacore, *Thunnus alalunga*, from tagging data. *Fish. Bull. U.S.* **97**, 421-433.

Bertrand, A. (1999) Le système {thon – environnement} en Polynésie Française : caractérisation de l'habitat pélagique, étude de la distribution et de la capturabilité des thons, par méthodes acoustiques et halieutiques. Thèse de Doctorat de l'ENSAR, N° d'ordre, 99-31, N° série, H43, 315 pp.

Bertrand, A., Josse, E., and Massé, J. (1999a) *In situ* acoustic target strength measurements of Bigeye (*Thunnus obesus*) and Yellowfin tuna (*Thunnus albacares*) by coupling split-beam echosounder observations and sonic tracking. *ICES J. Mar. Sci.* **56**, 54-60.

Bertrand, A., Le Borgne, R., and Josse, E. (1999b) Acoustic characterisation of micronekton in French Polynesia. *Mar. Ecol. Prog. Ser.* **191,** 127-140.

Bertrand, A., and Josse, E. (2000) Acoustic estimation of longline tuna abundance, *ICES J. Mar. Sci.* **57**(4), 919-926.

Block, B.A., Keen, J.E., Castillo, B., Dewar, H., Freund, E.V., Marcinek, D.J., Brill, R.W., and Farwell, C. (1997) Environmental preferences of yellowfin tuna (*Thunnus albacares*) at the northern extent of its range. *Mar. Biol.* **130**, 119-132.

Block, B.A., Dewar, H., Williams, T., Prince, E.D., Farwell, C., and Fudge, D. (1998a) Archival tagging of Atlantic bluefin tuna (*Thunnus thynnus thynnus*). *Mar. Technol. Soc. J.* **32**(1), 37-46.

Block, B.A., Dewar, H., Farwell, C., and Prince, E.D. (1998b) A new satellite technology for tracking the movements of Atlantic bluefin tuna. *Proc. Natl. Acad. Sci. USA* **95**, 9384-9389.

Bonabeau, E., Dagorn, L., and Fréon, P. (1999) Scaling in animal group-size distributions. *Proc. Natl. Acad. Sci. USA*. **96**, 4472-4477.

Brandt, S.B., and Mason, D.M. (1994) Landscape approaches for assessing spatial patterns in fish foraging and growth. In *Theory and Application in Fish Feeding Ecology* (ed. by D.J. Stouder, K.L. Fresh and R.J. Feller), GUTSHOP '92, San Juan Islands, WA, pp. 211-238.

Brill, R.W. (1996) Selective advantages conferred by the high performance physiology of tunas, billfishes, and dolphin fish. *Comp. Biochem. Physiol.* **113A**(1), 3-15.

Brill, R.W., Block, B.A., Boggs, C.H., Bigelow, K.A., Freund, E.V., and Marcinek, D.J. (1999) Horizontal movements and depth distribution of large adult yellowfin tuna (*Thunnus albacares*) near the Hawaiian Islands, recorded using ultrasonic telemetry: implications for the physiological ecology of pelagic fishes. *Mar. Biol.* **133**, 395-408.

Carey, F.G., and Olson, R.J. (1981) Sonic tracking experiments with tunas. *ICCAT Coll. Vol. Scient. Pap.* **XVII**, 458-466.

Cayré, P., and Chabanne, J. (1986) Marquage acoustique et comportement de thons tropicaux (albacore: *Thunnus albacares*, et listao: *Katsuwonus pelamis*) au voisinage d'un dispositif conentrateur de poissons. *Oceanogr. Trop.* **21**(2), 167-183.

Cayré, P., Amon Kothias, J.B., Stretta, J.M., and Diouf, T. (1988) Biology of tuna. In *Resources, Fishing and Biology of the Tropical Tunas of theEastern Central Atlantic*. Fonteneau A. and Marcille J., eds., FAO Fish. Tech. Pap. 292, 157-268.

Cayré, P. (1991) Behaviour of yellowfin tuna (*Thunnus albacares*) and skipjack tuna (*Katsuwonus pelamis*) around fish aggregating devices (FADs) in the Comoros Islands as determined by ultrasonic tagging. *Aquat. Living Resour.* **4**, 1-12.

Dagorn, L., Petit, M., Stretta, J.-M., Bernardet, X., and Ramos, A.G. (1995) Toward a synthetic eco-ethology of tropical tunas. *Sci. Mar.* **59**(3-4), 335-346.

Dagorn, L., Petit, M., Hallier, J.-P., Cayré, P., and Simier, M. (1997a) Does yellowfin tuna (Thunnus albacares) school size in the western Indian Ocean depend on individual size? ICCAT Coll. Vol. Scient. Pap. Vol. XLVI, 4, 122-128.

Dagorn, L., Petit, M., and Stretta, J.-M. (1997b) Simulation of large-scale tropical tuna movements in relation with daily remote sensing data: the artificial life approach. *BioSystems* **44**, 167-180.

Dagorn, L., and Fréon, P. (1999) Tropical tuna associated with floating objets: a simulation study of the meeting point hypothesis. *Can. J. Fish. Aquat. Sci.* **56**, 984-993.

Dagorn, L., Bach, P., and Josse, E. (2000a) Movement patterns of large bigeye tuna (*Thunnus obesus*) in the open ocean, determined using ultrasonic telemetry. *Mar. Biol.* **136**, 361-371.

Dagorn, L., Josse, E., and Bach, P. (2000b) Individual differences in horizontal movements of yellowfin tuna (*Thunnus albacares*) in nearshore areas in French Polynesia, determined using ultrasonic telemetry. *Aquat. Living Resour.* **13**(4), 193-202.

Dagorn, L., Menczer, F., Bach, P., and Olson, R.J. (2000c) Co-evolution of movement behaviours by tropical pelagic predatory fishes in response to prey environment: a simulation model. *Ecol. Model.* **134**(2-3), 325-341.

Dagorn, L., Josse, E., and Bach, P. (2001) Yellowfin tuna (*Thunnus albacares*) associated with tracking vessels during ultrasonic telemetry experiments: contribution to the tuna/floating objects issue. *Fish. Bull. U.S.* **99**(1), 40-48.

Deriso, R.B., Punsly, R.G., and Bayliff, W.H. (1991) A markov movement model of yellowfin tuna in the Eastern Pacific Ocean and some analyses for international management. *Fish. Res.* **11**, 375-395.

Fernandes, P.G., Brierly, A.S., Simmonds, E.J., Millard, N.W., McPhail, S.D., Armstrong, F., Stevenson, P., and Squires, M. (2000) Fish do not avoid survey vessels. *Nature* **404**, 35-36.

Fréon, P., and Misund, O.A. (1999) *Dynamics of Pelagic Fish Distribution and Behaviour: Effects on Fisheries and Stock Assessment.* Blackwell Science, Oxford, 348 pp.

Giske, J., Huse, G., and Fiksen, O. (1998) Modelling spatial dynamics of fish. *Rev. Fish Biol. Fish.* **8**, 57-91.

Goudey, C.A., Consi, T., Manley, J., Graham, M., Donovan, B., and Kiley, L. (1998) A robotic boat for autonomous fish tracking. *Mar. Technol. Soc. J.* **32**(1), 47-53.

Gunn, J.S., Polacheck, T., Davis, T., Sherlock, M., and Betlehem, A. (1994) The development and use of archival tags for studying the migration, behaviour and physiology of southern bluefin tuna, with an assessment of the potential for transfer of the technology to groundfish research. International Counc. for the Exploration of the Sea, Mini Symposium on Fish Migration, 21 pp.

Hampton, J. (1991) Estimation of southern bluefin tuna *Thunnus maccoyii* natural mortality and movement rates from tagging experiments. *Fish. Bull. US* **89**, 591-610.

Hampton, J. (2000) Natural mortality rates in tropical tunas: size really does matter. *Can. J. Fish. Aquat. Sci.* **57**(5), 1002-1010.

Hilborn, R. (1989) Models of tag dynamics with exchange between available and unavailable populations. *Can. J. Fish. Aquat. Sci.* **46**, 1356-1366.

Hilborn, R. (1991) Modeling the stability of fish schools: exchange of individual fish between schools of skipjack tuna (*Katsuwonus pelamis*). *Can. J. Fish. Aquat. Sci.* **48,** 1081-1091.

Hogeweg, P., and Hesper, B. (1990) Individual-oriented modelling in ecology. *Math. Comput. Model.* **13**, 83-90.

Holland, J.H. (1992) Genetic algorithms. *Sci. Am.*, July 1992, 66-72.

Holland, K.N., Brill, R.W., and Chang, R.K.C. (1990) Horizontal and vertical movements of yellowfin and bigeye tuna associated with fish aggregating devices. *Fish. Bull. U.S.* **88**, 493-507.

Holland, K.N., Brill, R.W., Sibert, J.R., and Fournier, D.A. (1992) Physiological and behavioural thermoregulation in bigeye tuna *Thunnus obesus*. *Nature* **358**, 410-412

Holland, K.N., Kleiber, P., and Kajiura, S.M. (1999) Different residence times of yellowfin tuna, *Thunnus albacares*, and bigeye tuna, *T. obesus*, found in mixed aggregations over a seamount. *Fish. Bull. U.S.* **97**, 392-395.

Humston, R., Ault, J.S., Lutcavage, M., and Olson, D.B. (2000) Schooling and migration of large pelagic fishes relative to environmental cues. *Fish. Oceanogr.* **9**(2), 136-146.

Hunter, J.R., Argue, A.W., Bayliff, W.H., Dizon, A.E., Fonteneau, A., Goodman, D., and Seckel, G.R. (1986) The dynamics of tuna movements: an evaluation of past and future research. FAO Fish. Tech. Pap. 277, 78 pp.

Huse, G., and Giske, J. (1998) Ecology in mare pentium: an individual based spatio-temporal model for fish with adapted behaviour. *Fish. Res.* **37**, 163-178.

Huse, G., Strand, E., and Giske, J. (1999). Implementing behaviour in individual-based models using neural networks and genetic algorithms. *Evol. Ecol.* **13**, 469-483.

Huston, M., DeAngelis, D., and Post, W. (1988) New computer models unify ecological theory. *BioScience* **38**, 682-691.

Huth, A., and Wissel, C. (1992) The simulation of the movement of fish schools. *J. Theor. Biol.* **156**, 365-385.

Josse, E, Bach, P, and Dagorn, L (1998) Simultaneous observations of tuna movements and their prey by sonic tracking and acoustic surveys. *Hydrobiologia* **371/372**, 61-69

Josse, E., Bertrand, A., and Dagorn, L. (1999) An acoustic approach to study tuna aggregated around fish aggregating devices in French Polynesia: methods and validation, *Aquat. Living Resour.* **12**, 303-313.

Josse, E., and Bertrand, A. (2000) *In situ* acoustic target strength measurements of tuna associated with a fish aggregating device, *ICES J. Mar. Sci.* **57**(4), 911-918.

Kleiber, P., Argue, A.W., and Kearney, R.E. (1987) Assessment of Pacific skipjack tuna (Katsuwonus pelamis) resources by estimating standing stock and components of population turnover from tagging data. *Can. J. Fish. Aquat. Sci.* **44**, 1122-1134.

Kleiber, P., and Hampton, J. (1994) Modeling effects of FADs and islands on movement of skipjack tuna (*Katsuwonus pelamis*): estimating parameters from tagging data. *Can. J. Fish. Aquat. Sci.* **51**, 2642-2653.

Klimley, A.P., and Holloway, C.F. (1999) School fidelity and homing synchronicity of yellowfin tuna, *Thunnus albacares*, *Mar. Biol.* **133**, 307-317.

Lehodey, P., Bertignac, M., Hampton, J., Lewis, A., and Picaut, J. (1997) El Nino southern oscillation and tuna in the western Pacific. *Nature*, **389**, 715-718.

Lehodey, P., André, J.-M., Bertignac, M., Hampton, J., Stoens, A., Menkes, C., Memery, L., and Grima, L. (1998) Predicting skipjack tuna forage distributions in the equatorial Pacific using a coupled dynamical bio-chemical model. *Fish. Oceanogr.* **7**(3-4), 317-325.

Levin, S.A. (1992) The problem of pattern and scale in ecology. *Ecology* **73**(6), 1943-1967.

Lutcavage, M.E., Brill, R.W., Skomal, G.B., Chase, B.C., and Howey, P.W. (1999) results of pop-up satellite tagging of spawning size class fish in the gulf of Maine: do North Atlantic bluefin tuna spawn in the mid-Atlantic? *Can. J. Fish. Aquat. Sci.* **56**, 173-177.

Magurran, A.E. (1993) Individual differences and alternative behaviours. In Pitcher T.J. (Ed.), *Behaviour of Teleost Fishes*, 2nd Edition, Chapman and Hall, pp. 441-477.

Marchal, E., and Lebourges, A. (1996) Acoustic evidence for unusual diel behaviour of a mesopelagic fish (*Vinciguerria nimbaria*), exploited by tuna. *ICES J. Mar. Sci.* **53**, 443-448.

Marsac, F., and Cayré, P. (1998) Telemetry applied to behaviour analysis of yellowfin tuna (*Thunnus albacares*, Bonnaterre, 1788) movements in a network of fish aggregating devices. *Hydrobiologia* **371/372** 155-171.

Maury, O., Gascuel, D., and Fonteneau, A. (2001) Spatial Modeling of Atlantic Yellowfin Tuna Population Dynamics: Application of a Habitat Based Advection-Diffusion-Reaction Model to the Study of Local Overfishing. *In* Spatial processes in marine fisheries, University of Alaska Sea Grant, Fairbanks, (in press).

Menczer, F., and Belew, R.K. (1996) From complex environments to complex behaviors. *Adapt. Behav.* **4**(3/4), 317-363.

Misund, O.A. (1993) Avoidance behaviour of herring (*Clupea harengus*) and mackerel (*Scomber scombrus*) in purse seine capture situations. *Fish. Res.* **16**, 179-194.

Parrish, F.A., Craig, M.P., Ragen, T.J., Marshall, G.J., and Buhleier, B.M. Identifying diurnal foraging habitat of endangered Hawaiian monk seals using a seal-mounted video camera. *Mar. Mamm. Sci.* **16**(2), 392-412. 2000.

Petit, M., Dagorn, L., Lena, P., Slepoukha, M., Ramos, A.G., and Stretta, J.M. (1994) Oceanic landscape concept and operational fisheries oceanography. Mémoires de l'Institut Océanographique, Monaco, n°18, 85-97.

Pitcher, T.J., Misund, O.A., Ferno, A., Totland, B., and Melle, V. (1996) Adaptive behaviour of herring schools in the Norwegian Sea as revealed by high-resolution sonar. *ICES J. Mar. Sci.* **53**, 449-452.

Reuter, H., and Breckling, B. (1994) Self-organization of fish schools: an object-oriented model. *Ecol. Model.* **75/76**, 147-159.

Romey, W.L. (1996) Individual differences make a difference in the trajectories of simulated schools of fish. *Ecol. Model.* **92**, 65-77.

Rumelhart, D.E., and McClelland (1986) *Parallel Distributed Processing.* MA: Bradford Books (PIT Press), Cambridge.

Shin, Y.-J., and Cury, P. (1999) OSMOSE: a multispecies individual-based model to explore the functional role of biodiversity in marine ecosystems. *In* Ecosystem approaches for fisheries management, University of Alaska Sea Grant, Fairbanks, pp. 593-607.

Shin, Y.-J., and Cury, P. (2001) Exploring fish community dynamics through size-dependent trophic interactions using a spatialized individual-based model. *Aquat. Living Resour.*, **14**(2), (in press).

Sibert, J.R., Hampton, J., Fournier, D.A., and Bills, P.J. (1999) An advection-diffusion-reaction model for the estimation of fish movement parameters from tagging data, with application to skipjack tuna (*Katsuwonus pelamis*). *Can. J. Fish. Aquat. Sci.* **56**, 925-938.

Stocker, S. (1999) Models for tuna school formation. *Math. Biosci.* **156**(1-2), 167-190.

Swartzman, G., Brodeur, R., Napp, J., Hunt, G., Demer, D., and Hewitt, R. (1999) Spatial proximity of age-0 walleye pollock (*Theragra chalcogramma*) to zooplankton near the Pribilof Islands, bering Sea, Alaska. *ICES J. Mar. Sci.*, **56**, 545-560.

Tyler, J.A., and Rose, K.A. (1994) Individual variability and spatialheterogeneity in fish population models. *Rev. Fish Biol. Fish.* **4**, 91-123.

Vabø, R., and Nøttestad, L. (1997) An individual based model of fish school reactions: predicting antipredator behaviour as observed in nature. *Fish. Oceanogr.* **6**, 155-171.

Welch, D.W., and Eveson, J.P. (1999) An assessment of light-based geoposition estimates from archival tags. *Can. J. Fish. Aquat. Sci.* **56**, 1317-1327.

White, G.C., and Garrott, R.A. (1990) *Analysis of Wildlife Radio-Tracking Data.* Academic Press, 383 pp.

Wilson, R.P. (1998) Following fishy things in far-off waters: are we sensing the right things? Marine Animal Telemetry Tags Report, New England Aquarium Forum Series, Report 98-3, pp. 37-42.

Wolcott, T.G. (1995) New options in physiological and behavioural ecology through multichannel telemetry. *J. Exp. Mar. Biol. Ecol.* **193**, 257-275.

Yuen, H.S.H. (1970) Behavior of skipjack tuna, *Katsuwonus pelamis*, as determined by tracking with ultrasonic devices. *J. Fish. Res. Bd Can.* **27**, 2071-2079.

On the Integrated Study of Tuna Behaviour and Spatial Dynamics: Tagging and Modelling as Complementary Tools

David S. Kirby[1,2,3]
[1]*Fisheries Research Group, Department of Biology, University of Leicester, UK.*
[2]*Earth Observation Science Group, Department of Physics, University of Leicester, UK.*
[3]*National Institute of Water and Atmospheric Research (NIWA) Ltd., Wellington, NZ*
Correspondence: < d.kirby@SCIENCEnet.com >

Key words: tuna, behaviour, tagging, modelling

Abstract: In this paper, modelling efforts that seek to describe, explain and predict the behaviour and spatial dynamics of tunas are reviewed and discussed in relation to tagging studies with the same goals. Tagging and tracking of fish with electronic devices can provide valuable observations of free-living animals, which may be used to help derive models and also to test their predictions. But we simply will not be able to derive and validate models for the fine-scale behaviour of tunas unless measurements are made of physiological and environmental variables, representing factors motivating behaviour, at the same time as position and activity are recorded. On longer time and space scales we must assess reproductive motivation, by identifying spawning grounds and times and measuring gonad state for individuals as they migrate throughout their range. Thermodynamics (through bioenergetics), fitness maximisation and adaptive behaviour with evolutionary motivation are appropriate paradigms for the derivation of models. But modelling will remain merely a technical exercise unless it is carried out as an integrated part of research programs pursuing the understanding of tuna behaviour and spatial dynamics as the ultimate goal. An observational framework that simultaneously measures environmental and physiological variables, with a complementary suite of statistical and theoretical models, will truly advance knowledge and enable us to understand these fish. This synthesis will only be achieved through collaboration between scientists with individual skills in field, laboratory and computational ecology and with innovative technical support. Some case studies are presented here to support this thesis, and it is hoped that this paper will stimulate discussion, collaboration and the development of new observational and computational methodologies.

J.R. Sibert and J.L. Nielsen (eds.), Electronic Tagging and Tracking in Marine Fisheries, 407–420.
© 2001 *Kluwer Academic Publishers. Printed in the Netherlands.*

1. INTRODUCTION

It has long been a goal of tuna ecologists to fully describe the behaviour of these magnificent fish in relation to their oceanic environment. Large scale movements have received much attention because of the complexities surrounding the management of highly migratory stocks (Block *et al.*, 1998a). Certainly these issues are important if seeking to apply traditional fisheries models for stock assessment and quota management. Mark and recapture methods have been used for various applications including large scale movements, growth rates, mortality, and transfer rates between stocks and gear types (see review by Hunter *et al.*, 1986). Acoustic telemetry has provided detailed data on tuna movements, often in relation to environmental variables measured from the recording platform (Holland *et al.*, 1990; Cayré and Marsac, 1993; Block *et al.*, 1997; Josse *et al.*, 1998; Brill *et al.*, 1999; Dagorn *et al.*, 2000). Archival tags have collected longer time series of data on position and water temperature, and sometimes body temperature (*e.g.* Block *et al.*, 1998b), without the need for a following platform, but the need to recapture the fish before data can be obtained is clearly limiting. Pop-up tags that relay their data via satellite have been used to effectively increase the return rate to near 100% (Block *et al.*, 1998a), but only position and daily mean water temperature have so far been recorded, limiting the utility of the results.

A common problem for all tagging studies to date is sample size; only a small number of fish may be tagged and tracked at any one time, and repeated sea-going experiments are costly. Biologists have not worried overly about this and for good reason — individual behaviour may be scaleable into populations and species as the same physiological and evolutionary imperatives apply to all. This doesn't negate the need to sample more than one fish at a time, but it does allow us to have confidence in good, comprehensive studies that have limited sampling size.

The common framework for understanding animal behaviour is behavioural ecology (Krebs *et al.*, 1993; Krebs and Davies, 1997). This is a unifying field that provides some powerful perspectives, paradigms and methods that can be applied across the animal kingdom. A conceptual model for the behavioural ecology of tunas is presented in Figure 1. Theoretical models in behavioural ecology (*e.g.* Kirby *et al.*, 2000) need observations for parameterisation of vital rates and for testing predictions. In turn they may contribute to experimental design by identifying key variables and parameters. Statistical models based solely on observations and without regard to mechanism or process can say nothing about causal links between variables, but may still have pragmatic value and find useful applications (*e.g.* Cayré and Marsac, 1993).

In this paper I discuss differences in modelling methodologies and emphasise the common ground that can exist between observational and computational studies. My contention is that we might learn so much more if we refine our observational methods and our modelling techniques to a point where they converge as valuable tools for the same task.

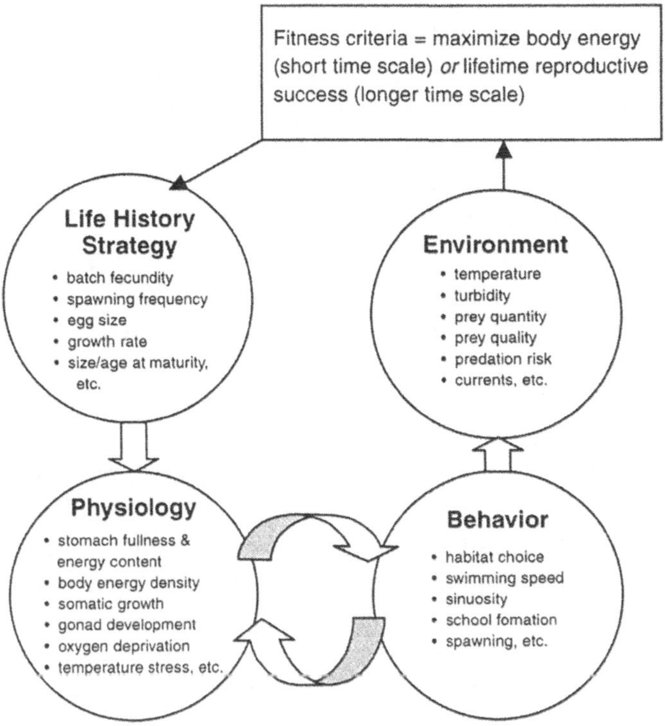

Figure 1. Conceptual model for the behavioural ecology of tunas. The properties of the oceanic environment interact with physiological constraints and evolutionary imperatives. In order to maximise some measure of fitness, a predatory fish comes to optimise its use of available habitat, through natural selection against sub-optimal strategies, and by making cognitive choices within the constraints of its sensory and learning abilities.

2. OBSERVATIONS OF TUNAS IN THE OCEAN AND IN THE LABORATORY

There is a long history of observations of tunas at sea and in the laboratory, and the ongoing development of dedicated facilities, research programs and databases will ensure that as long as we keep our minds open, we will continue to further our knowledge and understanding. This is not a comprehensive review of observational studies, but instead I have highlighted key studies that are central to my thesis *viz.* that the right variables must be measured at sea, in order to be able to derive and evaluate models that represent a true understanding of tuna behaviour and spatial dynamics.

Much observational work has focussed on the role of physical or abiotic aspects of the environment as potential limiting factors in horizontal and vertical range. Many physiological processes are temperature and/or oxygen limited, so this focus is understandable. Despite the mechanisms of heat conservation that are available to tunas (Kishinouye, 1923; Carey and Lawson, 1973; Holland *et al.*, 1992; Dewar *et al.*, 1994), temperature limitation of foraging range is suggested by laboratory experiment (Dizon *et al.*, 1977; Barkley *et al.*, 1978; Brill *et al.*, 1998) and apparent in field observations (Blackburn, 1965; Sund *et al.*, 1981; Brill, 1994; Brill *et al.*, 1999). Oxygen concentrations can be limiting in absolute terms, and even if they are not there may be clear preference for high oxygen depth strata (Block *et al.*, 1997). Cayré and Marsac (1993) tagged and tracked 3 yellowfin tuna, recording depth every 20 s and comparing vertical movements with profiles of temperature and dissolved oxygen. They fit modified normal distributions to the time series of data such that a depth-based catchability forecast could be made and fishing gear set depending on observed profiles of the physical variables. However, their conclusion that "...the vertical distributions of only two physical parameters (temperature and dissolved oxygen) explain the vertical distribution of yellowfin tuna" is somewhat exaggerated, particularly as they did not postulate any reason for the observed distributions, nor did they measure other potentially relevant variables such as light and/or turbidity. A statistical model of this nature may become a useful means of directing fishing effort but it does not advance our understanding of tuna behaviour. In the light of various experiments noting the dynamics of tuna prey (Marchal *et al.*, 1993; Josse *et al.*, 1998), conclusions constraining tuna behaviour by physical variables alone would seem overly simplistic. Even if we assume that adult tunas are apex predators and that their behaviour is not at all constrained by predation pressure, their behaviour is highly likely to be affected by the dynamics of their highly mobile and patchily distributed prey.

Tunas are known to have high metabolic energy demands (Brill, 1987) which necessitate a high energy intake. Given the need for tunas to keep swimming in order to maintain hydrostatic equilibrium (Magnuson, 1973), we might expect active foraging to occur whenever possible. However, it has been noted that the intensity of response of yellowfin tuna to prey odours varies with hunger state (Atema *et al.*, 1980); time is needed to recover from exercise and to digest and absorb food. Energy conserving 'dive and glide' behaviour has also been observed in more than one tuna species (Holland *et al.*, 1990; Block *et al.*, 1997), whereby rapid powered ascents are followed by a slow, lift-based glide. Both time- and state- dependent behaviour are predicted theoretically (Kirby *et al.*, 2000), which means that unless environmental and physiological variables are simultaneously measured, it is simply not possible to say what is controlling tuna behaviour.

Josse *et al.* (1998) acoustically tagged and tracked one yellowfin and two bigeye tuna, and simultaneously measured local prey density as indicated by a sound-scattering layer (SSL) on an echosounder. They obtained some good data illustrating the movements of these fish in relation to the SSL and noted that abiotic variables (temperature and oxygen) were not limiting. It is rather telling that they are able to conclude that they have observed a "...new (*sic*) explanatory factor of tuna behaviour: the biotic environment." It is new but it shouldn't be — as pelagic predators, tunas are ever likely to be affected by the dynamics of their prey. There has been too much focus on the relationship of tunas with physical environmental variables. These abiotic variables are still important for tunas and there may be real situations where they are limiting, but within the limits of these variables that directly affect physiological processes, it has long been recognised that the availability of forage will induce tuna distribution (Blackburn, 1965; Sund *et al.*, 1981).

Fine scale hunting behaviour is therefore also likely to be observed. Vertical movements in particular are common and frequent and the different tunas have different movement patterns. The reasons why these differences occur remain the subject of debate (Brill *et al.*, 1999) and morphological and/or biochemical adaptation to the different physical regimes above and below the thermocline are likely to be key (Lowe *et al.*, 2000).

The evolutionary advantage of these different behaviour patterns requires more holistic consideration of the trophic dynamics of the pelagic ecosystem. Trophic interactions are not just important for fine scale behaviour; even large scale movements and aggregations may be food-motivated. The gathering of young albacore at the fronts off the California upwelling is a case in point (Laurs *et al.*, 1984; Fielder and Barnard, 1987) — being sexually immature, these fish certainly aren't aggregating for reproduction. Assuming that tunas can occupy a broad oceanic niche, the proximate motivation for their dynamics within that space is likely to be food, with the ultimate motivation being to maximise reproductive success. Behavioural studies should therefore focus on estimating these factors as the free-living animal is observed.

3. MODELS FOR THE BEHAVIOUR AND SPATIAL DYNAMICS OF TUNAS

> Model: a simplified description of a system, process, *etc.*, put forward as a basis for theoretical or empirical understanding; a conceptual or mental representation of something. (Math.) A set of entities that satisfies all the formulae of a given formal or axiomatic system.
> (Oxford English Dictionary, 1993)

Biologists are often put off by the 'flute music' of calculus, the sophistry of statistics and the terse logic of programming code. When one's motivation for study comes from a deep appreciation of the beauty of nature it is easy to be put off by analytical methods that seem to grossly oversimplify or overcomplicate, sometimes simultaneously, a situation about which one already has an intuitive understanding. But such is the nature of science, and modelling is but another tool in the investigation of natural processes, and one that can contribute to knowledge at many different levels of understanding. While stock assessment may have "...degraded into mathematical games in which the object is to find best guesses and estimates for parameters that have little to do with any 'real' measured or measurable variables," (Sharp, 1995) the exercise of deriving a model can still help to identify and clarify the relative importance of different parameters and processes about which we previously had only a conceptual understanding. Simpler models often provide insight that is more valuable than accurate numerical fits, and the most influential models are often the ones where the numerical output is not needed to guide the qualitative understanding (Hilborn and Mangel, 1997).

Herein lies the truth that there are many different types of model that may be applied to any particular question. Most scientists will be familiar with the exercise of trying to find correlations between variables, and are happy when measurements of one variable can be used to reduce the variance in measurements of another (some use the term 'explain' — I prefer not to). But it is extremely difficult to find statistical relationships that hold in highly variable environments (Sharp, 1995; Mangel *et al.*, 2000; see Bigelow *et al.*, 1999; Andrade and Garcia, 1999). Furthermore, while statistical relationships may describe relationships well or badly, they cannot be used to show causality (Sharp, 1995; Brill, 1997; Hilborn and Mangel, 1997). The researcher may use statistical methods to identify relationships between variables but the ultimate questions regarding *why* such relationships exist cannot be answered in this way. Causality may be established through knowledge of specific obligate physiological responses and consequent behavioural decisions in a systems context (Sharp, 1995; Kirby *et al.*, 2000; Figure 1). This theoretical approach has its emphasis on identifying mechanisms of interaction between organism and environment, allowing cause and effect relationships to be established and used to make predictions that may be better founded than those based on projection of past trends into the future. The level of detail required to develop such process-oriented models is usually high, and simplifications and assumptions have to be made in order to progress. Nevertheless, even simple theoretical models that forgo mechanistic detail can still provide a better understanding of the system under study than may be obtained by statistical analyses alone, as they can also have explanatory power. There is a body of literature on this subject that is well worth reviewing before undertaking any kind of modelling exercise (*e.g.* citations above, plus Loehle, 1983 and reviews by Tyler and Rose, 1994, Giske *et al.*, 1998, Mangel *et al.*, 2000; Huse *et al.*, in press).

Models for the behaviour and spatial dynamics of tunas are many and varied (Deriso *et al.*, 1991; Cayré and Marsac, 1993; Dagorn *et al.*, 1995, 1997; Bertignac *et al.*, 1998; Dagorn and Freon, 1999; Sibert *et al.*, 1999; Stöcker, 1999; Humston *et al.*, 2000; Kirby *et al.*, 2000). Rule-based methods have been used to investigate both fine-scale behaviour of tunas (Dagorn *et al.*, 1995; Dagorn and Fréon, 1999) and larger-scale movements (Dagorn *et al.*, 1997; Humston *et al.*, 2000). The use of 'rules' is a prescriptive exercise and so the rules themselves must be simple, logical and defensible. Complex behaviour and spatial dynamics may then emerge in the model system that allows one to generate and evaluate hypotheses for real tuna. Large scale movements and population dynamics have been represented by advection-diffusion relations, with tagging data used either in model derivation (Sibert *et al.*, 1999) and/or evaluation (Bertignac *et al.*, 1998; Sibert *et al.*, 1999). Models of this type and at this scale ignore more fine-scale behaviour but may still incorporate interactions with environmental variability *e.g.* SST and forage density (Bertignac *et al.*, 1998).

The model of Bertignac *et al.* (1998) for the spatial population dynamics of Pacific skipack tuna (*Katsuwonus pelamis*) builds on earlier levels of modelling, covering general circulation (Blanke and Delecluse, 1993), biogeochemistry and new production (Stoens *et al.*, 1998), and tuna forage production (Lehodey *et al.*, 1998). In this way the model as a whole is prognostic for tuna. It is a bold attempt at modelling spatial population dynamics from a 'bottom-up' approach (*i.e.* from first principles — physics to fish) and is commendable in its endeavours to link biological oceanography with fisheries science. However, the use of differential equations implies that one has already identified the relevant dynamics and, in the case of fish movement models, the relationship between fish and environment (Eqs. 4, 9 and 10 in Bertignac *et al.*, 1998). Given the complexities of ecological interactions, and the different components of fitness that trade-off against each other in the course of an animal's lifetime, this may be somewhat premature.

A similar criticism would apply to the kinesis model of Humston *et al.* (2000) and, in fairness, to mathematical ecology in general. One can apply mathematical models of any particular functional form to any postulated relationship between variables, but there should be good justification for the choices made. Particular caution should also be exercised in the inferences drawn from the results. Humston *et al.* (2000) are successful in their aim to reproduce large-scale migration of Atlantic bluefin tuna from simple behavioural rules. These movement rules are formulated mathematically as functions that depend on the difference between actual and optimal temperature, the latter (18°C) being, "...chosen because it concurs with temperature data for those tuna," aerial survey data reported by Lutcavage *et al.* (1997), and because "...it is linked to the edge of Gulf Stream waters." In the absence of any other factors to trade off against this temperature preference it is then no surprise that the resulting distributions reflect those data used to derive the function, such that "...histograms of surface temperatures occupied at the end of model runs indicate highest concentrations of fish in surface waters of 18°C," and model results "also showed marked aggregations along the edges of sharp thermal fronts." This is skating on thin ice, below which lie the frigid waters of tautology. The model is attractive in its simplicity and forecasting skill, and in the aim to reflect observations the authors succeed admirably. But the assumptions made concern the mechanisms of interaction between fish and environment, as well as the motivations for action, upon which there is not yet convincing consensus. This is an area where future experimental research will be key.

It is well to remember that "...the realism of spatially resolved models cannot evolve faster than the acquisition of knowledge about the mechanisms governing the spatial behaviour of the constituents" (SERG, 2000). The level of realism that is incorporated into a model will also depend on its purpose and intended use. In a recent theoretical modelling exercise (Kirby *et al.*, 2000), where the aim was to be as true to mechanism and motivation as possible, it was not possible to simulate tuna behaviour unless a detailed representation of physiology (gastric evacuation, standard and active metabolic energy costs), sensory systems (visual range), and both biotic and abiotic characteristics of the environment (prey abundance and energy density, water temperature and turbidity) were included. We faced the same task as Bertignac *et al.* (1998) in trying to derive an equation to represent the effect of temperature stress on tuna, a task where we are totally dependent on experimental physiology to give us measurements of vital rates. By specifying a range of acceptable behaviours (swimming speeds and habitats) and a detailed representation of the state dynamics (*i.e.* physiology) the modelling technique calculates fitness values for all possible solutions and predicts optimal foraging behaviour (for more detail see Mangel and Clark, 1988; Clark and Mangel, 2000; Kirby *et al*., 2000). The exercise was started with the aim of deriving a means to predict the location of tunas in relation to ocean fronts visible in satellite imagery. But these data are of physical variables only, with the exception of chlorophyll concentration, and the importance of prey characteristics and state dependent behaviour became obvious only in the course of literature searching and model derivation and evaluation. Nonetheless, we succeeded in predicting behaviour from physiology in a complex environment, and inadvertently developed the optimal foraging model envisaged by Hunter *et al.* (1986; p. 30). The model itself still contains assumptions that may or may not be true, because our knowledge of various components is incomplete (*e.g.* sensory biology of tunas; physiological mechanisms and rates of accumulation of thermal stress; and the optical and nutritional properties of forage), but it is the first model for tunas that predicts behaviour from physiology and environment; it is this kind of model that is most closely related to tagging studies that seek to understand fine-scale behaviour.

For movements over larger time- and space-scales, if a mechanistic representation of reality is desired, a different approach again may be necessary. There are real issues regarding the scaling up of motivated individuals to the dynamics of populations but these may not be as formidable as they first seem. Using evolutionary motivation (*i.e.* some measure of reproductive success) spatial population dynamics has been modelled by both optimisation (Fiksen *et al*., 1995) and adaptation (Huse and Giske, 1998) approaches (see Giske *et al.,* 1998 for expansion and discussion of these terms). Adaptive models are well founded in evolutionary and life-history theory, and use computational methods inspired by biological processes (*i.e.* neural networks and genetic algorithms), which enable model agents to both learn and evolve just as with living creatures. They are also well suited to complex solution space (G. Huse, *pers comm.*) and may therefore be better suited to exploratory simulations of the effects of changes in exploitation patterns or ocean climate on fish population dynamics. This heuristic or 'black-box' approach has its detractors, usually amongst those more familiar with deterministic rather than adaptive processes, but it is conceptually satisfying to the biologist who is well aware of the complexities of living creatures and to whom adaptive behaviour and evolution are far from alien. The development of an adaptive model for the spatial dynamics of Pacific skipjack tuna is the subject of my current research.

4. MEANS TOWARDS A TRUE UNDERSTANDING OF TUNAS: A NEW SYNTHESIS

One of the greatest contributions that theoretical modelling can make in studies of behaviour is that various hypotheses can be jointly evaluated and refined, prior to field observation and statistical hypothesis testing. The use of models when planning an experiment may also help identify variables that may be confounded in the analysis of results (Hilborn and Mangel, 1997). This combined approach has practical as well as intellectual merit, as computational experiments are comparatively cheap to run, and may then allow field studies to focus on what is really important for enhancing understanding. This is the essential point that I want to press in this paper, with regard to the complementary roles of modelling and experimental studies of the behaviour of tunas in relation to their environment. Observations should be used to derive and evaluate models, which in turn may be used to guide investigations in the field through the generation of testable hypotheses. As mentioned above, in an optimal foraging model for tunas at fronts (Kirby *et al.*, 2000) we needed a detailed representation of environmental characteristics, prey characteristics, sensory systems and physiology in order to predict optimal habitat and swimming speed. Such detail can only come from experimental investigation in the laboratory and at sea. In addition, the model makes predictions, in particular regarding state-dependent behaviour, for which comparable observations are not presently available. We therefore hope that in future sea-going work, researchers will adopt the model, or at least the approach, and use it to guide their investigations.

Joseph and Wild (1984), summarising a meeting of the Inter-American Tropical Tuna Commission, noted that "...there is a need to organise more complete conceptual models on how environmental conditions and physiology can direct and limit tuna movements both vertically and horizontally...At-sea tagging operations should be accompanied by sampling to determine physiological state (energy storage, instantaneous growth rate, *etc.*) and recent reproductive history. Tuna stomachs can be used to monitor and assess the environment in terms of temporal and spatial food availability." From the literature it seems that this advice has not been taken. There are separate studies that tag fish and others that measure stomach contents but few that do both, let alone conduct the other physiological investigations suggested. In some biogeographical areas, we should have progressed our techniques by now such that we can identify forage fish by acoustic target strength. There may also be ways of non-destructively measuring hunger state through the use of chemical sensors on the fish. This would be vital information for a physiology based movement model, data that might be recorded by archival tag along with swimming speed and sinuousity. I made an assumption earlier that adult tunas are apex predators and are not themselves preyed upon. This is generally thought to be the case, but as Hampton (2000) has shown, natural mortality of small (21–30 cm) skipjack, yellowfin and bigeye tuna is an order of magnitude higher than that of mid-sized fish. To understand the behaviour of these fish we must then simultaneously measure or otherwise estimate predation risk for the environment where our tuna is under study. I am not aware of tagging methods that will record the presence of other fish, be they predators, prey or conspecifics, but it would be worthwhile considering how they might be developed, or at least how we might simultaneously measure the 'biotic environment' (*e.g.* Josse *et al.*, 1998) in terms of both predators and prey.

"The time and space scales of measurements of tuna movements is a critical issue in the design of future investigations. Tracking of individual fish over periods of hours or days is not equivalent to movements of groups or schools over months. The problem of using information from small-scale movements to model movements of large groups of tunas over weeks or months needs to be examined" (Joseph and Wild, 1984). Different models may be used to investigate these different aspects of movement. An optimal foraging model (*e.g*,. Kirby *et al.*, 2000) is a good paradigm for short time-scale behaviour but is not adequate for scales where motivation is different *i.e.* where reproductive activity must be considered. In this case, if the fitness-based modelling approach is still followed, a Darwinian fitness measure must be adopted (*e.g.* number of eggs laid per gram body mass above size at maturity—Fiksen *et al.*, 1995; Figure 1) and either an optimisation or an adaptation approach used (Giske *et al.*, 1998).

There are technological and logistical obstacles and constraints in the observational work suggested, and first we need to clarify which variables are most relevant to behaviour. Indices, proxies and vital rates for these variables may be identified in the laboratory, and then the technological development of new tools can begin. Modelling methods must be scrutinised, with methods used that are appropriate to the questions asked. Statistical models must not pretend to tell us why things happen, and theoretical models must be explicit in their assumptions and expand their scope from the artificial environments for which they are originally derived. Tremendous progress has been made in the physiological ecology of tunas, and in the development of computational methods; these fields must converge and be followed by behavioural and evolutionary studies that go beyond the descriptive and retrospective, and are focussed on understanding and prediction.

ACKNOWLEDGEMENTS

I would like to acknowledge all those whose work is referred to in this paper and I hope that this discussion will be a constructive stimulus towards further research. I am grateful to the reviewers for their comments and to John Sibert for encouraging me to commit my thoughts to paper.

REFERENCES

Andrade, H.A., and Garcia, C.A.E. (1999) Skipjack tuna fishery in relation to sea surface temperature off the southern Brazilian coast. *Fish. Oceanogr.* **8**(4), 245–254.

Atema, J., Holland, K. and Ikehara, W. (1980) Olfactory responses of Yellowfin tuna (*Thunnus albacares*) to prey odors: chemical search image. *J. Chem. Ecol.* **6**(2), 457–465.

Barkley, R.A., Neill, W.H., and Gooding, R.M. (1978) Skipjack tuna, *Katsuwonus pelamis*, habitat based on temperature and oxygen requirements. *Fish. Bull.* **76**(3), 653–662.

Bertignac, M., Lehodey, P., and Hampton, J. (1998) A spatial population dynamics model of tropical tunas using a habitat index based on environmental parameters. *Fish. Oceanogr.* **7**(3-4), 326–334.

Bigelow, K.A., Boggs, C.H., and He, X. (1999) Environmental effects on swordfish and blue shark catch rates in the US North Pacific longline fishery. *Fish. Oceanogr.* **8**(3), 178–198.

Blackburn, M. (1965) Oceanography and the ecology of tunas. *Oceanogr. Mar. Biol. Ann. Rev.* **3**, 299–322.

Blanke, B., and Delecluse P. (1993) variability of the tropical Atlantic Ocean simulated by a general circulation model with two different mixed-layer physics. *J. Phys. Oceanogr.* **23,** 1363–1388.

Block, B.A., Keen, J.E., Castillo, B., Dewar, H., Freund, E.V., Marcinek, D.J., Brill, R.W. and Farwell, C. (1997) Environmental preferences of yellowfin tuna (*Thunnus albacares*) at the northern extent of its range. *Mar. Biol.* **130,** 119-132.

Block, B.A., Dewar, H., Freund, Farwell, C. and Princes, E.D. (1998a) A new satellite technology for tracking the movements of Atlantic bluefin tuna. *Proc. Natl Acad. Sci. USA*, **95,** 9384–9389.

Block, B.A., Dewar, H., Williams, T., Princes, E.D. Farwell, C. and Fudge, D. (1998b) Archival Tagging of Atlantic Bluefin Tuna (*Thunnus thynnus thynnus*). *Mar. Technol. Soc. J.*, **32**(1), 37–46.

Brill, R.W. (1987) On the standard metabolic rate of tropical tunas, including the effect of body size and acute temperature change. *Fish. Bull.* **85,** 25-35.

Brill, R.W. (1997) How tuna physiology affects tuna movements and distribution. Pelagic Fisheries Research Program, University of Hawaii, Newsletter **2,** 4.

Brill, R.W. (1994) A review of temperature and oxygen tolerance studies of tunas pertinent to fisheries oceanography, movement models and stock assessments. *Fish. Oceanogr.* **3**(3), 204-216.

Brill, R.W., Lowe, T.E. and Cousins, K.L. (1998) *How water temperature really limits the vertical movements of tunas and billfishes - it's the heart stupid!* American Fisheries Society, International Congress on Biology of Fish, Towson University, Baltimore, MD. Abstract available: http://www.soest.hawaii.edu/PFRP/brill_afs.pdf

Brill, R.W., Block, B.A., Boggs, C.H., Bigelow, K.A., Freund, E.V. and Marcinek, D.J. (1999) Horizontal movements and depth distribution of large adult yellowfin tuna (*Thun-*

nus albacares) near the Hawaiian Islands recorded using ultrasonic telemetry: implications for the physiological ecology of pelagic fishes. *Mar. Biol.* **133,** 395-408

Carey, F.G. and Lawson, K.D. (1973) Temperature regulation in free swimming bluefin tuna. *Comp. Biochem. Physiol.* **44,** 375–392.

Cayré, P. and Marsac, F. (1993) Modelling the yellowfin tuna (*Thunnus albacares*) vertical distribution using sonic tagging results and local environmental parameters. *Aquatic Living Resources* **6**, 1–14.

Clark, C.W. and Mangel, M. (2000) Dynamic State Variable Models in Ecology. Oxford University Press Inc., USA, 302 pp.

Dagorn, L., Petit, M., Stretta, J-M., Bernadet, X. and Romos, A.G. (1995) Toward a synthetic eco-ethology of tropical tunas. *Sci. Mar.*, **59**(3–4), 335–346.

Dagorn, L., Petit, M., Stretta, J-M. (1997) Simulation of large-scale tropical tuna movements in relation with daily remote sensing data: the artificial life approach. *Biosystems*, **44,** 167–180.

Dagorn, L., and Fréon, P. (1999) Tropical tuna associated with floating objects: a simulation study of the meeting point hypothesis. *Can. J. Fish. Aquat. Sci.*, **56,** 984–993.

Dagorn, L., Bach, P., and Josse, E. (2000) Movement patterns of large bigeye tuna (*Thunnus obesus*) in the open ocean, determined using ultrasonic telemetry. *Mar. Biol.* **136,** 361–371.

Deriso, R.B., Punsly, R.G., and Bayliff, W.H. (1991) A Markov movement model of Yellowfin tuna in the Eastern Pacific Ocean and some analyses for international management. *Fish. Res.*, **11,** 375–395.

Dewar, H., Graham, J.B., and Brill, R.W. (1994) Studies of tropical tuna swimming performance in a large water tunnel. *J. Exp. Biol.* **192**, 33–44.

Dizon, A.E., Neill, W.H., and Magnuson, J.J. (1977) Rapid temperature compensation of volitional swimming speeds and lethal temperatures in tropical tunas (Scombridae). *Env. Biol. Fishes* **2,** 83–92.

Fielder, P., and Bernard, H. (1987) Tuna aggregation and feeding near fronts observed in satellite imagery. *Cont. Shelf Res.* **7**, 871–881.

Fiksen, Ø., Giske, J., and Slagstad, D. (1995) A spatially explicit fitness-based model of capelin migrations in the Barents Sea. *Fish. Oceanogr.* **4**(3), 193–208.

Giske, J., Huse, G., and Fiksen, Ø. (1998) Modelling spatial dynamics of fish. *Rev. Fish Biol. Fisher.*, **8**, 57–91.

Hampton, J. (2000) Natural mortality rates in tropical tunas: size really does matter. *Can. J. Fish. Aquat. Sci.*, **57,** 1002–1010

Hilborn, R., and M. Mangel (1997) The Ecological Detective: Confronting Models with Data. Princeton University Press, USA, 315 pp.

Holland, K.N., Brill, R.W., and Chang, R.K.C. (1990) Horizontal and vertical movements of Yellowfin and Bigeye tuna associated with fish aggregating devices. *Fish. Bull.* **88**, 493–507.

Holland, K.N., Brill, R.W., Chang, R.K.C., Sibert, J.R., and Fournier, D.A. (1992) Physiological and behavioural thermoregulation in bigeye tuna (*Thunnus obesus*). *Nature* **358,** 410–411.

Humston, R., Ault, J.S., Lutcavage, M., and Olson, D.B. (2000) Schooling and migration of large pelagic fishes relative to environmental cues. *Fish. Oceanogr.*, **9**(2), 136–146.

Hunter, J.R., Argue, A.W., Bayliff, W.H., Dizon, A.E., Fonteneau, A., Goodman, D., and Seckel, G.R. (1986) The dynamics of tuna movements: an evaluation of past and future research. *FAO Fisheries Technical Paper* **277**, 78 pp.

Huse G., and Giske, J. (1998) Ecology in Mare Pentium: An individual-based spatio-temporal model for fish with adapted behaviour. *Fish. Res.* **37**(1–3), 163–178.

Huse, G., Giske, J., and Salvanes, A.G.V. (in press) Individual-based modelling. *In* Hart, P.J.B., and J.D. Reynolds (eds.), *Fish and Fisheries Handbook*, Blackwell Science, UK.

Joseph, J., and Wild, A. (1984) Considerations for Tuna Research to the Year 2000. Inter-American Tropical Tuna Commission. La Jolla, CA.

Josse, E., Bach, P., and Dagorn, L. (1998) Simultaneous observations of tuna movements and their prey by sonic tracking and acoustic surveys. *Hydrobiologia*, **371/372,** 61–69.

Kirby, D.S., Fiksen, Ø., and Hart, P.J.B. (2000) A dynamic optimisation model for the behaviour of tunas at ocean fronts. *Fish. Oceanogr.* **9,** 4., 328–342.

Kishinouye, E. (1923) Contributions to the comparative study of the so-called scombroid fishes. *J. Coll. Aric. Imp. Univ. Tokyo.* **8**, 293-475.

Krebs, J.R., Davies, N.B., and Parr, J. (1993) *Introduction to Behavioural Ecology.* Blackwell Science, UK, 432 pp.

Krebs, J.R., and Davies, N.B. (1997) *Behavioural Ecology.* Blackwell Science, UK, 464 pp.

Laurs, R.M., Fielder, P.C., and Montgomery, D.R. (1984) Albacore tuna catch distributions relative to environmental features observed from satellites. *Deep-Sea Res.* **31**(9), 1085-1099.

Lehodey, P., Andre, J-M., Bertignac, M., Hampton, J., Stoens, A., Menkes, C., Memery, L., and Grima, N. (1998) Predicting skipjack tuna forage distributions in the equatorial Pacific using a coupled dynamical bio-geochemical model. *Fish. Oceanog.* **7**(3/4), 317–325.

Loehle, C. (1983) Evaluation of theories and calculation tools in ecology. *Ecol. Model.*, **19**, 239–247.

Lowe (2000) Blood oxygen-binding characteristics of Bigeye tuna (*Thunnus obesus*) , a high-energy-demand teleost that is tolerant of low ambient oxygen. *Mar. Biol.* **136,** 1087–1098.

Lutcavage, M. (1997) Distribution, relative abundance, and behavior of giant bluefin tuna in New England waters, 1995. Coll. Vol XLVI, **2,** 332–347. *Int. Comm. Conserv. Atlantic Tunas*, Madrid.

Mangel, M., and Clark, C.W. (1988) *Dynamic Modeling in Behavioural Ecology.* Monographs in Behaviour and Ecology, Princeton University Press, Princeton, New Jersey, 308 pp.

Mangel, M., Fiksen, Ø., and Giske, J. (2000) Logical, statistical and theoretical models in natural resource management and research. *In* Franklin, A.. and Schenk, T. (Eds.), *How to Practice Safe Modelling.* Island Press, Boulder, CO.

Magnuson, J.J. (1973) Comparative study of adaptations for continuous swimming and hydrostatic equilibrium of scombroid and xiphoid fishes. *Fish. Bull.* **71**, 337–356.

Marchal, E., Gerlotto, F. and Stequert, B. (1993) On the relationship between scattering layer, thermal structure and tuna abundance in the Eastern Atlantic equatorial current system. *Oceanol. Acta,* **16**(3), 261–272.

SERG (2000) URL: http://www.ifm.uib.no/Staff/giske/SERG2000.htm

Sharp, G.D. (1995) It's about time: new beginnings and old good ideas in fisheries science. *Fish. Oceanogr.* **4**(4), 324–341.

Sibert, J.R., Hampton, J., Fournier, D.A., and Bills, P.J. (1999) An advection-diffusion-reaction model for the estimation of fish movement parameters from tagging data, with application to skipjack tuna (*Katsuwonus pelamis*). *Can. J. Fish. Aquat. Sci.*, **56,** 6, 925-938.

Stöcker, S. (1999) Models for tuna school formation. *Math. Biosci.* **156**, 167–190.

Stoens A., Menkes C., Dandonneau, Y., and Memery, L. (1998) New production in the equatorial Pacific: a coupled dynamical-biogeochemical model. *Fish. Oceanogr.* **7**(3/4) 311–316.

Sund, P.N., Blackburn, M., and Williams, F. (1981) Tunas and their environment in the Pacific Ocean: a review. *Oceanogr. Mar. Biol. Ann. Rev.* **19,** 443–512.

Tyler, J.A., and Rose, K.A. (1994) Individual variability and spatial heterogeneity in fish population models. *Rev. Fish Biol.Fisher.,* **4,** 91–123.

From Individuals to Local Population Densities: Movements of North Atlantic Bluefin Tuna (*Thunnus Thynnus*) in the Gulf of Maine/Northwestern Atlantic

Nathaniel Newlands[1] and Molly Lutcavage[2]
[1] *Fisheries Center, University of British Columbia, 2204 Main Mall, Vancouver, British Columbia, Canada, V6R 1Z4*
[2] *New England Aquarium, Edgerton Research Laboratory, Central Wharf, Boston, MA, U.S.A. 02110-3399*

Key words: behaviours, bluefin tuna, movement models, population density, schooling

Abstract: North Atlantic bluefin tuna have high rates of spatial redistribution and display complex aggregation behavior in response to changes in their environment. This paper presents preliminary results obtained from an analysis of individual movements of giant bluefin tuna derived from tracking studies conducted in association with aerial surveys in the Gulf of Maine (NW Atlantic). From the perspective of a correlated and biased random walk movement model, we examined how alternative behavioral modes of tuna movement affect estimation of local population density. We summarize information needed to link movements and behavior of individuals and schools (i.e., depth and surfacing behaviour, spatial mixing, residency, and school and size distribution) in order to portray spatial and temporal dynamics of the bluefin assemblage. Our goals are to integrate hydroacoustic tracking and aerial survey movement data in order formulate a modeling and simulation framework that will generate predictions and estimates of local population densities. This will support the development of fishery-independent assessments of adult bluefin tuna that are crucial to accurate monitoring, management, and rebuilding of the Atlantic population.

1. INTRODUCTION

The Atlantic bluefin tuna (*Thunnus thynnus*) is a long-lived, highly migratory species that attains sizes of 300 kg or more. Adults undertake cyclic migrations between coastal feeding zones and offshore wintering areas and

J.R. Sibert and J.L. Nielsen (eds.), Electronic Tagging and Tracking in Marine Fisheries, 421–441.
© 2001 *Kluwer Academic Publishers. Printed in the Netherlands.*

spawning grounds (Mather *et al.*, 1995). Bluefin tuna are intensively pursued by commercial and recreational fisheries because of their high commercial value. Their population is currently believed to have plummeted to 20% of 1970's levels, yet there is significant uncertainty in the status and size of the current population (Sissenwine *et al.*, 1998; ICCAT, 1999). A recent scientific review called for the development of alternative approaches that might improve this situation (NRC, 1994). In an attempt to develop new, fishery independent population assessments, a series of aerial surveys were conducted for giant bluefin tuna in the Gulf of Maine (NW Atlantic) to study the spatial and temporal patterns and relative abundance of surface schools, which are easily detected and photographed by spotter pilots (Lutcavage and Kraus, 1995; Lutcavage *et al.*, 1997).

In general, schooling is believed to improve search and foraging strategies and defense against predators, to faciliate social communication and learning, and to increase mating success (Pitcher and Parrish, 1993). The motivations and movements of individuals subsequently influence the behavior of their schools, and the process of joining, leaving or staying within schools may offer mechanisms for balancing continuous risks to survival (Pitcher and Parrish, 1993). The collective result of individual decisions influences the spatial and temporal patterns of distribution, size and mixing of schools, and ultimately, affects population-scale dynamics (Mogliger *et al.*, 2000).

Previous studies have shown that bluefin tuna form dynamic, highly polarized schools (Partridge *et al.*, 1983; Lutcavage and Kraus, 1995; Lutcavage *et al.*, 1997), but further investigation of the responses of individuals in schools of various sizes and structures is required in order to undertand their movements in time and space. Individual-based models have generated important predictions for population-scale dynamics, and have shown that large spatial and temporal variations within a population can occur when individuals possess differences in relative survival 'fitness' (DeAngelis and Gross, 1992). These models have shown that the mixing and size variability of schools depends upon the benefits schooling offers the individual. The net effect is fluctuation in both the school size and structure as a function of prey density, physiological regulation, activity, or limits of sensory detection systems (Parrish and E-Keshet., 1999). For example, school members benefit from the collective knowledge of refuge and foraging arenas, and through social facilitation, can select optimal search paths, which would then reduce energetic costs and survival risks of migrations (Kieffer and Colgan, 1992).

This paper presents preliminary results obtained from an analysis of individual movement of giant bluefin tuna derived from hydro-acoustic tracking studies conducted in association with aerial surveys in the Gulf of Maine (NW Atlantic). We first examine how correlated and biased random walk models of movement trajectories affect estimation of local population density.

Information on depth and surfacing behaviour, mixing, residency, and the structure and size distribution of schools is then examined to see how behaviors evident from individual fish and school movements are linked. The application of appropriate models from real time data on fish movements and behavior are important steps in the development of sampling frameworks for fishery independent aerial surveys and other direct census methods (e.g. Light Detection and Ranging (LIDAR), hydroacoustics) (Lutcavage and Newlands, 1999).

2. MATERIAL AND METHODS

The horizontal and vertical movement and behavior of twelve giant bluefin tuna (113-340 kg) tracked by hydroacoustic telemetry for up to 48 hours in three regions in the Gulf of Maine (Stellwagan Bank, Great South Channel, Cape Cod Bay) were recently reported by Lutcavage *et al.* (2000). This size-class is the primary seasonal assemblage documented in the fishery-linked aerial surveys conducted in the Gulf of Maine from 1993-96 (Lutcavage and Kraus, 1995; Lutcavage *et al.*, 1997). Individuals traveling in schools were tracked with an ultrasonic transmitter and single-directional hydrophone (V-10/V-11, Vemco, Inc.) that had a detection zone of approximately 1.7 km. (Lutcavage *et al.*, 2000). Fish positions were approximated as the position of the tracking vessel, which was determined with GPS (geographical position system) coordinates recorded every 60 seconds. Fish depth was sampled every 15 seconds with a Vemco VR-60 receiver. This study identified three consistent movement modes in the observed tracks: repetitive travel through the thermocline with small displacements (<5-40 km d^{-1}), travel primarily in the surface layer with large displacements (40-76 km d^{-1}), and large, single vertical displacement/dives at dusk and dawn. In our current analysis we compared individual fish travel patterns to two existing theoretical movement models in order to classify their movement modes. This was essential for extrapolating from the small sample size of hydroacoustic tracks to a large number of individuals in order to simulate the spatial consequences of selected movement modes for estimating local population densities.

2.1 Movement Trajectories in Three-Dimensions

2.1.1 Individual movements as correlated random walks

Patlak's model addresses three important limitations of standard diffusion models for representing population dispersal (Patlak, 1953a,b; Okubo, 1980; McCulloch and Cain, 1989). This model assumes (i) dispersal in more than one spatial dimension for the purposes of characterizing marine habitats in three-dimensions, (ii) that serial correlation in move direction of an individual at any given time has a relative influence on the direction of movement in the near future, and (iii) variations in speed and turning frequency can be linked to external factors producing changes in their spatial distribution (Turchin, 1991). This modeling approach characterizes observed movement paths according to move speed, duration and turning angle frequency distributions, from which repeated random and independent (Monte-Carlo) sampling may be used to predict the probability of an individual at any point in space and time (population density). This means that each movement path in the model is represented as a stochastic, Markov chain process (i.e., correlated random-walk, CRW).

Turchin (1991) derived the spatial *equilibrium* solution to Patlak's model assuming all individual movements are spatially contained by specified linear boundary conditions (no immigration/emmigration) with the spatial extent of movement divided into two regions j=1,2. In addition, the variance about the mean move length for the observed tracks within each region is assumed to be proportional to its mean. Under these assumptions, the ratio of population densities between two regions is expressed as a ratio of *residency indices*, ρ_j for each region j,

$$\rho_j = \tau_j \varphi_j^{\alpha} \mu_j^{-\left(\frac{2-2\gamma}{2-\gamma}+\alpha\right)}, \gamma \neq 2 \tag{1}$$

where τ is the mean move duration (s), $\gamma = M/\mu^2$ a function of mean-squared move length (m^2) M, and mean move length (m), μ. The angle, φ [0, 2π], is related to the two variables, γ and the mean cosine of the turning angle ψ [0,π], through the substitution, $\varphi = \gamma + \psi(2-\gamma)$. The additive term in the exponent, α, is related to the values of CRW parameters for each region according to

$$\alpha = \frac{2\gamma(\mu_2 - \mu_1)}{(2-\gamma)(\varphi_1\mu_2 - \varphi_2\mu_1)} \tag{2}$$

Equations used in calculating parameters associated with equation (2) are provided in the Results section. Low values in the residency index imply

Equations used in calculating parameters associated with equation (2) are provided in the Results section. Low values in the residency index imply rapid and/or directed movement within a region, while high values imply low movement speeds and/or frequent turning.

Equilibrium residency estimation is applicable to marine habitat patches with sizes that are similar to the average move length of individuals (Turchin, 1991). Patlak's correlated random walk (CRW) model utilizes static distributions of movement parameters. Frequency distributions of its parameters are taken as empirical mean values across all observed paths, and movement stochasticity is introduced in the random sampling from these distributions. Therefore, this model does not attempt to explain the uncertainty or variability of movements caused by each parameter being behaviorally or spatially dependent. Given the high mobility of bluefin tuna, time-dependent residency estimation is required to specify the fluctuations in local equilibrium residency over time. However, equilibrium residency estimates may still provide indications of bluefin tuna habitat preference, and inter-specific competition (Brown, 1988). For example, a spatial model proposed for yellowfin tuna (*Thunnus albacores*) in the eastern Pacific when the fishery was restricted in its range (late 1960's) showed that catches in the coastal fishery were sustained by immigration from larger offshore regions (Mullen, 1994). Even though individuals are highly mobile, this study suggests that the *fine*-scale characteristics of tuna movement paths (i.e. diffusivity or directionality) provide key information for identifying potential impacts of fishing dynamics at the mesoscale (10-100) km. In contrast to yellowfin tuna movements, which are usually diffusive (Sibert *et al.*, 1999; Bertignac *et al.*, 1998), bluefin tuna movements have both diffusive and directed movement modes (Humston *et al.*, 2000; Lutcavage *et al.*, 2000).

2.1.2 Movements as biased random walks

A model recently developed by Grünbaum (1999) provides a generalized framework that facilitates systematic study of behavioral variation and the consequences of variation across different environments. This model contains a key biological feature of non-random taxis through the dependency of turning rate on external cues, without which the model reduces to a CRW process. The added realism of Grünbaum's model addresses a current criticism of stock-assessment models for tunas because it allows inclusion of crucial biological and environmental information necessary for developing better indices of abundance (Brill, 1994; NRC, 1994), and relaxes the assumption that fish are randomly and uniformly distributed. This model is particularly applicable to the bluefin tuna, an apex predator, where the distribution of its prey scales at slower rates than a fixed number of successive search steps or moves of individual fish. The successive displacements may then be associated with the time required by the fish to sample external cues across sensory detection zones. Similar zones can be conceived where a fish would transit and sample a gradient zone in order to modify its behavior in response to changes in physiological regulation. In comparison with equation (1), the mesoscale residency or equilibrium searcher density for this model is given by

$$\rho(\varepsilon(x,y,z)) = \left\{ \mathrm{F}(\varepsilon)\exp\left[-\int_{\varepsilon_0}^{\varepsilon} \mathrm{G}(\varepsilon')d\varepsilon' \right] \right\}^{1-\varphi} \tag{3}$$

where F(ε) is a function that describes the dependence of turning rate on local environmental potential (prey concentration), ε, and G(ε) modulates the sensitivity of movement to the rate of environmental change perceived by an individual as

$$\mathrm{G}(\varepsilon) = -\frac{d}{d\varepsilon}\ln\left(\mathrm{F}(\varepsilon)\right) \tag{4}$$

Grünbaum (1999) describes a variety of functional forms for F(ε) and G(ε) according to several possible search strategies, ranging from area-restricted search behavior to directional taxis. These alternative movements taken from the bluefin tuna tracks (Figure 1) are depicted in Figure 2.

Behaviors where turning *distribution* is a function of integrated density cannot result in taxis (Grünbaum, 1997; 1999), while behavioral turning *rates* are functions of prey density. Therefore, predicted turning distributions and rates can be generated in order to simulate shifts in local population densities. By identifying the relationship of various movement parameters associated with behavioral modes, the effect of environmental gradients on turning rate can be determined. This biased random walk model is under development to quantify and integrate alternative individual and schooling movement behaviors observed in the bluefin hydroacoustic tracks and aerial surveys.

3. RESULTS

Geolocation (X, Y) and depth measurement (Z) records for 10 selected individuals were time-correlated using a linear-interpolation scheme. Conversion of positions to unequal, linear move displacements enabled three-dimensional reconstruction of each path, shown here for individual 9705 (Figure 3), which was followed in a school of approximately σ=200 individuals (Lutcavage *et al.*, 2000). The cartesian coordinates $(X,Y,Z)_n$ following k = (1,...,n) successive moves were converted into spherical coordinates $(l, \theta, \varphi)_k$ using the following transformation equations

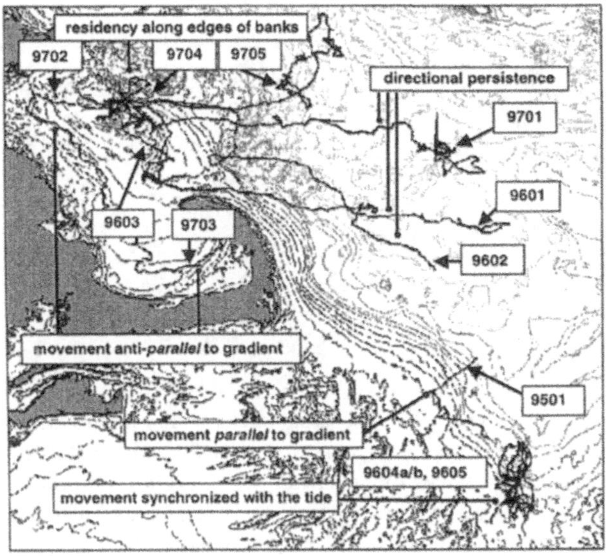

Figure 1. Observed hydro-acoustic tracks of bluefin in the Gulf of Maine (1996-97, N=12) contained within the region (69-71 °W, 41-43 °N). Several key observations for inclusion into individual-based or schooling models for bluefin tuna are shown. The tracks are shown over a bathymetric grid and associated contour lines at a resolution of 15 arc-second grid (~1/2 km grid cell size) compiled by the Research Environmental Data and Information Management System (REDIMS) for the Gulf of Maine. (http://oracle.er.usgs.gov/gomaine/)

$$X_n = \sum_{k=1}^{n} l_k \sin\theta_k \cos\varphi_k$$
$$Y_n = \sum_{k=1}^{n} l_k \sin\theta_k \sin\varphi_k \qquad (5)$$
$$Z_n = \sum_{k=1}^{n} l_k \cos\theta_k$$

where θ_k is the polar angle and, φ_k, the azimuthal angle of the $k^{th.}$ movement measured from the z-axis and x axes respectively. The azimuthal angle, φ_k, of the $k^{th.}$ move is related to the sum of previous turning angles, ψ_j (j=1,...,k-1) which represent the angular change in successive move direction by

$$\varphi_k = \varphi_1 + \sum_{j=1}^{k-1} \psi_j \qquad (6)$$

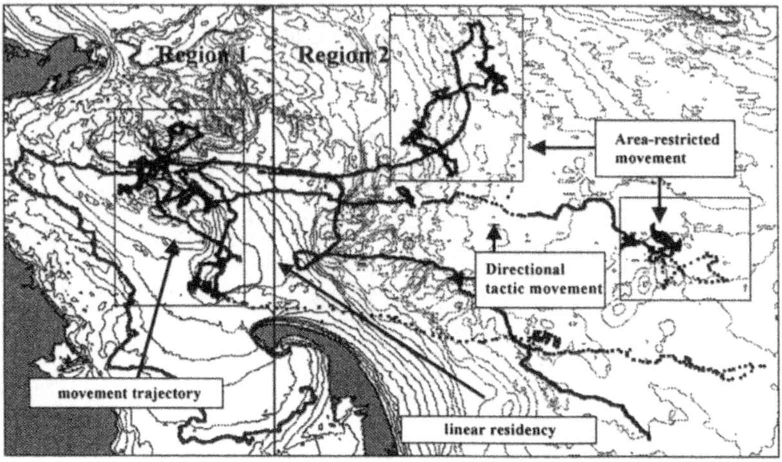

Figure 2. A linear boundary is used to separate tracks within mesoscale regions 1 and 2 for illustrating the calculation of spatial population density following Patlak's correlated random walk model for the bluefin tracks of Lutcavage *et. al.* (2000). Directional tactic and area-restricted type movements may follow external taxis response by bluefins to local oceanic bathymetric or thermal gradients. This chart was generated using Arcview© Movement Analysis extension (USGS) (Hooge *et al.*, 1997, 1999).

The mean move length (μ), mean-squared move length (M), mean move duration (τ), and mean cosine of the turning angle ($<\cos\psi>$) used to characterize their trajectories are given by

$$\begin{aligned} \mu &= \frac{1}{n}\sum_{k=1}^{n} l_k \\ M &= \frac{1}{n}\sum_{k=1}^{n} l_k^2 \\ \tau &= \frac{1}{n}\sum_{k=1}^{n}\left(\frac{l_k}{v_k}\right) \\ <\cos\psi> &= \frac{1}{n}\sum_{k=1}^{n}\cos\psi_k \end{aligned} \tag{7}$$

Values for these CRW parameters were estimated from the empirical movement data for each of the ten individuals and are provided in Table 1. Associated distributions of move-length, turning angle and speed are shown in Figures 3a-d. The turning angle distribution (Figure 3b) shows slight asymmetry with respect to a circular (Von Mises) normal distribution. As expected, the move-length and speed distributions are gamma-distributed. Profiles showing the successive alterations in the spherical coordinate angles $(\theta,\varphi)_k$ of their move displacements, used to calculate turning angle, ψ_k, are shown (k=1,…,n) in Figure 4. Each path contains successive displacements occurring at unequal time-intervals, whereby movement patterns can be examined according to both their *time* dimension: the turning frequency (mean move duration), and *spatial* dimension (mean cosine of the turning angle). Variation in each of these dimensions is expected to produce very different predictions for their mesoscale spatial distribution (Turchin, 1998).

Table 1. Mean values of move length (m), duration(s), and cosine of turning angle (dimensionless) for observed trajectories of tracked individuals. The number of mode-switching events, s, where the number of modes is m=(s-1). For the individual labeled as 9705, m=2 where modes m1 and m2 were identified from the spectral analysis.

Track Id.	Total moves (depth-correlated record) n_i	Mean move-length $\mu(\varepsilon_i)$ (m)	Mean move duration $\tau(\varepsilon_i)$ (s)	Mean cosine of turning angle $<\cos(\psi)>$	Mode-switching events s=(m+1), for m modes
9601	205	1030	1267	0.872	2
9602	2787	140.6	138.7	0.902	4
9603	1752	100.9	59.47	0.877	1
9604a	2477	95.57	161.0	0.899	4
9604b	1183	124.8	163.0	0.913	0
9701	1888	256.2	202.5	0.901	4
9702	1820	122.1	156.4	0.906	3
9703	352	118.3	254.2	0.881	1
9704	180	131.4	412.9	0.924	2
9705	2567	117.0	153.6	0.886	2

Extracting the best resolution of behavioral information from observed movement paths requires that both over- and under-sampling be avoided when segmenting a path into equal move-displacements. In our analysis, move displacements detected at high resolution (location and depth were recorded at 60-second intervals) were aggregated and kept as variable move lengths in order to analyze behavioral adjustments of move-length, duration, speed and turning angle. In this way, high frequency alterations detected between right and left turns or 'wobble' (see Figure 4) are characterized as a short-term (small lag) negative auto-correlation in turning angle. Such segments are typically 'tuned' by shifting to an appropriate lag (time-integration of successive positions), thereby reducing serial correlation. Here, a spectral method was used so that the unequal move displacements could be analyzed, enabling auto-correlations in the turning angle and other movement variables to be identified and associated with alternative behaviors (Lomb, 1976; Press *et al.*, 1999).

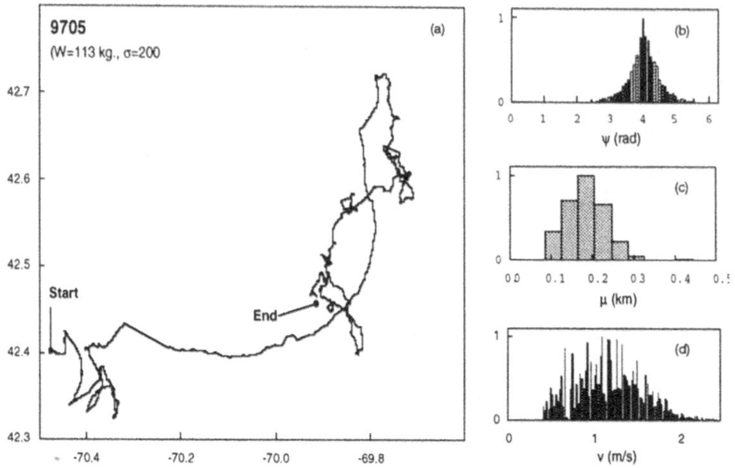

Figure 3. Left: (a) Track 9705 for an individual of estimated weight, W=113 kg. and school size, σ=200 individuals is shown in two-dimensions. *Right:* (b) Probability distributions for turning angle, ψ (rad.), (c) move-length, μ (rad.), and (d) speed, v(m/s) obtained from our analysis of this trajectory are shown.

In order to compare bluefin tuna movements to correlated random walk predictions, removal of serial-correlation in move duration, speed, and turning angle is required (Turchin, 1998). Fast computation of the Lomb periodogram was used in translating the movement trajectory of each individual from the time-domain to frequency domain for the purposes of identifying the relative contribution and significance of separate frequency components for turning angle and speed (Table 1). The so-called *Nyquist* frequency, representing equal temporal partitioning of the trajectory, occurs at $\omega=2\pi f=0.5$ Hz. Above Gaussian distributed noise, a second peak is evident in the power spectrum of the turning angle (see Figure 5b), below the Nyquist frequency, at approximately 0.38 Hz. Associated peaks in the speed profile of 9705 occur at much smaller frequencies, with a prominent peak at approximately $\omega=0.01$ Hz (Figure 5c). Figure 5a shows spectral power, $P(\omega)$, as a function of angular frequency, ω, and variation of the amplitude of the maximum signal, $P(\omega)_{max}$, across its entire movement trajectory (fish 9705). This plot depicts two movement modes identified within the trajectory of 9705, denoted m1 and m2, where during mode m1, second-order autocorrelation of turning angle and move speed is evident, and during m2, turning angle and speed appear uncorrelated. Over the duration of mode m1, this fish and its school were near actively feeding whales, seabirds and

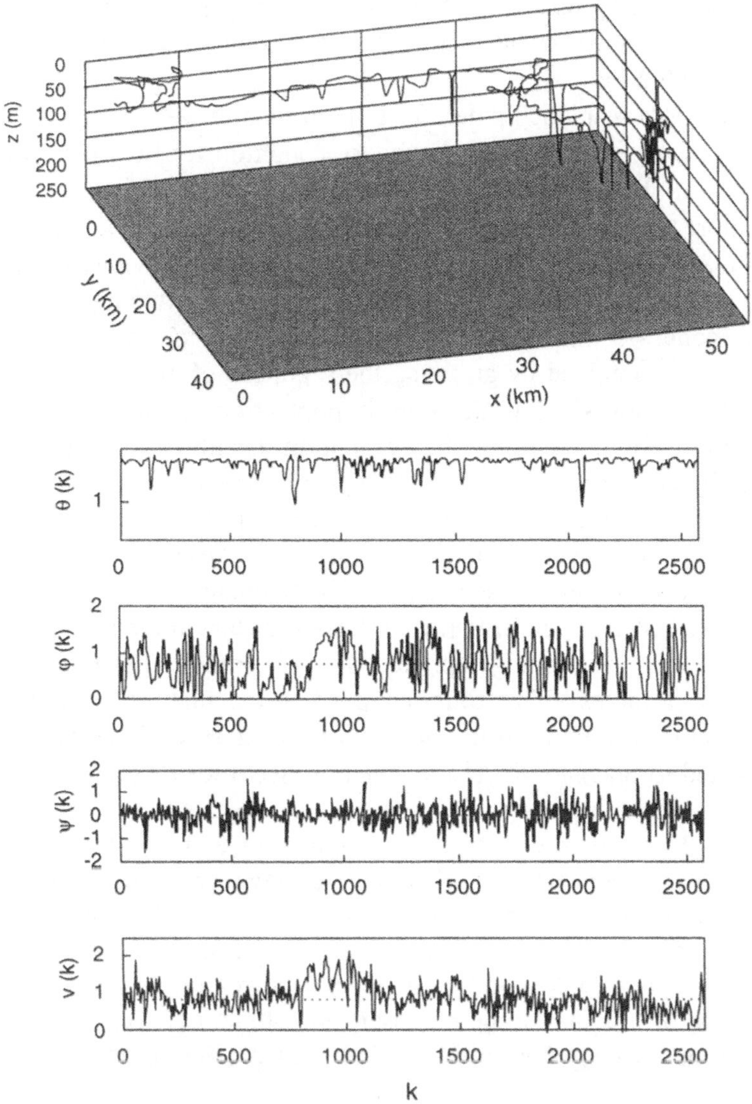

Figure 4. Three-dimensional reconstruction of movement path (9705) using a linear-interpolative scheme to correlate depth (z) with latitude and longitude readings with two movement modes (m1,m2) (see also Fig. 3). Top to bottom: the variation in spherical movement angles (θ,φ)(rad.), turning angle, ψ (rad.) for 60 second intervals, and move speed, v(m/s) calculated for this track are profiled over k=(1,..,n), where the total number of move intervals n=2567.

basking sharks, suggesting that this movement mode may be associated with feeding. Similarly, a number of distinct modes were found in our analysis of other fish tracks (Table 1).

Calculation of Patlak's residency index for bluefin tuna movement paths using equation (1) is shown for a linear boundary condition for two regions at 70.2 °W (Figure 2). This boundary was chosen to delineate region (1), enclosing an inshore foraging area associated with the topographical feature Stellwagan Bank (Lutcavage *et al.*, 2000). The estimate provides an expected ratio of equilibrium population densities between the two regions for Patlak's CRW model. In this estimation, values of calculated random-walk parameters (Table 2) obtained by grouping the segments of all the observed tracks within each region yield an equilibrium population density for the inshore region (1) of about six times the population density of the offshore region (2). Individuals shift behaviors abruptly, since there are distinct changes in their movement parameters as they cross the boundary. If relevant physical and/or biological perturbations are small, these factors may be considered homogeneous across each region. The boundary subdividing the regions would then represent an instantaneous gradient that alters individual-level movements without invoking explicit spatial or temporal heterogeneity of environmental variables. This time-independent calculation shows that under an assumption of random-walk movement, the effects of school-level dynamics expected to bias individual movement trajectories on the expected ratio of population densities is significant at the spatial scale of our observations. However, the additive effects of environment or extrinsic factors are clearly needed to generate meaningful population-level predictions. Time-dependent estimation of residency dynamics calls for path observations of sufficiently large sample size, spatial extent, and short time span so that appreciable mixing of unobserved individuals between regions is unlikely to occur. This condition might also be necessary to estimate accurate feeding rates from the time-varying distribution of individuals (Mason and Patrick, 1993).

Table 2. Correlated random walk (CRW) parameters for n^i discrete moves for the observed trajectories of individuals for the two regions delineated as region 1, ε1 (inshore) and region 2, ε2 (offshore) shown in Figure 1.

Region	Sample size of moves n_i	Mean move length $\mu(\varepsilon_i)$ (m)	Mean move duration $\tau(\varepsilon_i)$ (s)	Mean cosine of turning angle $\langle\cos(\psi)\rangle$	Scaling constant γ (m^{-1})	Residency index $\rho(\varepsilon_i)$	Ratio of population densities $\rho(\varepsilon_1)/\rho(\varepsilon_2)$
ε_1	10120	107.1	56.8	0.276	0.00930	0.511	6.03
ε_2	5576	971.9	86.7	0.319	0.00100	0.0850	

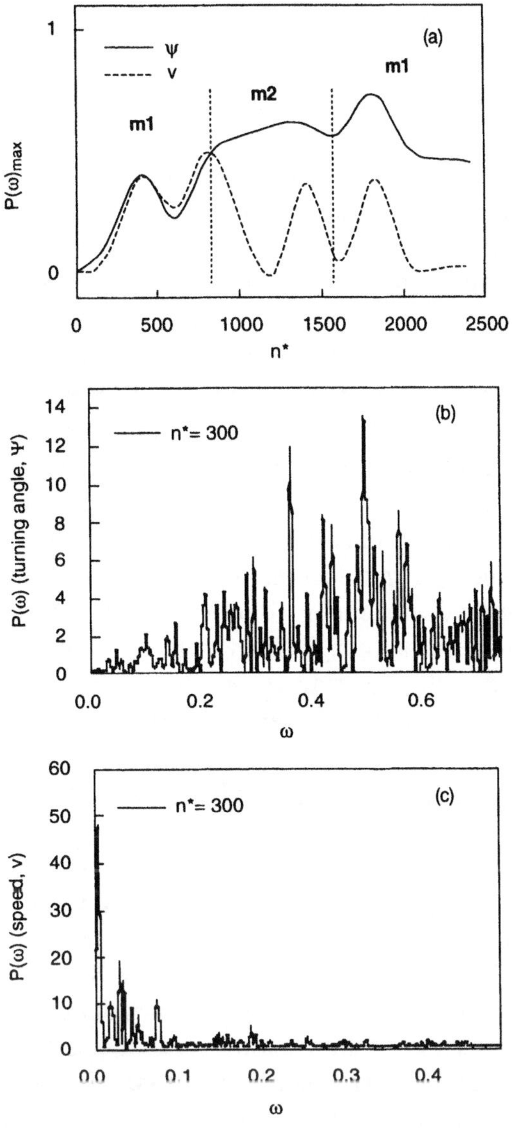

Figure 5. (Top) (a) The amplitude of the maximum signal, $P(\omega)_{max}$, with greater than 99% significance above Gaussian white noise is plotted versus n*=(1,..n) (for a total of n=2567 successive moves). This plot reveals that alternation in the trajectory follows two movement modes (m1, m2): m1—second-order autocorrelation of turning angle with move speed, m2—uncorrelated turning angle and speed. (Bottom) Spectral analysis of the unevenly sampled moves within the three-dimensional track shown in Figure 4. Translation of the profiles for turning angle and speed from the time domain to frequency domain was performed by calculating the Lomb (normalized) periodogram, P(ω) (spectral power as a function of angular frequency, ω) as shown in (b) and (c) respectively, for n*=300 total moves within the trajectory of 9705.

4. DISCUSSION

Identification of travel paths of individuals within schools offers the following advantages for the analysis and prediction of school movements (Turchin, 1998): (1) samples of real-time travel paths are not biased to fishing effort or gear, (2) tracks of individual fish provide important data for testing mechanisms of physiological responses (Brill, 1994) or environmental cues that may influence diffusivity and directionality of school movements (Bertignac *et al.*, 1999; Sibert *et al.*, 1999; Humston *et al.*, 2000), (3) re-construction of independent path observations can identify consistent behavioral movement modes and spatial cues for temporal 'mode-switching' events, and (4) spatial population shifts can be identified because tracks contain time step information needed to model the attraction and repulsion of individuals/schools within a heterogeneous environment (Dagorn and Fréon, 1999). Bluefin tuna movement responses that fit within realistic individual-based and dynamic schooling models (see Figure 1) include diffusive and rapid vertical movements during residency near topographic features, directed movement to regions of high prey concentration, direction shifts synchronized with tidal flows, and movement relative to bathymetric or topographic gradients (Lutcavage *et al.*, 2000).

4.1 Structure and Size Distribution of Observed Schools

In a model of the stability of skipjack tuna (*Katsuwonus pelamis*) schools from mark-recapture data, 16-63% of individuals left one school and joined another over 24 hours (Hilborn, 1991). From this analysis, two alternative hypotheses of school-school interactions were proposed for calculating school stability resulting from the net effect of individual decision schemes. One hypothesis for school stability is that individuals leave schools according to exponential decay over time. Simulations of school size distribution according to fission and fusion rates also yield an exponential distribution for school splitting as a function of its size, consistent with an associated maximum entropy limit (Flierl *et al.*, 1999). Analysis of school size statistics for mixed schools of yellowfin *(Thunnus albacares*), skipjack (*Katsuwonus pelamis*) and bigeye tuna (*Thunnus obesus*) confirms a power-law relationship between the distribution and size of observed schools (Niwa, 1998; Bonabeau *et al.*, 1999). Another hypothesis, that schools exchange individuals based on proximity (Hilborn, 1991), predicts an increased likelihood that individuals within separate schools will later be found traveling together. Observations of the internal structure of schools, movement characteristics of schools of various sizes, and school residency within distinct spatial regions of high school densities may provide crucial estimates of school integrity and spatial mixing.

Mesoscale frontal regions (where coastal upwelling aggregates primary producers and attracts biomass) are also expected to lead to higher probabilities for mixing because these regions concentrate tuna schools (Laurs *et al.*, 1977; Fiedler and Bernard, 1987; Podesta *et al.*, 1993; Marchal *et al.*, 1993; Olson *et al.*, 1994; Swartzman *et al.*, 1994; Maravelias, 1999). In the Gulf of Maine, bluefin movements have been linked to thermal stratification and bathymetric structure (Lutcavage *et al.*, 1997; Lutcavage *et al.*, 2000). Identification of these well known frontal features (Ullman and Cornillon, 1999) may also lead to a high probability of detection, and consequently, may improve regional abundance estimates.

4.2 Depth, Travel, and Behavioral Patterns

In the Gulf of Maine, bluefin tuna schools travel primarily in the shallow mixed layer (>90% of time within the uppermost 30m), but their depth and horizontal displacements are highly variable (Lutcavage *et al.*, 2000). The probability of two schools converging and exchanging individuals may depend upon only the distance between them, if this distance is small. This may be the case for movements that follow mode m1, where individuals within schools exhibit area-restricted behavior (Figure 4). During directed movements (such as mode m2), school structure may best determine potential for mixing. The specific structure adopted by group members then affects the relationship between school size and velocity, and may allow individuals to maintain specific individual movement modes across different spatial regions.

In an individual-based simulation whereby individuals possess different sets of decision rules, Romey (1996) showed that school–level adjustments were necessary for maintaining specific searching behaviors, and that the relationship between school formation and its trajectory (independent of external environmental effects) is inherently nonlinear. In schools comprising individuals with high nearest-neighbor attraction (increasing integrity) group size had a large effect on school trajectory, since larger groups move more slowly. Dynamic structural adjustments (i.e., alternative movement modes) may account for the fact that bluefin tuna experience maximal thermal gradients of approximately 14°C during vertical excursions, yet manage to spend the majority of their time occupying the warmest water available (Brill and Lutcavage, 2000).

Partridge *et al.* (1983) examined two hypotheses for the formation of schools of less than 25 giant bluefin tuna in Cape Cod Bay based on a two-dimensional analysis of their structures from aerial photographs. They suggested that predators receive cooperative hunting and hydrodynamic benefits by schooling in parabolic formations, although no evidence was provided that these individuals actually engaged in feeding. We are conducting analyses on several thousand bluefin tuna schools photographed in recent aerial surveys (1993-97) that cover a much broader range of school size, structure, and geographic locations than those documented by Partridge *et al.* (1983). The static observational data from aerial surveys, used in conjunction with identification of movement dynamics of individuals within schools, will allow development of time-dependent calculations examining how bluefin school trajectories change as a function of a network of real time biotic and abiotic variables.

4.3 Prey Distribution and Mesoscale Foraging Regions

Our current modeling work aims to quantitatively test different movement hypotheses, such as the hypothesis that distinct assemblages of bluefin tuna reside in the Gulf of Maine. Support for this comes from historic observations of schools of similar size, shape and behavior in subregions of the New England and Canadian Atlantic shelf (Crane, 1936; Porter and Hogans, 1991) and recent aerial survey data showing clumped distribution of schools within these areas (Lutcavage *et al.*, 1997). Although aerial surveys provide only a static view of bluefin distribution, our modeling shows that spatially distinct assemblages may arise dynamically from schools that have different foraging strategies for exploiting their spatially-distinct prey (e.g. sandlance, squid, mackerel, and herring).

4.4 Spatial Mixing and Residency

Bluefin tuna enter the Gulf of Maine to feed, consuming an estimated 1-10 % of their biomass per day (Crane, 1936; Chase, 1995). If bluefin tuna feeding patterns are a function of the relative concentrations of their prey (see Cosner *et al.*, 1999), the spatial mixing of schools associated with specific movement behaviors would alter school characteristics. For example, predator-prey encounter rates would improve with frequent residency in distinct, productive foraging regions. If certain movement modes are consistently adopted within certain regions, schooling enhances information transfer and improves the decision-making process of individuals across space and time. When individuals can elect optimal durations and times for switching between alternative movement modes, schools might then possess a spatial memory mapping ability that would allow them to learn changes in prey concentrations (Dill, 1983; Milinski, 1994; Boyd, 1996) via possible mechanisms such as forage area copying (FAC) (Pitcher and Parrish, 1993), or "joining" strategies to improve foraging benefits (Giraldeau, 1984; Metcalfe and Thomson, 1995).

Equilibrium residency calculations (described in the Methods section) are justified under the assumption that mesoscale movement dynamics follow the ideal free distribution (IFD). The IFD assumes that schools of bluefin tuna have equal competitive abilities and that competition among schools of different sizes scales in proportion to their local density. If the prey are fixed, schools are then considered free to move to locations that maximize feeding rates (Rose and Leggett, 1990; Gillis *et al.*, 1993; Pitcher and Parrish, 1993). To maximize their feeding rates with respect to spatially distributed prey, schools of bluefin tuna may utilize distinct responses that lead to increasing prey encounter and capture rates, and responding effectively to changes in prey distribution and density (Belovsky *et al.*, 1989; Getty and Pulliam, 1991; Fahrig, 1992; Cosner *et al.*, 1999). This might also explain why adjustments in school size and structure are necessary and support modeling the nonlinearities of school dynamics along alternative school movement trajectories, as noted by Romey (1996).

5. CONCLUSIONS

We have discussed how the analysis of alternative movement modes within individual fish trajectories, when integrated with information on static characteristics of their schools from aerial surveys, will help identify movement trends of bluefin tuna. This approach seeks to integrate observations of individuals and schools, conducted at various time and spatial scales. The dependence of movement trajectories on school-level properties examined under two theoretical models suggests that individual paths may be better understood by examining how changes in schooling behaviors produce variability in movement (Grünbaum, 1997). Work is in progress to apply appropriate models to improve understanding of the sampling process, schooling dynamics, and biotic and abiotic factors that influence the detection and behavior of bluefin tuna schools. A comprehensive modeling approach is crucial for the development of an integrated, direct-assessment of regional biomass, leading to the development of fishery-independent estimates of bluefin tuna population size and spatial trends.

ACKNOWLEDGEMENTS

This work was funded in part by the Office of Naval Research (ONR), National Satellite Information Distribution Service (NESDIS), and National

Marine Fisheries Service (NMFS) and a Research Fellowship from the University of British Columbia, Vancouver, Canada to N. Newlands. Special thanks to Dr. Richard Brill, Jennifer Goldstein, Brad Chase, Greg Skomal, Jeff. Tutein, and Dr. T.J. Pitcher, and to the reviewers who helped improve this manuscript.

REFERENCES

Belovsky, G.E., Ritchie, M.E., and Moorehead, J. (1989) Foraging in complex environments: when prey availability varies over time and space. *Theor. Pop. Biol.* **36**, 144-160.

Bertignac, M, Lehodey, P., and Hampton, J. (1998) A spatial population dynamics simulation model of tropical tunas using a habitat index based on environmental parameters. *Fish. Oceanogr.* **7**(3/4), 326-334.

Bonabeau, E., Dagorn, L., and Freon, P. (1999) Scaling in animal group-size distributions. *Proc. Natl. Acad. Sci. USA. (Ecology, Physics)* **96**, 4472-4477

Boyd, I.L. (1996) Temporal scales of foraging in a marine predator. *Ecology* **77**(2), 426-434.

Brill, R., and Lutcavage. M. (2001) Understanding environmental influences on movements and depth distributions of tuna and billfishes can significantly improve population assessments. Charleston Bump Symposium. Trans. Am. Fish. Soc. Sp. Publ. (in press).

Brill, R.W. (1994) A review of temperature and oxygen tolerance studies of tunas pertinent to fisheries oceanography, movement models and stock assessments. *Fish. Oceanogr.* **3**(3), 204-216.

Brown, J.S. (1988) Patch use as an indicator of habitat preference, predation risk, and competition. *Behav. Ecol. Sociobiol.* **22**, 37-47.

Chase, B.C. (1995) The diet of bluefin tuna (*Thunnus thynnus*) off the coast of Massachussetts. Mass. Div. Of Marine Fisheries, Salem, MA 01970. June, 1995. Unpubl. Report.

Cosner, C., DeAngelis, D.L., Ault, J.S., and Olson, D.B. (1999). Effects of spatial grouping on the functional response of predators. *Theor. Pop. Biol.* **56**(1), 65-75.

Crane, J. (1936) Notes on the biology and ecology of giant tuna, *Thunnus thynnus* Linneaus, observed at Portland, Maine. *Zoologica* (NY Zoological Soc.) **21**, 207-211.

Dagorn, L., and Freon, P. (1999) Tropical tuna associated with floating objects: a simulation study of the meeting point hypothesis. *Can. J. Fish. Aquat. Sci.* **56**, 984-993.

DeAngelis, D.L., and Gross, L.J. (1992) Individual-Based Models and Approaches in Ecology: Populations, Communities and Ecosystems. Chapman and Hall, New York.

Dill, L.M. (1983) Adaptive flexibility in the foraging behaviour of fishes. *Can. J. Fish. Aquat. Sci.* **40**, 398-408.

Fahrig, L. (1992) Relative importance of spatial and temporal scales in a patchy environment. *Theor. Pop. Biol.* **41**, 300-314.

Fiedler, P.C., and Bernard, H.J. (1987) Tuna aggregation and feeding near fronts observed in satellite imagery. *Cont. Shelf Res.* **7**(8) 871-881.

Flierl, G., Grünbaum, D., Levin, S., and Olson, D. (1999) From individuals to aggregations: the interplay between behavior and physics. *J. Theor. Biol.* **196**, 397-454.

Getty, T. and Pulliam, H.R. (1991) Random prey detection with pause-travel search. *Am. Nat.* **138**(6), 1459-1477.

Gillis, D.M., Peterman, R.M., and Tyler, A.V. (1993) Movement dynamics in a fishery: application of the ideal free distribution to spatial allocation of effort. *Can. J. Fish. Aquat. Sci.* **50**, 323-333.

Giraldeau, L.-A. (1984). Group foraging: the skill pool effect and frequency-dependent learning. *Am. Nat.* **124**, 77-79.

Grünbaum, D. (1999) Advection-diffusion equations for generalized tactic search behaviors. *J. Math. Biol.* **38**, 169-194.

Grünbaum, D. (1997) Schooling as a strategy for taxis in a noisy environment, 257-281. *In* Parrish, J.K., and Hamner, W.H. (eds.) Animal Groups in Three Dimensions. Cambridge University Press, New York. 378 pp.

Hilborn, R. (1991) Modeling the stability of fish schools: exchange of individual fish between schools of skipjack tuna (*Katsuwonus pelamis*). *Can. J. Fish. Aquat. Sci.* **48**, 1081-1091.

Hooge, P.N., Eichenlaub, W.M., and Solomon, E.K. (1999) Using GIS to Analyze Animal Movements in the Marine Environment, http://www.absc.usgs.gov/glba/gistools/.

Hooge, P.N., and Eichenlaub, W.M. (1997) Animal movement extension to Arcview. Ver. 1.1. Alaska Biological Science Center, U.S. Geological Survey, Anchorage, AK, USA.

Humston, R., Ault, J.S., Lutcavage, M., and Olson, D.B. (2001) Schooling and migration of large pelagic fishes relative to environmental cues. *Fish. Oceanogr.* (in press)

ICCAT. (1999) Report of the Standing Committee on Research and Statistics. International Commission for the Conservation of Atlantic Tunas.

Kieffer, J.D., and Colgan, P.W. (1992) The role of learning in fish behavior. *Rev. Fish. Biol. Fish.* **2**, 125-143.

Laurs, R.M., Yeun, H.S.H., and Johnson, J.H. (1977) Small-scale movements of albacore, *Thunnus alalunga*, in relation to ocean features as indicated by ultrasonic tracking and oceanographic sampling. *Fish. Bull., U.S.* **75**(2), 347-355.

Lomb, N.R. (1976) Least-squares frequency analysis of unequally spaced data. *Astrophys. Space Sci.* **39**, 447-462.

Press, W.H., Teukolsky, S.A., Vetterling, W.T., and Flannery, B.P. (1999) Numerical Recipes in C: The Art of Scientific Computing. 2nd. Ed., Cambridge University Press, New York. 994 pp.

Lutcavage, M., Brill, R.W., Goldstein, J.L., Skomal, G.B., Chase, B.C., and J. Tutein. (2001) Movements and behaviour of adult North Atlantic bluefin tuna (*Thunnus thynnus*) in the northwest Atlantic determined using ultrasonic telemetry. *Mar. Biol.* (in press)

Lutcavage, M., and Newlands, N. (1999) A strategic framework for fishery-independent aerial assessment of bluefin tuna. Int. Comm. Conserv. Altantic Tunas Coll. Vol. Scient. Pap. SCRS/98/75, 1-5.

Lutcavage, M., Goldstein, J.L., Kraus, S. (1997) Distribution, relative abundance, and behavior of giant bluefin tuna in New England waters, 1995. Int. Comm. Conserv. Atlantic Tunas Coll. Vol. Sci. SCRS/96/129.

Lutcavage, M. and S. Kraus. (1995) The feasibility of direct photographic assessment of giant bluefin tuna (*Thunnus thynnus*) in New England waters. *Fish. Bull., U.S.* **95**(2), 300-310.

Lutcavage, M., Goldstein, J., and Kraus, S. (1997) Distribution, relative abundance, and behavior of giant bluefin tuna in New England waters, 1995. Int. Comm. Conserv. Atlantic Tunas Coll. Vol. Sci. **XLVI** (2), 332-347.

Maravelias, C.D. (1999) Habitat selection and clustering of a pelagic fish: effects of topography and bathymetry on species dynamics. *Can. J. Fish. Aquat. Sci.* **56**, 437-450.

Marchal, E, Gerlotto, F. and Stequert, B. (1993) On the relationship between scattering layer and tuna abundance in the Eastern Atlantic equatorial current system. *Oceanologica Acta* **16**(3), 261-272.

Mason, D.M. and Patrick, E.V. (1993) A model for the space-time dependence of feeding for pelagic fish populations. *Trans. Am. Fish. Soc.* **122**, 884-901.

Mather, F.J., III, Mason, J.M., Jr., and Jones, A.C. (1995) Historical Document: Life history and fisheries of Atlantic bluefin tuna. NOAA Technical Memorandum NMFS-SEFSC-370.

McCulloch, C.E., and Cain, M.L. (1989) Analyzing discrete movement data as a correlated random walk. *Ecology* **70**(2), 383-388.

Metcalfe, N.B., and Thomson, B.C. (1995) Fish recognize and prefer to shoal with poor competitors. *Proc. R. Soc. Lond. B.* **259**, 207-210.

Milinski, M. (1994) Long-term memory for food patches and implications for ideal free distributions in sticklebacks. *Ecology* **75**(4), 1150-1156.

Mogilner, A., Bent, L., Edelstein-Keshet, L. (2001) An individual-based model for aggregation in social organisms: attraction-repulsion interactions that lead to a well-spaced group. (in press)

Mullen, A.J. (1994) Effects of movement on stock assessment in a restricted-range fishery. *Can. J. Fish. Aquat. Sci.* **51**, 2027-2033.

National Research Council (NRC). (1994) An Assessment of Atlantic Bluefin Tuna. National Academy Press, Washington, D.C.

Niwa, H.-S. (1998) School size statistics of fish. *J. Theor. Biol.* **195**, 351-361.

Okubo, A. (1980) Diffusion and Ecological Problems: Mathematical Models. Springer-Verlag, New York. 254 pp.

Olson, D.B., Hitchcock, G.L., Mariano, A.J., Ashjian, C.J., Peng, G., Nero, R.W., and Podesta, G.P. (1994) Life on the edge: marine life and fronts. *Oceanography* 7(2), 52-60.

Parrish, J.K., and Edelstein-Keshet, L. (1999) Complex systems, Viewpoint: Complexity, pattern and evolutionary trade-off in animal aggregation. *Science*, **284**, April, 99-101.

Partridge, B.L., Johansson, J., and Kalish, J. (1983) The structure of schools of giant bluefin tuna in Cape Cod Bay. *Env. Biol. Fish.* **9**(3/4), 253-262.

Patlak, C.S. (1953a) Random walk with persistence and external bias. *Bull. Math. Biophys.* **15**, 311-337.

Patlak, C.S. (1953b) A mathematical contribution to the study of orientation or organisms. *Bull. Math. Biophys.* **15**, 431-476.

Pitcher, T.J., and Parrish, J.K. (1993) Functions of shoaling behavior in teleosts. pp. 363-439, In Pitcher, T.J. (ed.) Behavior of Teleost Fishes. 2nd ed., Chapman and Hall, New York.

Podesta, G.P., Browder, J.A., and J.J. Hoey. (1993) Exploring the association between swordfish catch rates and thermal fronts on U.S. longline grounds in the western North Atlantic. *Cont. Shelf Res.* **13**(2/3), 253-277.

Porter, J.M., and Hogans, W.E. (1991) A mark-recapture experiment on bluefin tuna (*Thunnus thynnus*) from the Browns-Georges Banks region of the Canadian Atlantic. *ICCAT Coll. Vol. Sci. Pap. Madrid*, **35**, 253-256.

Romey., W.L. (1996) Individual differences make a difference in the trajectories of simulated schools of fish. *Ecol. Modell.* **92**, 65-77.

Rose, G.A., and Leggett, W.C. (1990) The importance of scale to predator-prey spatial correlations: an example of Atlantic fishes. *Ecology* **71**(1), 33-43.

Sibert, J.R., Hampton, J., Fournier, D.A., and Bills, P.J. (1999) An advection-diffusion-reaction model for the estimation of fish movement parameters from tagging data, with application to skipjack tuna (*Katsuwonus pelamis*).*Can J. Fish. Aquat. Sci.* **56**, 925-938.

Sissenwine, M.P., Mace, P.M., Powers, J.E., and Scott, G.P. (1998) A commentary on Western Atlantic bluefin tuna assessments. *Trans. Am. Fish. Soc.* **127**, 838-855.

Swartzman, G., Stuetzle, W., Kulman, K., and Powojowski, M. (1994) Relating the distribution of pollock schools in the Bering Sea to environmental factors. *ICES J. Mar. Sci.*, **51**, 481-492.

Turchin, P. (1998) Quantitative Analysis of Movement: Measuring and Modeling Population Re-distribution in Animals and Plants. Sinauer Associates, Inc., Sunderland, 396pp.

Turchin, P. (1991) Translating foraging movements in heterogeneous environments into the spatial distribution of foragers. *Ecology* **72**(4), 1253-1266.

Ullman, D.S., and Cornillon, P. (1999) Satellite-derived sea surface temperature fronts on the continental shelf off the northeast U.S. coast. *J. Geophys. Res.* **104**(C10) 23,459-478.

Possible Models for Combining Tracking Data with Conventional Tagging Data

John Sibert[1] and David Fournier[2]
[1] *University of Hawaii, Honolulu, e-mail jsibert@soest.hawaii.edu*
[2] *Otter Research Ltd., Sidney, British Columbia, e-mail otter@otter-rsch.com*

Key words: Kalman Filter, random walk, fish track, archival tag

Abstract: Advection-diffusion models have been used successfully to describe the time and place of recapture of tuna tagged with conventional dart tags. Such models are the continuous analogs of a biased random walk. This paper demonstrates how biased random walks can be used to simulate large scale movements of tunas as recorded by archival tags in a way that captures all of the major characteristics of the tracks. The parameters of the biased random walk model are identical to the parameters of the advection diffusion model, suggesting that a joint parameter estimation procedure might be feasible. Finally, the potential application of the Kalman filter to the analysis of tracking data is discussed. This statistical model has the potential to increase the accuracy of geoposition estimates from tracking devices as well as to estimate biased random walk parameters from tracking data.

1. INTRODUCTION

Acoustic transponders and archival tags provide large amounts of data pertaining to the vertical and horizontal position of individual fish. Ancillary sensors provide additional information such as internal body temperature and ambient water temperature. In contrast, conventional "dart" or "spaghetti" tags provide a small amounts of data pertaining to the horizontal position of a large number of fish and no data at all pertaining to their vertical position. Whether the data derived from the use of these devices are informative depends on the scientific questions being addressed and the means used to analyze the data. Tracking data on vertical distribution and apparent temperature preferenda have been particularly informative when analyzed in the context of attempting to quantitatively define the habitat of fish species in terms of depth, temperature and oxygen preferences. The results have found ready application in the analysis of catch and effort data from longline fisheries. Indices of

J.R. Sibert and J.L. Nielsen (eds.), Electronic Tagging and Tracking in Marine Fisheries, 443–456.
© 2001 *Kluwer Academic Publishers. Printed in the Netherlands.*

abundance have been refined by estimating the number of longline hooks deployed in the depth strata most likely to be frequented by the species of interest (Hinton and Nakano, 1996; Hampton *et al.*, 1998). It remains to be seen whether the accuracy of these indices of abundance is improved.

Data pertaining to horizontal movements have proven similarly useful when analyzed in the context of determining the prefered habitat of tunas, but have been slower to penetrate the quantitative fisheries stock assessment and management mainstream. While there are undoubtedly many reasons for this lack of application, one major barrier is the lack of a generally accepted context in which to apply data on the movement of individual fish. Stock assessment is concerned with estimation of population-level processes, such as mortality, that cannot easily be estimated from a track of an individual fish. Furthermore, unknown position errors in data from archival tags makes the application of these data problematical.

In this paper, we introduce the related concepts of a biased random walk and the diffusion-advection equations as a consistent theoretical framework for combining tracking data with conventional with tag release and recapture data. A simple biased random walk simulation of a tuna track is presented to demonstrate that this framework may be suitable for analysis of tracking data. The results of a preliminary statistical model for estimating diffusion-advection parameters from long-distance tracks is then presented.

2. BIASED RANDOM WALKS

For simplicity, the following discussion is restricted to one dimensions; the ideas are easily extended to two dimension. In a simple random walk, an animal is assumed to move a constant small distance, λ, in a constant small interval of time, τ, as shown schematically below.

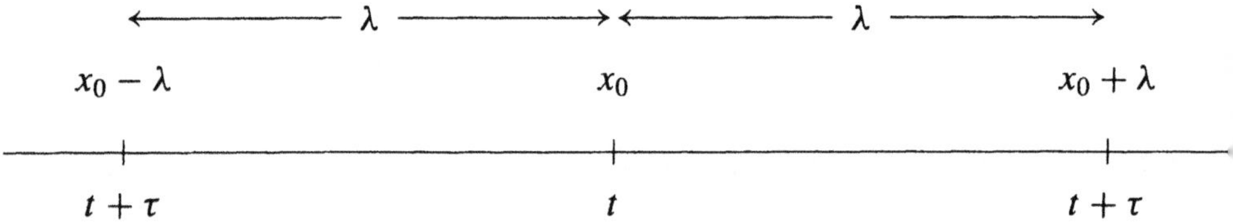

At time 0, the animal is at position x_0 and can move either to the right with probability P_1 or to the left with probability P_2. At time $t + \tau$, the animal is at position $x_0 + \lambda$ with probability P_1 or at position $x_0 - \lambda$ with probability P_2; $P_1 + P_2 = 1$. The difference in probability of moving to the left or to the right, the "bias", is $\varepsilon = P_1 - P_2$. Although a random walk is a stochastic process and the position of an individual fish cannot be predicted exactly, the fish's movement is not haphazard. The fish moves at constant speed λ/τ, but the direction of movement is not constant. Therefore it can never move farther than λ in any time step. Furthermore, it is possible to imagine that the bias or the distance moved in a time step might be variable in time and space and may be dependent on characteristics of the environment or may reflect some inherent tendency to prefer one direction over another.

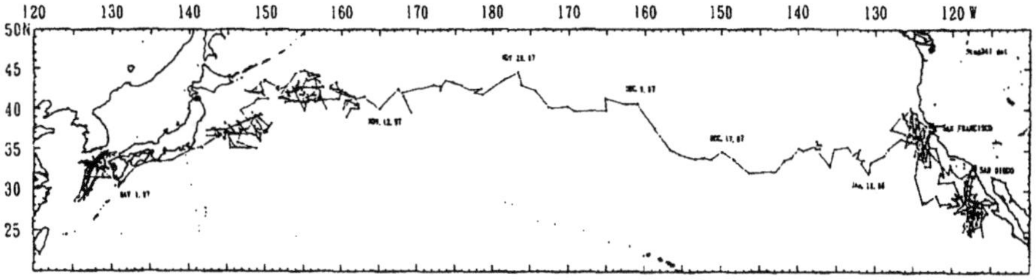

Figure 1. Track of juvenile bluefin tuna from Tsuji et al. 1999.

The advection-diffusion equation is closely related to a biased random walk. Okubo (1980) derives the advection-diffusion equation from a random walk by showing that

$$\lim_{\lambda,\tau,\epsilon\to 0} \epsilon \cdot \frac{\lambda}{\tau} = u \qquad \lim_{\lambda,\tau\to 0} \frac{\lambda}{2} \cdot \frac{\lambda}{\tau} = D \tag{1}$$

and that the rate of change in density of animals at a point can be expressed as

$$\frac{\partial N}{\partial t} = -u\frac{\partial N}{\partial x} + D\frac{\partial^2 N}{\partial x^2} \tag{2}$$

In other words, the advection-diffusion equation is the limiting case of a biased random walk and there is a straightforward interpretation of the parameters of one model in the context of the other.

Advection-diffusion models are finding increasing application to the analysis of large-scale movement and distribution of tunas (Lehodey *et al.*, 1998; Bertignac *et al.*, 1998; Maury and Gascuel, 1999). Advection-diffusion models, in which the parameters vary regionally and seasonally, can predict the time and place of recapture of tagged skipjack tuna (*Katsuwonus pelamis*) with reasonable accuracy (Sibert *et al.*, 1999). Therefore it should be possible to model the track of a singe tuna using a biased random walk parameterized to be consistent with an advection-diffusion model of a population of conventionally tagged tunas.

3. RANDOMLY WALKING ACROSS THE PACIFIC

Tsuji *et al.* (1999) presented preliminary data from an archival tag recovered from a Pacific northern bluefin tuna, *Thunnus thynnus*. The animal was tagged in November, 1996 near Tsushima Island and recaptured in August 1998 off California. The track presented by Tsuji *et al.* (1999) is reproduced in Figure 1. As the authors note in their paper, the animal appeared to change its behavior several times during the period it was at liberty. Whether these changes are due to season, geography or local environmental conditions must await a more complete analysis.

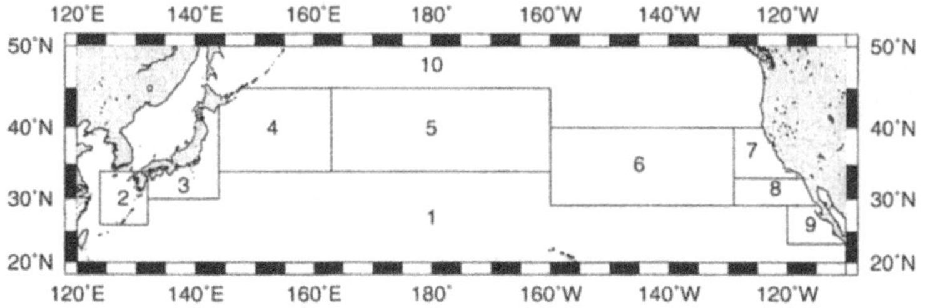

Figure 2. Regions used for bluefin tuna random walk simulations.

Sibert *et al.* (1999) superimposed an array of regions over the model domain, in which the parameters of the advection-diffusion model were assumed constant. The same principle can be applied to the tuna track in Figure 1. Eight model regions were designated (Figure 2), corresponding to regions where the tagged bluefin appeared to have uniform behavior. Two additional model regions were added to the north of the track and the south of the track. Values of the advection-diffusion parameters, u, v, and D, were assigned to each region. A characteristic speed, $\mu = \lambda/\tau$, was assigned to be similar to daily movement speed estimated from the bigeye track, approximately 100 Nmi da^{-1}. The time step was fixed at 1 hour. Biases in the north-south (ε_y) and east-west (ε_x) dimensions were calculated from the parameters u and v using the relationships in equation (1). The distance moved at each time step was calculated from the parameter D and the characteristic speed, $\lambda = \frac{D}{2\mu}$. Two uniform random numbers between 0 and 1 were generated at each time step and compared with ε_y and ε_x to determine the direction of the jump in the north-south and east-west dimensions. The simulation was run for twenty months and the positions of simulated fish were recorded once per day. The numerical values of the parameters in each region are shown in Table 1.

Figure 3 shows two representative tracks generated by two different series of random numbers. These two simulated tracks display all of the important features of the observed bluefin track. There are places where the simulated fish appeared to have no prefered direction of movement and places where it moved consistently towards the East. The two fish did not arrive at the West coast of North America simultaneously, and both show a "bimodal" distribution after arriving off the West Coast. These results suggest that a statistical model of a tuna track based on a biased random walk is feasible.

Table 1. Advection-diffusion parameters used for biased random walk simulation of trans-Pacific bluefin tuna track. u and v in Nmi da^{-1}; D in Nmi^{2}da^{-1}.

Region	u	v	D
1	2.0	10.0	100
2	4.0	3.0	100
3	10.0	6.0	33
4	2.0	2.0	33
5	15.0	0.0	23
6	15.0	0.0	23
7	0.0	0.0	100
8	0.0	0.0	167
9	0.0	0.0	100
10	0.0	0.0	100

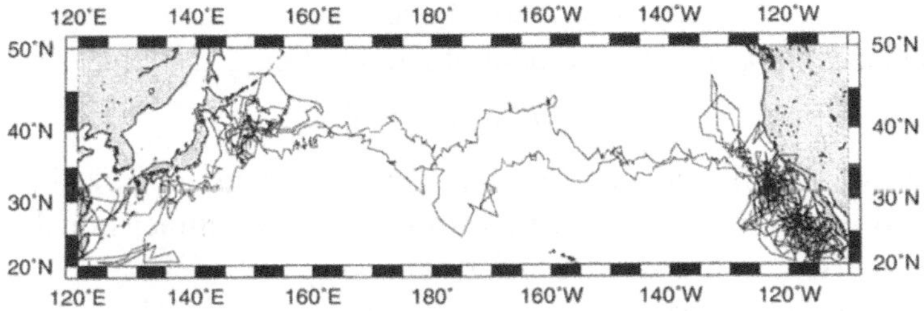

Figure 3. Two simulated bluefin tuna tracks.

4. PRELIMINARY STATISTICAL MODEL

A useful statistical model for application to tracking data would estimate the parameters of a biased random walk parameterized consistently with statistical models of the advection-diffusion equation applied to conventional tags. In addition, the model would estimate the errors in the position estimates from the tracking device so that a "corrected" track could be estimated.

The advection-diffusion equation is closely related to the normal probability distribution. Feller (1968) shows that if the probability of observing an animal at point x at time t is given as a normal probability density function,

$$p(t, x) = \frac{1}{\sqrt{4\pi Dt}} e^{-\frac{1}{2}\frac{(x-ut)^2}{2Dt}} \tag{3}$$

the function p uniquely satisfies equation (2). In other words, if animals are dispersing according to equation (2), after a period of time t their mean position will be given by ut and the variance of their position will be $2Dt$. Thus, u can be considered to be the mean rate of displacement (which may be zero) and D a measure of the rate at which the uncertainty of the position increases with time (Feller, 1966).

The notions of a state space model and the Kalman filter (Harvey, 1990) form the basis for a statistical model of a track. Let

$$y_t = \alpha_t + d_t + \epsilon_t, \quad t = 1, \ldots, T \tag{4}$$

where y_t is a two dimensional vector representing the observed position of the tagged fish at time t, α_t its true position, d_t a two dimensional vector the bias in observing the position, and ε_t a serially uncorrelated 2 dimensional random vector with mean 0 and 2 2 covariance matrix, H_t,

$$d_t = \begin{pmatrix} 0 \\ 0 \end{pmatrix} \quad \text{and} \quad H_t = \begin{pmatrix} \sigma_x^2 & 0 \\ 0 & \sigma_y^s \end{pmatrix}. \tag{5}$$

In other words, we assume that there is no systematic bias in the tag's position estimate and that there are independent random errors in the estimate of longitude and latitude, where σ_x^2 and σ_y^2 are the mean square errors of the longitude and latitude estimates.

The true position of the tagged fish is not observed, but is assumed to be generated by a biased random walk

$$\alpha_t = \alpha_{t-1} + c_t + \eta_t, \quad t = 1, \ldots, T \tag{6}$$

where c_t is a 2 dimensional vector representing the bias of the random walk, and η_t is a 2 dimensional vector of serially uncorrelated random variables with mean 0 and covariance matrix, Q_t. For the biased random walk in equation (6)

$$c_t = \begin{pmatrix} u \\ v \end{pmatrix} \quad \text{and} \quad Q_t = \begin{pmatrix} 2D & 0 \\ 0 & 2D \end{pmatrix}. \tag{7}$$

The Kalman filter consists of a set of recursive relations that update the estimated position of the tagged fish and the components of the variance of the estimated position.

$$
\begin{aligned}
a_{t|t-1} &= a_{t-1} + c_t \\
P_{t|t-1} &= P_{t-1} + Q_t \\
w_t &= y_t - a_{t|t-1} - d_t \\
F_t &= P_{t|t-1} + H_t \\
a_t &= a_{t|t-1} + P_{t|t-1} F_t^{-1} w_t \\
P_t &= P_{t|t-1} - P_{t|t-1} F_t^{-1} P_{t|t-1}
\end{aligned}
\qquad (8)
$$

where P_t is a 2×2 covariance matrix of the estimation error and $a_{t|t-1}$ can be interpreted as an estimate of the "true" position of the tagged fish, α_t.

The parameters to be estimated are u, v, D, σ_x^2, and σ_y^2. The estimates of these parameters are the values that maximize the log likelihood function

$$
\ln L = -T \ln 2\pi - 0.5 \sum_{t=1}^{T} \ln |F_t| - 0.5 \sum_{t=1}^{T} w_t' F_t^{-1} w_t \qquad (9)
$$

The performance of this model was examined using a series of simulations. For each series, one thousand replicate simulated archival tag tracks were generated for assumed values of the model parameters u, v, D, σ_x^2, and σ_y^2. The RMS error of the geolocation estimates produced by currently available archival tags is approximately $\frac{1}{2}$ degree of longitude and $1\frac{1}{2}$ degrees of latitude (R. Hill, personal communication). These errors were fixed for all simulations at 30 and 90 nautical miles (Nmi) respectively assuming the tracks to be near the equator. Three behavior scenarios were simulated: (1) net eastward movement with movement variability about the same order of magnitude as geolocation error; (2) net eastward movement with movement variability about half of the geolocation error; and (3) no net movement and movement variability about half of the geolocation error. The numerical values of the input parameters are given in Table 2.

Simulation results were evaluating by computing the bias as the difference between the input value of a parameter and the value estimated by the Kalman filter. The bias was "standardized" by dividing by the standard deviation of the estimates computed over all 1000 simulations so that parameters with different ranges could be easily compared. This comparison evaluates the ability of the Kalman filter estimator to correctly recover the input parameters given a large number of replicate tracks.

The worst (lowest value of the likelihood function, equation (9) and best (highest value of the likelihood function) fitting simulations were examined by Markov-chain Monte Carlo exploration of the likelihood surface. This

Table 2. Input parameter values for simulations of Kalman filter estimation procedure. u and v in Nmi da^{-1}; D in Nmi^{2}da^{-1}; σ_x and σ_y in Nmi. The column headed 'sd' is the standard deviation of the movement uncertainty in Nmi.

Simulation	σ_x	σ_y	u	v	D	sd
1	30	90	50	0	1800.0	60
2	30	90	50	0	112.5	15
3	30	90	0	0	112.5	15

analysis treats the parameter estimates as random variables and computes the posterior probability distribution of the estimates given the observations (i.e., the simulated track) and the constraints on the parameters. It evaluates the ability of the Kalman filter estimator to recover the parameters from a single track.

5. RESULTS

Figure 4 shows "box and whisker" diagrams of the standardized bias for all three simulations. The mean bias, given by the horizontal line within the boxes, is close to zero for all variables and the estimates appear to be approximately normally distributed. The model appears to slightly overestimate diffusivity, D.

Figure 5 compares the likelihood profiles of all the parameter estimates for a "good" fit, L1754 ($\ln L = -1753.94$) and a "bad" fit, L2053 ($\ln L = -2053.14$). For both cases, the estimates of the random walk bias parameters u and v are well behaved, their estimates are very close to the true values, and the posterior probability of the estimates given the data appears normal. The posterior distribution of the estimates of diffusivity, D, are asymmetric with the point estimates higher than the "true" value. Nevertheless, the true value of D lies within two standard deviations of the estimated value. In the case of the "good" fit, which happens to be a model with diffusivity less than geolocation error, the true value of the longitude error is more than two standard deviations higher than the estimate.

The magnitude of the diffusivity in relation to the latitude error appears to influence the bias in the estimates of diffusivity and both components of the geolocation error. The high diffusivity case, L2053, appears to cause overestimation of both diffusivity and the geolocation errors. On the other hand, the low diffusivity case, L1754, produces an accurate estimate of diffusivity, but underestimates the geolocation error.

These relationships are reflected in the correlations coefficient between the parameter estimates, Table 3. The estimates of diffusivity and

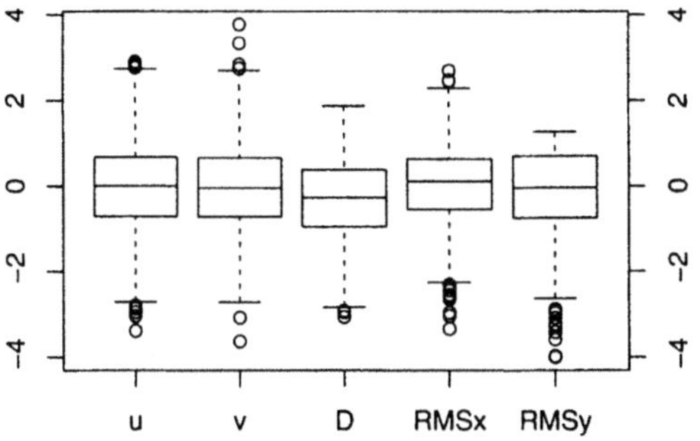

Simulation 1 $u = 50$, $D = 1800$

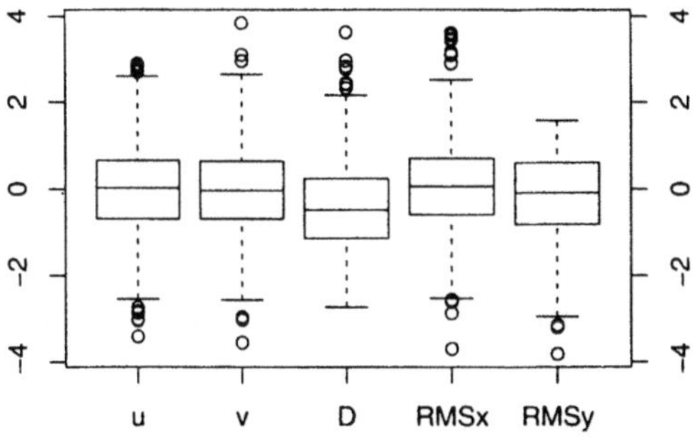

Simulation 2, $u = 50$, $D = 112.5$

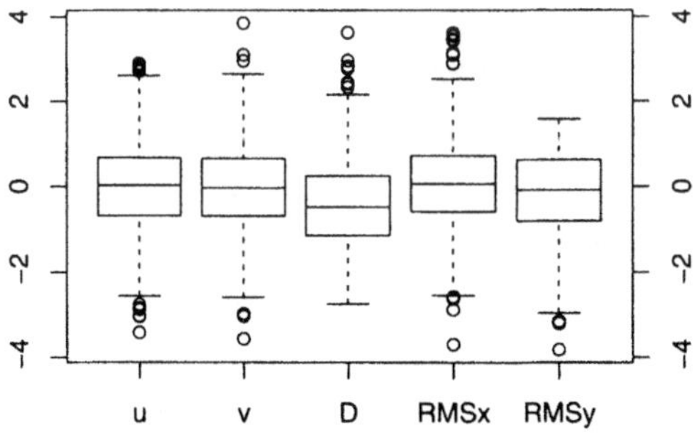

Simulation 3, $u = 0$, $D = 112.5$

Figure 4. Standardized bias plots. The boxes indicate the interquartile range, i.e., the area encompassing the central 50% of the estimates. The range bars extend outside of the boxes to the most extreme data point which is no more than 1.5 times the interquartile range. The model parameters, u, v, D, σ_x and σ_y are indicated on the abscissa as u, v, D, RMSx and RMSy repectively. The scale on the ordinate in standard deviations.

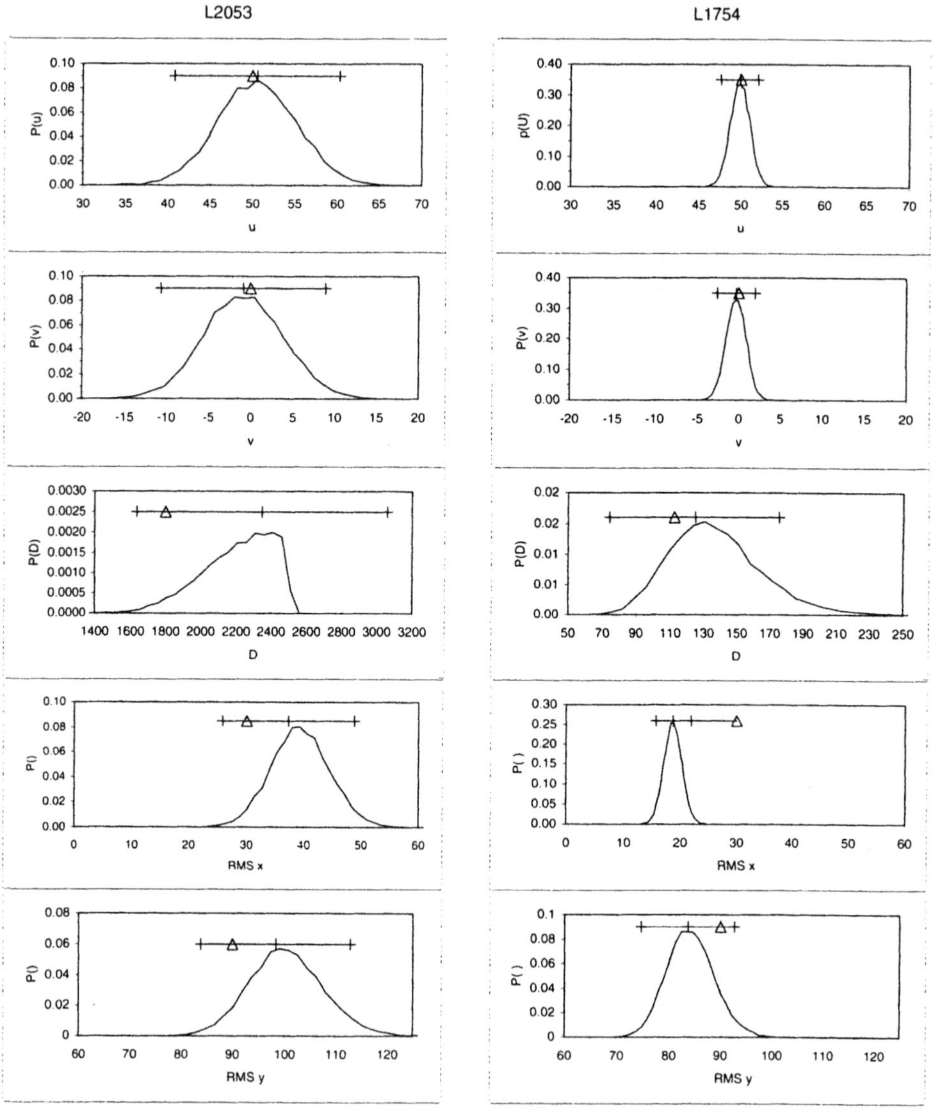

Figure 5. Likelihood profiles for two realizations of simulation. The horizontal bar in each plot represents the point estimate of the parameter two standard deviations. The triangle represents the "true" value of the parameter used in the simulations. L2053 pertains to simulation 1 (high diffusivity). L1754 pertains to simulations 2 and 3 (low diffusivity).

Table 3. Correlation coefficients between parameter estimates. L2053 pertains to simulation 1 (high diffusivity). L1754 pertains to simulations 2 and 3 (low diffusivity).

Parameter	u	v	D	σ_x
		L1754		
v	0			
D	0.0034	0.0032		
σ_x	-0.0031	0.0013	-0.4201	
σ_y	-0.0002	0.0018	-0.0674	0.0283
		L2053		
v	-0.0001			
D	0.0089	-0.0126		
σ_x	-0.0108	0.0069	-0.5499	
σ_y	-0.0028	0.0124	-0.3136	0.1724

geolocation error are negatively correlated and the magnitude of the correlation is larger in the high diffusivity case.

6. DISCUSSION

The state space Kalman filter model appears to perform reasonably well as a means to estimate population movement parameters from a track. The model has some difficulty in estimating both diffusivity and geolocation error, however. This difficulty, while potentially troublesome, is not unexpected. In its simplest form, the Kalman filter simply estimates the true position as the weighted mean of the position predicted by the model and the position estimated by the tag with weights inversely proportional to the errors in the two position estimates (Maybeck, 1979). In other words, the Kalman filter is interpreting diffusivity as process error and giving the model prediction less weight in a high-diffusivity track. In spite of this problem, the predicted track is closer to the actual track in all simulations (Figure 6).

The ability of the estimation procedure to correctly evaluate the geolocation error and estimate the true positions of the tagged fish depends on the magnitude of the diffusivity relative to the geolocation error. The geolocation errors used in these simulations are similar to the geolocation errors of the current generation of archival tags operating under optimal conditions and are equivalent to diffusivities of approximately 2000 $\mathrm{Nmi^2da^{-1}}$. During sub-optimal conditions, e.g., around the time of the equinoxes, the geolocation errors would be expected to be much higher (Hill, 2000). Estimates of diffusivity from tagged skipjack in the tropical western Pacific (Sibert *et al.*, 1999) range from 250 to 800 $\mathrm{Nmi^2da^{-1}}$. Similar values were used by Bertignac *et al.* (1998) for ocean-basin scale of skipjack population dynamics. At the higher range of published tuna diffusivity, Maury *et al.* (1999) appear

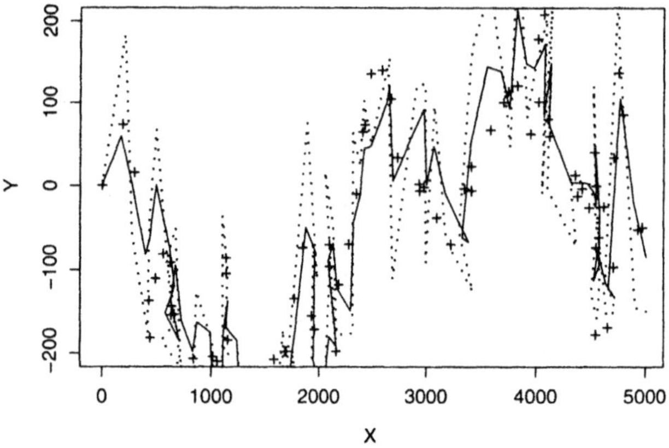

Simulation 1 $u = 50$, $D = 1800$

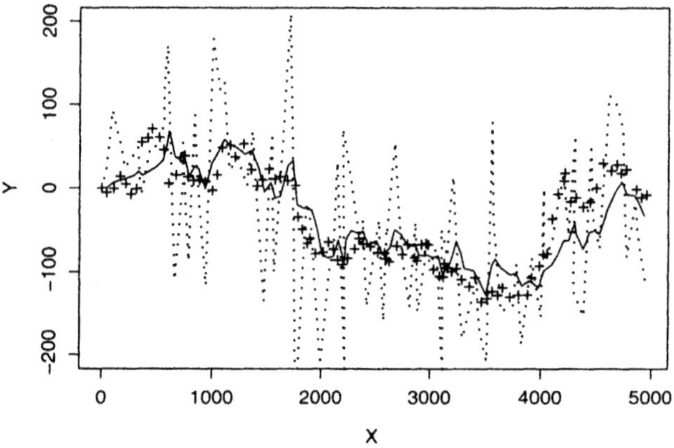

Simulation 2 $u = 50$, $D = 112.5$

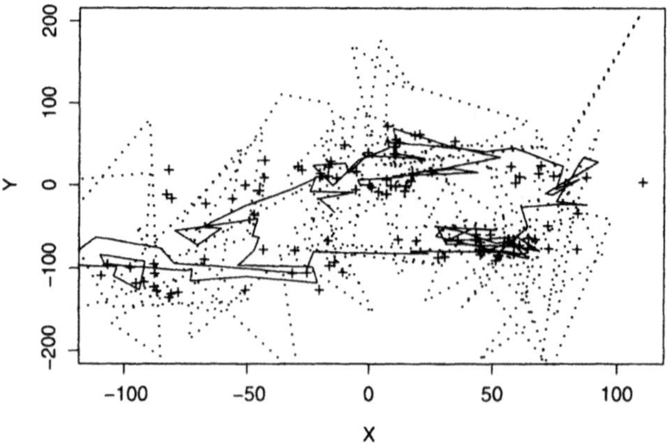

Simulation 3 $u = 0$, $D = 112.5$

Figure 6. True (+), observed (dotted line) and "corrected" (solid line) position of tags in simulations.

to have used values near 9000 Nmi^2da^{-1} in simulations yellowfin tuna populations. These rough comparisons indicate that the diffusivity expected for tunas is probably less than the geolocation errors of archival tags suggesting that the Kalman filter would be applicable to real data from archival tags.

In situations where fish with archival tags and fish with conventional tags are simultaneously at liberty, it is possible to envisage a model which estimates movement and mortality parameters as well as geolocation error in a completely integrated fashion. Such an estimation procedure would link estimation of advection-diffusion and mortality parameters from conventional tags to the estimation of random walk parameters and geolocation errors via a composite likelihood function. Thus, information on the horizontal movements from tracking data would be integrated into large scale population models.

7. ACKNOWLEDGEMENTS

The work described in this paper was sponsored by the University of Hawaii Pelagic Fisheries Research Program under Cooperative Agreement number NA67RJ0154 from the National Oceanic and Atmospheric Administration. We would like to thank Pierre Kleiber for the suggestion to apply the Kalman filter to archival tag position estimates. We are grateful to Dr. Tomoyuki Itoh of the National Research Institute of Far Seas Fisheries, Japan, for permission to reproduce Figure 1.

REFERENCES

Bertignac, M., P. Lehodey, and J. Hampton. (1998) A spatial population dynamics simulation model of tropical tunas using a habitat index based on environmental parameters. *Fish. Oceanogr.* **7**, 326-334.

Feller, W. 1966. *An Introduction to Probability Theory and its Applications.* Volume II. John Wiley & Sons, N.Y.

Feller, W. 1968. *An Introduction to Probability Theory and its Applications.* Volume I. (Third Edition). John Wiley & Sons, N.Y. 509 pp.

Hampton, J., K. Bigelow and M. Labelle. 1998. A summary of current information on the biology, fisheries and stock assessment of bigeye tuna (*Thunnus obesus*) in the Pacific Ocean, with recommendations for data requirements and future research. OFP Tech. Rep. No. 56. Secretariat of the Pacific Community, Noumea.

Harvey, A. C. 1990. *Forecasting, Structural Time Series Models and the Kalman Filter.* Cambridge University Press.

Hill, R. 2000. Geolocation by Light-Level. The next Step: Latitude. Symposium on Tagging and Tracking Marine Fish with Electronic Devices, February 7-11, 2000, Honolulu.

Hinton, M. G. and H. Nakano. 1996. Standardizing catch and effort statistics using physiological, ecological, or behavioral constraints and environmental data, with an application to blue marlin (*Makaira nigricans*) catch and effort data from Japanese longline fisheries in the Pacific. *Bull. Int. Am. Trop. Tuna Comm.* **21**(4), 171-200.

Lehodey, P., J. Andre, M. Bertignac, J. Hampton, A. Stoens, C. Menkes, L. Mamery and N. Grima. 1998. Predicting skipjack tuna forage distributions using a coupled dynamical biogeochemical model. *Fish. Oceanogr.* **7**, 317-325.

Maybeck, P. S. 1979. *Stochastic Models, Estimation, and Control.* Volume 1. Academic Press. Republished 1994 by Navtech Book and Software Store, Arlington, VA.

Maury, O. and D. Gascuel. 1999. SHADYS, a GIS based numerical model of fisheries. Example application: the study of a marine protected area. *Aquat. Living Resour.* **12**, 77-88.

Maury, O., D. Gascuel and A. Fonteneau. 2001. Spatial modeling of Atlantic yellowfin tuna population dynamics: a habitat based advection-diffusion-reaction approach with application to the local overfishing study. "Spatial processes and management of fish populations", Anchorage October 27 - 39, 1999. Alaska Sea Grant, in press.

Okubo, A. 1980. *Diffusion and Ecological Problems:Mathematical Models.* Springer, N.Y. 254 pp.

Sibert, J. R., J. Hampton, D. A. Fournier, and P. J. Bills. 1999. An advection-diffusion-reaction model for the estimation of fish movement parameters from tagging data, with application to skipjack tuna (*Katsuwonus pelamis*). *Can. J. Fish. Aquat. Sci.* **56**, 925-938.

Tsuji, S., T. Itoh, A. Nitta, and S. Kume. 1999. The trans-Pacific migration of a young bluefin tuna, *Thunnus thynnus*, recorded by an archival tag. Working Paper ISC2/99/15. Interim Scientific Committee for Tuna and Tuna-Like Species in the North Pacific Ocean, January 15 -23, 1999, Honolulu.

PARTICIPANTS
Symposium on Tagging and Tracking Marine Fish with Electronic Devices
February 7-11, 2000
East-West Center, University of Hawaii

Jill Aitken
Biology Department
Dalhousie University
1545 Oxford Street, Apt. 42
Halifax, NS B3H 3Z3
CANADA
jpaitken@is2.dal.ca

Franco Andaloro
Istituto Centrale per la Ricera
Applicata al Mare (ICRAM
Via E. Amari, 124
90139 Palermo, Sicily
ITALY
Ph. 39-091-6114044
Fax 39-091-6114060
andalorf@tin.it

Alex Antoniou
Shark Research Institute
P.O. Box 40
Princeton, NJ 08540
USA
antoniou@rci.rutgers.edu

Nobuaki Arai
Graduate School of Informatics
Kyoto University
606-8501
Kyoto,
JAPAN
arai@i.kyoto-u.ac.jp

Michael Arendt
Virginia Institute of Marine Science
College of William and Mary
P.O. Box 334
Gloucester Point, VA 23062
USA
Ph. 804-684-7647
mda@vims.edu

Geoff Arnold
The Centre for Environment,
Fisheries and Aquaculture Science (CEFAS)
Pakefield Road
Lowestoft Laboratory
Lowestoft, Suffolk NR33 0HT
UNITED KINGDOM
G.P.Arnold@cefas.co.uk

Pascal Bach
IRD
Centre de Montpellier
B.P. 5045
34032 Montpellier Cedex
FRANCE
bach@mpl.ird.fr

Sallie Beavers
P.O. Box 4992
Kailua-Kona, HI 96745
USA
Ph. 808-325-0630
sbeavers@nethawaii.net

Susanna Blackwell
Hopkins Marine Station
Stanford University
120 Ocean View Blvd.
Pacific Grove, CA 93950
USA
Ph. 831-655-6237
apoflav@leland.stanford.edu

Barbara Block
Hopkins Marine Station
Stanford University
120 Ocean View Blvd.
Pacific Grove, CA 93950
USA
bblock@leland.stanford.edu

Scott Bloom
Oceanic Institute
Makapuu Point
41-202 Kalanianaole Hwy.
Waimanalo, HI 96759-1820
USA
sbloom@oceanicinstitute.org

Christofer Boggs
National Marine Fisheries Service
Honolulu Lab
2570 Dole Street
Honolulu, HI 96822
USA
Ph. 808-983-5370
Fax 808-983-2902
cboggs@honlab.nmfs.hawaii.edu

Stephania Bolden
National Marine Fisheries Service
Southeast Fisheries Science Center
75 Virginia Beach Drive
Miami, FL 33133
USA
Ph. 305-361-4293
Fax 305-361-4499
stephania.bolden@noaa.gov

Junio Fabrizio Borsani
Istituto Centrale per la Ricera Applicata al Mare
ICRAM
via di Casalotti 300
00166
Rome,
ITALY
Ph. 39-0661570407
Fax 39-0661550581
jfaber@tin.it

Andre Boustany
Hopkins Marine Station
Stanford University
120 Ocean View Blvd.
Pacific Grove, CA 93950
USA
Ph. 831-655-6237
andre27@leland.stanford.edu

Russell Brainard
National Marine Fisheries Service
Honolulu Lab
2570 Dole Street
Honolulu, HI 96822
USA
Ph. 808-983-5392
Fax 808-982-2902
Rusty.Brainard@noaa.gov

Christopher Bridges
Institute for Zoophysiologie
Heinrich-Heine-Universitat
UniversitatsstraBe 1/26 12
D-40225
Dusseldorf,
GERMANY
bridge@uni-duesseldorf.de

Richard Brill
National Marine Fisheries Service
Honolulu Lab
2570 Dole Street
Honolulu, HI 96822
USA
Ph. 808-592-8304
Fax 808-983-2902
rbrill@honlab.nmfs.hawaii.edu

Raffaele Brunetti
via Ottaviano 9
00192
Rome,
ITALY
Ph. 39-06-39729989
Fax 39-06-39726129
b.b@iol.it

Aaron Bush
University of Hawaii
Hawaii Institute of Marine Biology
P.O. Box 1346
Kaneohe, HI 96744
USA

Sally Campen
1800 Birch Rd.
McClean, VA 22101
USA
Ph. 703-847-3143
David Cote
Department of Biology
University of Waterloo
Waterloo, Ontario N2L 3G1
CANADA

Laurent Dagorn
IRD
Centre de Montpellier
B.P. 5045
34032 Montpellier Cedex
FRANCE
dagorn@mpl.ird.fr

Gregorio De Metrio
Department of Animal Production
University of BARI - Italy
via Amendola, 165\A
70125
Bari,
ITALY
Ph. 39-80-4670604
Fax 39-80-4670283
G.Demetrio@tno.it

Heidi Dewar
Pleger Institute of Environmental Research
1400 N. Pacific St.
Oceanside, CA 92054
USA
Ph. 760-721-2531
Fax 760-721-1475
Heidi_Dewar@alumni.ucsd.edu

Michael Domeier
Pleger Institute of Environmental Research
1400 N. Pacific St.
Oceanside, CA 92054
USA
DomeierML@cs.com

Kanesa Duncan
University of Hawaii
Hawaii Institute of Marine Biology
P.O. Box 1346
Kaneohe, HI 96744
USA

Anne-Marie Eklund
National Marine Fisheries Service
Southeast Fisheries Science Center
75 Virginia Beach Drive
Miami, FL 33133
USA
Ph. 305-361-4271
Fax 305-365-4102
anne.marie.eklund@noaa.gov

Tim Fitzgerald
University of Hawaii
Hawaii Institute of Marine Biology
P.O. Box 1346
Kaneohe, HI 96744
USA

Malcom Francis
Natl. Institute of Water & Atmospheric Research (NIWA)
P.O. Box 14901
Wellington,
NEW ZEALAND
m.francis@niwa.cri.nz

Alan Friedlander
Oceanic Institute
Makapuu Point
41-202 Kalanianaole Hwy.
Waimanalo, HI 96759-1820
USA
afriedlander@oceanicinstitute.org

Rachel Graham
Environment Department
University of York
York, Y010 5DD
UNITED KINGDOM
Ph. 44-1904-434-068
Fax 44-1904-432-998
rtg@btl.net

John Graves
Virginia Institute of Marine Science
College of William and Mary
P.O. Box 1346
Gloucester Point, VA 23062
USA
Ph. 804-642-7352
Fax 804-642-7186
graves@vims.edu

John Gunn
CSIRO
Division of Marine Research
GPO 1538
Hobart, TAS 7001
AUSTRALIA
Ph. 61-3-6232-5222
Fax 61-3-6232-5000
John.Gunn@marine.csiro.au

Jason Hartog
CSIRO
Division of Marine Research
GPO 1538
Hobart, TAS 7001
AUSTRALIA
Jason.Hartog@marine.csiro.au

Michelle Heupel
Mote Marine Laboratory
1600 Ken Thompson Pkwy
Sarasota, FL 34236
USA
mheupel@mote.org

William Heyman
The Nature Conservancy
62 Front Street
Punta Gorda,
BELIZE
Ph. 501-7-22503
will@btl.net

Roger Hill
Wildlife Computers
16150 NE 85th Street, #226
Redmond, WA 98052
USA
Ph. 425-881-3048
Fax 425-881-3405
tags@wildlifecomputers.com

Kim Holland
University of Hawaii
Hawaii Institute of Marine Biology
P.O. Box 1346
Kaneohe, HI 96744
USA
kholland@hawaii.edu

John Hoolihan
Environmental Research &
Wildlife Development Agency
Chamber of Commerce Tower,
10th floor
P.O. Box 45553
Abu Dhabi,
UNITED ARAB EMIRATES
Ph. 011-9712-4414441
Fax 011-9712-4414131
jph@erwda.gov.ae

Marianne Holm
Institute of Marine Research
P.O. Box 1870 Nordnes
N-5817
Bergen,
NORWAY
Ph. 47-55-23-85-00
Fax 47-55-23-85-31
marianne.holm@imr.no

Paul Howey
Microwave Telemetry
10280 Old Columbia Rd., Ste. 260
Columbia, MD 21046
USA
Ph. 410-290-8672
Fax 410-290-8847
Microwt@aol.com

Robert Humphreys
National Marine Fisheries Service
Honolulu Lab
2570 Dole Street
Honolulu, HI 96822
USA
rhumphre@honlab.nmfs.hawaii.edu

Ewan Hunter
The Centre for Environment, Fisheries and Aquaculture Science (CEFAS)
Pakefield Road
Lowestoft Laboratory
Lowestoft, Suffolk NR33 0HT
UNITED KINGDOM
Ph. 44-1502-524226
Fax 44-1502-513865
e.hunter@cefas.demon.co.uk

Walter Ikehara
Hawaii Dept. Land & Natural Resources
Division of Aquatic Resources
1151 Punchbowl St., Rm. 330
Honolulu, HI 96813
USA
walter_n_ikehara@exec.state.hi.us

David Itano
University of Hawaii
JIMAR, Pelagic Fisheries Research Program
1000 Pope Road, MSB 312
Honolulu, HI 96822
USA
Ph. 808-956-4108
Fax 808-945-4104
ditano@soest.hawaii.edu

Adam Jaffe
University of Hawaii
Hawaii Institute of Marine Biology
P.O. Box 1346
Kaneohe, HI 96744
USA
jaffe@hawaii.edu

Erwan Josse
IRD
Centre de Bretagne
B.P. 70
29280 Plouzane,
FRANCE
Ph. 33-2-98-22-45-03
Fax 33-2-98-22-45-14
Erwan.Josse@ird.fr

Stephen Kajiura
University of Hawaii
Hawaii Institute of Marine Biology
P.O. Box 1346
Kaneohe, HI 96744 USA
kajiura@hawaii.edu

Shinichiro Kakuma
Okinawan Prefectural Fisheries Experiment Station
1 Nishizaki
Itoman City, Okinawa 901-03
JAPAN
skakuma@fish.pref.okinawa.jp

David Kirby
National Institute of Water and Atmospheric Research
(NIWA)
Remote Sensing for Fisheries Project
P.O. Box 14-901
Greta Point
Wellington,
NEW ZEALAND
Ph. 64-4-386-0487
Fax 64-4-386-2153
d.kirby@niwa.cri.nz

Pierre Kleiber
National Marine Fisheries Service
Honolulu Lab
2570 Dole Street
Honolulu, HI 96822
USA
Ph. 808-983-5399
Fax 808-983-2902
pkleiber@honlab.nmfs.hawaii.edu

Gilles Lacroix
Dept. of Fisheries and Oceans,
Ocean Sector
Government of Canada
Biological Station
531 Brandy Cove Road
St. Andrews, NB E5B 2L9
CANADA
Ph. 506-529-8854
Fax 506-529-5862
lacroixg@mar.dfo-mpo.gc.ca

Michael Laurs
National Marine Fisheries Service
Honolulu Lab
2570 Dole Street
Honolulu, HI 96822
USA
Ph. 808-983-5301
Fax 808-983-2901
mlaurs@honlab.nmfs.hawaii.edu

Bruno Leroy
Secretariat of the Pacific
Community (SPC)
B.P. D5
Noumea Cedex,
NEW CALEDONIA
brunol@spc.org.nc

James Lotimer
Lotek Marine Technologies, Inc.
115 Pony Drive
Newmarket, Ontario L3Y 7B5
CANADA
Ph. 905-836-6680
Fax 905-836-6455
jim.lotimer@lotek.com

Christopher Lowe
Dept. of Biological Sciences
California State University,
Long Beach
1250 Bellflower Blvd.
Long Beach, CA 90804
USA
clowe@csulb.edu

Jon Lucy
Virginia Institute of Marine
Science
College of William and Mary
Gloucester Point, VA 23062
USA
Ph. 804-684-7161
Fax 804-684-7166
lucy@vims.edu

Molly Lutcavage
Edgerton Laboratory
New England Aquarium
Central Wharf
Boston, MA 02110-3399
USA
mlutcavg@neaq.org

Greg McKinnon
VEMCO Limited
100 Osprey Drive
Shad Bay, NS B3T 2C1
CANADA

Julie Meka
USGS Biological Resources Division
Alaska Biological Science Center
1011 E. Tudor Rd.
Anchorage, AK 99503
USA
Ph. 907-786-3917
julie_meka@brd.usgs.gov

Carl Meyer
University of Hawaii
Hawaii Institute of Marine Biology
P.O. Box 1346
Kaneohe, HI 96744
USA
carlm@hawaii.edu

Julian Metcalfe
The Centre for Environment, Fisheries and Aquaculture Science (CEFAS)
Pakefield Road
Lowestoft Laboratory
Lowestoft, Suffolk NR33 0HT
UNITED KINGDOM
J.D.Metcalfe@cefas.co.uk

Hiroshi Nakajima
Fisheries Research Institute of Mie
3564-3 Hamajima
Hamajimacho Simagun
Mie,
JAPAN
fishmie@lilac.ocn.ne.jp

Nathaniel Newlands
Edgerton Laboratory
New England Aquarium
Central Wharf
Boston, MA 02110-3399
USA
nnewland@fisheries.com

Jennifer Nielsen
U.S. Geological Survey
Biological Resources Division
1011 E. Tudor Rd.
Anchorage, AK 99503-6199
USA
Jennifer_Nielsen@usgs.gov

Ron O'Dor
Biology Department
Dalhousie University
Halifax, NS B3H 4J1
CANADA
odor@is.dal.ca

Itaru Ohta
Okinawan Prefectural Fisheries Experiment Station
1 Nishizaki
Itoman City, Okinawa 901-03
JAPAN
i_ohta@fish.pref.okinawa.jp

Francois Poisson
IFREMER
Delegation de La Reunion
B.P. 60
97822 Le Port Cedex FRANCE
Ph. 0262-42-03-40
Fax 0262-43-36-84
fpoisson@guetali.fr

Julie Porter
Dept. of Fisheries and Oceans, Ocean Sector
Government of Canada
Biological Station
531 Brandy Cove Road
St. Andrews, NB E0G 2X0
CANADA
Ph. 506-529-5902/8854
Fax 506-529-5862
porterjm@mar.dfo-mpo.gc.ca

Richard Radtke
University of Hawaii
Hawaii Institute of Geophysics and Planetology
2525 Correa Road
Honolulu, HI 96822 USA
Ph. 808-956-7498
Fax 808-956-8516
radtke@hawaii.edu

Erin Rechisky
1194 Boston Neck Road
Narragansett, RI 2992 USA
erechisky@yahoo.com

Attilio Rinaldi
Istituto Centrale per la Ricera
Applicata al Mare
ICRAM
via di Casalotti 300
00166
Rome,
ITALY

Wataru Sakamoto
Graduate School of Agriculture
Kyoto University
606-8502 Kitashirakawa
Oiwakecho
Sakyo-ku
Kyoto,
JAPAN
sakamoto@kais.kyoto-u.ac.jp

Kurt Schaefer
Inter-American Tropical Tuna Commission
8604 La Jolla Shores Drive
La Jolla, CA 92037-1508
USA
kschaefer@iattc.ucsd.edu

David Scruton
Dept. of Fisheries and Oceans
Government of Canada
P.O. Box 5667
St. John's, NF A1C 5X1
CANADA
Ph. 709-772-2007
Fax 709-772-5315
scrutond@dfo-mpo.gc.ca

Michael Seki
National Marine Fisheries Service
Honolulu Lab
2570 Dole Street
Honolulu, HI 96822
USA
Ph. 808-983-5393
Fax 808-983-2902
mseki@honlab.nmfs.hawaii.edu

Chugey Sepulveda
12582 Hill Street
Orange, CA 92869
USA
Chuggee@aol.com

Michael Shane
San Diego State University
Hubbs-Sea World Research Institute
2595 Ingraham Street
San Diego, CA 92109 USA
Ph. 619-226-3946
Fax 619-226-3944
mshane@hswri.org

John Sibert
University of Hawaii
JIMAR, Pelagic Fisheries Research Program
1000 Pope Road, MSB 312
Honolulu, HI 96822 USA
Ph. 808-956-8083
Fax 808-956-4104
jsibert@soest.hawaii.edu

Colin Simpfendorfer
Mote Marine Laboratory
1600 Ken Thompson Pkwy
Sarasota, FL 34236 USA
colins@mote.org

Richard Starr
University of California
Sea Grant Extension Program
P.O. Box 440
Moss Landing, CA 95039
USA
starr@mlml.calstate.edu

John Stevens
CSIRO
Division of Marine Research
GPO 1538
Hobart, TAS 7001
AUSTRALIA
Ph. 61-3-6232-5222
Fax 61-3-6232-5000
John.Stevens@marine.csiro.au

Johannes Sturlaugsson
Institute of Freshwater Fisheries
Vagnhofdi 7, 112
Reykjavik,
ICELAND
Ph. 354-567-64-00
Fax 354-567-64-20
johannes@veidimal.is

Greg Stutzer
6551 Edmonton Ave.
San Diego, CA 92122
USA
gstutzer@hswri.org

Vito Susca
Institute for Zoophysiologie
Heinrich-Heine-Universitat
UniversitatsstraBe 1/26 12
D-40225
Dusseldorf,
GERMANY
Ph. 0049-211-8114847
Fax 0049-211-8113897
susca@uni-duesseldorf.de

Marc Taquet
IFREMER - Delegation Antilles
Laboratoire Ressources Halieutiques
Pointe Fort 97231
Le Robert, MARTINIQUE
FRENCH W. INDIES
Ph. 596-65-79-09
Fax 596-65-11-56
Marc.Taquet@ifremer.fr

Elizabeth Tyler
University of London
7 Westcombe Park Road
London, UNITED KINGDOM

Fred Voegeli
VEMCO Limited
100 Osprey Drive
Shad Bay, NS B3T 2C1
CANADA
Ph. 902-852-3047
Fax 902-852-4000
voegeli@istar.ca

Michael Walker
University of Auckland
School of Biological Sciences
Private Bag 98019
Auckland, NEW ZEALAND
Ph. 64-9-373-7599 ext. 7054
Fax 64-9-393-7417
m.walker@auckland.ac.nz

William Walsh
Hawaii Dept. Land & Natural Resources
Division of Aquatic Resources
P.O. Box 936
Hilo, HI 96721-0936 USA
darkona@interpac.net

David Welch
Ocean Sciences & Productivity Division
Pacific Biological Station
Nanaimo, B.C. V9R 5K6
CANADA
WelchD@pac.dfo-mpo.gc.ca

Kevin Weng
Hopkins Marine Station
120 Ocean View Blvd.
Pacific Grove, CA 93950
USA
Ph. 831-601-7719
kcmw@stanford.edu

Andrew West
75-5863 Kuakini Hwy.
Kailua-Kona, HI 96740 USA
ajwest4@ilhawaii.net

Grant West
CSIRO
Division of Marine Research
GPO 1538
Hobart, TAS 7001
AUSTRALIA
Ph. 61-3-6232-5222
Fax 61-3-6232-5000
Grant.West@marine.csiro.au

Hakan Westerberg
National Board of Fishery
Nya Varvet 31
SE-426 71 Va Frolunda
SWEDEN
hakan.westerberg@fiskeriverket.se

Brad Wetherbee
National Marine Fisheries Service
Northeast Fisheries Science Center
28 Tarzell Drive
Narragansett, RI 02882
USA
Brad.Wetherbee@noaa.gov

Bill Woodward
Service ARGOS, Inc.
1801 McCormick Drive, Ste. 10
Largo, MD 20774
USA
Ph. 301-925-4411
Fax 301-925-8995
woodward@argosinc.com

VENDOR PARTICIPANTS

Alec Electronics Co., Ltd.
7-2-3 Ibukidai-Higashi
Nishi-ku
Kobe,
JAPAN
Ph. 078-997-8686
Fax 078-997-8609
fabian.wolk@alec-electronics.co.jp
contact: Fabian Wolk

Lotek Marine Technologies, Inc.
114 Cabot Street
St. John's, NF A1C 1Z8
CANADA
marine.telemetry@lotek.com
contact: Keith Stoodley

VEMCO Limited
100 Osprey Drive
Shad Bay, NS B3T 2C1
CANADA
Ph. 902-852-3047
Fax 902-852-4000
voegeli@istar.ca
contact: Fred Voegeli

Wildlife Computers
16150 NE 85th Street, #226
Redmond, WA 98052
USA
Ph. 425-881-3048
Fax 425-881-3405
tags@wildlifecomputers.com
contact: Melinda Braun, Roger Hill

GPSR Compliance

The European Union's (EU) General Product Safety Regulation (GPSR) is a set of rules that requires consumer products to be safe and our obligations to ensure this.

If you have any concerns about our products, you can contact us on ProductSafety@springernature.com

In case Publisher is established outside the EU, the EU authorized representative is:

Springer Nature Customer Service Center GmbH
Europaplatz 3
69115 Heidelberg, Germany

Batch number: 10823957

Printed by Printforce, the Netherlands